TEXTBOOK

OF POLYMER SCIENCE

Second Edition

Wiley-Interscience, a Division of John Wiley and Sons, Inc.

New York | London | Sydney | Toronto

Library of Congress Catalog Card Number: 78-142713

ISBN   0   471   07296   6

Printed in the United States of America

10   9   8   7   6   5   4   3   2   1

Preface

"Dear Colleague, Leave the concept of large molecules well alone . . . there can be no such thing as a macromolecule."

It is said* that this advice was given to Hermann Staudinger just 45 years ago, after a major lecture devoted to his evidence in favor of the macromolecular concept. Today it seems almost impossible that this violent opposition to the idea of the existence of polymer molecules could have existed in relatively recent times. Now we take for granted not only the existence of macromolecules but their value to us in food, clothing, shelter, transportation, communication, most other aspects of modern technology, and last but far from least the muscles, sinews, genes, and chromosomes that constitute our bodies and intellect.

Even within the years since the first edition of *Textbook of Polymer Science* (1962) was written, the use of synthetic polymers has proliferated, as discussed in Chapter 7E. Not only has the annual production of plastics (for example) increased some 250% in the last eight years, but on a volume basis it has already exceeded that of copper and aluminum, and is expected to surpass the production of steel by the mid-1980's. One consequence of this widening use of polymeric materials is that a substantial if not major fraction of all chemists and chemical engineers, to say nothing of those in other disciplines, is employed in industry related in some way to polymers. Estimates vary, but this fraction appears to be one-third to one-half or higher.

Education in polymer science has not kept pace. By far the majority of colleges and universities in the United States have no courses in polymer

* Robert Olby, "The Macromolecular Concept and the Origins of Molecular Biology", *J. Chem. Educ.* **47**, 168–174 (1970)

science, no staff member conducting research in this area, and only cursory mention of polymers in other courses. There are, needless to say, many exceptions, ranging from the isolated effort of a single staff member to such major centers for polymer research as those at Case-Western Reserve, the Universities of Akron and Massachusetts, and Rensselaer, where ten or more staff members constitute a formal or informal polymer research center.

Fortunately, this scarcity of education in polymer science is slowly diminishing, but it is still evident in many areas. What is most unfortunate is that it appears to exist, not because of a lack of awareness, but rather a lack of interest. For example, on several occasions a graduating Ph.D. student in my own institution has dropped by to say that he was trained in another area of chemistry and was vaguely aware of our polymer science program, but he had now accepted employment with a large company where he was told that he would be working in the polymer field, and please, was there any way he could learn all about polymers in the short time remaining before he started the job?

Perhaps we could overlook a few individual instances of this sort, but it is more disturbing to note that, while polymeric materials are widely used as previously indicated, most "materials" curricula are merely renamed metallurgy departments giving only lip service to major families of materials other than metals, such as ceramics and polymers. Again, there are many exceptions, and I do not wish to leave the impression that this is always the case. I have, however, yet to see an introductory materials course or textbook that treats polymers fairly in accord with their wide usage throughout the world.

Another dichotomy which deserves mention, and which I wish I had the ability to overcome, is exemplified by the communications gap between the polymer scientist as now trained in the university or in industry and the biologist or biomedical scientist, whose concepts of macromolecules are vastly different. Many of my colleagues seem to share my feeling that major advances can be made in the next few years by the application of polymer physics and physical chemistry to biological materials.

The foregoing commentary emphasizes my feeling that the need for education in polymer science exists more than ever today, and that it must be filled by teaching at several levels in several disciplines. Clearly, no single book can serve all needs in this field, but I hope that in its new edition *Textbook of Polymer Science* will continue to be valuable in many ways. In the revision I have attempted to keep in mind its use as supplemental reading material in such undergraduate chemistry courses as physical chemistry, organic chemistry, and instrumental analysis, where an effort is made to introduce polymer science into the chemistry curriculum at appropriate places; as supplemental reading in other curricula such as biology and its interdisciplinary offshoots, materials in the broad sense, and the environmental sciences growing in popularity today; as the textbook in polymer science and engineering courses at the undergraduate and first-year graduate levels; as supplementary reading to broaden the background of the student in advanced polymer courses using appropriate specialized texts; in continuing education at the post-graduate level, in universities but particularly in industry; and as a reference and guide to the literature for the practicing polymer scientist and engineer.

Just as the use of polymers has proliferated in the past decade, so has its literature. The leading journal in the field, *Journal of Polymer Science*, has subdivided into several largely independent parts. New journals have appeared, some with national society sponsorship, such as *Macromolecules*, sponsored by the American Chemical Society; others as commercial ventures. Polymer articles, and even polymer sections, appear in other journals, new and old. *POST—J*, described below, abstracts approximately 500 journals for articles on polymers.

Abstract journals devoted solely to polymers have appeared. *POST—J*, acronym for Polymer Science and Technology—Journals, and its companion *POST—P* (Patents) are published bi-weekly, at a rate of 500–700 abstracts per issue in mid-1970, by the Chemical Abstracts Service of the American Chemical Society. Other such services exist, sponsored by universities and as commercial ventures.

Not only has the number of general and specialized books on polymers increased at a tremendous rate in recent years, but a number of valuable compilations has appeared. Outstanding among these are the *Encyclopedia of Polymer Science and Technology* and the *Modern Plastics Encyclopedia*, both referenced in this book.

One result of the appearance of many specialized books and encyclopedias is that it is no longer necessary for a textbook to cite the original literature in detail; indeed, it would be impossible to do so today. Therefore I have limited references to specialized books and compilations where possible, on the assumption that most of them will be as readily available as the original literature and can in their turn cover the subject more completely than is possible in the present volume. The exceptions, aside from some key references of historical value, lie in the areas where adequate specialized coverage has not yet appeared. The reader will recognize these areas by the extent and content of the bibliography at the end of each chapter.

In revising the *Textbook of Polymer Science*, I have reached the conclusion that many of the basic principles of polymer science are now well established. Examples are the kinetics of condensation and free-radical addition polymerization. In areas such as these, the reader will find only minor changes from the 1962 edition. Elsewhere, the revision has been more extensive. Much new material has been added, particularly in areas still under rapid development. This has had to be counterbalanced by the omission of an equivalent amount of material that no longer represents the current state of our knowledge, or was of lesser or only historical interest, in order to keep the length (and price) of the volume in hand.

The actual arrangement of material has undergone little change. Chapters 1 to 4 now comprise Part I, dealing with introductory concepts and the characterization of macromolecules. Important additions in this section include discussion of solubility parameters, free-volume theories of polymer solution thermodynamics, gel permeation chromatography, vapor-phase osmometry, and scanning electron microscopy, with extensive revision of many other sections.

Part II (Chapters 5 to 7) deals with the structure and properties of bulk polymers, and includes considerable revision of parts of Chapter 5, where a few of the concepts of crystallinity in polymers, new in 1962, have had to be modified as our knowledge in this area has grown. Chapter 7 has been revised in order of presentation, with considerable new material added.

The format and content of Part III, concerned with polymerization kinetics, have been revised primarily for the citation of recent advances and new references. The exception lies in Chapter 10, whose topic is ionic and coordination polymerization, in which field many of the concepts new in 1962 have now reached a stage of further elucidation and acceptance.

As in the earlier edition, the chapters of Part IV describe the polymerization, structure, properties, fabrication, and applications of commercially important polymers, including those used as plastics, fibers, and elastomers. The reader comparing old and new chapter titles will find that considerable rearrangement has been made, and the content is likewise extensively revised. Of particular note are new sections in Chapter 15 on aromatic heterochain, heterocyclic, ladder, and inorganic polymers.

Part V, dealing with polymer processing, has in contrast been revised primarily by the addition of new references.

It is my hope that the many readers whose kind comments on earlier editions have given me pleasure will continue to find this revision useful.*

FRED W. BILLMEYER, JR.
Troy, New York
September, 1970

---

*I regret that the name of Sidney Gross, the Editor of the *Modern Plastics Enclyopedia*, has inadvertently been misspelled as Gruss.

Acknowledgments

The assistance of my many colleagues, at Rensselaer, in industry, and editors and fellow authors, has been most valuable in this revision. Special thanks go to Professor Sonja Krause for helpful suggestions, to Dr. Norbert Bikales and his authors of yet-unpublished articles for the *Encyclopedia of Polymer Science and Technology* for kindly placing their manuscripts at my disposal, and to the McGraw-Hill Book Company for furnishing a copy of the *Modern Plastics Encyclopedia*.

The manuscript was typed by Patricia D'Angelo and indexed in part by Eleanor Ann Billmeyer.

My present research is carried out in Rensselaer's Materials Research Centre, a facility supported by the National Aeronautics and Space Administration.

I am deeply indebted for the indulgence of my graduate students, and particularly my wife Annette and our children, for whom this undertaking has resulted in many hours when I was unavailable for tasks more important and pleasant to them.

FRED W. BILLMEYER, JR.
Troy, New York
September, 1970

Contents

Textbook

of Polymer Science

I

Polymer Chains and their Characterization

I

The Science of Large Molecules

A. Basic Concepts of Polymer Science

Over half a century ago, Wilhelm Ostwald (1914)* coined the phrase "the land of neglected dimensions" to describe the range of sizes between molecular and macroscopic within which occur most colloidal particles. The term "neglected dimensions" might have been applied equally well to the world of polymer molecules, the high-molecular-weight compounds so important to man and his modern technology. It was not until the 1930's that the science of high polymers began to emerge, and the major growth of the technology of these materials came even later. Yet today polymer dimensions are neglected no more, for industries associated with polymeric materials employ more than half of all American chemists and chemical engineers.

The science of macromolecules is divided between biological and nonbiological materials. Each is of great importance. Biological polymers form the very foundation of life and intelligence, and provide much of the food on which man exists. This book, however, is concerned with the chemistry, physics, and technology of nonbiological polymers. These are primarily the synthetic materials used for plastics, fibers, and elastomers, but a few naturally occurring polymers, such as rubber, wool, and cellulose, are included. Today, these substances are truly indispensable to mankind, being essential to his clothing, shelter, transportation, and communication, as well as to the conveniences of modern living.

A *polymer* is a large molecule built up by the repetition of small, simple chemical units. In some cases the repetition is linear, much as a chain is

* Parenthetical years or names and years refer to items in the bibliography at the end of the chapter.

built up from its links. In other cases the chains are *branched* or inter-connected to form *three-dimensional networks*. The *repeat unit* of the polymer is usually equivalent or nearly equivalent to the *monomer*, or starting material from which the polymer is formed. Thus (Table 1-1) the repeat unit of poly(vinyl chloride) is —CH_2CHCl—; its monomer is vinyl chloride, CH_2=$CHCl$.

The length of the polymer chain is specified by the number of repeat units in the chain. This is called the *degree of polymerization* (DP). The molecular weight of the polymer is the product of the molecular weight of the repeat unit and the degree of polymerization. Using poly(vinyl chloride) as an example, a polymer of degree of polymerization 1000 has a molecular weight of $63 \times 1000 = 63,000$. Most high polymers useful for plastics, rubbers, or fibers have molecular weights between 10,000 and 1,000,000.

Unlike many products whose structure and reactions were well known before their industrial application, some polymers were produced on an industrial scale long before their chemistry or physics was studied. Empiricism in recipes, processes, and control tests was usual.

Gradually the study of polymer properties began. Almost all were first called anomalous because they were so different from the properties of low-

TABLE 1-1. *Some linear high polymers, their monomers, and their repeat units*

Polymer	Monomer	Repeat Unit
Polyethylene	CH_2=CH_2	—CH_2CH_2—
Poly(vinyl chloride)	CH_2=$CHCl$	—CH_2CHCl— 63
Polyisobutylene	CH_2=$\overset{\displaystyle CH_3}{\underset{\displaystyle CH_3}{C}}$	—CH_2—$\overset{\displaystyle CH_3}{\underset{\displaystyle CH_3}{C}}$—
Polystyrene*	CH_2=CH ⬡	—CH_2—CH— ⬡
Polycaprolactam (6 nylon)	H—$N(CH_2)_5C$—OH with H below N and O below C	—$N(CH_2)_5C$— with H below N and O below C
Polyisoprene (natural rubber)	CH_2=CH—$\overset{\displaystyle}{\underset{\displaystyle CH_3}{C}}$=$CH_2$	—CH_2CH=$\overset{\displaystyle}{\underset{\displaystyle CH_3}{C}}$—$CH_2$—

* By convention, the symbol ⬡ is used throughout to represent the benzene ring, double bonds being omitted.

molecular-weight compounds. It was soon realized, however, that polymer molecules are many times larger than those of ordinary substances. The presumably anomalous properties of polymers were shown to be normal for such materials, as the consequences of their size were included in the theoretical treatments of their properties.

Primary chemical bonds along polymer chains are entirely satisfied. The only forces between molecules are secondary bond forces of attraction, which are weak relative to primary bond forces. The high molecular weight of polymers allows these forces to build up enough to impart excellent strength, dimensional stability, and other mechanical properties to the substances.

Polymerization processes   The processes of polymerization were divided by Flory (1953) and Carothers (Mark 1940) into two groups known as *condensation* and *addition* polymerization or, in more precise terminology (Chapter 8*A*), *step-reaction* and *chain-reaction* polymerization.

Condensation or *step-reaction polymerization* is entirely analogous to condensation in low-molecular-weight compounds. In polymer formation the condensation takes place between two polyfunctional molecules to produce one larger polyfunctional molecule, with the possible elimination of a small molecule such as water. The reaction continues until almost all of one of the reagents is used up; an equilibrium is established which can be shifted at will at high temperatures by controlling the amounts of the reactants and products.

Addition or *chain-reaction polymerization* involves chain reactions in which the chain carrier may be an ion or a reactive substance with one unpaired electron called a *free radical*. A free radical is usually formed by the decomposition of a relatively unstable material called an *initiator*. The free radical is capable of reacting to open the double bond of a vinyl monomer and add to it, with an electron remaining unpaired. In a very short time (usually a few seconds or less) many more monomers add successively to the growing chain. Finally two free radicals react to annihilate each other's growth activity and form one or more polymer molecules.

With some exceptions, polymers made in chain reactions often contain only carbon atoms in the main chain (*homochain polymers*), whereas polymers made in step reactions may have other atoms, originating in the monomer functional groups, as part of the chain (*heterochain polymers.*)

Molecular weight and its distribution   In both chain and stepwise polymerization, the length of a chain is determined by purely random events. In step reactions, the chain length is determined by the local availability of reactive groups at the ends of the growing chains. In radical polymerization, chain length is determined by the time during which the chain grows before it diffuses into the vicinity of a second free radical and the two react.

In either case, the polymeric product contains molecules having many different chain lengths. For some types of polymerization the resulting *distribution of molecular weights* can be calculated statistically. It can be illustrated by plotting the weight of polymer of a given size against the chain length or molecular weight (Fig. 1-1).

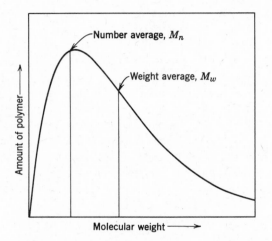

Fig. 1-1.   Distribution of molecular weights in a typical polymer.

Since a distribution of molecular weights exists in any finite sample of polymer, the experimental measurement of molecular weight can give only an average value. Several different averages are important. For example, some methods of molecular-weight measurement in effect count the number of molecules in a known mass of material. Through knowledge of Avogadro's number this information leads to the *number-average molecular weight* $\overline{M}_n$ of the sample. For typical polymers the number average lies near the peak of the weight-distribution curve or the most probable molecular weight.

In other experiments, such as light scattering, the contribution of a molecule to the observed effect is a function of its mass. Heavy molecules are favored in the averaging process; a *weight-average molecular weight* $\overline{M}_w$ results. $\overline{M}_w$ is equal to or greater than $\overline{M}_n$. The ratio $\overline{M}_w/\overline{M}_n$ is sometimes used as a measure of the breadth of the molecular weight distribution. Values of $\overline{M}_w/\overline{M}_n$ for typical polymers range from 1.5–2.0 to 20–50.

The molecular weight averages shown in Fig. 1-1 are defined mathematically in Chapter 3.

Branched and network polymers   In contrast to the linear-chain molecules discussed so far, some polymers have branched chains, often as a result of side reactions during polymerization (Fig. 1-2*a*). The term *branching* implies that the individual molecules are still discrete; in still other cases *crosslinked* or *network structures* are formed (Fig. 1-2*b*), as in the use of monomers containing more than two reactive groups in stepwise polymerization. If, e.g.,

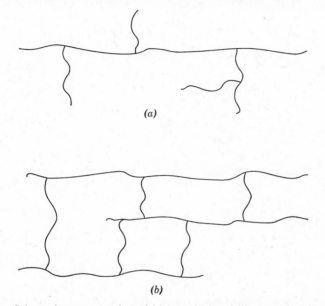

(*a*)

(*b*)

Fig. 1-2.   Schematic representation of (*a*) branched and (*b*) network polymers.

glycerol is substituted for ethylene glycol in the reaction with a dibasic acid, a three-dimensional network polymer results. In recent years, a variety of branched polymer structures, some with outstanding high-temperature properties, has been synthesized (Chapter 15, Fig. 1-3).

    In commercial practice crosslinking reactions may take place during the fabrication of articles made with *thermosetting* resins. The crosslinked network extending throughout the final article is stable to heat and cannot be made to flow or melt. In contrast, most linear polymers can be made to soften and take on new shapes by the application of heat and pressure. They are said to be *thermoplastic.*

The texture of polymers   The geometrical arrangement of the atoms in a polymer chain can be divided conveniently into two categories:

    a. Arrangements fixed by the chemical bonding in the molecule, such as *cis* and *trans* isomers, or *d-* and *l-*forms. Throughout this book, such

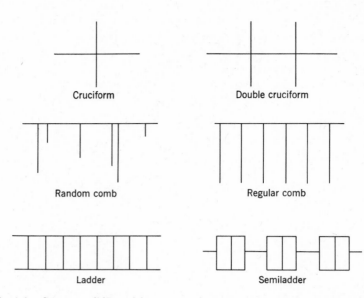

Fig. 1-3. Some possible model structures for branched polymers (Tawn 1969).

arrangements are described as *configurations*. The configuration of a polymer chain cannot be altered unless chemical bonds are broken and reformed.

b. Arrangements arising from rotation about single bonds. These arrangements, including the manifold forms that the polymer chain may have in solution, are described as *conformations*.

In dilute solution, where the polymer chain is surrounded by small molecules, or in the melt, where it is in an environment of similar chains, the polymer molecule is in continual motion because of its thermal energy, assuming many different *conformations* in rapid succession. As a polymer melt is cooled, or as this molecular motion so characteristic of polymers is restrained through the introduction of strong interchain forces, the nature of the polymer sample changes systematically in ways that are important in determining its physical properties and end uses (Fig. 1-4).

In the molten state, polymer chains move freely, though often with enormous viscosity, past one another if a force is applied. This is the principle utilized in the fabrication of most polymeric articles, and is the chief example of the plasticity from which the very name plastics is derived. If the irreversible flow characteristic of the molten state is inhibited by the introduction of a tenuous network of primary chemical-bond crosslinks in the process commonly called *vulcanization* (Chapter 19), but the local freedom of motion of the polymer chains is not restricted, the product shows the

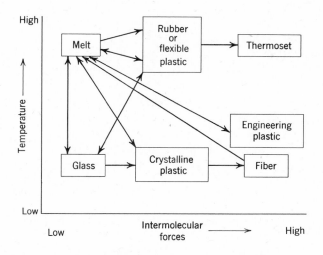

Fig. 1-4. The interrelation of the states of bulk polymers. The arrows indicate the directions in which changes from one state to another can take place (Billmeyer 1969).

elastic properties we associate with typical rubbers. If, however, the interchain forces result from secondary bonds, such as the interaction of polar groups, rather than primary chemical bonds, the rubber is not one of high elasticity but has the properties of limpness and flexibility: a familiar example is the vinyl film widely used alone or in coated fabrics. Secondary bond forces are capable of forming and breaking reversibly as the temperature is changed, as indicated by the arrows in Fig. 1-4.

Continued primary-bond crosslinking in the postpolymerization step of vulcanization converts rubber into hard rubber or ebonite, while crosslinking concurrent with polymerization produces a wide variety of thermosetting materials. Common examples are the phenol-formaldehyde and amine-formaldehyde families widely used as plastics.

As the temperature of a polymer melt or rubber is lowered, a point known as the *glass-transition temperature* is reached where polymeric materials undergo a marked change in properties associated with the virtual cessation of molecular motion on the local scale. Thermal energy is required for segments of a polymer chain to move with respect to one another; if the temperature is low enough, the required amounts of energy are not available. Below their glass-transition temperature, polymers have many of the properties associated with ordinary inorganic glasses including hardness, stiffness, brittleness, and transparency.

In addition to undergoing a glass transition as the temperature is lowered, some polymers can *crystallize* at temperatures below that designated as their crystalline melting point. Not all polymers are capable of crystallizing; to oversimplify somewhat, the requirements for crystallizability in a polymer are that it have either a geometrically regular structure, or that any substituent atoms or groups on the backbone chain be small enough so that, if irregularly spaced, they can still fit into an ordered structure by virtue of their small size (see Chapter 5).

The properties of crystalline polymers are highly desirable. Crystalline polymers are strong, tough, stiff, and generally more resistant to solvents and chemicals than their noncrystalline counterparts. Further improvements in these desirable properties can be brought about in at least two ways:

First, by increasing intermolecular forces through the selection of highly polar polymers, and by using inherently stiff polymer chains, crystalline melting points can be raised so that the desirable mechanical properties associated with crystallinity are retained to quite high temperatures. There is a large research effort in this direction at the present time, leading to plastics capable of competing with metals and ceramics in engineering applications (Chapters 7 and 15).

Second, the properties of crystalline polymers can be improved for materials in fiber form by the process of orientation or drawing. The result is the increased strength, stiffness, and dimensional stability associated with synthetic fibers (Chapter 18).

GENERAL REFERENCES:

Mark 1940; Flory 1953, Chaps. II-1, II-2; Mark 1966; Margerison 1967.

B. History of Macromolecular Science

Early investigations

Natural polymers Natural polymers have been utilized throughout the ages. Since his beginning man has been dependent on animal and vegetable matter for sustenance, shelter, warmth, and other requirements and desires. Natural resins and gums have been used for thousands of years. Asphalt was utilized in pre-Biblical times; amber was known to the ancient Greeks; and gum mastic was used by the Romans.

About a century ago the unique properties of natural polymers were recognized. The term *colloid* was proposed to distinguish polymers as a class from materials which could be obtained in crystalline form. The concept was later broadened to that of the "colloidal state of matter," which was considered to be like the gaseous, liquid, and solid states. Although useful for describing many colloidal substances, such as gold sols and soap solutions, the concept of a reversibly attainable colloidal state of matter has no validity.

The hypothesis that colloidal materials are very high in molecular weight is also quite old, but before the work of Raoult and van't Hoff in the 1880's no suitable methods were available for estimating molecular weights. When experimental methods did become available, molecular weights ranging from 10,000 to 40,000 were obtained for such substances as rubber, starch, and cellulose nitrate. The existence of large molecules implied by these measurements was not accepted by the chemists of the day for two reasons.

First, true macromolecules were not distinguished from other colloidal substances that could be obtained in noncolloidal form as well. When a material of well-known structure was seen in the colloidal state, its apparent high molecular weight was considered erroneous. Thus it was assumed that Raoult's solution law did not apply to any material in the colloidal state. Second, coordination complexes and the association of molecules were often used to explain polymeric structures in terms of physical aggregates of small molecules.

For example, the empirical formula $C_5 H_8$ was found for rubber as early as 1826, and isoprene was obtained on destructive distillation of the polymer in 1860. The presence of the repeating unit

$$CH_2-\overset{\overset{\textstyle CH_3}{|}}{C}=CH-CH_2-$$

was demonstrated in the early 1900's. At that time rubber was thought to consist of short sequences of this unit arranged in either chains or cyclic structures. Obvious difficulties regarding end groups, which could not be found chemically, favored ring structures, leading to the concept of the rubber molecule being a ring like dimethylcycooctadiene. Large numbers of these were considered held together by "association" to give the colloidal material:

$$\left[\begin{array}{c} \overset{\textstyle CH_3}{|} \\ CH_2-C=CH-CH_2 \\ | \qquad\qquad | \\ CH_2-C=CH-CH_2 \\ | \\ CH_3 \end{array} \right]_x$$

Synthetic polymers   In the search by the early organic chemists for pure compounds in high yields, many polymeric substances were discovered and as quickly discarded as oils, tars, or undistillable residues. A few of these materials, however, attracted interest. Poly(ethylene glycol) was prepared about 1860; the individual polymers with degree of polymerization up to 6 were isolated and their structures correctly assigned. The concept of extending the structure to very high molecular weights by continued condensation was understood.

Other condensation polymers were prepared in succeeding decades. As the molecular aggregation theories gained in popularity, structures involving small rings held together by secondary bond forces were often assigned to these products.

Some vinyl polymers were also discovered. Styrene was polymerized as early as 1839, isoprene in 1879, and methacrylic acid in 1880. Again cyclic structures held together by "partial valences" were assigned.

The rise of polymer science

Acceptance of the existence of macromolecules   Acceptance of the macromolecular hypothesis came about in the 1920's, largely because of the efforts of Staudinger (1920), who received the Nobel Prize in 1953 for his championship of this viewpoint. He proposed long-chain formulas for polystyrene, rubber, and polyoxymethylene. His extensive investigations of the latter polymers left no doubt as to their long-chain nature. More careful molecular-weight measurements substantiated Staudinger's conclusions, as did x-ray studies showing structures for cellulose and other polymers which were compatible with chain formulas. The outstanding series of investigations by Carothers (1929, 1931) supplied quantitative evidence substantiating the macromolecular viewpoint.

The problem of end groups   One deterrent to the acceptance of the macromolecular theory was the problem of the ends of the long-chain molecules. Since the degree of polymerization of a typical polymer is at least several hundred, chemical methods for detecting end groups were at first not successful. Staudinger (1925) suggested that no end groups were needed to saturate terminal valences of the long chains; they were considered to be unreactive because of the size of the molecules. Large ring structures were also hypothesized (Staudinger 1928), and this concept was popular for many years. Not until Flory (1937) elucidated the mechanism for chain-reaction polymerization did it become clear that the ends of long-chain molecules consist of normal, satisfied valence structures. The presence and nature of end groups have since been investigated in detail by chemical methods (Price 1942, Joyce 1948, Bevington 1954).

Molecular weight and its distribution   Staudinger (1928) was among the first to recognize the large size of polymer molecules, and to utilize the dependence on molecular weight of a physical property, such as dilute-solution viscosity (Staudinger 1930), for determining polymer molecular weights. He also understood clearly that synthetic polymers are polydisperse (Staudinger 1928). A few years later, Lansing (1935) distinguished unmistakably among the various average molecular weights obtainable experimentally.

Configurations of polymer chain atoms   Staudinger's name is also associated with the first studies (1935) of the configuration of polymer chain atoms. He showed that the phenyl groups in polystyrene are attached to alternate chain carbon atoms. This regular head-to-tail configuration has since been established for most vinyl polymers. The mechanism for producing branches in normally linear vinyl polymers was introduced by Flory (1937), but such branches were not adequately identified and characterized for another decade (see Chapter 3D). Natta (1955a,b) first recognized the presence of stereospecific regularity in vinyl polymers.

Early industrial developments

Rubber   The modern plastics industry began with the utilization of natural rubber for erasers and in rubberized fabrics a few years before Goodyear's discovery of vulcanization in 1839. In the next decade the rubber industry arose both in England and in the United States. In 1851 hard rubber, or ebonite, was patented and commercialized.

Derivatives of cellulose   Cellulose nitrate, or nitrocellulose, discovered in 1838, was successfully commercialized by Hyatt in 1870. His product, Celluloid, cellulose nitrate plasticized with camphor, could be formed into a wide variety of useful products by the application of heat and pressure. Nitrocellulose found application in the manufacture of explosives, photographic film, synthetic fibers (Chardonnet silk), airplane dopes, automobile lacquers, and automobile safety glass. Nitrocellulose in turn has been superseded in almost all these uses by more stable and more suitable polymers.

Cellulose acetate, discovered in 1865, was not used commercially for several decades because of the low solubility and lack of dyeability of the early cellulose triacetate products. In the early 1900's the manufacture of less highly substituted, more readily soluble compositions opened the way for the commercial development of acetate rayon fibers and cellulose acetate plastics.

Later still, processes were developed whereby cellulose itself could be dissolved and reprecipitated by chemical treatment. These processes led to the production of viscose rayon fiber and cellophane film.

Synthetic polymers   The oldest of the purely synthetic plastics is the family of phenol-formaldehyde resins, of which Baekeland's <u>Bakelite</u> was the first commercial product. Small-scale production of phenolic resins and varnishes was begun in 1907.

The first commercial use of styrene was in synthetic rubbers made by copolymerization with dienes in the early 1900's. Polystyrene was produced commercially in Germany about 1930 and successfully in the United States in 1937. Large-scale production of vinyl chloride-acetate resins began in the early 1920's also.

Thus the last quarter-century has seen the development of all but a handful of the wide variety of synthetic polymers now in common use.

GENERAL REFERENCES

Mark 1940; Flory 1953, Chap. I; Purves 1954.

C. Molecular Forces and Chemical Bonding in Polymers

The nature of the bonds that hold atoms together in molecules is explained by quantum mechanics in terms of an atom consisting of a small nucleus, concentrating the mass and positive charge, surrounded by clouds or shells of electrons relatively far away. It is among the outermost, more loosely bound electrons, called *valence electrons*, that chemical reactions and primary bond formation take place.

Primary bonds

Ionic bond   The most stable electronic configuration for all but very light atoms is a complete outer shell of eight electrons, called an *octet*. This structure may be obtained by the donation of an electron by one atom to another:

$$\text{Na}\cdot + \cdot\overset{\cdot\cdot}{\underset{\cdot\cdot}{\text{Cl}}}: \longrightarrow \text{Na}^+ + :\overset{\cdot\cdot}{\underset{\cdot\cdot}{\text{Cl}}}:^-$$

This results in electrostatic charges on the atoms, creating the attractive forces of these *ionic bonds*. Ionic bonds lead to the buildup of large salt crystals. While these bonds are not usually found in macromolecular substances, divalent ions have long been used to provide "crosslinks" between carboxyl groups in natural resins. The same type of bonding has recently been reintroduced in the class of materials called *ionomers* (Chapter 13).

Covalent bond   These bonds are formed when one or more pairs of valence electrons are shared between two atoms, again resulting in stable electronic shells:

$$\cdot \overset{\cdot\cdot}{\underset{\cdot\cdot}{C}} \cdot \ + \ 4H\cdot \ \longrightarrow \ H\!:\!\overset{\overset{\textstyle H}{\cdot\cdot}}{\underset{\underset{\textstyle H}{\cdot\cdot}}{C}}\!:\!H$$

The *covalent bond* is the predominant bond in polymers.

Coordinate bond   This bond is similar to the covalent bond in that electrons are shared to produce stable octets; but in the coordinate bond both of the shared electrons come from one atom, called the *donor*. Addition compounds of boron trichloride are common examples:

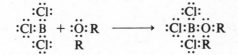

where R is an organic group. The donor atom (O) effectively loses an electron and carries a positive charge, whereas the *acceptor* (B) becomes negative. Thus, the *coordinate* or *semipolar bond* has properties between those of the ionic and the covalent bonds. Although no polymers containing true coordinate bonds have reached commercialization, the use of this type of bonding in inorganic (Chapter 15) and semi-organic polymers is being actively explored.

Metallic bond   In the *metallic bond*, the number of valence electrons is far too small to provide complete shells for all the atoms. The resulting bonds involve the concept of positively charged atoms embedded in a permeating "gas" of electrons free to move about at will. The features of high electrical conductivity, nonsaturation of valence, and lack of direction in the bonding forces are accounted for. The metallic bond is perhaps the least understood of all the primary-bond types. Metal-to-metal atom bonds are not yet utilized in polymeric systems, but metal-organic bonds form the basis of the new class of *metallocene* polymers (Chapter 15).

Typical primary-bond distances and energies   From studies of the positions of atoms in molecules and the energetics of molecular formation and dissociation, it is possible to assign typical energies and lengths to primary bonds. Table 1-2 lists some of these properties of interest in polymeric systems. The angles between successive single bonds involving the atomic arrangements usual in polymers range between $105°$ and $113°$, not far from the tetrahedral angle of $109° \ 28'$.

TABLE 1-2 *Typical primary bond lengths and energies*

Bond	Bond Length, A	Dissociation Energy, kcal/mole
C—C	1.54	83
C=C	1.34	146
C—H	1.10	99
C—N	1.47	73
C≡N	1.15	213
C—O	1.46	86
C=O	1.21	179
C—F*	1.32–1.39	103–123
C—Cl	1.77	81
N—H	1.01	93
O—H	0.96	111
O—O	1.32	35

* The bond length decreases and the dissociation energy increases as additional fluorine atoms are substituted on the same carbon atom (Bryant 1962).

Primary bonding and polymer topology   In recent years more detailed consideration has been given to the topology of crosslinked polymer networks than is apparent from the early classification of polymers as linear, branched, or three-dimensional network structures. Among the newer types of polymers under intense investigation and in some cases already of commercial interest are the various types of branched polymers depicted in Fig. 1-3, the ladder polymers (Chapter 15) noted for their outstanding thermal properties and the semiconducting capability demonstrated by some structures (Baker 1964, De Winter 1966), and two-dimensional sheet polymers (Blumstein 1966).

Secondary bond forces

Experience shows that, even after all the primary valences within covalent molecules are saturated, there are still forces acting between the molecules. These are generally known as *secondary valence* or *intermolecular forces*. The first quantitative statements about them were made by van der Waals in his studies of the equation of state of real gases; hence they are frequently called *van der Waals forces*.

Dipole forces When different atoms in a molecule carry equal and opposite electric charges, the molecule is said to be *polar* or to have a *dipole moment*. At large distances such a molecule acts like an electrically neutral system, but at molecular distances the charge separation becomes significant and leads to a net intermolecular force of attraction. The magnitude of the interaction energy depends on the mutual alignment of the dipoles. Molecular orientation of this sort is always opposed by thermal agitation; hence the dipole force is strongly dependent upon temperature.

Induction forces A polar molecule also influences surrounding molecules that do not have permanent dipoles. The electric field associated with a dipole causes slight displacements of the electrons and nuclei of surrounding molecules, which lead to induced dipoles. The intermolecular force between the permanent and induced dipoles is called the *induction force*. The ease with which the electronic and nuclear displacements are made is called the *polarizability* of the molecule. The energy of the induction force is always small and is independent of temperature.

Dispersion forces The existence of intermolecular forces in nonpolar materials, plus the small temperature dependence of intermolecular forces even where the dipole effect is known to far outweigh the induction effect, suggests the presence of a third type of intermolecular force. All molecules have time-varying dipole moments which average out to zero and which arise from different instantaneous configurations of the electrons and nuclei. These fluctuations lead to perturbations of the electronic clouds of neighboring atoms and give rise to attractive forces called *dispersion forces*. They are present in all molecules and make up a major portion of the intermolecular forces unless very strong dipoles are present. In nonpolar materials only the dispersion forces exist. They are independent of temperature.

Interrelation of intermolecular forces The energy of the intermolecular attractive forces varies as the inverse sixth power of the intermolecular distance. As with primary bond forces, repulsion arises when the atoms approach more closely than the equilibrium bond distance of 3–5 A. The energy of typical secondary bond attractive forces is 2–10 kcal/mole, divided among the three secondary bond types according to the polarizability and dipole moment of the bonding molecules.

The hydrogen bond The bond in which a hydrogen atom is associated with two other atoms is particularly important in many polymers, including proteins, and is held by many to be essential to life processes. Since the classical concepts of chemical bonding allow hydrogen to form only one covalent bond, the hydrogen bond can be considered electrostatic or ionic in character. This model does not, however, account for all the properties of the hydrogen bond; it is appealing to consider the bond covalent in some cases. The hydrogen bond occurs between two functional groups in the same or different molecules. The hydrogen is usually attached to an acidic group (a proton donor), typically a hydroxyl, carboxyl, amine, or amide group. The other group must be basic, usually oxygen as in carbonyls, ethers, or hydroxyls; nitrogen as in amines and amides; and occasionally halogens. The association of such polar liquid molecules as water, alcohols, and hydrofluoric acid, the formation of dimers of simple organic acids, and important structural effects in polar polymers such as nylon, cellulose, and proteins, are due to hydrogen bonding.

Typically, hydrogen bonds range between 2.4 and 3.2 A in length and between 3 and 7 kcal/mole in dissociation energy. Only fluorine, nitrogen, oxygen, and (occasionally) chlorine are electronegative enough to form hydrogen bonds.

Intermolecular forces and physical properties

Secondary bond forces are not of great importance in the formation of stable chemical compounds. They lead rather to the aggregation of separate molecules into solid and liquid phases. As a result many physical properties such as volatility, viscosity, surface tension and frictional properties, miscibility, and solubility are determined largely by intermolecular forces.

The *cohesive energy* is the total energy necessary to remove a molecule from a liquid or solid to a position far from its neighbors. This is approximately equal to the heat of vaporization or sublimation at constant volume and can be estimated from thermodynamic data. The cohesive energy per unit volume, sometimes called the *specific cohesive energy* or *cohesive energy density*, and its variation with molecular structure, illustrate the effects of intermolecular forces on the physical properties of matter.

Volatility and molecular weight The tendency of a molecule to volatilize from its liquid is a function of its total translational energy and therefore of the temperature. The boiling point depends on the relation of the translational energy to the cohesive energy, and thus is a function of molecular weight in a homologous series. At high molecular weights the total cohesive energy per molecule becomes greater than primary-bond energy, and the molecules decompose before they volatilize. This point is reached at molecular weights far below those of typical polymers.

The melting point is also related to the cohesive energy, but here another important factor comes into play. This is the influence of molecular order or entropy. In thermodynamic terms changes of state take place only when the free energy* change in the process

GIBBS

$$\Delta G = \Delta H - T \Delta S$$

is favorable, and the energy term ΔH may easily be outweighed by the entropy term $T \Delta S$ whenever a radical change in molecular configurations occurs in the process. Thus in general a high boiling point is associated with a high melting point, but the relation between melting point and molecular structure is fairly complicated. Symmetrical molecules, which have low entropies of fusion, melt at higher temperatures than do similar but less symmetrical molecules.

Effect of polarity   A molecule containing strongly polar groups exerts correspondingly strong attractive forces on its neighbors. This is reflected in higher boiling and melting points and other manifestations of higher cohesive energy density.

Miscibility and solubility   These properties are also determined by the intermolecular forces. The thermal effect on mixing or solution is the difference between the cohesive energy of the mixture and that of the individual pure components. Again entropy considerations are important, but in general a negative heat of mixing favors solubility and a positive heat of mixing favors immiscibility. The intermolecular forces therefore lead directly to the solubility law of "like dissolves like."

The role of the intermolecular forces and the cohesive energy density in determining the solubility of polymers is discussed further in Chapter 2A.

Intermolecular forces and polymer types   Table 1-3 lists the cohesive energy densities of some typical polymers. These data corroborate the conclusions of Section A regarding the texture of polymers for, in the absence of primary-bond crosslinks, it is the intermolecular forces that provide the restraints on molecular motion which, as illustrated in Fig. 1-4, are a major determinant of the nature of bulk polymers.

If the intermolecular forces are small and the cohesive energy is low, and the molecules have relatively flexible chains, they comply readily to applied stresses and have properties usually associated with elastomers. Somewhat higher cohesive energy densities, accompanied in some cases by bulky side groups giving stiffer chains, are characteristic of typical plastics. If the cohesive energy is higher still, the materials exhibit the high resistance

* This book follows the convention of defining the (Gibbs) *free energy* as $G = H - TS$, and the *work content* or Helmholtz free energy as $A = E - TS$.

TABLE 1-3. *Cohesive energy densities of linear polymers (Walker 1952, Small 1953)*

Polymer	Repeat Unit	Cohesive Energy Density, cal/cm^3
Polyethylene	$-CH_2CH_2-$	62
Polyisobutylene	$-CH_2C(CH_3)_2-$	65
Polyisoprene	$-CH_2C(CH_3)=CHCH_2-$	67
Polystyrene	$-CH_2CH(C_6H_5)-$	74
Poly(methyl methacrylate)	$-CH_2C(CH_3)(COOCH_3)-$	83
Poly(vinyl acetate)	$-CH_2CH(OCOCH_3)-$	88
Poly(vinyl chloride)	$-CH_2CHCl-$	91
Poly(ethylene terephthalate)	$-CH_2CH_2OCOC_6H_4COO-$	114
Poly(hexamethylene adipamide)	$-NH(CH_2)_6NHCO(CH_2)_4CO-$	185
Polyacrylonitrile	$-CH_2CHCN-$	237

to stress, high strength, and good mechanical properties typical of fibers, especially where molecular symmetry is favorable for crystallization.

GENERAL REFERENCES

Ketelaar 1953; Cottrell 1958; Pauling 1960, 1964; Pimentel 1960; Chu 1967; Phillips 1970.

BIBLIOGRAPHY

Baker 1964. W. O. Baker, "Structure and Electrical Properties of some Synthetic Solid Polymers," *J. Polymer Sci.* **C4,** 1633–1650 (1964).

Bevington 1954. J. C. Bevington, H. W. Melville, and R. P. Taylor, "The Termination Reaction in Radical Polymerization. Polymerizations of Methyl Methacrylate and Styrene at 25°," *J. Polymer Sci.* **12,** 449–459 (1954).

Billmeyer 1969. Fred W. Billmeyer, Jr., "Molecular Structure and Polymer Properties," *J. Paint Technol.* **41,** 3–16; erratum, 209 (1969).

Blumstein 1966. Alexandre Blumstein and Fred W. Billmeyer, Jr., "Polymerization of Adsorbed Monolayers. III. Preliminary Structure Studies in Dilute Solution of the Insertion Polymers," *J. Polymer Sci.* **A-2 4,** 465–474 (1966).

Bryant 1962. W. M. D. Bryant, "Free Energies of Formation of Fluorocarbons and their Radicals. Thermodynamics of Formation and Depolymerization of Polytetrafluoroethylene," *J. Polymer Sci.* **56,** 277–296 (1962).

Carothers 1929. W. H. Carothers, "An Introduction to the General Theory of Condensation Polymers," *J. Am. Chem. Soc.* **51,** 2548–2559 (1929).

Carothers 1931. Wallace H. Carothers, "Polymerization," *Chem. Revs.* **8**, 353–426 (1931).

Chu 1967. Benjamin Chu, *Molecular Forces: Based on the Baker Lectures of Peter J. W. Debye*, Interscience Div., John Wiley and Sons, New York, 1967.

Cottrell 1958. Tom L. Cottrell, *The Strengths of Chemical Bonds*, 2nd ed., Academic Press, New York, 1958.

DeWinter 1966. W. deWinter, "Double-Strand Polymers," *Revs. Macromol. Chem.* 1966–1967, 329–353 (1966).

Flory 1937. Paul J. Flory, "Mechanism of Vinyl Polymerizations," *J. Am. Chem. Soc.* **59**, 241–253 (1937).

Flory 1953. Paul J. Flory, *Principles of Polymer Chemistry*, Cornell University Press, Ithaca, N.Y., 1953.

Joyce 1948. R. M. Joyce, W. E. Hanford, and J. Harmon, "Free Radical-Initiated Reaction of Ethylene with Carbon Tetrachloride," *J. Am. Chem. Soc.* **70**, 2529–2532 (1948).

Ketelaar 1953. J. A. A. Ketelaar, *Chemical Constitution*, Elsevier Publishing Co., New York, 1953.

Lansing 1935. W. D. Lansing and E. O. Kraemer, "Molecular Weight Analysis of Mixtures by Sedimentation Equilibrium in the Svedberg Ultracentrifuge," *J. Am. Chem. Soc.* **57**, 1369–1377 (1935).

Margerison 1967. D. Margerison and G. C. East, *An Introduction to Polymer Chemistry*, Pergamon Press, New York, 1967.

Mark 1940. H. Mark and G. Stafford Whitby, eds., *Collected Papers of Wallace Hume Carothers on High Polymeric Substances*, Interscience Publishers, New York, 1940.

Mark 1966. Herman F. Mark and the Editors of Life, *Giant Molecules*, Time, Inc., New York, 1966.

Natta 1955a. G. Natta, Piero Pino, Paolo Corradini, Ferdinando Danusso, Enrico Mantica, Giorgio Mazzanti, and Giovanni Moranglio, "Crystalline High Polymers of α-Olefins," *J. Am. Chem. Soc.* **77**, 1708–1710 (1955).

Natta 1955b. G. Natta, "A New Class of α-Olefin Polymers with Exceptional Regularity of Structure" (in French), *J. Polymer Sci.* **16**, 143–154 (1955).

Ostwald 1914. Wilhelm Ostwald, *Die Welt der Vernachlässigten Dimensionen (The World of Neglected Dimensions)*, T. Steinkopff, Dresden, 1914.

Pauling 1960. Linus Pauling, *The Nature of the Chemical Bond*, 3rd ed., Cornell University Press, Ithaca, N.Y., 1960.

Pauling 1964. Linus Pauling and Roger Hayward, *The Architecture of Molecules*, W. H. Freeman and Co., San Francisco, Calif., 1964.

Phillips 1970. James C. Phillips, *Covalent Bonding in Crystals, Molecules and Polymers*, Chicago University Press, Chicago, Ill., 1970.

Pimentel 1960. George C. Pimentel and Aubrey L. McClellan, *The Hydrogen Bond*, W. H. Freeman and Co., San Francisco, Calif., 1960.

Price 1942. Charles C. Price, Robert W. Kell, and Edwin Krebs, "Addition Polymerization Catalyzed by Substituted Acyl Peroxides," *J. Am. Chem. Soc.* **64**, 1103–1106 (1942).

Purves 1954. Clifford B. Purves, "Historical Survey," Chap. III A in Emil Ott, Harold M. Spurlin and Mildred W. Grafflin, eds., *Cellulose and Cellulose Derivatives*, 2nd ed., Part I, Interscience Publishers, New York, 1954.

Small 1953. P. A. Small, "Some Factors Affecting the Solubility of Polymers," *J. Appl. Chem.* **3**, 71–80 (1953).

Staudinger 1920. H. Staudinger, "Polymerization" (in German), *Ber.* **53B**, 1073–1085 (1920).

Staudinger 1925. H. Staudinger, "The Constitution of Polyoxymethylenes and Other High-Molecular Compounds" (in German), *Helv. Chim. Acta* **8**, 67–70 (1925).

Staudinger 1928. H. Staudinger, "The Constitution of High Polymers. XIII" (in German), *Ber.* **61B**, 2427–2431 (1928).

Staudinger 1930. H. Staudinger and W. Heuer, "Highly Polymerized Compounds. XXXIII. A Relation Between the Viscosity and the Molecular Weight of Polystyrenes" (in German), *Ber.* **63B**, 222–234 (1930).

Staudinger 1935. H. Staudinger and A. Steinhofer, "Highly Polymerized Compounds. CVII. Polystyrenes" (in German), *Ann.* **517**, 35–53 (1935).

Tawn 1969. A. R. H. Tawn, "New Concepts in Polymer Architecture," *J. Oil Colour Chem. Assoc.* **52**, 600–622 (1969).

Walker 1952. E. E. Walker, "The Solvent Action of Organic Substances on Polyacrylonitrile," *J. Appl. Chem.* **2**, 470–481 (1952).

2

Polymer Solutions

A. Criteria for Polymer Solubility

The solution process Dissolving a polymer is a slow process that occurs in two stages. First, solvent molecules slowly diffuse into the polymer to produce a swollen gel. This may be all that happens—if, for example, the polymer-polymer intermolecular forces are high because of crosslinking, crystallinity, or strong hydrogen bonding. But if these forces can be overcome by the introduction of strong polymer-solvent interactions, the second stage of solution can take place. Here, the gel gradually disintegrates into a true solution. Only this stage can be materially speeded by agitation. Even so, the solution process can be quite slow (days or weeks) for materials of very high molecular weight.

The degree of swelling of a lightly crosslinked polymer can be measured and related to the thermodynamic properties of the system, but this pheno-menon is not widely utilized at the present time.

Polymer texture and solubility Solubility relations in polymer systems are more complex than those among low-molecular-weight compounds, because of the size difference between polymer and solvent molecules, the viscosity of the system, and the effects of the texture and molecular weight of the polymer. In turn, the presence or absence of solubility as conditions (such as the nature of the solvent or the temperature) are varied can give much information about the polymer; this is in fact the topic of most of this chapter.

From what has already been said, it is clear that the topology of the polymer is highly important in determining its solubility. Crosslinked polymers do not dissolve but only swell if indeed they interact with the solvent at all. In part, at least, the degree of this interaction is determined by the extent of crosslinking: lightly crosslinked rubbers swell extensively in solvents in which the unvulcanized material would dissolve, but hard rubbers, like many thermosetting resins, may not swell appreciably in contact with any solvent.

The absence of solubility does not imply crosslinking, however. Other features may give rise to sufficiently high intermolecular forces to prevent solubility. The presence of crystallinity is the common example. Many crystalline polymers, particularly nonpolar ones, do not dissolve except at temperatures near their crystalline melting points. Because crystallinity decreases as the melting point is approached (Chapter 5) and the melting point is itself depressed by the presence of the solvent, solubility can often be achieved at temperatures significantly below the melting point. Thus linear polyethylene, with crystalline melting point $T_m = 135°C$, is soluble in many liquids at temperatures above 100°C, while even polytetrafluoroethylene, $T_m = 325°C$, is soluble in some of the few liquids that exist above 300°C. More polar crystalline polymers, such as 66 nylon, $T_m = 265°C$, can dissolve at room temperature in solvents that interact strongly with them (for example, to form hydrogen bonds).

There is little quantitative information about the influence of branching on solubility; in general, branched species appear to be more readily soluble than their linear counterparts of the same chemical type and molecular weight.

Of all these systems, the theory of solubility, based on the thermodynamics of polymer solutions, is highly developed only for linear polymers in the absence of crystallinity. This theory is described in Sections C and D. Here, the chemical nature of the polymer is by far the most important determinant of solubility, as is elucidated in the remainder of this section. The influence of molecular weight (within the polymer range) is far less, but it is of great importance to fractionation processes (Sections D and E) which yield information about the distribution of molecular weights in polymer samples.

Solubility parameters   Solubility occurs when the free energy of mixing

$$\Delta G = \Delta H - T \, \Delta S$$

is negative. It was long thought that the entropy of mixing ΔS was always positive, and therefore the sign of ΔG was determined by the sign and magnitude of the heat of mixing ΔH. For reasonably nonpolar molecules and in the absence of hydrogen bonding, ΔH is positive and was assumed to be the same as that derived rigorously for the mixing of small molecules. For this case, the heat of mixing per unit volume is (Hildebrand 1950)

$$\Delta H = v_1 v_2 (\delta_1 - \delta_2)^2$$

where v is volume fraction, and subscripts 1 and 2 refer to solvent and polymer, respectively. The quantity δ^2 is the cohesive energy density or, for small molecules, the energy of vaporization per unit volume. The quantity δ is known as the *solubility parameter*. (This expression for the heat of mixing is one of several alternates used in theories of the thermodynamics of polymer solutions; in Section $C, \Delta H$ is written in a different but equivalent way.)

The value of the solubility-parameter approach is that δ can be calculated for both polymer and solvent. As a first approximation, and in the absence of strong interactions such as hydrogen bonding, solubility can be expected if $\delta_1 - \delta_2$ is less than 1.7–2.0, but not if it is appreciably larger.

This approach to polymer solubility, pioneered by Burrell (1955), has been extensively used, particularly in the paint industry. A few typical values of δ_1 and δ_2 are given in Table 2-1; for polymers, they are the square roots

TABLE 2-1 *Typical values of the solubility parameter* δ *(Walker 1952, Small 1953, Bristow 1958 a, b)*

Solvent	δ_1, $(cal/cm^3)^{1/2}$	Polymer	δ_2, $(cal/cm^3)^{1/2}$
n-Hexane	7.24	Polyethylene	7.9
Carbon tetrachloride	8.58	Polystyrene	8.6
2-Butanone	9.04	Poly(methyl methacrylate)	9.1
Benzene	9.15	Poly(vinyl chloride)	9.5
Chloroform	9.24	Poly(ethylene terephthalate)	10.7
Acetone	9.71	66 Nylon	13.6
Methanol	14.5	Polyacrylonitrile	15.4

of the cohesive-energy densities of Table 1-3. Extensive tabulations have been published (Burrell 1966, Hoy 1970). Perhaps the easiest way to determine δ_2 for a polymer of known structure is by the use of the molar-attraction constants E of Table 2-2;

$$\delta_2 = \frac{\rho \sum E}{M}$$

where values of E are summed over the structural configuration of the repeating unit in the polymer chain, with repeat molecular weight M and density ρ.

The solubility-parameter approach is useful only in the absence of strong polymer-solvent interactions, and a number of modifications have been proposed to account for these cases. Hydrogen bonding is often used as a second parameter in addition to δ and is incorporated in some of Burrell's

TABLE 2-2   *Molar attraction constants E, (cal cm³)$^{1/2}$/mole (Hoy 1970)*

Group	E	Group	E
—CH$_3$	148	NH$_2$	226.5
—CH$_2$—	131.5	—NH—	180
>CH—	86	—N—	61
>C<	32	C≡N	354.5
CH$_2$=	126.5	NCO	358.5
—CH=	121.5	—S—	209.5
>C=	84.5	Cl$_2$	342.5
—CH= aromatic	117	Cl primary	205
—C= aromatic	98	Cl secondary	208
—O— ether, acetal	115	Cl aromatic	161
—O— epoxide	176	F	41
—COO—	326.5	Conjugation	23
>C=O	263	*cis*	−7
—CHO	293	*trans*	−13.5
(CO)$_2$O	567	6-Membered ring	−23.5
—OH→	226	*ortho*	9.5
OH aromatic	171	*meta*	6.5
—H acidic dimer	−50.5	*para*	40

tabulations (1966), dielectric constant or some related quantity can be used as a third (Crowley 1966), but the point of diminishing returns is soon reached.

It is now known, as discussed in Section C, that both ΔH and ΔS can be negative under some circumstances. While this obviously invalidates the conditions for which the solubility-parameter concept is appropriate, that concept is nevertheless still extremely useful and should not be abandoned without test, even in these unusual cases.

In contrast to the above considerations of the thermodynamics of dis-solution of polymers, the rate of this step depends primarily on how rapidly the polymer and the solvent diffuse into one another (Ueberreiter 1962, Asmussen 1963). Solvents that promote rapid solubility are usually small, compact molecules, but these kinetically good solvents need not be thermo-dynamically good as well. Mixtures of a kinetically good and a thermodyna-mically good liquid are often very powerful and rapid polymer solvents.

GENERAL REFERENCES

Hildebrand 1950; Allen 1959; Gardon 1965; Burrell 1966, 1970; Hoy 1970.

B. *Conformations of Dissolved Polymer Chains*

As specified in Chapter 1, those arrangements of the polymer chain differing by reason of rotations about single bonds are termed *conformations*.* In solution, a polymer molecule is a randomly coiling mass most of whose conformations occupy many times the volume of its segments alone. The average density of segments within a dissolved polymer molecule is of the order of 10^{-4}–10^{-5} g/cm^3. The size of the molecular coil is very much influenced by the polymer-solvent interaction forces. In a thermodynamically "good" solvent, where polymer-solvent contacts are highly favored, the coils are relatively extended. In a "poor" solvent, they are relatively contracted. It is the purpose of this section to describe the conformational properties of both ideal and real polymer chains.

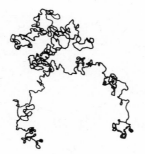

Fig. 2-1. Model of one of the many conformations of a random-coil chain of 1000 links (Treloar 1958).

The importance of the random-coil nature of the dissolved, molten, amorphous, and glassy states of high polymers cannot be overemphasized. As the following chapters show, many important physical as well as thermodynamic properties of high polymers result from this characteristic structural feature. The random coil (Fig. 2-1) arises from the relative freedom of rotation associated with the chain bonds of most polymers and the formidably large number of conformations accessible to the molecule.

One of these conformations, the fully extended chain (often an all-*trans* planar zigzag carbon chain, Fig. 2-2) has special interest because its length,

---

* Here we differ from some authorities, notably Flory (1969), who prefers the use, well established in statistical mechanics, of the term *configuration* here. We follow the convention of organic chemistry in which configuration designates stereochemical arrangement (see Chapter 5*A*).

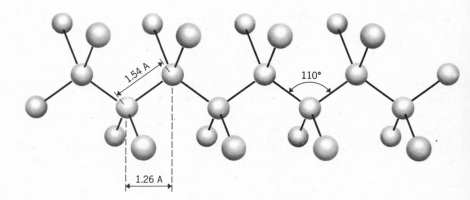

Fig. 2-2. The fully extended all-*trans* conformation of the carbon-carbon chain (Billmeyer 1969).

the *contour length* of the chain, can be calculated in a straightforward way. In all other cases, the size of the random coil must be expressed in terms of statistical parameters such as the root-mean-square distance between its ends, $(\overline{r^2})^{1/2}$, or its *radius of gyration*, the root-mean-square distance of the elements of the chain from its center of gravity, $(\overline{s^2})^{1/2}$. For linear polymers the mean-square end-to-end distance and the square of the radius of gyration are simply related: $\overline{r^2} = \overline{6s^2}$.

The freely jointed chain A simple model of a polymer chain consists of a series of x links of length l joined in a linear sequence with no restrictions on the angles between successive bonds. The probability W that such an array has a given end-to-end distance r can be calculated by the classical random-flight method (Rayleigh 1919, Chandrasekhar 1943). The most important result of the calculation is that the end-to-end distance is proportional to the square root of the number of links:

$$(\overline{r_f^2})^{1/2} = lx^{1/2}$$

(The subscript f indicates the random-flight end-to-end distance.) Thus $(\overline{r_f^2})^{1/2}$ is proportional to $M^{1/2}$, or $\overline{r_f^2}/M$ is a characteristic property of the polymer chain structure, independent of molecular weight or length.

The distribution of end-to-end distances over the space coordinates $W(x, y, z)$ is given by the Gaussian distribution function shown graphically in Fig. 2-3. This is the density distribution of end points and shows that if one end of the chain is taken at the origin, the probability is highest of finding the other end in a unit volume near the origin. This probability decreases continuously with increasing distance from the origin. On the other hand,

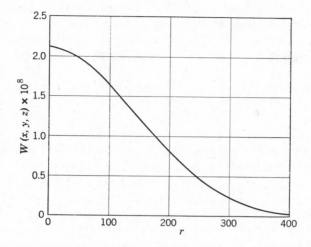

Fig. 2-3. Gaussian distribution of end-to-end distances in a random coil of 10^4 links each 2.5 A long. $W(x,y,z)$ is expressed in A^{-3} and r in A.

the probability of finding a chain end within the volume of a spherical shell between distances r and $r + dr$ from the origin has a maximum, as shown in Fig. 2-4. It should be noted that the Gaussian distribution does not fall to zero at large extensions and so must fail to describe the actual conformations of the chain where r nears the contour length. Over most of the range of interest for the dilute solution properties of polymers this is of little consequence; better approximations are available where needed, as in the treatment of rubber elasticity (Chapter 6B).

The freely jointed or random-flight model seriously underestimates the true dimensions of real polymer molecules for two reasons: first, restrictions to completely free rotation, such as fixed bond angles and steric hindrances,

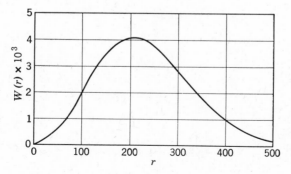

Fig. 2-4. Radial distribution of end-to-end distances for the coil of Fig. 2-3. $W(r)$ is expressed in A^{-1}.

provide short-range interactions leading to larger dimensions than calculated above; and second, long-range interactions resulting from the inability of chain atoms far removed from one another to occupy the same space at the same time result in a similar effect.

Short-range interactions   One of the triumphs of modern polymer chemistry is the extent to which Flory and his collaborators (Flory 1969), among others, have calculated completely and accurately the effect of short-range interactions on the dimensions of random-coil polymers. Several effects are involved. The restriction to fixed bond angle θ expands the chain by a factor of $[(1 - \cos\theta)/(1 + \cos\theta)]^{1/2}$, equal to $\sqrt{2.0}$ for carbon-carbon bonds. Restricted rotation, whether resulting from steric hindrances and resulting potential-energy barriers or from resonance leading to rigid planar conformations, increases dimensions still more. Finally, conformations that would place two atoms close together along the chain too close to one another are not allowed, leading to further expansion. The most important of these is the so-called *pentane interference* (Taylor 1948) between the first and fifth chain atoms in a sequence.

The net results of these short-range interactions can be expressed as a *characteristic ratio* of the square of the actual chain dimensions in the absence of long-range interactions (called the *unperturbed dimensions* and given the symbol $\overline{r_0^2}$) and the square of the random-flight end-to-end distance, $l^2 x$. Typical values of the characteristic ratio for some common polymers are given in Table 2-3.

TABLE 2-3.   The characteristic ratio $\overline{r_0^2}/l^2 x$ for some common polymers, evaluated at the limit of high chain length (Flory 1969)

Polymer	$\dfrac{\overline{r_0^2}}{l^2 x}$
Polyethylene	6.7
Polystyrene (atactic)	10.0
Polypropylene (isotactic)	5.7
Poly(methyl methacrylate)	
atactic	6.9
isotactic	9.3
syndiotactic	~ 7
Poly(ethylene oxide)	4.0
66 Nylon	5.9

Long-range interactions and the excluded volume Correction of polymer chain dimensions for short-range interactions still fails to eliminate conformations in which two widely separated chain segments occupy the same space. Each segment of a real chain exists within a volume from which all other segments are excluded. The theoretical calculation of the *excluded volume* and its effect on the dimensions of the polymer chain has remained a major unsolved problem in polymer science for many years (Berry 1970). A statistical approach has been more fruitful and has allowed the simulation of chain conformations by digital computer up to chain lengths of several hundred segments (Smith 1963). From these studies has come the result that, at sufficiently high chain length, proportionality between the end-to-end distance and the square root of the number of segments is retained.

Practically, the effect of the long-range interactions is to cause a further expansion of the chain over its unperturbed dimensions, since more of the compact conformations, with small values of $\overline{r^2}$, must be excluded. The actual dimensions of the real chain can be expected to exceed its unperturbed dimensions by an expansion factor α, thus $(\overline{r^2})^{1/2} = \alpha(\overline{r_0^2})^{1/2}$. The value of α depends on the nature of the solvent: a solvent for which α is large is said to be a thermodynamically "good" solvent for that polymer, and vice versa.

In a sufficiently poor solvent, or at a sufficiently low temperature (since solvent power and α vary with the temperature), it is possible to achieve the condition $\alpha = 1$, where the chain attains its unperturbed dimensions. This special point is called the *Flory temperature* Θ; a solvent used at $T = \Theta$ is called a Θ-*solvent*. The calculation of α and Θ and their relation to thermodynamic quantities is treated in Section *C*.

Nonlinear chains Since branched chains possess a multiplicity of ends it is more suitable to speak in terms of the radius of gyration $(\overline{s^2})^{1/2}$. A branched molecule occupies a smaller volume than a linear one with the same number of segments, that is, the same molecular weight. It is convenient to express this diminution of size as a factor $g = \overline{s^2}(\text{branched})/\overline{s^2}(\text{linear})$, which can be calculated statistically for various degrees and types of branching (Zimm 1949). For random branching, five trifunctional branch points per molecule reduce g to about 0.70. The change in size is the basis of a method to measure branching (Chapter 3*D*).

Preferred conformations Even at $T = \Theta$ and for solvents with about the same Θ-temperatures, it is found that some polymers do not have the same unperturbed dimensions in different solvents. From this and other evidence, the belief is arising that portions of dissolved polymer chains may at times exist in preferred nonrandom conformations, such as helices (see also Chapter 5*B*).

GENERAL REFERENCES

Flory 1953; Tompa 1956; Volkenstein 1963; Birshtein 1966; Opschoor 1966; Flory 1969; Lowry 1970.

C. Thermodynamics of Polymer Solutions*

The behavior of polymers toward solvents is characteristic and different from that of low-molecular-weight substances. The size and conformations of dissolved polymer molecules require special theoretical treatment to explain their solution properties. Conversely, it is possible to obtain information about the size and shape of polymer molecules from studies of their solution properties.

The first group of properties of interest includes those depending upon equilibrium between two phases, one or both of which is a solution of the polymer. In this section we discuss situations in which one phase is pure solvent; in Section *D*, solubility phenomena in which both phases contain polymer.

Thermodynamics of simple liquid mixtures

From the condition for equilibrium between two phases may be derived relations such as that for the free energy of dilution of a solution:

$$\Delta G_A = kT \ln \left(\frac{p_A}{p_A{}^0} \right) \tag{2-1}$$

where ΔG is the free energy of dilution resulting from the transfer of one molecule of liquid A from the pure liquid state with vapor pressure $p_A{}^0$ to a large amount of solution with vapor pressure p_A. This expression is written in terms of one of the *colligative properties* of solutions: vapor-pressure lowering, freezing-point depression, boiling-point elevation, and osmotic pressure. The values of these properties cannot be related to the composition of the system by pure thermodynamic reasoning. It is necessary to know the type of variation of one property with concentration; the variation of the others may then be deduced.

Ideal solutions   In the simplest type of mixing, the molecules of components A and B have roughly the same size and shape and similar force fields. They

---

* For reviews of thermodynamics as background to this subject see Hildebrand 1950, Lewis 1961.

may then form an *ideal solution*, defined as one in which Raoult's law is obeyed. This law states that the partial vapor pressure of each component in the mixture is proportional to its mole fraction. Therefore

$$p_A = p_A{}^0 \frac{N_A}{N_A + N_B} = p_A{}^0 n_A \qquad (2\text{-}2)$$

where n denotes mole fraction, and Eq. 2-1, for example, becomes

$$\Delta G_A = kT \ln n_A \qquad (2\text{-}3)$$

The total free energy of mixing is

$$\begin{aligned}
\Delta G &= N_A \Delta G_A + N_B \Delta G_B \\
&= kT(N_A \ln n_A + N_B \ln n_B)
\end{aligned} \qquad (2\text{-}4)$$

The conditions for ideal mixing imply that the heat of mixing $\Delta H = 0$, i.e., the components mix without change in energy. Since $\Delta G = \Delta H - T \Delta S$, the entropy of mixing is given by

$$\Delta S = -k(N_A \ln n_A + N_B \ln n_B) \qquad (2\text{-}5)$$

which is positive for all compositions, so that, by the second law, spontaneous mixing occurs in all proportions.

Other types of mixing   In practice few liquid mixtures obey Raoult's law. Three types of deviations are distinguished:

    a. "Athermal" solutions, in which $\Delta H = 0$ but ΔS is no longer given by Eq. 2-5.
    b. "Regular" solutions, in which ΔS has the ideal value but ΔH is finite.
    c. "Irregular" solutions, in which both ΔH and ΔS deviate from the ideal values.

It is usually found in systems of similar-sized molecules that ΔS is nearly ideal when $\Delta H = 0$; therefore athermal solutions are nearly ideal. Many mixtures are found for which ΔH is finite, however. Such cases arise when the intermolecular force fields around the two types of molecule are different. Expressions for the heat of mixing may be derived in terms of the cohesive energy density of these force fields (Section A).

Entropy and heat of mixing of polymer solutions

Deviations from ideal behavior   Polymer solutions invariably exhibit large deviations from Raoult's law except at extreme dilutions, where ideal behavior is approached as an asymptotic limit. At concentrations above a few per cent, deviations from ideality are so great that the ideal law is of little value for predicting or correlating the thermodynamic properties of polymer solutions. Even if the mole fraction is replaced with the volume fraction, in view of the different sizes of the polymer and solvent molecules, there is not a good correlation with the experimental results.

Entropy of mixing Deviations from ideality in polymer solutions arise largely from small entropies of mixing. These are not abnormal but are the natural result of the large difference in molecular size between the two components. They can be interpreted in terms of a simple molecular model. The molecules in the pure liquids and the mixture are assumed to be representable without serious error by a lattice. A two-dimensional representation of such an arrangement for nonpolymer liquids is shown in Fig. 2-5a. Whereas the molecules of a pure component can be arranged in only one way on such a lattice, assuming that they cannot be distinguished from one another, the molecules of a mixture of two components can be arranged on a lattice in a large but calculable number of ways, W. By the Boltzmann relation the entropy of mixing $\Delta S = k \ln W$. Equation 2-5 results for the case of molecules which can replace one another indiscriminately on the lattice.

It is assumed that the polymer molecules consist of a large number x of chain segments of equal length, flexibly joined together. Each link occupies one lattice site, giving the arrangement of Fig. 2-5b. The solution is assumed to be concentrated enough that the occupied lattice sites are distributed at random rather than lying in well-separated regions of x occupied sites each. It can now be seen qualitatively why the entropy of mixing of polymer solutions is small compared to that with normal solutes. There are fewer ways

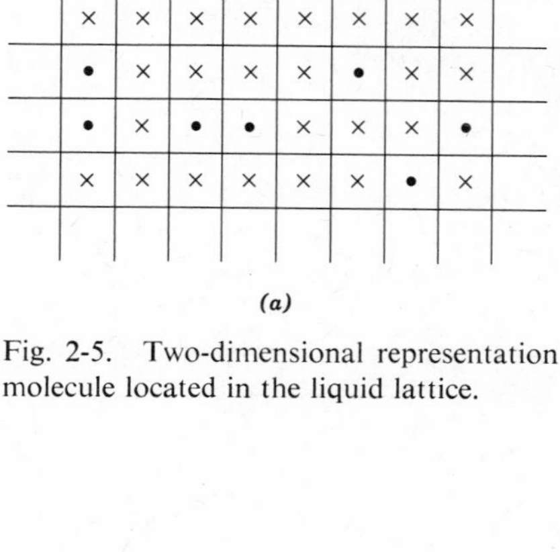

(a) 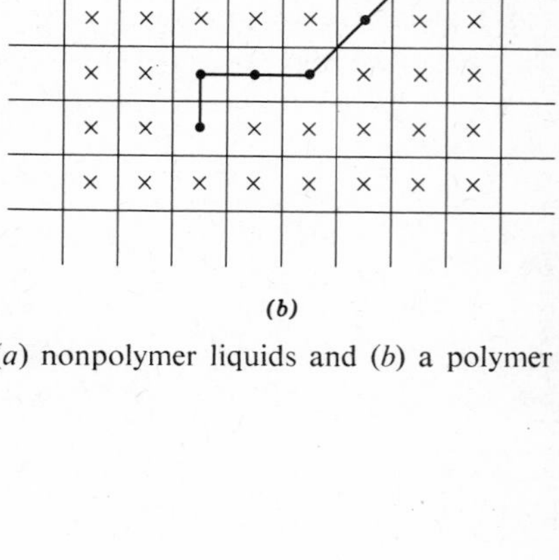   (b)

Fig. 2-5. Two-dimensional representation of (a) nonpolymer liquids and (b) a polymer molecule located in the liquid lattice.

in which the same number of lattice sites can be occupied by polymer segments: fixing one segment at a lattice point severely limits the number of sites available for the adjacent segment. The approximate calculation of W for such a model is due separately to Flory (1942) and Huggins (1942a,b,c); their results (which differ only in minor detail) are known as the *Flory-Huggins theory* of polymer solutions. The entropy of mixing is analogous to that given in Eq. 2-5 for simple liquids: for polymer solutions,

$$\Delta S = -k(N_1 \ln v_1 + N_2 \ln v_2) \tag{2-6}$$

where subscript 1 denotes the solvent and 2 the polymer, and v_1 and v_2 are *volume fractions* defined as

$$v_1 = \frac{N_1}{N_1 + xN_2}$$
$$v_2 = \frac{xN_2}{N_1 + xN_2} \tag{2-7}$$

Heat and free energy of mixing   The heat of mixing of polymer solutions is analogous to that of ordinary solutions:*

$$\Delta H = \chi_1 k T N_1 v_2 \tag{2-8}$$

where χ_1 characterizes the interaction energy per solvent molecule divided by kT. Combining Eqs. 2-6 and 2-8 gives the Flory-Huggins expression for the free energy of mixing of a polymer solution with normal heat of mixing:

$$\Delta G = kT(N_1 \ln v_1 + N_2 \ln v_2 + \chi_1 N_1 v_2) \tag{2-9}$$

From this expression may be derived many useful relations involving experimentally obtainable quantities. For example, the partial molar free energy of mixing is

$$\Delta \bar{G}_1 = kT\left[\ln(1 - v_2) + \left(1 - \frac{1}{x}\right)v_2 + \chi_1 v_2^2\right] \tag{2-10}$$

and from it is obtained the osmotic pressure

$$\pi = -\frac{kT}{V_1}\left[\ln(1 - v_2) + \left(1 - \frac{1}{x}\right)v_2 + \chi_1 v_2^2\right] \tag{2-11}$$

where V_1 is the molecular volume of the solvent. If the logarithmic term in Eq. 2-11 is expanded and only low powers of v_2 are retained, the working equation for molecular weight determination by osmotic pressure measurement (Eq. 2-17) is obtained.

---

* The nomenclature of Flory is used throughout. Huggins writes $\mu_h°$ instead of χ_1, and Hildebrand expresses ΔH in terms of the solubility parameter as in Section A.

Experimental results with polymer solutions

The first system for which accurate experimental results were compared with the Flory-Huggins theories was that of rubber in benzene (Gee 1946, 1947). Free energies for the system were calculated from vapor-pressure measurements by Eq. 2-1, and heats of solution were measured in a calorimeter. The predicted concentration dependence of the partial molar heat of dilution

$$\Delta \bar{H}_1 = kT\chi_1 v_2{}^2 \qquad (2\text{-}12)$$

was not observed. Despite this fact, the entropy of mixing, calculated from the heat and free energy, was in fair agreement with theory except for the dilute solution region.

The properties of other systems do not in general conform as well to the predictions of the theory. For example, the anticipated independence of χ_1

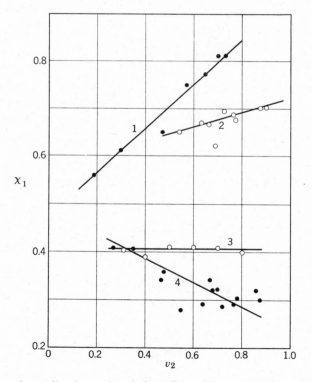

Fig. 2-6. Experimentally observed variation of χ_1 with concentration. Curve 1, poly-(dimethyl siloxane) in benzene (Newing 1950); curve 2, polystyrene in benzene (Bawn 1950); curve 3, rubber in benzene (Gee 1946, 1947); curve 4, polystyrene in toluene (Bawn 1950).

of v_2 is usually not observed (Fig. 2-6) (but see the free-volume theories described below).

Flory-Krigbaum theory   The lattice model used in the Flory-Huggins treatment neglects the fact that a very dilute polymer solution must be discontinuous in structure, consisting of domains or clusters of polymer chain segments separated on the average by regions of polymer-free solvent. Flory and Krigbaum (1950) assume such a model, in which each cloud of segments is approximately spherical, with a density that is a maximum at the center and decreases in an approximately Gaussian function with distance from the center. Within the volume occupied by the segments of one molecule, all other molecules tend to be excluded. Within such an excluded volume (Section *B*) occur long-range intramolecular interactions whose thermodynamic functions can be derived. The partial molar heat content, entropy, and free energy of these interactions are

$$\Delta \bar{H}_1 = kT\kappa_1 v_2{}^2$$
$$\Delta \bar{S}_1 = k\psi_1 v_2{}^2 \qquad (2\text{-}13)$$
$$\Delta \bar{G}_1 = kT(\kappa_1 - \psi_1)v_2{}^2$$

where, by comparison with the expansion of Eq. 2-10,

$$\kappa_1 - \psi_1 = \chi_1 - \tfrac{1}{2} \qquad (2\text{-}14)$$

It is convenient to define as a parameter the *Flory temperature* Θ such that $\Theta = \kappa_1 T/\psi_1$. It follows that

$$\psi_1 - \kappa_1 = \psi_1\left(1 - \frac{\Theta}{T}\right) = \tfrac{1}{2} - \chi_1 \qquad (2\text{-}15)$$

and that at the temperature $T = \Theta$ the partial molar free energy due to polymer-solvent interactions is zero and deviations from ideal solution behavior vanish. The excluded volume becomes smaller as the solvent becomes poorer and vanishes at $T = \Theta$, where the molecules interpenetrate one another freely with no net interactions. At temperatures below Θ they attract one another and the excluded volume is negative. If the temperature is much below Θ, precipitation occurs (Section *D*).

The parameter α can be evaluated in terms of thermodynamic quantities:

$$\alpha^5 - \alpha^3 = 2C_m \psi_1\left(1 - \frac{\Theta}{T}\right)M^{1/2} \qquad (2\text{-}16)$$

where C_m lumps together numerical and molecular constants. This equation, which involves the assumption of validity of theories involving intermolecular interactions as well as intramolecular interactions, must be accepted with some reservations. It predicts that α increases without limit with increasing molecular weight; since $\overline{r_0^2}$ is proportional to M, $\overline{r^2}/M$ should increase with M. It also leads to the fact that, at $T = \Theta$, $\alpha = 1$ and molecular dimensions are unperturbed by intramolecular interactions. Since α depends on the entropy parameter ψ, it is larger in better solvents.

The *second virial coefficient* A_2 is related to the coefficient of v_2^2 in the expansion of Eq. 2-11,

$$\pi = \frac{kT}{V_1}\left(\frac{v_2}{x} + (\tfrac{1}{2} - \chi_1)v_2^2 + \ldots\right) \tag{2-17}$$

and is defined as

$$A_2 = \frac{\bar{v}_2^2}{N_0 V_1}(\tfrac{1}{2} - \chi_1) \tag{2-18}$$

consistent with Eq. 3-6. Here $\bar{v}_2$ is the specific volume of the polymer. This simple treatment does not describe the temperature and molecular-weight dependence of A_2 in good agreement with experiment.

Free-volume theories

From what has been said, it can be seen that both the Flory-Huggins and Flory-Krigbaum theories have serious shortcomings. Both are based on, and conserve the important features of, the theories of " regular " solutions of small molecules. Only the appropriate combinatorial entropy of mixing has been modified to fit the polymer case. The most important assumption which is retained is that there is no volume change on mixing.

The deviations between theory and experiment for the interaction parameter χ_1 led to its reinterpretation as a combined entropy and enthalpy parameter, as in Eqs. 2-13 and 2-14. The entropy term was considered to be a small negative correction, corresponding to the plausible idea of a small increase in order as polymer-polymer and solvent-solvent interactions were replaced with polymer-solvent interactions. Typical values of ψ_1 and κ_1 (for example, Tompa 1956, p. 170) show that this is not the case, for the entropic contribution to χ_1 is positive and much larger than the enthalpic term. Still other failures of the traditional theories of the thermodynamics of polymer solutions are discussed in Section D.

These difficulties have been overcome in powerful new theories first developed around 1952 by Prigogine (1957) and his co-workers, and extended and put into practice by Flory (Flory 1968, Eichinger 1968*a,b,c,d*) and Patterson (1967, 1969) among others. The major new factor in the theories is the recognition of the dissimilarity in the free volumes of the polymer and the solvent as a result of their great difference in size, the usual solvent being

much more expanded than the polymer. Mixing is not unlike the condensation of a gas (the solvent) into a dense medium (the polymer). The total
volume change on mixing is usually negative and is accompanied with a
negative (exothermic) ΔH and a negative contribution to ΔS. Thus, the total
ΔH on mixing consists of the usual positive interaction term (Eq. 2-8) and
the new negative term; both contribute to a new interaction parameter like
χ_1. Similarly, the entropy of mixing consists of the Flory-Huggins combinatorial term (Eq. 2-6) plus the new negative term; only the latter contributes
to the new χ. The relative magnitude of these contributions is discussed
further in Section *D*. Their exact mathematical formulation is beyond the
scope of this section.

It can be said, however, that a basic assumption of the free-volume
theories is that all liquids, polymers when noncrystalline, and mixtures of
these follow the same reduced equation of state. In Flory's formulation (see,
for example, Carpenter 1970) this can be written

$$\frac{\tilde{p}\tilde{V}}{\tilde{T}} = \frac{\tilde{V}^{1/3}}{\tilde{V}^{1/3}-1} - \frac{1}{\tilde{V}\tilde{T}} \tag{2-19}$$

which simplifies further at low pressures, including atmospheric, where
$\tilde{p} \to 0$. Here $\tilde{p}$, $\tilde{V}$, $\tilde{T}$ are the ratios of the real pressure, volume, and temperature to reduction parameters which can be evaluated from the thermal
coefficient of expansion and the isothermal compressibility. It is possible
to calculate reduced parameters for mixtures as well as for the pure components; from the resulting equations of state we can evaluate the volume
change on mixing, the heat and entropy of mixing, and such details as the
dependence of the new χ on concentration. It has been found that volume
changes on mixing can be predicted to within 10–15% of the experimental
values, and the correct variation of χ with concentration is predicted, in
contrast to the result of the Flory-Huggins theory discussed above.

GENERAL REFERENCES

Flory 1953; Tompa 1956; Huggins 1958; Miller 1966; Patterson 1967; Doolittle 1969;
Patterson 1969; Berry 1970; Carpenter 1970.

D. Phase Separation in Polymer Solutions

This section is concerned with the equilibrium between two liquid
phases, both of which contain amorphous polymer and one or more solvents.
The treatment of cases involving a crystalline polymer phase is more complex and is given in part in Section *E* and in part in Chapter 5E.

When the temperature of a polymer solution is raised or lowered, the solvent eventually becomes thermodynamically poorer. Finally a temperature is reached beyond which polymer and solvent are no longer miscible in all proportions. At more extreme temperatures, the mixture separates into two phases. Such phase separation may also be brought about by adding a nonsolvent liquid to the solution. In either case, it takes place when the interaction parameter χ_1 exceeds a critical value near $\frac{1}{2}$ (see below).

It was shown in Section C that the Flory-Huggins theory attributes χ_1 to polymer-solvent interactions alone, and predicts it to increase monotonically as the temperature decreases (Fig. 2-7). Thus, phase separation is predicted to take place only on lowering the temperature, the phase diagram looking like that in the lower portion of Fig. 2-8. The maximum temperature for phase separation is designated the *upper critical solution temperature*.

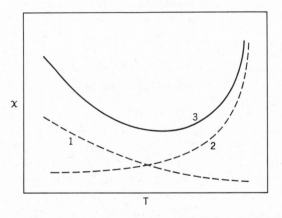

Fig. 2-7.   Qualitative behavior of the parameter χ_1 according to the free-volume theories. Curve 1, contribution from enthalpic polymer-solvent interactions (as predicted by the Flory-Huggins theory); curve 2, contribution due to free-volume dissimilarity between polymer and solvent; curve 3, total χ_1, sum of 1 and 2 (Patterson 1969).

Although it was not frequently observed until recent years (Freeman 1960), it is now recognized that phase separation invariably occurs also when the temperature is raised until a *lower critical solution temperature** (Fig. 2-8, upper area) is reached. This phenomenon is explained by the free-volume theories of polymer solutions described qualitatively in Section C. The contribution to χ_1 from the free-volume dissimilarity between polymer and

* The perverse names of the upper and lower critical solution temperatures relate more directly to an alternate case (the common example is nicotine in water) in which the upper critical solution temperature is higher than the lower and the phase diagram consists of a region of immiscibility completely surrounded by a one-phase region.

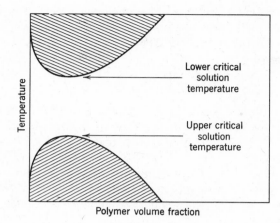

Fig. 2-8. Qualitative phase diagram of a polymer solution showing phase separation both on heating (at the lower critical solution temperature) and on cooling (at the upper critical solution temperature) (Patterson 1969).

solvent is an increasing function of the temperature (Fig. 2-7). The total interaction parameter thus goes through a minimum and two critical values of χ_1 are accessible. Each has the character of a Θ-temperature (see below); at each the second virial coefficient approaches zero and the polymer chain approaches its unperturbed dimensions (Delmas 1966).

For the remainder of this section reference is made only to phase separation at an upper critical solution temperature. The predictions of the Flory-Huggins theory remain unchanged except for the sign of the temperature dependence in the case of a lower critical solution temperature.

Binary polymer-solvent systems   The condition for equilibrium between two phases in a binary system is that the partial molar free energy of each component be equal in each phase. This condition corresponds to the requirement that the first and second derivatives of $\Delta \bar{G}_1$ (Eq. 2-10) with respect to v_2 be zero. Application of this condition leads to the critical concentration at which phase separation first appears:

$$v_{2c} = \frac{1}{1 + x^{1/2}} \simeq \frac{1}{x^{1/2}} \qquad (2\text{-}20)$$

This is a rather small volume fraction; for a typical polymer ($x \simeq 10^4$), $v_{2c} \simeq 0.01$. The critical value of χ_1 is given by

$$\chi_{1c} = \frac{(1 + x^{1/2})^2}{2x} \simeq \frac{1}{2} + \frac{1}{x^{1/2}} \qquad (2\text{-}21)$$

The critical value of χ_1 exceeds $\frac{1}{2}$ by a small increment depending on molecular weight and at infinite molecular weight equals $\frac{1}{2}$. The temperature at which phase separation begins is given by

$$\frac{1}{T_c} = \frac{1}{\Theta}\left[1 + \frac{1}{\psi_1}\left(\frac{1}{x^{1/2}} + \frac{1}{2x}\right)\right] \simeq \frac{1}{\Theta}\left(1 + \frac{C}{M^{1/2}}\right) \qquad (2\text{-}22)$$

where C is a constant for the polymer-solvent system. Thus $1/T_c$ (K) varies linearly with the reciprocal square root of molecular weight. The Flory temperature Θ is the critical miscibility temperature in the limit of infinite molecular weight.

The qualitative features of the theory are in agreement with experiment. The dependence of precipitation temperature on polymer concentration is shown in Fig. 2-9. The critical values T_c and v_{2c} correspond to the maxima

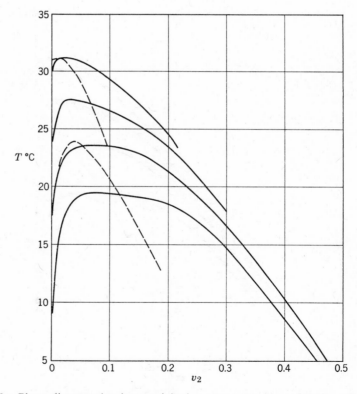

Fig. 2-9. Phase diagram showing precipitation temperature as a function of polymer concentration for fractions of polystyrene in diisobutyl ketone (Shultz 1952). Higher-molecular-weight fractions precipitate at higher temperatures. Solid lines, experimental; dotted lines, theoretical.

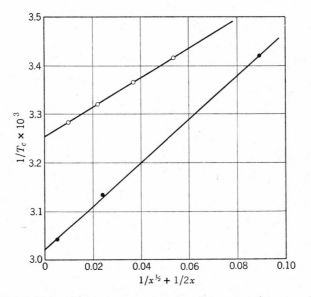

Fig. 2-10. Dependence of critical temperature for phase separation on molecular weight for (○) polystyrene in cyclohexane, and (●) polyisobutylene in diisobutyl ketone (Shultz 1952).

in the curves. In the neighborhood of the critical point the polymer concentration in the dilute phase is extremely small. The phenomenon of the coexistence of two liquid phases, one of which is a dilute solution and the other nearly pure solvent, is called *coacervation* (Section *E*).

The discrepancy between theory and experiment results (Koningsveld 1968) from the effect of finite molecular-weight distribution breadth in the polymer samples used.

The dependence of T_c upon molecular weight is shown in Fig. 2-10. The curves are accurately linear, and the Flory temperatures Θ obtained from the intercepts agree within experimental error ($<1°$) with those derived from osmotic measurements, taking Θ to be the temperature where A_2 is zero. Precipitation measurements on a series of sharp fractions offer perhaps the best method of determining Θ.

Ternary systems   Although a thorough discussion of ternary systems is outside the scope of this book, a few cases have particular interest. The most commonly encountered system is that of polymer, solvent, and nonsolvent. The phase relations are conveniently displayed in the usual ternary diagram (Fig. 2-11). The position of the *binodal curve*, along which two phases are in equilibrium, depends upon molecular weight; the limiting critical point at infinite molecular weight is the analog of Θ in a two-component system.

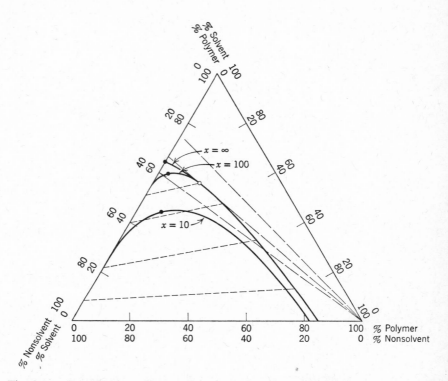

Fig. 2-11. Ternary phase diagram for a polymer–solvent–nonsolvent system, showing phase separation boundaries for the indicated values of x. Dotted lines are tie lines for x = 100, connecting points representing the compositions of pairs of phases in equilibrium. Dashed lines represent constant solvent–nonsolvent ratios. ●, critical points; ○, precipitation threshold for x = 100.

With appropriate conditions, solubility may be achieved for some polymers in mixtures of liquids, neither of which is a solvent for that polymer. A well-known example is the solubility of cellulose acetate in mixtures of ethanol and chloroform but not in either pure liquid (Bamford 1950). In contrast, mixtures of two different polymer types in a single solvent almost invariably separate into two phases, each of which contains practically all of one of the polymers.* An example is rubber and polystyrene in benzene (Dobry 1947).

* This general incompatibility also applies to mixtures of two polymers without solvent, for which the critical-point equations for phase separation predict that, for miscibility over the entire composition range, the difference in solubility parameters of the two polymers cannot exceed 0.35 at $M = 10,000$ or 0.1 at $M = 100,000$ (S. Krause, private communication).

Multicomponent systems The theory of phase separation in systems comprising a heterogeneous polymer in a single solvent is developed with the simplifying assumption that the interaction constants χ for all values of x are identical; only the size parameter x itself varies from one molecular species to another. In theory all details of the system could be calculated from a knowledge of the size distribution of the polymer and the equilibrium conditions. In practice this would be an enormous task; but the principal point of interest, namely the efficiency with which molecular species are separated between the two phases, is simply derived. It is given by the equation

$$\ln\left(\frac{v_x'}{v_x}\right) = \sigma x \tag{2-23}$$

where v_x' and v_x are, respectively, the concentrations of species x in the precipitated or more concentrated phase (primed) and the dilute phase (unprimed), and

$$\sigma = v_2\left(1 - \frac{1}{\bar{x}_n}\right) - v_2'\left(1 - \frac{1}{\bar{x}_n'}\right) + \chi_1[(1 - v_2)^2 - (1 - v_2')^2] \tag{2-24}$$

Here v_2' and v_2 are the total polymer concentrations in the two phases, and $\bar{x}_n$ is the number average of x. The parameter σ thus depends in a complicated way on the relative amounts of all the polymer species in each phase. The following conclusions may be drawn from the theory:

a. A part of any given polymer species is always present in each phase: in fact, every species is actually more soluble in the precipitated phase, i.e., $v_x' > v_x$ for all values of x.

b. Since σ depends on the concentrations of all species and on the details of their weight distribution, the result of a fractionation cannot be predicted in advance unless the details of the distribution are known. The Flory-Huggins theory predicts that only the number average $\bar{x}_n$ for each phase appears in σ, while an alternate analysis (Stockmayer 1949) expresses σ in terms of higher moments of the distribution for the whole polymer.

c. Fractionation is significantly more efficient in very dilute solutions: if $R = v'/v$, the ratio of the volumes of the precipitated and dilute phases, the fraction f_x of the constituent x in the dilute phase is

$$f_x = \frac{1}{1 + Re^{\sigma x}} \tag{2-25}$$

and that in the precipitated phase is

$$f_x' = \frac{Re^{\sigma x}}{1 + Re^{\sigma x}} \tag{2-26}$$

If $R \ll 1$, most of the smaller species remain in the dilute phase because of its greater volume. The larger partition factor v_x'/v_x for the higher-molecular-weight species causes them to appear selectively in the precipitated phase despite its smaller volume.

The resulting inefficient nature of the fractionation process is illustrated in Figs. 2-12 and 2-13. Figure 2-12 shows, in the three lower curves, the distribution of polymer species left in the dilute phase after precipitation at various values of R. The initial distribution is shown in the upper curve. The increase in efficiency as R becomes smaller is apparent. Figure 2-13 shows the overlapping distribution curves resulting even when R is unusually small. It is clear that separation into fractions which are close to monodisperse is a most difficult and time-consuming process. It should be noticed that the distribution curves of the fractions in Fig. 2-13 become narrower at lower molecular weights.

The foregoing discussion refers to fractionation by cooling from a single solvent. Similar considerations apply to fractionation at constant temperature in a solvent-nonsolvent system. A difference arises in that the

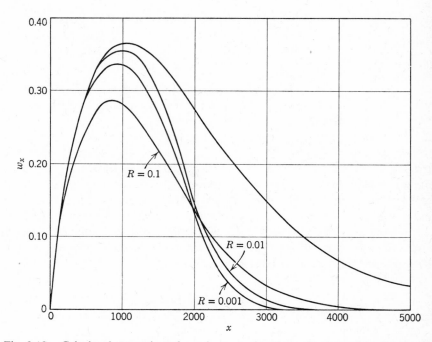

Fig. 2-12. Calculated separation of a polymer having the molecular weight distribution shown in the upper curve with $R = 0.1$, 0.01, and 0.001, f_x chosen to be $\frac{1}{2}$ at $x = 200$ in each case (Flory 1953).

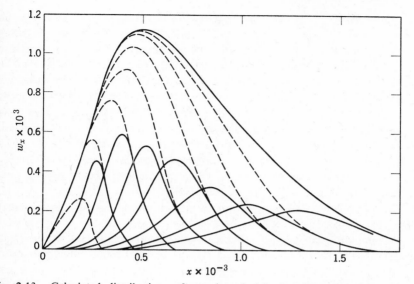

Fig. 2-13. Calculated distributions of a series of eight fractions separated from the distribution of the upper curve (Flory 1953). $R = 0.001$ in each case. Dashed curves represent the polymer remaining in the dilute phase after each successive precipitation.

solvent composition is different in the two phases. This difference should lead to more efficient fractionation in solvent-precipitant mixtures. Experimental evidence for such an effect has not been reported.

GENERAL REFERENCES

Flory 1953; Tompa 1956; Huggins 1967; Koningsveld 1968, 1970.

E. Fractionation of Polymers by Solubility

In this section are discussed only the most important and widely practiced techniques for fractionating polymers by solubility differences. Many variations and less useful techniques exist, and are reviewed by Hall (1959), Guzmán (1961), and Cantow (1967).

Bulk fractionation by nonsolvent addition

Fractional precipitation in bulk (Kotera 1967) is carried out by adding the nonsolvent to a dilute solution of the polymer until a slight turbidity develops at the temperature of fractionation. To ensure the establishment of equilibrium, the mixture may be warmed until it is homogeneous and allowed to cool slowly back to the required temperature, which should thereafter be carefully maintained. The precipitated phase is allowed to settle to a coherent layer, and the supernatant phase is removed. A further increment of precipitant is added and the process repeated. The polymer is isolated from the precipitated phase, which may still be relatively dilute (perhaps 10% polymer).

Refractionation is often used to achieve better separation. A more efficient procedure is to restrict the initial experiments to very dilute solutions, or to refine each initial fraction by a single reprecipitation at higher dilution, returning the unprecipitated polymer from this separation to the main solution before the next fraction is removed.

The solvent and the precipitant should be chosen so that precipitation occurs over a relatively wide range of solvent composition, yet is complete before too high a ratio of precipitant to solvent is reached. Other important considerations are the stability and volatility of the liquids, and their ability to form a highly swollen, mobile gel phase. Relatively simple equipment, for example a three-neck flask of several liters capacity (Fig. 2-14), is adequate for bulk precipitation fractionation.

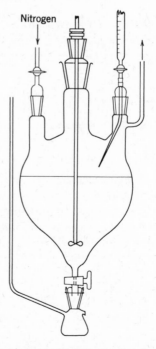

Fig. 2-14.   An apparatus suitable for bulk fractional precipitation (Hall 1959).

Solvent-gradient elution In elution methods (Elliott 1967), polymer is placed in contact with a series of liquids of gradually increasing solvent power. Species of lowest molecular weight, and thus highest solubility, dissolve in the first liquid, and successively higher molecular-weight fractions in subsequent liquids. To ensure rapid equilibration, the polymer must be present as a very thin film. For convenience, the polymer is often applied to a finely divided substrate such as sand or glass beads, which is subsequently used to pack a column (Fig. 2-15). By the use of mixing vessels, a continuous gradient of solvent-nonsolvent composition can be produced and eluted through the column. There is evidence both for (Kenyon 1960) and against (Traskos 1968) the suggestion that the efficiency of the subsequent extraction is improved if the polymer is selectively precipitated, with the lower-molecular-weight species precipitating last. Regardless of the mechanism, it is true that slow deposition of the polymer improves the resolution of the subsequent fractionation.

Thermal-gradient elution A modification of the solvent-gradient method utilizes a small temperature gradient from end to end of the column in addition to the gradient of solvent composition (Baker 1956, Porter 1967). The sample is initially confined to a small zone of the packing at the warm end of the column. Each species undergoes a series of solution and precipitation steps as it progresses down the column. Despite this apparent advantage, there seems to be little difference in efficiency between solvent-gradient and thermal-gradient elution methods (Schneider 1961a,b).

Analytical precipitation techniques

Analytical fractionation techniques are those in which fractions are not isolated, in contrast to the methods discussed above. They may be, but are not necessarily, carried out on a smaller scale than their preparative counterparts.

Summative fractionation In this method (Billmeyer 1950, Battista 1967), a series of polymer solutions is used. In each, part of the polymer is precipitated, the point of fractionation being varied throughout the series. Weight fractions w_x and molecular weights M_x of the precipitates are obtained; the product $w_x M_x$ may be related to the *integral* of the cumulative distribution curve. Since the two differentiations required to obtain the differential distribution would introduce intolerably large errors into the resulting differential curve, the data are properly analyzed to provide only information

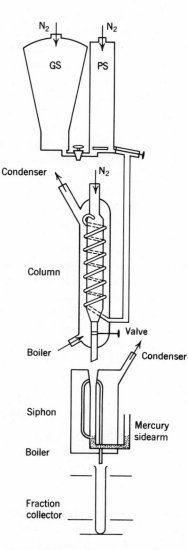

Fig. 2-15. Apparatus for column elution fractionation of polyethylene by the solvent–gradient method (Kokle 1965). Mixing flasks at the top provide a gradient of concentration by mixing good solvent (GS) and poor solvent (PS). This is prewarmed and injected at the top of the thermostatted column. The siphon delivers constant-volume fractions to the collector.

50

equivalent to the position of the maximum and breadth of the molecular-weight distribution curve. The method is advantageous for obtaining limited information about the molecular-weight distribution with a minimum of time, preliminary calibration effort, and equipment expense.

Turbidimetric titration In this technique (Giesekus 1967) a precipitant is added slowly to a polymer solution, and the turbidity due to precipitated polymer is measured by the decrease in intensity of a beam of transmitted light, or by the increase in intensity of scattered light. Selection of an appropriate solvent-nonsolvent system for a given polymer requires much preliminary study, and it is usually not possible to obtain accurate information on molecular-weight distribution from the experimental data. Nonetheless, the method is attractive for control purposes once the proper operating conditions are established.

Effect of polymer structure on solubility-based fractionation

Effect of chemical type It must be reemphasized (Section *A*) that the solubility of polymers is determined primarily by their chemical composition, and in the usual case molecular weight is only a secondary variable. If chemical composition differences exist among the polymer species in a sample, fractionation is likely to occur primarily as a result of these differences and to yield information on them rather than on the distribution of molecular weights (Fuchs 1967). Turbidimetric titration has often been used to explore qualitatively the effects of chemical composition in copolymer systems (Giesekus 1967).

Effect of chain branching On the basis of limited information (Schneider 1968), it appears that branching increases the solubility of high polymers. Thus, at any given stage in a solubility-based fractionation process, the branched polymer being separated will consist of a mixture of species, some with low branching and low molecular weight, others more branched but with compensating higher molecular weight, all having the same solubility. Additional information is needed to characterize such a system in terms of either variable separately.

Effect of crystallinity Precipitation of polymer to a crystalline, rather than an amorphous, phase can lead to fractionation by molecular weight (with control of the other variables just mentioned) only if the (depressed) crystalline melting point is a strong function of molecular weight. This is the case only for polymers of rather low molecular weight (say, less than 20,000), but in these cases the technique works rather well for the production of narrow fractions in large amounts (Pennings 1968).

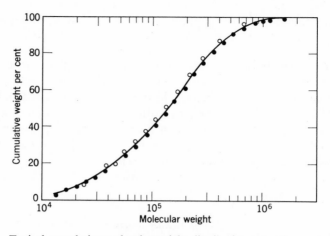

Fig. 2-16.   Typical cumulative molecular-weight distribution curve for a sample of poly-propylene (Crouzet 1969). $\bigcirc$, gradient-elution data; $\bullet$ data from gel permeation chroma-tography. Molecular weight is plotted logarithmically because of the very broad distribu-tion in this sample.

Treatment of data

Information about the distribution of molecular weights in a polymer is often the goal of a fractionation experiment. Such information requires not only that the variables mentioned in the previous paragraphs be con-trolled, but also requires the measurement of the weight and molecular weight of each fraction. For convenience, intrinsic viscosity is often used for the molecular weight measurement, but this must be limited to experiments with linear polymers.

The data are treated (Tung 1967) by plotting against molecular weight the cumulative or integral distribution curve of the combined weight of all fractions having molecular weight up to and including M. A typical curve of this type is shown in Fig. 2-16. The differential distribution curve is obtained by differentiation. Experimental errors, indicated by the scatter of points in the integral curve, are magnified in this process. Only the gross features of the differential curve, such as the position of the peak and the breadth of the curve, are experimentally significant. Unless the fractionation is performed with extreme care, spurious features, such as double peaks in the differential distribution curve, may appear.

GENERAL REFERENCES

Hall 1959; Guzmán 1961; Schneider 1965; Cantow 1967; Johnson 1967.

F. Gel Permeation Chromatography

Gel permeation chromatography is a powerful new separation technique (similar in principle to but advanced in practice (Maley 1965) over gel filtration as practiced by biochemists) which has found wide acceptance (Cazes 1966, 1970) in the polymer field since its discovery (Moore 1964) in 1961. The separation takes place in a chromatographic column filled with beads of a rigid porous "gel"; highly crosslinked porous polystyrene and

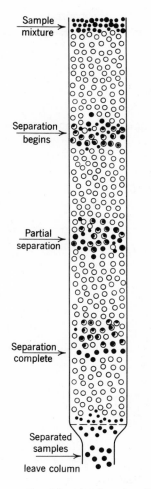

Sample mixture

Separation begins

Partial separation

Separation complete

Separated samples

leave column

Fig. 2-17.   Principle of the separation of molecules according to size by gel permeation chromatography (after Cazes 1966 and Billmeyer 1969).

porous glass are preferred column-packing materials. The pores in these gels are of the same size as the dimensions of polymer molecules.

A sample of a dilute polymer solution is introduced into a solvent stream flowing through the column. As the dissolved polymer molecules flow past the porous beads (Fig. 2-17), they can diffuse into the internal pore structure of the gel to an extent depending on their size and the pore-size distribution of the gel. Larger molecules can enter only a small fraction of the internal portion of the gel, or are completely excluded; smaller polymer molecules penetrate a larger fraction of the interior of the gel.

The larger the molecule, therefore, the less time it spends inside the gel, and the sooner it flows through the column. The different molecular species are eluted from the column in order of their molecular size (Benoit 1966) as distinguished from their molecular weight, the largest emerging first.

A complete theory predicting retention times or volumes as a function of molecular size has not appeared for gel permeation chromatography. A specific column or set of columns (with gels of differing pore size) is calibrated empirically to give such a relationship, by means of which a plot of amount of solute versus retention volume (the chromatogram, Fig. 2-18) can be converted into a molecular-size distribution curve. For convenience, commercially available narrow-distribution polystyrenes (Pressure) are often used. If the calibration is made in terms of a molecular-size parameter, for example $[\eta]M$ (whose relation to size is given by Eq. 3-29), it can be applied to a wide variety of both linear and branched polymers (Fig. 2-19).

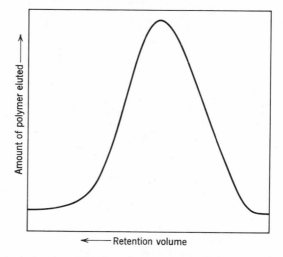

Fig. 2-18. A typical gel permeation chromatogram. Polystyrene in tetrahydrofuran $\overline{M}_w/\overline{M}_n = 2.9$.

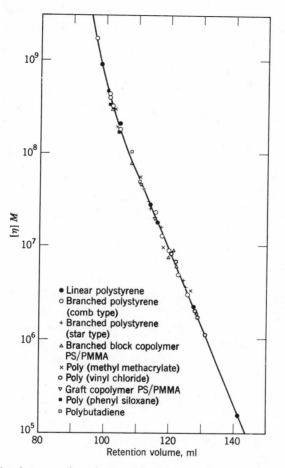

Fig. 2-19. Calibration curve for gel permeation chromatography based on hydrodynamic volume as expressed by the product $[\eta]M$ (Grubisic 1967). Among the polymer types shown are linear polystyrene, two types of branched polystyrene, poly(methyl methacrylate), poly(vinyl chloride), polybutadiene, poly(phenyl siloxane), and two types of copolymer.

As in all chromatographic processes, the band of solute emerging from the column is broadened by a number of processes, including contributions from the apparatus, flow of the solution through the packed bed of gel particles, and the permeation process itself (Billmeyer, 1968*a,b*, Kelley 1969, 1970*a,b*). Corrections for this zone broadening can be made empirically (Duerksen 1968); it usually becomes unimportant when the sample has $\overline{M}_w/\overline{M}_n > 2$.

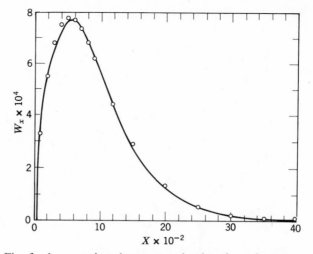

Fig. 2-20. Fit of gel permeation chromatography data for polystyrene to a molecular-weight distribution curve calculated from polymerization kinetics (May 1968).

Gel permeation chromatography has proved extremely valuable for both analytical and preparative work with a wide variety of systems ranging from low to very high molecular weights. The method can be applied to a wide variety of solvents and polymers, depending on the type of gel used. With polystyrene gels, relatively nonpolar polymers can be measured in solvents such as tretrahydrofuran, toluene, or (at high temperatures) *o*-dichlorobenzene; with porous glass gels, more polar systems including aqueous solvents can be used. A few milligrams sample suffices for analytical work, and the determination is complete in 2 to 4 hours in typical cases. An extensive bibliography on gel permeation chromatography has been compiled (Waters).

The results of careful gel permeation chromatography experiments for molecular-weight distribution agree so well with results from other techniques that there is serious doubt as to which is correct when residual discrepancies occur. Figure 2-16 shows the extent of agreement between this method and a solvent-gradient elution fractionation, while Fig. 2-20 demonstrates the degree of fit between the experiment and a distribution curve calculated from polymerization kinetics.

GENERAL REFERENCES

Cazes 1966, 1970; Altgelt 1967, 1968.

BIBLIOGRAPHY

Allen 1959. P. W. Allen, "Introductory (Solubility and the Choice of Solvent; the Preparation and Handling of Polymer Solutions)," Chap. 1 in P. W. Allen, ed., *Techniques of Polymer Characterization*, Butterworths, London, 1959.

Altgelt 1967. K. H. Altgelt and J. C. Moore, "Gel Permeation Chromatography," Chap. B.4 in Manfred J. R. Cantow, ed., *Polymer Fractionation*, Academic Press, New York, 1967.

Altgelt 1968. K. H. Altgelt, "Theory and Mechanics of Gel Permeation Chromatography," pp. 3–46 in J. Calvin Giddings and Roy A. Keller, eds., *Advances in Chromatography*, Vol. 7, Marcel Dekker, New York, 1968.

Asmussen 1962. Frithjof Asmussen and Kurt Ueberreiter, "Velocity of Dissolution of Polymers. Part II," *J. Polymer Sci.* **57**, 199–208 (1962).

Baker 1956. C. A. Baker and R. J. P. Williams, "A New Chromatographic Procedure and its Application to High Polymers," *J. Chem. Soc.* **1956**, 2352–2362.

Bamford 1950. C. H. Bamford and H. Tompa, "The Theory of Coacervation," *Trans. Faraday Soc.* **46**, 310–316 (1950).

Battista 1967. O. A. Battista, "Summative Fractionation," Chap. C.4 in Manfred J. R. Cantow, ed., *Polymer Fractionation*, Academic Press, New York, 1967.

Bawn 1950. C. E. H. Bawn, R. F. J. Freeman, and A. R. Kamaliddin, "High Polymer Solutions. Part I. Vapour Pressure of Polystyrene Solutions," *Trans. Faraday Soc.* **46**, 677–684 (1950).

Benoit 1966. Henri Benoit, Zlatka Grubisic, Paul Rempp, Danielle Decker, and Jean-Georges Zilliox, "Study by Liquid-Phase Chromatography of Linear and Branched Polystyrenes of Known Structure" (in French), *J. Chem. Phys.* **63**, 1507–1514 (1966).

Berry 1970. Guy C. Berry and Edward F. Casassa, "Thermodynamic and Hydrodynamic Behavior of Dilute Polymer Solutions," *Macromol. Revs.* **4**, 1–66 (1970).

Billmeyer 1950. F. W. Billmeyer, Jr., and W. H. Stockmayer, "Method of Measuring Molecular Weight Distribution," *J. Polymer Sci.* **5**, 121–137 (1950).

Billmeyer 1968a. F. W. Billmeyer, Jr., G. W. Johnson, and R. N. Kelley, "Evaluating Dispersion in Gel Permeation Chromatography. I. Theoretical Analysis," *J. Chromatog.* **34**, 316–321 (1968).

Billmeyer 1968b. F. W. Billmeyer, Jr., and R. N. Kelley, "Evaluating Dispersion in Gel Permeation Chromatography. II. The Injection—Detection System," *J. Chromatog.* **34**, 322–331 (1968).

Billmeyer 1969. Fred W. Billmeyer, Jr., "Molecular Structure and Polymer Properties," *J. Paint Technol.* **41**, 3–16, 209 (1969).

Birshtein 1966. T. M. Birshtein and O. B. Ptitsyn, *Conformations of Macromolecules*, translated by Serge N. Timasheff and Marina J. Timasheff, John Wiley and Sons, New York, 1966.

Bristow 1958a. G. M. Bristow and W. F. Watson, "Cohesive Energy Densities of Polymers. Part I. Cohesive Energy Densities of Rubbers by Swelling Measurements," *Trans. Faraday Soc.* **54**, 1731–1741 (1958).

Bristow 1958b. G. M. Bristow and W. F. Watson, "Cohesive Energy Densities of Polymers. Part 2. Cohesive Energy Densities from Viscosity Measurements," *Trans Faraday Soc.* **54**, 1742–1747 (1958).

Burrell 1955. Harry Burrell, "Solubility Parameters for Film Formers," *Official Digest* **27**, 726–758 (1955).

Burrell 1966. H. Burrell and B. Immergut, "Solubility Parameter Values," pp. IV-341–IV-368 in J. Brandrup and E. H. Immergut, eds., with the collaboration of H.-G. Elias, *Polymer Handbook*, Interscience Div., John Wiley and Sons, New York, 1966.

Cantow 1967. Manfred J. R. Cantow, ed., *Polymer Fractionation*, Academic Press, New York, 1967.

Carpenter 1970. D. K. Carpenter, "Solution Properties," in press in Herman F. Mark, Norman G. Gaylord, and Norbert F. Bikales, eds., *Encyclopedia of Polymer Science and Technology*, Vol. 12, Interscience Div., John Wiley and Sons, New York, 1970.

Cazes 1966. Jack Cazes, "Topics in Chemical Instrumentation. XXIX. Gel Permeation Chromatography," *J. Chem. Educ.* **43**, "Part One" A567–A582, and "Part Two" A625–A642 (1966).

Cazes 1970. Jack Cazes, "Current Trends in Gel Permeation Chromatography 1970," *J. Chem. Educ.* **47**, "Part one: Theory and Experiment" A461–A471 and "Part Two: Methodology" A505–A514 (1970).

Chandrasekhar 1943. S. Chandrasekhar, "Stochastic Problems in Physics and Astronomy," *Revs. Modern Phys.* **15**, 1–89 (1943).

Crouzet 1969. P. Crouzet, F. Fine, and P. Magnin, "Comparison of the Gel Permeation Chromatographic and Gradient Elution Fractionation Methods for the Determination of the Molecular Weight Distribution of Polypropylene" (in French), *J. Appl. Polymer Sci.* **13**, 205–213 (1969).

Crowley 1966. James D. Crowley, G. S. Teague, Jr., and Jack W. Lowe, Jr., "A Three-Dimensional Approach to Solubility," *J. Paint Technol.* **38**, 269–280 (1966).

Delmas 1966. G. Delmas and D. Patterson, "The Temperature Dependence of Polymer Chain Dimensions and Second Virial Coefficients," *Polymer* **7**, 513–524 (1966).

Dobry 1947. A. Dobry and F. Boyer-Kawenoki, "Phase Separation in Polymer Solution," *J. Polymer Sci.* **2**, 90–100 (1947).

Doolittle 1969. Arthur K. Doolittle, "Polymer Solution Thermodynamics, State-of-the-Art Survey," *J. Paint Technol.* **41**, 483–488 (1969).

Duerksen 1968. J. H. Duerksen and A. E. Hamilec, "Polymer Reactors and Molecular Weight Distribution. Part III. Gel Permeation Chromatography, Methods of Correcting for Imperfect Resolution," *J. Polymer Sci.* **C21**, 83–103 (1968).

Eichinger 1968a. B. E. Eichinger and P. J. Flory, "Thermodynamics of Polymer Solutions. Part 1. Natural Rubber and Benzene," *Trans. Faraday Soc.* **64**, 2035–2052 (1968).

Eichinger 1968b. B. E. Eichinger and P. J. Flory, "Thermodynamics of Polymer Solutions. Part 2. Polyisobutylene and Benzene," *Trans. Faraday Soc.* **64**, 2053–2060 (1968).

Eichinger 1968c. B. E. Eichinger and P. J. Flory, "Thermodynamics of Polymer Solutions. Part 3. Polyisobutylene and Cyclohexane," *Trans. Faraday Soc.* **64**, 2061–2065 (1968).

Eichinger 1968d. B. E. Eichinger and P. J. Flory, "Thermodynamics of Polymer Solutions. Part 4. Polyisobutylene and *n*-Pentane," *Trans. Faraday Soc.* **64**, 2066–2072 (1968).

Elliott 1967. John H. Elliott, "Fractional Solution," Chap. B.2 in Manfred J. R. Cantow, ed., *Polymer Fractionation*, Academic Press, New York, 1967.

Flory 1942. Paul J. Flory, "Thermodynamics of High Polymer Solutions," *J. Chem. Phys.* **10**, 51–61 (1942).

Flory 1950. P. J. Flory and W. R. Krigbaum, "Statistical Mechanics of Dilute Polymer Solutions," *J. Chem. Phys.* **18**, 1086–1094 (1950).

Flory 1953. Paul J. Flory, *Principles of Polymer Chemistry*, Cornell University Press, Ithaca, N.Y., 1953.

Flory 1968. P. J. Flory, J. L. Ellenson, and B. E. Eichinger, "Thermodynamics of Mixing of *n*-Alkanes with Polyisobutylene," *Macromolecules* **1**, 279–284 (1968).

Flory 1969. Paul J. Flory, *Statistical Mechanics of Chain Molecules*, Interscience Div., John Wiley and Sons, New York, 1969.

Freeman 1960. P. I. Freeman and J. S. Rowlinson, "Lower Critical Points in Polymer Solutions," *Polymer* **1**, 20–26 (1960).

Fuchs 1967. O. Fuchs and W. Schmieder, "Chemical Inhomogeneity and its Determination," Chap. D. in Manfred J. R. Cantow, ed., *Polymer Fractionation*, Academic Press, New York, 1967.

Gardon 1965. J. L. Gardon, "Cohesive-Energy Density," pp. 833–862 in Herman F. Mark, Norman G. Gaylord, and Norbert M. Bikales, eds., *Encyclopedia of Polymer Science and Technology*, Vol. 3, Interscience Div., John Wiley and Sons, New York, 1965.

Gee 1946. Geoffrey Gee and W. J. C. Orr, "The Interaction between Rubber and Liquids. VIII. A New Examination of the Thermodynamic Properties of the System Rubber + Benzene," *Trans. Faraday Soc.* **42**, 507–517 (1946).

Gee 1947. Geoffrey Gee, Part 5, "Equilibrium Properties of High Polymers and Gels," in "Discussion on 'Some Aspects of the Chemistry of Macromolecules,'" *J. Chem. Soc.* **1947**, 280–288.

Giesekus 1967. Hanswalter Giesekus, "Turbidimetric Titration," Chap. C.1 in Manfred J. R. Cantow, ed., *Polymer Fractionation*, Academic Press, New York, 1967.

Grubisic 1967. Z. Grubisic, P. Rempp, and H. Benoit, "A Universal Calibration for Gel Permeation Chromatography," *J. Polymer Sci.* **B5**, 753–759 (1967).

Guzmán 1961. G. M. Guzmán, "Fractionation of High Polymers," pp. 113–183 in J. C. Robb and F. W. Peaker, eds., *Progress in High Polymers*, Vol. 1, Academic Press, New York, 1961.

Hall 1959. R. W. Hall, "The Fractionation of High Polymers," Chap. 2 in P. W. Allen, ed., *Techniques of Polymer Characterization*, Butterworths, London, 1959.

Hildebrand 1950. Joel H. Hildebrand and Robert L. Scott, *The Solubility of Nonelectro-lytes*, 3rd ed., Reinhold Publishing Corp., New York, 1950 (reprinted by Dover Publications, Inc., New York, 1964).

Hoy 1970. K. L. Hoy, "New Values of the Solubility Parameters from Vapor Pressure Data," *J. Paint Technol.* **42**, 76–118 (1970).

Huggins 1942a. Maurice L. Huggins, "Thermodynamic Properties of Solutions of Long-Chain Compounds," *Ann. N. Y. Acad. Sci.* **43**, 1–32 (1942).

Huggins 1942b. Maurice L. Huggins, "Some Properties of Solutions of Long-Chain Compounds," *J. Phys. Chem.* **46**, 151–158 (1942).

Huggins 1942c. Maurice L. Huggins, "Theory of Solutions of High Polymers," *J. Am. Chem. Soc.* **64**, 1712–1719 (1942).

Huggins 1958. Maurice L. Huggins, *Physical Chemistry of High Polymers*, John Wiley and Sons, New York, 1958.

Huggins 1967. Maurice L. Huggins and Hiroshi Okamoto, "Theoretical Considerations," Chap. A in Manfred J. R. Cantow, ed., *Polymer Fractionation*, Academic Press, New York, 1967.

Johnson 1967. Julian F. Johnson, Manfred J. R. Cantow, and Roger S. Porter, "Fractionation," pp. 231–260 in Herman F. Mark, Norman G. Gaylord, and Norbert M. Bikales, eds., *Encyclopedia of Polymer Science and Technology*, Vol. 7, Interscience Div., John Wiley and Sons, New York, 1967.

Kelley 1969. R. N. Kelley and F. W. Billmeyer, Jr., "Evaluating Dispersion in Gel Permeation Chromatography. Axial Dispersion of Polymer Molecules in Packed Beds of Non-Porous Beads," *Anal. Chem.* **41**, 874–879 (1969).

Kelley 1970a. Richard N. Kelley and Fred W. Billmeyer, Jr., "Evaluating Dispersion in Gel Chromatography. Dispersion Due to Permeation," *Anal. Chem.* **42**, 399–403 (1970).

Kelley 1970b. Richard N. Kelley and Fred W. Billmeyer, Jr., "A Review of Peak Broadening in Gel Chromatography," *Sep. Sci.* **5**, 291–316 (1970).

Kenyon 1960. A. S. Kenyon and I. O. Salyer, "Elution Fractionation of Crystalline and Amorphous Polymers," *J. Polymer Sci.* **43**, 427–444 (1960).

Koningsveld 1968. R. Koningsveld, "Liquid-Liquid Separation of Polymer Solutions," pp. 172–213 in Donald McIntyre, ed., *Characterization of Macromolecular Structure*, National Academy of Sciences Publication No. 1573, Washington, D.C., 1968.

Koningsveld 1970. R. Koningsveld, "Preparative and Analytical Aspects of Polymer Fractionation," *Fortschr. Hochpolym.-Forsch. (Advances in Polymer Science)* **7**, 1–69 (1970).

Kokle 1965. V. Kokle and F. W. Billmeyer, Jr., "The Molecular Structure of Polyethylene. XVI. Analytical Column Fractionation," *J. Polymer Sci.* **C8**, 217–232 (1965).

Kotera 1967. Akira Kotera, "Fractional Precipitation," Chap. B.1 in Manfred J. R. Cantow, ed., *Polymer Fractionation*, Academic Press, New York, 1967.

Lewis 1961. Gilbert N. Lewis and Merle Randall, *Thermodynamics*, revised by Kenneth S. Pitzer and Leo Brewer, 2nd ed., McGraw-Hill Book Co., New York, 1961.

Lowry 1970. George G. Lowry, ed., *Markov Chains and Monte Carlo Calculations in Polymer Science*, Marcel Dekker, New York, 1970.

Maley 1965. L. E. Maley, "Application of Gel Permeation Chromatography to High and Low Molecular Weight Polymers," *J. Polymer Sci.* **C8**, 253–268 (1965).

May 1968. James A. May, Jr., and William B. Smith, "Polymer Studies by Gel Permeation Chromatography. II. The Kinetic Parameters for Styrene Polymerization," *J. Phys. Chem.* **72**, 216–221 (1968).

Miller 1966. M. L. Miller, *The Structure of Polymers*, Reinhold Publishing Corp., New York, 1966.

Moore 1964. J. C. Moore, "Gel Permeation Chromatography. I. A New Method for Molecular Weight Distribution of High Polymers," *J. Polymer Sci.* **A2**, 835–843 (1964).

Newing 1950. M. J. Newing, "Thermodynamic Studies of Silicones in Benzene Solution," *Trans. Faraday Soc.* **46**, 613–620 (1950).

Opschoor 1966. A. Opschoor, *Conformations of Polyethylene and Polypropylene*, Gordon and Breach, New York, 1966.

Patterson 1967. D. Patterson, "Thermodynamics of Non-Dilute Polymer Solutions," *Rubber Chem. Tech.* **40**, 1–35 (1967).

Patterson 1969. D. Patterson, "Free Volume and Polymer Solubility: A Qualitative View," *Macromolecules* **2**, 672–679 (1969).

Pennings 1968. A. J. Pennings, "Liquid-Crystal Phase Separation in Polymer Solutions," pp. 214–244 in Donald McIntyre, ed., *Characterization of Macromolecular Structure*, National Academy of Sciences Publication No. 1573, Washington, D.C., 1968.

Pressure. Pressure Chemicals Co., 3419–25 Smallman St., Pittsburgh, Pa., 15201.

Porter 1967. Roger S. Porter and Julian F. Johnson, "Chromatographic Fractionation," Chap. B.3 in Manfred J. R. Cantow, ed., *Polymer Fractionation*, Academic Press, New York, 1967.

Prigogine 1957. I. Prigogine (with the collaboration of A. Bellemans and V. Mathot), *The Molecular Theory of Solutions*, Interscience Publishers, New York, 1957.

Rayleigh 1919. Lord Rayleigh, "On the Problem of Random Vibrations, and of Random Flights in One, Two, or Three Dimensions," *Phil Mag.* **37**, 321–347 (1919).

Schneider 1961a. N. S. Schneider, J. D. Loconti, and L. G. Holmes, "A Comparison of the Thermal Gradient and Elution Methods of Polymer Fractionation," *J. Appl. Polymer Sci.* **5**, 354–363 (1961).

Schneider 1961b. N. S. Schneider, "Polymer Fractionation by Column Methods," *Anal. Chem.* **33**, 1829–1831 (1961).

Schneider 1965. Nathaniel S. Schneider, "Review of Solution Methods and Certain Other Methods of Polymer Fractionation," *J. Polymer Sci.* **C8**, 179–204 (1965).

Schneider 1968. N. S. Schneider, R. T. Traskos, and A. S. Hoffman, "Characterization of Branched Polyethylene Fractions from the Elution Column," *J. Appl. Polymer Sci.* **12**, 1567–1587 (1968).

Shultz 1952. A. R. Shultz and P. J. Flory, "Phase Equilibria in Polymer-Solvent Systems," *J. Am. Chem. Soc.* **74**, 4760–4767 (1952).

Small 1953. P. A. Small, "Some Factors Affecting the Solubility of Polymers," *J. Appl. Chem.* **3**, 71–80 (1953).

Smith 1963. Richard P. Smith, "Excluded Volume Effect in Polymer Chains," *J. Chem. Phys.* **38**, 1463–1470 (1963).

Stockmayer 1949. W. H. Stockmayer, "Solubility of Heterogeneous Polymers," *J. Chem. Phys.* **17**, 588 (1949).

Taylor 1948. William J. Taylor, "Average Length and Radius of Normal Paraffin Hydrocarbon Molecules," *J. Chem. Phys.* **16**, 257–267 (1948).

Tompa 1956. H. Tompa, *Polymer Solutions*, Butterworths, London, 1956, Chap. 8.

Traskos 1968. R. T. Traskos, N. S. Schneider, and A. S. Hoffman, "Elution Column Fractionation of Branched Polyethylene," *J. Appl. Polymer Sci.* **12**, 509–525 (1968).

Treloar 1958. L. R. G. Treloar, *The Physics of Rubber Elasticity*, 2nd ed., Clarendon Press, Oxford, 1958.

Tung 1967. L. H. Tung, "Treatment of Data," Chap. E in Manfred J. R. Cantow, ed., *Polymer Fractionation*, Academic Press, New York, 1967.

Ueberreiter 1962. Kurt Ueberreiter and Frithjof Asmussen, "Velocity of Dissolution of Polymers. Part I," *J. Polymer Sci.* **57**, 187–198 (1962).

Volkenstein 1963. M. V. Volkenstein, *Configurational Statistics of Polymeric Chains*, transl. by Serge N. Timasheff and M. J. Timasheff, Interscience Div., John Wiley and Sons, New York, 1963.

Walker 1952. E. E. Walker, "The Solvent Action of Organic Substances on Polyacrylonitrile," *J. Appl. Chem.* **2**, 470–481 (1952).

Waters. Waters Associates, Inc., 61 Fountain Street, Framingham, Mass., 01701.

Zimm 1949. Bruno H. Zimm and Walter H. Stockmayer, "The Dimensions of Chain Molecules Containing Branches and Rings," *J. Chem. Phys.* **17**, 1301–1314 (1949).

3

Measurement of Molecular Weight and Size

The molecular weights of polymers can be determined by chemical or physical methods of functional group analysis, by measurement of the colligative properties, light scattering, or ultracentrifugation, or by measurement of dilute solution viscosity. All these methods except the last are, in principle, absolute: molecular weights can be calculated without reference to calibration by another method. Dilute solution viscosity, however, is not a direct measure of molecular weight. Its value lies in the simplicity of the technique and the fact that it can be related empirically to molecular weight for many systems.

With the exception of some types of end group analysis, all molecular weight methods require solubility of the polymer, and all involve extrapolation to infinite dilution or operation in a Θ-solvent in which ideal solution behavior is attained.

GENERAL REFERENCES

Billmeyer 1965, 1966a, 1969.

A. End Group Analysis

Molecular weight determination through group analysis requires that the polymer contain a known number of determinable groups per molecule. The long-chain nature of polymers limits such groups to end groups. Thus the method is usually referred to as *end group analysis*. Since methods of end group analysis count the number of molecules in a given weight of sample, they yield the number-average molecular weight (Section *B*) for polydisperse materials. The methods become insensitive at high molecular weight, as the fraction of end groups becomes too small to be measured with precision. Although published procedures often indicate that loss of precision occurs at molecular weights above 25,000, there is no doubt that the adoption of more sensitive techniques, such as gas chromatography or mass spectrometry to estimate reaction products, can raise this limit by at least a factor of 10. Similar considerations apply to the use of infrared spectroscopy to measure end groups.

The discussion of end group methods is conveniently divided to cover condensation and addition polymers because of the difference in the types of end groups usually found.

Condensation polymers   End group analysis in condensation polymers usually involves chemical methods of analysis for functional groups. Carboxyl groups in polyesters (Pohl 1954) and in polyamides (Waltz 1947) are usually titrated directly with base in an alcoholic or phenolic solvent, while amino groups in polyamides may be titrated with acid under similar conditions (Waltz 1947). Hydroxyl groups are usually determined by reacting them with a titratable reagent (Ogg 1945, Conix 1958), but infrared spectroscopy has been used (Ward 1957, Daniels 1958). The chemical methods are often limited by insolubility of the polymer in solvents suitable for the titrations.

Addition polymers   No general procedures can be given for end group analysis in addition polymers because of the variety of type and origin of the end groups. When the polymerization kinetics is well known, analysis may be made for initiator fragments containing identifiable functional groups (Evans 1947), elements (Price 1943), or radioactive atoms (Bevington 1954*a*, *b*); or for end groups arising from transfer reactions with solvent (Mayo 1943, Joyce 1948); or for unsaturated end groups in linear polyethylene and poly-α-olefins, as in the infrared spectroscopic analysis for vinyl groups (Billmeyer 1964*b*).

GENERAL REFERENCES

Price 1959; Hellman 1962.

B. Colligative Property Measurement

The relations between the colligative properties and molecular weight for infinitely dilute solutions rest upon the fact that the activity of the solute in a solution becomes equal to its mole fraction as the solute concentration becomes sufficiently small. It can be shown that the activity of the solvent must equal its mole fraction under these conditions, and it follows that the depression of the activity of the solvent by a solute is equal to the mole fraction of the solute.

The colligative property methods are based on vapor-pressure lowering, boiling-point elevation (*ebulliometry*), freezing-point depression (*cryoscopy*), and the osmotic pressure (*osmometry*). The working equations for these methods are derived from Eq. 2-10. When Boltzmann's constant is replaced by the gas constant throughout to convert to a molar basis, the equations for the methods of interest for polymer solutions are:

$$\lim_{c \to 0} \frac{\Delta T_b}{c} = \frac{RT^2}{\rho \Delta H_v} \frac{1}{M}$$

$$\lim_{c \to 0} \frac{\Delta T_f}{c} = -\frac{RT^2}{\rho \Delta H_f} \frac{1}{M} \tag{3-1}$$

$$\lim_{c \to 0} \frac{\pi}{c} = \frac{RT}{M}$$

where ΔT_b, ΔT_f, and π are the boiling-point elevation, freezing-point depression, and osmotic pressure, respectively, ρ is the density of the solvent, ΔH_v and ΔH_f are the latent heats of vaporization and fusion of the solvent per gram, and c is the solute concentration in grams per cubic centimeter.

The relative applicability of the colligative methods to polymer solutions is demonstrated in Table 3-1. It is clear that direct measurement of vapor-

TABLE 3-1. *Colligative properties of a solution of a polymer of $M = 20,000$ at $c = 0.01 g/cm^3$*

Property	Value
Vapor-pressure lowering	4×10^{-3} mm Hg
Boiling-point elevation	1.3×10^{-3} °C
Freezing-point depression	2.5×10^{-3} °C
Osmotic pressure	15 cm solvent

pressure lowering in polymer solutions is unrewarding. It is possible, however, to utilize vapor-pressure lowering indirectly through the technique of *vapor-phase osmometry*, in which one measures a temperature difference relatable to vapor-pressure lowering through the Clapeyron equation. This temperature difference is the same order of magnitude as those observed in cryoscopy and ebulliometry.

While such differences, of the order of $1 \times 10^{-3}°C$, can currently be measured with considerable precision, the larger effect observed in the osmotic experiment suggests that this technique is the most fruitful for polymer solutions. Although other considerations, such as the perfection of available osmotic membranes, are important, the osmotic method has been more widely used than other colligative techniques for polymer systems. To distinguish it from vapor-phase osmometry, we refer hereafter to membrane osmometry as the specific name for the last-named technique in Table 3-1.

Number-average molecular weight

Since typical polymers consist of mixtures of many molecular species, molecular weight methods always yield average values. The measurement of the colligative properties in effect counts the number of moles of solute per unit weight of sample. This number is the sum over all molecular species of the number of moles N_i of each species present:

$$\sum_{i=1}^{\infty} N_i$$

The total weight w of the sample is similarly the sum of the weights of each molecular species,

$$w = \sum_{i=1}^{\infty} w_i = \sum_{i=1}^{\infty} N_i M_i \tag{3-2}$$

The average molecular weight given by these methods is known as the *number-average molecular weight* $\overline{M}_n$. By this definition of molecular weight as weight of sample per mole,

$$\overline{M}_n = \frac{w}{\sum_{i=1}^{\infty} N_i} = \frac{\sum_{i=1}^{\infty} M_i N_i}{\sum_{i=1}^{\infty} N_i} \tag{3-3}$$

It is easy to show that the colligative methods measure $\overline{M}_n$. Taking membrane osmometry as an example, we may write Eq. 3-1 separately for each species i in the mixture present. Omitting the limit sign for convenience, and transposing,

$$\pi_i = RT\frac{c_i}{M_i}$$

Summing over i, we may replace $\sum \pi_i$ by π since each solute molecule contributes independently to the osmotic pressure, and note that $c = \sum c_i$. The mixture is then described by an unspecified average molecular weight $\overline{M}$:

$$\pi = RT\sum\left(\frac{c_i}{M_i}\right) = RT\frac{c}{M} \qquad (3\text{-}4)$$

Solving, $\overline{M} = c/\sum(c_i/M_i)$. For unit volume, the c's may be replaced by w's. Noting that $w_i = M_i N_i$ shows that $\overline{M} = \overline{M}_n$.

Equations 3-1 may then be put in the form applicable to heterogeneous solutes:

$$\lim_{c \to 0}\frac{\Delta T_b}{c} = \frac{RT^2}{\rho \Delta H_v}\frac{1}{\overline{M}_n}$$

$$\lim_{c \to 0}\frac{\Delta T_f}{c} = -\frac{RT^2}{\rho \Delta H_f}\frac{1}{\overline{M}_n} \qquad (3\text{-}5)$$

$$\lim_{c \to 0}\frac{\pi}{c} = \frac{RT}{M_n}$$

End group methods as well as colligative methods give the number-average molecular weight. The number average is very sensitive to changes in the weight fractions of low-molecular-weight species, and relatively insensitive to similar changes for high-molecular-weight species.

Concentration dependence of the colligative properties

For convenience, the concentration dependence of the colligative properties is developed in terms of the osmotic pressure; the results are directly applicable to cryoscopy and ebulliometry by appropriate substitution in Eq. 3-5. By Eq. 2-11 or 2-17, a general expression for the osmotic pressure at a finite concentration is

$$\frac{\pi}{RTc} = (A_1 + A_2 c + A_3 c^2 + \cdots)$$

$$\frac{\pi}{RTc} = \frac{1}{M_n}(1 + \Gamma c + g\Gamma^2 c^2 + \cdots) \qquad (3\text{-}6)$$

where $\Gamma = A_2/A_1$, and g is a slowly varying function of the polymer-solvent interaction with values near zero for poor solvents and near 0.25 for good solvents (Krigbaum 1952, Stockmayer 1952).

In most cases, the term in c^2 may be neglected; when dependence on c^2 is significant, it may be convenient to take $g = 0.25$ and write

$$\frac{\pi}{RTc} = \frac{1}{\overline{M}_n}\left(1 + \frac{\Gamma}{2}c\right)^2 \qquad (3\text{-}7)$$

In terms of the polymer-solvent interaction constant χ_1 of the Flory-Huggins theory, the osmotic pressure is given by

$$\frac{\pi}{RTc} = \frac{1}{M_2} + \frac{\rho_1}{M_1\rho_2^2}\left(\frac{1}{2} - \chi_1\right)c + \cdots \qquad (3\text{-}8)$$

where subscript 1 indicates the solvent, and 2 the polymer.

As Eqs. 3-6 and 3-8 suggest, it is usual to plot π/c vs. c. In general a straight line results whose intercept at $c = 0$ is $A_1 = 1/\overline{M}_n$ and whose slope is the second virial coefficient A_2 and allows evaluation of the polymer-solvent interaction constant χ_1. If the solvent is good enough or the concentration high enough that the c^2 term is significant, the points may deviate from a straight line. In such cases, it is useful to plot $(\pi/RTc)^{1/2}$ vs. c as suggested by Eq. 3-7. Typical data are discussed under *Membrane Osmometry*.

Vapor-phase osmometry

An indirect measurement of vapor-pressure lowering by the technique known as vapor-phase osmometry (Pasternak 1962) has become popular in recent years, and is now the method of choice for measuring $\overline{M}_n$ of samples too low in molecular weight to be measured in a membrane osmometer. In this method, the property measured, in a so-called "vapor-pressure osmometer" (Fig. 3-1), is the small temperature difference resulting from different rates of solvent evaporation from and condensation onto droplets of pure solvent and polymer solution maintained in an atmosphere of solvent vapor. An important assumption of the method is that this temperature difference is proportional to the vapor-pressure lowering of the polymer solution at equilibrium and thus to the number-average molecular weight: because of heat losses, the full temperature difference expected from theory is not attained. Measurements must be made at several concentrations and extrapolated to $c = 0$. Like ebulliometry and cryoscopy, the method is calibrated with low-molecular-weight standards and is useful for values of $\overline{M}_n$ at least up to 40,000 in favorable cases. It is rapid and has the additional advantage that only a few milligrams of sample are required. A series of articles describing the technique and instrumentation has appeared (Simon 1960; Wegmann 1962; Tomlinson 1963; Chylewski 1964; Dohner 1967). Successful measurement has been reported of number-average molecular weights as high as 160,000 (Simon 1966; Wachter 1969), but with commercial equipment the upper limit remains near 25,000. The lower limit is that at

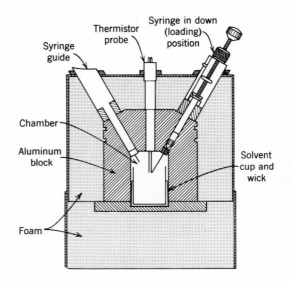

Fig. 3-1.   Measurement chamber of a vapor-phase osmometer (Pasternak 1962). Droplets of solvent and polymer solution are placed, with the aid of hypodermic syringes, on the "beads" of two thermistors used as temperature-sensing elements and maintained in equilibration with an atmosphere of solvent vapor. The lower activity of the solvent in the solution droplet leads to an excess of solvent condensation over evaporation there compared to the droplet of pure solvent. The excess heat of vaporization thus liberated leads to a rise in temperature, usually somewhat less than that predicted from thermodynamic considerations because of heat losses.

which the solute becomes appreciably volatile. Since the method does not measure equilibrium vapor-pressure lowering but depends on the development of quasi-steady-state phenomena, care must be taken to standardize such variables as time of measurement and drop size between calibration and sample measurement. Significant nonlinearity has been observed (Sharma 1967) in the extrapolation of data from relatively high concentrations to infinite dilution. The method is finding increasing use in the study of association (Elias 1966, 1967) and in the rapid estimation of osmotic coefficients of electrolytes (Burge 1963; Kratohvil 1966b).

Ebulliometry

In the boiling-point elevation experiment (Glover 1966, Ezrin 1968) the boiling point of the polymer solution is compared directly with that of the (condensing) pure solvent in a vessel known as an ebulliometer. Sensing devices include differential thermometers and multijunction thermocouples or thermistors arranged in a Wheatstone bridge circuit. It is customary to calibrate the ebulliometer with a substance of known molecular weight, e.g., octacosane ($M = 396$) or tristearin ($M = 892$).

Although with care accurate molecular weights can be obtained up to $\overline{M}_n = 30,000$ or more, the ebulliometric experiment occasionally suffers from a limitation in the tendency of polymer solutions to foam on boiling. Not only may this lead to unstable operation, but also the polymer may concentrate in the foam because of its greater surface, thus rendering uncertain the actual concentration of the solution. No equipment suitable for ebulliometry in the high-polymer range is commercially available, and this method remains largely in the category of a referee technique.

Cryoscopy

The freezing-point depression, or cryoscopic, method (Vofsi 1957; Bonnar 1958; Newitt 1966) is similar to the ebulliometric technique in several respects. The preferred temperature-sensing element is the thermistor used in a bridge circuit. The freezing points of solvent and solution are often compared sequentially. Calibration with a substance of known molecular weight is customary. Although the limitations of the method appear to be somewhat less severe than those of ebulliometry, care must be taken that supercooling is controlled. The use of a nucleating agent to provide controlled crystallization of the solvent is helpful in this respect. Reliable results are obtained for molecular weights at least as high as 30,000. Again, lack of commercial equipment relegates this method largely to referee status despite the appeals of simplicity and ease of operation.

Membrane osmometry

The principle of membrane osmometry is illustrated in Fig. 3-2. The two compartments of an *osmometer* are separated by a *semipermeable membrane*, through which only solvent molecules can penetrate, and which is closed except for capillary tubes. With the polymer solute confined to one side of the osmometer, the activity of the solvent is different in the two compartments. The thermodynamic drive towards equilibrium results in a

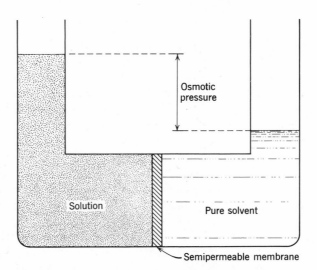

Fig. 3-2. The principle of operation of the membrane osmometer (Billmeyer 1966*b*).

difference in liquid level in the two capillaries. The resulting hydrostatic pressure increases the solvent activity on the solution side until, when the applied pressure equals the osmotic pressure, equilibrium is achieved.

Osmometers Two types of osmometer have been used extensively in the past. One, the block osmometer (Fuoss 1943; Krigbaum 1953), is a relatively large and bulky metal instrument. The membrane area is large and the solution volume small—advantages for rapid equilibration. More popular are relatively small and simple osmometers based on the Zimm-Myerson design (Zimm 1946), in which two membranes are held against a glass solution cell by means of perforated metal plates, as shown in Fig. 3-3. The assembled instrument is suspended in a large tube partly filled with solvent. In the Stabin (1954) modification, the membranes are supported on both sides by channeled metal plates to ensure their rigidity. Advantages of these osmometers are small size and cost, making multiple installations feasible, the possibility of complete immersion in constant-temperature baths, and ease of filling and adjusting osmotic height.

In the last few years rapid automatic osmometers have become available [Reiff 1959 (Melabs); Steele 1963 (F&M); Rolfson 1964 (Hallikainen)], in which the solvent compartment is completely closed and fitted with a sensitive pressure-sensing device rather than a capillary (Fig. 3-4). A servo system rapidly adjusts the liquid level in the solvent compartment to balance the osmotic pressure before an appreciable amount of solvent has to pass through the membrane. As a result of this rapid action, the osmometers come to equilibrium in 1 to 5 minutes instead of 10 to 20 hours as typically required in conventional instruments.

Membranes Osmotic membranes fall into two classes: (*a*) nonswellable inorganic substances such as porous glass and metal; and (*b*) swellable materials such as cellulose and rubber. Although some theories of membrane action suggest that a "sieve" theory is not adequate and that the membrane material must interact with or be swollen by the solvent, the successful use of glass membranes has been reported (Elias 1958).

Common organic membrane materials (Immergut 1954; Patat 1956, 1959) include collodion (nitrocellulose of 11–13.5% nitrogen); regenerated cellulose, made by denitration of collodion; gel cellophane that has never

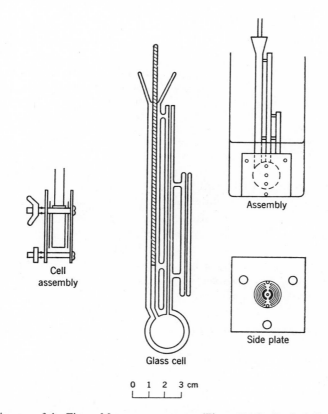

Fig. 3-3. Diagram of the Zimm–Myerson osmometer (Zimm 1946). Typical diameter for the measuring and reference (solvent cell) capillary is 0.5–1 mm. The closure of the filling tube is a 2-mm metal rod. A mercury seal is used at the top to ensure tightness.

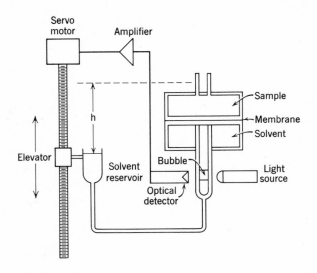

Fig. 3-4. Mode of operation of a high-speed automatic membrane osmometer (F&M). In a glass capillary in the solvent chamber, below the horizontal membrane, a bubble of air is trapped. Motion of this bubble as solvent flows through the membrane is detected photoelectrically, and provides a signal to a servomechanism. The height of a reservoir of solvent is adjusted to provide the hydrostatic pressure required to maintain the bubble at its original position. This is just the osmotic pressure. Other osmometers of this type use capacitance detection and a similar servomechanism (Rolfson 1964) and direct pressure detection with a strain gauge (Reiff 1959).

been allowed to dry after manufacture (this material is available from commercial cellophane manufacturers in several thicknesses); bacterial cellulose, made by the action of certain strains of bacteria; rubber; poly(vinyl alcohol); polyurethanes; poly(vinyl butyral); and polychlorotrifluoroethylene. Of these, gel cellophane is probably most widely used at present. Commercially available membranes are described by Armstrong (1968), and their preparation for use with organic solvents (*conditioning*) is discussed by Chiang (1964) and Coll (1968).

Permeability of solute through the membrane   The success of the osmotic experiment depends on the availability of a membrane through which solvent molecules pass freely, but through which solute molecules cannot pass. Existing membranes only approximate ideal semipermeability, and the chief limitation of the osmotic method is the diffusion of low-molecular-weight species through the membrane. As currently practiced, the method is reliable and accurate only for experiments in which diffusion is absent, such as measurement of unfractionated polymers with $\overline{M}_n > 50,000$, or polymers from which low-molecular-weight species have been removed by fractionation or extraction, with $\overline{M}_n > 20,000$. The upper limit to molecular weights that can be measured by the osmotic method is about 1 million, set by the precision with which small osmotic heights can be read.

Theoretical treatments of the diffusion of low-molecular-weight species through the osmotic membrane have been reviewed by Elias (1968). Theory and experiment show that the apparent osmotic pressure is always less than the true osmotic pressure and falls with time. Extrapolation back to zero time still gives too low an osmotic pressure and hence too high a molecular weight; however, the correct osmotic pressure may be calculated from the data from independent dialysis and diffusion experiments (Hoffman 1965). In principle, measurement of the osmotic pressure after a very long time should give the true molecular weight of the nondiffusing species, but it seems likely that equilibrium is difficult to reach.

The magnitude of the error resulting from solute diffusion and its dependence on the type of measurement are illustrated in Table 3-2.

TABLE 3-2. *Apparent molecular weights of branched polyethylenes by membrane osmometry (Holleran 1968)*

	Apparent $\overline{M}_n$ of Sample		
Method	75	76	77
Average of cryoscopy, ebulliometry, and vapor-phase osmometry (Table 3-3)	11,000	15,300	18,500
Rapid membrane osmometer, first observable values (5–7 min after sample introduction)	19,900	22,100	27,900
Rapid membrane osmometer, 36 min after sample introduction	29,400	25,000	32,800
Conventional membrane osmometer, 1000 min	39,500	30,300	36,400

Typical data Figures 3-5 and 3-6 show typical osmotic data. In both cases concentrations were low enough so that dependence of π/RTc on c^2 was negligible. The parallelism of the lines in Fig. 3-5, representing samples of different molecular weight in the same solvent, illustrates the usual result that the polymer-solvent interactions are a slowly varying function of molecular weight. The constant intercept of the lines in Fig. 3-6 obtained for one sample in several different solvents illustrates the independence of $(\pi/RTc)_{c=0} = 1/\overline{M}_n$ of solvent type and the range of A_2 or χ_1 which can be obtained with a given polymer and a range of solvents.

The advantage of selecting a poor solvent for osmotic measurements in order to minimize the error in extrapolation is evident. At the Flory

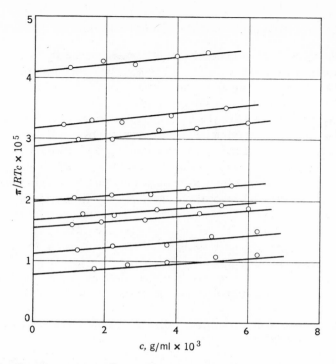

Fig. 3-5.   Plots of $\pi/RTc = 1/\bar{M}_n$ vs. c for cellulose acetate fractions in acetone solution (Badgley 1949).

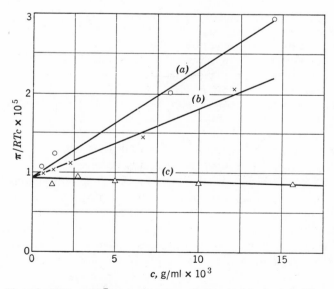

Fig. 3-6.   Plot of $\pi/RTc = 1/\bar{M}_n$ vs. c for nitrocellulose in (a) acetone, (b) methanol, and (c) nitrobenzene (Gee 1944, data of Dobry 1935).

temperature Θ the slope of π/RTc vs. c is zero to concentrations of several per cent.

Comparison of data

Although it is conservative to state that most methods for measuring polymer molecular weights have an accuracy of only 5–10%, significantly closer agreement among methods can be achieved in favorable cases. An extensive intercomparison of methods for determining $\overline{M}_n$ (Billmeyer 1964*b*) gave the data in Table 3-3. Here, despite additional difficulties imposed by

TABLE 3-3.   Number-Average Molecular Weight of Polyethylene by Various Methods (Billmeyer 1964b)

	Sample						
	Branched Polyethylene			Linear Polyethylene			
Method	75	76	77	99	99H	99L	101
End groups by infrared	—	—	—	8,100	37,400	4,800	11,700
Ebulliometry	11,150	—	18,400	—	—	—	—
Cryoscopy	10,700	13,300	19,100	8,100	—	3,320	11,800
Membrane osmometry	(30,300)	(26,600)	(31,400)	—	40,600	—	—
Vapor-phase osmometry	11,000	16,300	18,800	8,600	38,400	3,500	11,900

high-temperature operation, several methods gave excellent agreement for samples of polyethylene. The osmotic molecular weights in parentheses show the effects of diffusion of low-molecular-weight species through the membranes. They were obtained with conventional osmometers; data on these samples obtained with rapid automatic osmometers are given in Table 3-2.

GENERAL REFERENCES

Flory 1953; Bonnar 1958; Allen 1959; Krigbaum 1967; McIntyre 1968.

C. Light Scattering

The scattering of light occurs whenever a beam of light encounters matter. The nuclei and electrons undergo induced vibrations in phase with the incident light wave, and act as sources of light which is propagated in all directions, aside from a polarization effect, with the same wavelength as the exciting beam. Light scattering accounts for many natural phenomena, including the colors of the sky and the rainbow, and of most white materials.

Light scattering in gases and liquids

Lord Rayleigh (1871a,b) applied classical electromagnetic theory to the problem of the scattering of light by the molecules of a gas. He showed that the molecules act as sources of light whose intensity per unit volume of scattering material is i_θ when observed at a distance r and at an angle θ with respect to the incident beam:

$$\frac{i_\theta r^2}{I_0} = R_\theta = \frac{2\pi^2(n-1)^2}{\lambda^4}\frac{1}{v}(1+\cos^2\theta) \tag{3-9}$$

where I_0 is the intensity of the primary beam, and n, v, and λ are, respectively, the refractive index of the gas, the number of scattering particles per cubic centimeter, and the wavelength of the light.

The quantity R_θ, called the Rayleigh ratio, is often replaced by the turbidity, τ, the total scattering integrated over all angles: $\tau = \int_\theta i_\theta\, d\theta$. In the absence of absorption, τ is related to the primary beam intensities before and after passing through the scattering medium. If the incident intensity I_0 is reduced to I in a length l of sample, then

$$\frac{I}{I_0} = e^{-\tau l} \tag{3-10}$$

For particles small compared to λ, $\tau = (8\pi/3)R_{(\theta=0)}$, whence

$$\tau = \frac{32\pi^3}{3}\frac{(n-1)^2}{\lambda^4}\frac{1}{v} \tag{3-11}$$

The term $(1 + \cos^2\theta)$ in Eq. 3-9 arises from the fact that light can be propagated only at right angles to the direction of oscillation of the electric moment.

From the number of scattering particles per cubic centimeter, v, either the molecular weight of the gas or Avogadro's number, N_0, may be calculated if the other is known.

In Rayleigh's treatment it was assumed that each particle scattered as a point source independent of all others. This is equivalent to assuming that the relative positions of the particles are random. In liquids this is not the case, and the scattered light intensity is reduced about fiftyfold because of destructive interference of the light scattered from different particles. In the calculation of scattered intensity in liquids (Einstein 1910; Smoluchowski

1908, 1912), the scattering is considered due to local thermal fluctuations in density which make the liquid optically inhomogeneous. The fluctuations in the density can be related to the compressibility κ of the liquid by comparing the thermal energy per mole RT with the work required to cause a change in density through the application of a hydrostatic pressure:

$$\tau = \frac{8\pi^3}{3\lambda^4}(n^2 - 1)^2 RT\kappa \tag{3-12}$$

Light scattering from solutions of particles small compared to the wavelength

In solutions and in mixtures of liquids, additional light scattering arises from irregular changes in density and refractive index due to fluctuations in composition. Debye (1944, 1947) calculated the effect of these fluctuations, relating them to the change in concentration c associated with the osmotic pressure π per mole of solute:

$$\tau = \frac{32\pi^3}{3\lambda^4}\frac{RTc}{N_o}\left(n\frac{dn}{dc}\right)^2 \bigg/ \left(\frac{d\pi}{dc}\right) \tag{3-13}$$

Inserting the relation between osmotic pressure and molecular weight yields the *Debye equation*:

$$K\frac{c}{R_{90}} = H\frac{c}{\tau} = \frac{1}{M} + 2A_2 c + \cdots$$

where

$$\tag{3-14}$$

$$K = \frac{2\pi^2 n^2}{N_o \lambda^4}\left(\frac{dn}{dc}\right)^2 \quad \text{and} \quad H = \frac{32\pi^3 n^2}{3N_o \lambda^4}\left(\frac{dn}{dc}\right)^2$$

Equation 3-14 forms the basis of the determination of polymer molecular weights by light scattering. Beyond the measurement of τ or R_θ, only the refractive index n and the *specific refractive increment* dn/dc require experimental determination. The latter quantity is a constant for a given polymer, solvent, and temperature, and is measured with an interferometer or a differential refractometer (Brice 1951) or in the analytical ultracentrifuge (Section *E*).

Equation 3-14 is correct only for vertically polarized incident light and for optically isotropic particles. The use of unpolarized light requires that τ be multiplied by $1 + \cos^2 \theta$. Whether polarized or unpolarized light is used, the scattered intensity is symmetrical about the angle of observation 90°. If the scattering particles are anisotropic, a correction for depolarization is required (Cabannes 1929).

In the derivation of Eq. 3-14 it is shown that the amplitude of the scattered light is proportional to the polarizability and hence to the mass of the scattering particle. Thus the intensity of scattering is proportional to the square of the particle mass. If the solute is polydisperse, the heavier molecules contribute more to the scattering than the light ones. The total scattering at zero concentration is

$$\tau = \sum_{i=1}^{\infty} \tau_i = H \sum_{i=1}^{\infty} c_i M_i = Hc\overline{M}_w \tag{3-15}$$

defining the *weight-average molecular weight* according to any of the relationships

$$\overline{M}_w = \frac{\sum_{i=1}^{\infty} c_i M_i}{c} = \sum_{i=1}^{\infty} w_i M_i = \frac{\sum_{i=1}^{\infty} N_i M_i^{2}}{\sum_{i=1}^{\infty} N_i M_i} \tag{3-16}$$

Here w_i is the weight fraction, and

$$c = \sum_{i=1}^{\infty} c_i$$

as in the corresponding definition of number-average molecular weight.

The significance of the weight- and number-average molecular weights is noteworthy. $\overline{M}_w$ is always greater than $\overline{M}_n$ except for a monodisperse system. The ratio $\overline{M}_w/\overline{M}_n$ is a measure of the *polydispersity* of the system. $\overline{M}_w$ is particularly sensitive to the presence of high-molecular-weight species, whereas $\overline{M}_n$ is influenced more by species at the lower end of the molecular weight distribution: if equal weights of molecules with $M = 10,000$ and $M = 100,000$ are mixed, $\overline{M}_w = 55,000$ and $\overline{M}_n = 18,200$; if equal numbers of each kind of molecule are mixed, $\overline{M}_w = 92,000$ and $\overline{M}_n = 55,000$.

Light scattering in solutions of larger particles

When the size of a scattering particle exceeds about $\lambda/20$, different parts of the particle are exposed to incident light of different amplitude and phase. The scattered light is made up of waves coming from different parts of the particle and interfering with one another. The general solution describing the distribution of scattered light for large particles is complex and has been solved only for the case of a homogeneous sphere (Mie 1908). However, an approximate solution applies to most polymer systems. In this treatment, the intensity of light scattered from the elements making up the particle is shown to be proportional to

$$\sum\sum\left[\sin\left(\frac{4\pi r}{\lambda_s}\sin\frac{\theta}{2}\right)\Big/\left(\frac{4\pi r}{\lambda_s}\sin\frac{\theta}{2}\right)\right] \qquad (3\text{-}17)$$

where $\lambda_s = \lambda/n$ is the wavelength of light in the medium of refractive index n, and the double summation is carried out over all pairs of scattering elements separated by a distance r.

For spherical particles, the scattering elements are symmetrically distributed, and the radial density function of Eq. 3-17 is a constant. Solution of the integral form of Eq. 3-17 leads to the following equation for the intensity of scattered light $P(\theta)$ (Rayleigh 1881, 1914; Gans 1925; Debye 1944, 1947):

$$P(\theta) = \left[\left(\frac{3}{u^3}\right)\left(\sin u - u\cos u\right)\right]^2 \qquad (3\text{-}18)$$

where

$$u = 2\pi\left(\frac{d}{\lambda_s}\right)\sin\frac{\theta}{2}$$

and d is the diameter of the spheres.

For a monodisperse system of randomly coiling polymers the radial density function is the Gaussian distribution, and the intensity of scattering is given (Debye 1944, 1947) by

$$P(\theta) = \left(\frac{2}{v^2}\right)[e^{-v} - (1 - v)] \qquad (3\text{-}19)$$

where

$$v = (16\pi^2)\left(\frac{\overline{s^2}}{\lambda_s^2}\right)\sin^2\frac{\theta}{2}$$

and $\overline{s^2}$ is the mean-square radius of gyration of the molecule.

Similar functions can be derived for other models, such as rods and discs. The functions $P(\theta)$ are plotted in Fig. 3-7. It can be seen from the curves that the scattered intensity is greater in the forward than in the backward direction, falling off smoothly with angle except for rather large spheres. It is sometimes convenient to characterize the size of the scattering particles by the *dissymmetry* of light scattering, $z = i_{45°}/i_{135°}$, measured at two angles symmetrical about 90°. The dissymmetry increases rapidly with increasing particle size.

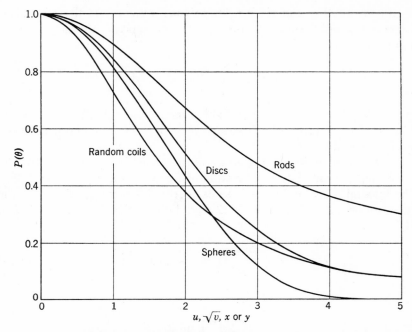

Fig. 3-7.   The variation of the particle scattering function $P(\theta)$ with the size parameter u (spheres), v (monodisperse random coils), x (rods), or y (discs).

Molecular weight determination from light scattering
in solutions of larger particles

Equation 3-14 is derived with the assumption that the scattering particles are small compared to λ. If this is not true, dissymmetry of scattering exists, and the turbidity of the solution is reduced by the interference between waves from different parts of the same particle. This effect is taken into account by modifying Eq. 3-14 to read

$$K\frac{c}{R_{90}} = H\frac{c}{\tau} = \frac{1}{\overline{M}_w P(\theta)} + 2A_2 c + \cdots \tag{3-20}$$

where M is replaced by $\overline{M}_w$ for application to polydisperse systems. It has been shown (Zimm 1948a,b) that a z-average dimension is obtained from $P(\theta)$ for polydisperse systems.

In practice the correction for dissymmetry can be made in two ways. In Debye's original method, the dissymmetry was measured and compared with the functions of Fig. 3-7 to evaluate particle size and thence $P(\theta)$. In more common current use is Zimm's method, in which the left-hand side of Eq. 3-20 is plotted (*Zimm plot*) against $\sin^2 \theta/2 + kc$ where k is an arbitrary constant. A rectilinear grid results (Fig. 3-8), allowing extrapolation to both $c = 0$ and $\theta = 0$, where $P(\theta) = 1$.

Determination of molecular weight by light scattering

Equipment   Early light-scattering turbidimeters, measuring scattered intensity photographically or visually, have been almost entirely superseded by photoelectric instruments. Representative instruments are described in the literature (Brice 1950; Wippler 1954; Baum 1961) and are commercially available (Aminco, Bausch & Lomb, Wood). Briefly, in the widely used Brice-Phoenix instrument (Fig. 3-9) light from a mercury-arc source (S) passes through a lens (L), a polarizer (P_1) and a monochromatizing filter (F). It then strikes either a calibrated reference standard or a glass cell (C) holding the polymer solution. After passing through the cell, the primary light beam is absorbed in a light-trap tube (T).

Scattered light from cell or standard is viewed by a multiplier phototube (R) after passing through a slit system D_1, D_2 and polarizer (P_2). The

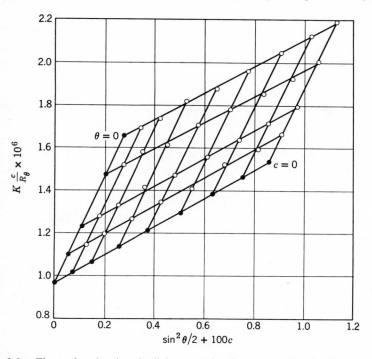

Fig. 3-8.   Zimm plot showing the light scattering from a sample of polystyrene in butanone (Billmeyer 1955). Molecular parameters are $\bar{M}_w = 1,030,000$; $A_2 = 1.29 \times 10^{-4}$ ml mole g^{-2}; $(\bar{s_z^2})^{1/2} = 460$ A; $z = 1.44$.

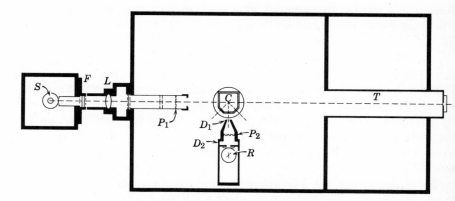

Fig. 3-9.   Sketch of the essential components of a light-scattering photometer.

phototube is powered by high voltage from a regulated electronic power supply. Its output signal is transmitted either to a galvanometer or to a strip chart recorder.

A typical scattering cell consists of a glass cylinder having flat entrance and exit windows. The scattering cell is centered on the axis of rotation of the receiver phototube. The receiver is rotated manually, its angular position being controlled and observed by means of a calibrated disc on the outside of the scattering compartment.

Sample preparation   The proper preparation of a sample for light-scattering measurement is of major importance. Polymer molecules are usually small compared to particles of dust or other extraneous material. The scattering of these extraneous substances, while usually confined to low angles, may easily outweigh the scattering from the polymer solution. Consequently, solvents and polymer solutions must be clarified by filtration or by ultracentrifugation. Proper clarification, as evidenced by obtaining a rectilinear Zimm plot, is essential for the unambiguous interpretation of light-scattering data (Muus 1957; Bryant 1959).

Accuracy in light-scattering measurement is favored by the proper choice of solvent. The difference in refractive index between polymer and solvent should be as large as possible. In addition, the solvent should itself have relatively low scattering, and the polymer-solvent system should not have too high a second virial coefficient, since the extrapolation to zero polymer concentration becomes less certain for high A_2. Mixed solvents must be avoided unless both components have the same refractive index.

Calibration   Although in principle observed scattered intensity can be related to turbidity through knowledge of the geometry of the photometer, calibration with substances of known turbidity is common practice. These substances include reflecting standards (Brice 1950), colloidal suspensions (Maron 1954), and simple liquids (Carr 1950). Tungstosilicic acid, $H_4SiW_{12}O_{40}$, $M = 2879$, has been recommended as a primary calibration standard in preference to pure liquids (too sensitive to impurities) or uniform-particle-size latexes (too sensitive to residual polydispersity) (Kratohvil 1966*b*). (See also Kratohvil 1962, 1965, 1966*a*; Kerker 1964.)

Treatment of data The weight-average molecular weight is the inverse of the intercept at $c = 0$ and $\theta = 0$ in the Zimm plot (Fig. 3-8). The second virial coefficient is calculated from the slope of the lines at constant angle by Eq. 3-20, and to a first approximation is independent of angle. The radius of gyration is derived from the slope of the zero concentration line as a function of angle. The appropriate relationship, derived from Eq. 3-9, is

$$\overline{s_z^2} = \frac{3\lambda_s^2}{16\pi^2} \times \frac{\text{initial slope}}{\text{intercept}} \tag{3-21}$$

Range of applicability Light scattering from solutions has been used to measure molecular weights as low as that of sucrose and as high as those of proteins. In practice, polymer molecular weights in the order of 10,000–10,000,000 can readily be measured, with the possibility of extending the range in either direction in favorable cases. Applications include determination of absolute values of $\overline{M}_w$, especially for proteins and other biological materials, and calibration of intrinsic viscosity-molecular weight relations (Section *D*).

It has long been assumed that information on particle shape could be determined from measurement of the angular dissymmetry of light scattering. It has been demonstrated (Benoit 1968) that this viewpoint should be abandoned except for data obtained under very favorable circumstances. Likewise, the assumption that information on polydispersity can be obtained from similar data has been shown (Carpenter 1966) to be untenable except under most unusual conditions. Finally, the severe restrictions on the applicability of the light-scattering method to copolymer systems (Bushuk 1958, Krause 1961) should be mentioned: the measurements must be made in a solvent selected for the highest possible value of dn/dc (for example, water, methanol, or acetone for most organic polymers) if values of $\overline{M}_w$ which are nearly correct are to be obtained. Use of the complete theory for binary copolymers is complex, but yields parameters describing the distribution of compositions (for example, which monomer may be concentrated in the high- or low-molecular-weight fractions) as well as correct values of $\overline{M}_w$.

The measurement by light scattering of the size of particles considerably larger than polymer molecules is discussed by van de Hulst (1957), Billmeyer (1964*a*), Kerker (1969), and Livesey (1969).

GENERAL REFERENCES

Peaker 1959; Billmeyer 1964*a*; Eskin 1964; Kratohvil 1964, 1966*c*; McIntyre 1964; Kerker 1969.

D. Solution Viscosity and Molecular Size

The usefulness of solution viscosity as a measure of polymer molecular weight has been recognized ever since the early work of Staudinger (1930). Solution viscosity is basically a measure of the size or extension in space of polymer molecules. It is empirically related to molecular weight for linear polymers; the simplicity of the measurement and the usefulness of the viscosity-molecular weight correlation are so great that viscosity measurement constitutes an extremely valuable tool for the molecular characterization of polymers.

Experimental methods

Measurements of solution viscosity are usually made by comparing the *efflux time t* required for a specified volume of polymer solution to flow through a capillary tube with the corresponding efflux time t_0 for the solvent. From t, t_0, and the solute concentration are derived several quantities whose defining equations and names are given in Table 3-4. Two sets of nomenclature are in use for these quantities; one (Cragg 1946) has had long and

TABLE 3-4. *Nomenclature of solution viscosity*

Common Name	Recommended Name	Symbol and Defining Equation
Relative viscosity	Viscosity ratio	$\eta_r = \eta/\eta_0 \simeq t/t_0$
Specific viscosity	—	$\eta_{sp} = \eta_r - 1 = (\eta - \eta_0)/\eta_0 \simeq (t - t_0)/t_0$
Reduced viscosity	Viscosity number	$\eta_{red} = \eta_{sp}/c$
Inherent viscosity	Logarithmic viscosity number	$\eta_{inh} = (\ln \eta_r)/c$
Intrinsic viscosity	Limiting viscosity number	$[\eta] = (\eta_{sp}/c)_{c=0} = [(\ln \eta_r)/c]_{c=0}$

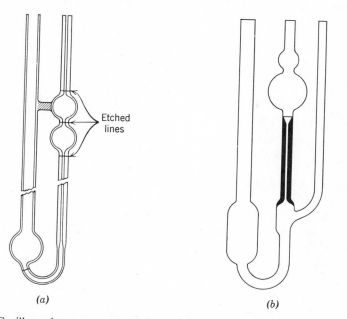

Fig. 3-10. Capillary viscometers commonly used for measurement of polymer solution viscosities: (*a*) Ostwald-Fenske; (*b*) Ubbelohde.

widespread application, the other (International Union 1952) was proposed for greater clarity and precision. In this book the common nomenclature is retained. In this system, the concentration *c* is expressed in grams per deciliter (g/dl, g/100 ml).

The intrinsic viscosity [η] is independent of concentration by virtue of extrapolation to $c = 0$, but is a function of the solvent used. The inherent viscosity at a specified concentration, usually 0.5 g/dl, is sometimes used as an approximation to [η].

Dilute solution viscosity is usually measured in capillary viscometers of the Ostwald-Fenske or Ubbelohde type (Fig. 3-10*a* and *b*, respectively). The latter has the advantage that the measurement is independent of the amount of solution in the viscometer; measurement at a series of concentrations can easily be made by successive dilution.

For highest precision, the following precautions are usually observed. The viscosity measurement is made in a constant-temperature bath regulated to at least $\pm 0.02°C$. The efflux time is kept long (preferably greater than 100 sec) to minimize the need for applying corrections to the observed data. For accuracy in extrapolating to $c = 0$, the solution concentration is restricted to the range that gives relative viscosities between 1.1 and 1.5.

Treatment of data   Viscosity data as a function of concentration are extrapolated to infinite dilution by means of the Huggins (1942) equation

$$\frac{\eta_{sp}}{c} = [\eta] + k'[\eta]^2 c \tag{3-22}$$

where k' is a constant for a series of polymers of different molecular weights in a given solvent. The alternate definition of the intrinsic viscosity leads to the equation (Kraemer 1938):

$$\frac{\ln \eta_r}{c} = [\eta] + k''[\eta]^2 c \tag{3-23}$$

where $k' - k'' = \frac{1}{2}$. Typical data are shown in Fig. 3-11. The slopes of the lines vary as $[\eta]^2$ as required by Eqs. 3-22 and 3-23 for polymer species of different molecular weight in the same solvent.

At intrinsic viscosities above about 2, and even lower in some cases, there may be an appreciable dependence of viscosity on rate of shear in the viscometer. This dependence is not eliminated by extrapolation to infinite dilution; measurement as a function of shear rate and extrapolation to zero shear rate is also required. Special apparatus and extreme care is required (Zimm 1962; Van Oene 1968).

Empirical correlations between intrinsic viscosity and molecular weight for linear polymers

Staudinger's prediction in 1930 that the reduced viscosity of a polymer is proportional to its molecular weight has needed only slight modification: the intrinsic viscosity has been substituted for the reduced viscosity, and it

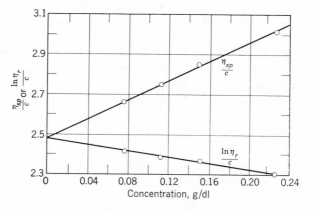

Fig. 3-11. Reduced and inherent viscosity-concentration curves for a polystyrene in benzene (Ewart 1946).

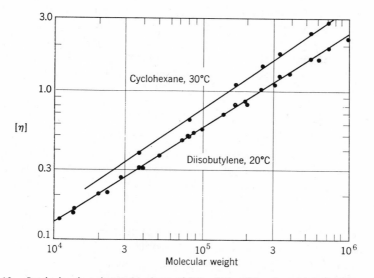

Fig. 3-12. Intrinsic viscosity-molecular weight relationships for polyisobutylene in diisobutylene and cyclohexane (Flory 1943; Krigbaum 1953).

has been recognized that the proportionality is to a power of the molecular weight somewhat less than 1. The relation is expressed in the equation

$$[\eta] = K'M^a \tag{3-24}$$

where K' and a are constants determined from a double logarithmic plot of intrinsic viscosity and molecular weight (Fig. 3-12). Such plots are usually found to be straight lines within experimental error over large ranges of the variables. For randomly coiled polymers, the exponent a varies from 0.5 in a Θ-solvent to a maximum of about 1.0. For many systems, a lies between 0.6 and 0.8. Typical values of K' range between 0.5 and 5×10^{-4}. Both K' and a are functions of the solvent as well as of the polymer type. This empirical relation between viscosity and molecular weight is valid only for linear polymers.

Intrinsic viscosity measurement leads to the *viscosity-average molecular weight,* defined by the equation (Schaefgen 1948)

$$\overline{M}_v = \left[\sum_{i=1}^{\infty} w_i M_i^a \right]^{1/a} = \left[\frac{\sum_{i=1}^{\infty} N_i M_i^{1+a}}{\sum_{i=1}^{\infty} N_i M_i} \right]^{1/a} \tag{3-25}$$

Thus $\overline{M}_v$ depends on a as well as on the distribution of molecular species. For many polymers, $\overline{M}_v$ is 10–20 % below $\overline{M}_w$ (Fig. 1-1). For $a = 1, \overline{M}_v = \overline{M}_w$.

If fractionated polymers, for which $\overline{M}_n \simeq \overline{M}_v \simeq \overline{M}_w$, are available, any absolute molecular weight measurement may be combined with a viscosity measurement to evaluate the constants in Eq. 3-24. If the molecular weight distribution of the samples is known in sufficient detail to calculate $\overline{M}_v$ from a molecular weight average which has been measured, the constants in the equation

$$[\eta] = K' \overline{M}_v^{\,a} \tag{3-26}$$

may be evaluated. Since $\overline{M}_v$ is nearer to $\overline{M}_w$ than to $\overline{M}_n$, the weight-average molecular weight is preferred for this calibration. It can be shown that polymers with fairly broad but well-known distributions are preferred to fractions for this purpose. A less precise relation,

$$[\eta] = K' \overline{M}_w^{\,a} \tag{3-27}$$

is sometimes used to relate weight-average molecular weight directly to intrinsic viscosity for a limited series of samples. Extensive tabulations of K' and a are available (Meyerhoff 1961, Kurata 1963, 1966).

Equation 3-27 is known to become inaccurate for molecular weights below about 50,000 because deviations from the straight-line relationship set in. For better results in this region, and with some theoretical justification (Stockmayer 1963), use of one of several expressions of the form

$$[\eta] = K M^{1/2} + K'' M \tag{3-28}$$

is recommended. Here the first term is determined by short-range interactions, as in Eq. 3-31, and the second term by long-range interactions. The form and value of K'' vary among the several theories available (Cowie 1966).

Intrinsic viscosity and molecular size

Theories (Debye 1948; Kirkwood 1948; Flory 1949, 1950, 1951) of the frictional properties of polymer molecules in solution show that the intrinsic viscosity is proportional to the effective hydrodynamic volume of the molecule in solution divided by its molecular weight. The effective volume is proportional to the cube of a linear dimension of the randomly coiling chain. If $(\overline{r^2})^{1/2}$ is the dimension chosen,

$$[\eta] = \Phi \frac{(\overline{r^2})^{3/2}}{M} \tag{3-29}$$

where Φ is a universal constant. Replacing $(\overline{r^2})^{1/2}$ by $\alpha(\overline{r_0^2})^{1/2}$ and recalling that r_0^2/M is a function of chain structure independent of its surroundings or molecular weight, it follows that

$$[\eta] = \Phi\left(\frac{r_0{}^2}{M}\right)^{3/2} M^{1/2}\alpha^3 = KM^{1/2}\alpha^3 \qquad (3\text{-}30)$$

where $K = \Phi(\overline{r_0{}^2}/M)^{3/2}$ is a constant for a given polymer, independent of solvent and molecular weight. The dimensions $\overline{(r^2)}^{1/2}$ and $\overline{(r_0{}^2)}^{1/2}$ and the expansion factor α used here are usually identified with those discussed in Chapter 2. Extensions of the Flory-Krigbaum treatment (Chapter 2C) imply that Φ is not a universal constant, and that the power of α in Eq. 3-30 is somewhat less than 3 (Kurata 1960, 1963).

It follows from the properties of α that, at $T = \Theta$, $\alpha = 1$ and

$$[\eta]_\Theta = KM^{1/2} \qquad (3\text{-}31)$$

There is ample experimental evidence for the validity of this equation. Values of K are near 1×10^{-3} for a number of polymer systems.

An important feature of Flory's viscosity theory is that it furnishes confirmation (if not evaluation) of the Θ-temperature as that at which $a = \frac{1}{2}$, and it allows the determination of the unperturbed dimensions of the polymer chain. Even if a Θ-solvent is not available, several extrapolation techniques (summarized by Cowie 1966) are available for estimating the unperturbed dimensions from viscosity data in good solvents. The best and simplest of these techniques appears to be that of Stockmayer (1963) based on Eq. 3-28.

Determination of chain branching

The size of a dissolved polymer molecule, which can be estimated by means of its intrinsic viscosity, depends upon a number of factors. These include the molecular weight, the local chain structure as reflected in the factor $\overline{r_0{}^2}/M$, the perturbations reflected in α, and finally the gross chain structure, in which the degree of long-chain branching is an important variable. [The presence of short branches, such as ethyl or butyl groups in low-density polyethylene (Chapter 13A), has little effect on solution viscosity.] By evaluating or controlling each of the other factors, viscosity measurements may be used to estimate the degree of long-chain branching of a polymer.

The relation between molecular size and the number and type of branch points has been calculated (Zimm 1949), leading to a relation g between the radii of gyration for branched and linear chains of the same number of segments:

$$g = \frac{\overline{s^2} \text{ (branched)}}{\overline{s^2} \text{ (linear)}} \qquad (3\text{-}32)$$

Several theories have been proposed relating g to $[\eta]$(branched)/$[\eta]$(linear). Only recently has it been possible to decide which of these yields the best fit to experimental data; the situation is well summarized by Graessley (1968). At present it still seems appropriate to consider the ratio $[\eta]$(branched)/$[\eta]$(linear) evaluated at constant molecular weight as a qualitative indication of chain branching rather than to attempt to assign numerical values to the degree of chain branching in a given polymer sample.

GENERAL REFERENCES

Flory 1953; Onyon 1959; Meyerhoff 1961; Kurata 1963; Lyons 1967; Moore 1967.

E. Ultracentrifugation

Ultracentrifugation techniques are the most intricate of the existing methods for determining the molecular weights of high polymers. They are far more successful in application to compact protein molecules than to random coil polymers, where extended conformations increase deviations from ideality and the chances of mechanical entanglement of the polymer chains. Despite recent theoretical and experimental developments which increase the utility of the techniques for random-coil polymer systems, the methods are used overwhelmingly for biological materials.

Experimental techniques

The ultracentrifuge consists of an aluminum rotor several inches in diameter which is rotated at high speed in an evacuated chamber. The solution being centrifuged is held in a small cell within the rotor near its periphery. The rotor may be driven electrically or by an oil or air turbine. The cell is equipped with windows, and the concentration of polymer along its length is determined by optical methods based on measurements of refractive index or absorption. A schematic view of a commercially available ultracentrifuge is shown in Fig. 3-13. The instrument can be used over a range of speeds suitable for all types of ultracentrifugation, and at temperatures up to about 100°C.

Solvents for ultracentrifugation experiments must be chosen for difference from the polymer in both density (to ensure sedimentation) and refractive index (to allow measurement). In addition, mixed solvents must usually be avoided, and low solvent viscosity is desirable. It is customary to use *schlieren optics* in the ultracentrifuge; experimental data are obtained as a (photographed) curve of dn/dr vs. r, where n is refractive index and r is the distance

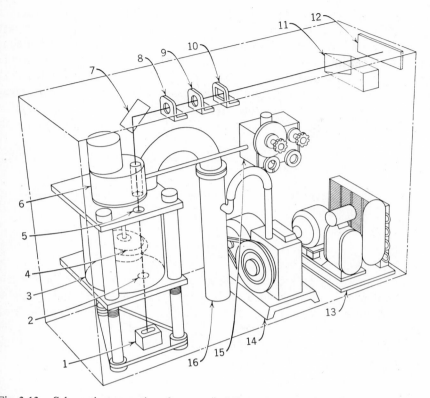

Fig. 3-13.   Schematic perspective of a commercially available, electrically driven analytical ultracentrifuge. Numbered parts include: 1, lamp; 2, 5, 9, lenses; 3, vacuum chamber; 4, rotor; 6, drive unit; 7, 11, mirrors; 8, 10, diaphragm and cylindrical lens of schlieren optics; 12, photographic plate; 13, refrigeration unit; 14, 16, vacuum pumps; 15, speed-control unit (Courtesy Spinco).

from the center of rotation to the point of observation in the cell. These data are converted to a curve of dc/dr vs. r through knowledge of the specific refractive increment dn/dc. A *synthetic boundary cell*, in which pure solvent is layered onto a polymer solution during ultracentrifugation to form a sharp boundary, can be used to calibrate the optics of the ultracentrifuge or to determine dn/dc after calibration.

Sedimentation equilibrium

In the sedimentation equilibrium experiment, the ultracentrifuge is operated at a low speed of rotation for times up to 1 or 2 weeks under constant conditions. A thermodynamic equilibrium is reached in which the polymer is distributed in the cell solely according to its molecular weight and molecular weight distribution, the force of sedimentation on each species being just balanced by its tendency to diffuse back against the concentration gradient resulting from its movement in the centrifugal field. The force on a particle of mass m at a distance r from the axis of rotation is $\omega^2 r(1 - \bar{v}\rho)m$, where ω is the angular velocity of rotation, $\bar{v}$ the partial specific volume of the polymer, and ρ the density of the solution. Writing the partial molar free energy in terms of this force and applying the conditions for equilibrium gives the result

$$\frac{M(1 - \bar{v}\rho)\omega^2 r}{RT} = \frac{1}{c}\frac{dc}{dr}\left(1 + c\frac{d\ln\gamma}{dc}\right) \qquad (3\text{-}33)$$

where the second term corrects for nonideality, γ being the activity coefficient. For ideal solutions, this equation integrates to

$$M = \frac{2RT\ln(c_2/c_1)}{(1 - \bar{v}\rho)\omega^2(r_2^2 - r_1^2)} \qquad (3\text{-}34)$$

where c_1 and c_2 are the concentrations at two points r_1 and r_2 in the cell. If the polymer is heterogeneous, M is the weight-average molecular weight; r_1 and r_2 must be taken at the meniscus and the cell bottom to include all the molecular species in the average. If the data are treated in terms of the refractive increment $\tilde{n}$, the difference between the refractive indices of solution and solvent, it can be shown that

$$\bar{M}_z = \frac{RT}{(1 - \bar{v}\rho)\omega^2}\frac{\left(\dfrac{1}{r}\dfrac{d\tilde{n}}{dr}\right)_2 - \left(\dfrac{1}{r}\dfrac{d\tilde{n}}{dr}\right)_1}{\tilde{n}_2 - \tilde{n}_1} \qquad (3\text{-}35)$$

where $\bar{M}_z$ is the *z-average molecular weight*

$$\bar{M}_z = \frac{\displaystyle\sum_{i=1}^{\infty} N_i M_i^3}{\displaystyle\sum_{i=1}^{\infty} N_i M_i^2} = \frac{\displaystyle\sum_{i=1}^{\infty} w_i M_i^2}{\displaystyle\sum_{i=1}^{\infty} w_i M_i} \qquad (3\text{-}36)$$

and again positions 1 and 2 must be the ends of the cell.

The problem of nonideality is best met by working at the Θ-temperature. In nonideal solutions, the apparent molecular weight is a linear function of concentration at temperatures near Θ, the slope depending primarily on the second virial coefficient.

The major disadvantage to the sedimentation equilibrium experiment lies in the fact that several days may be required to reach equilibrium, even if the time is shortened by the use of short cells or a synthetic boundary cell to reduce the distance the various species must travel as equilibrium is established.

The distribution of molecular weights can be obtained directly from sedimentation equilibrium measurements (Scholte 1968a,b, 1970). The solution of an integral equation is involved, however, and the analysis remains a difficult one.

Equilibrium sedimentation in a density gradient   In this approach, a mixed solvent is chosen so that the relative sedimentation of the two components gives rise to a density gradient. The solute forms a band centering at the point where its effective density equals that of the solvent mixture. The band is Gaussian in shape with respect to solute concentration, the half-width being inversely proportional to the solute molecular weight. A feature which is of major importance is the sensitivity of the method to small differences in effective density among the solute species. Density-gradient ultracentrifugation has been used to separate ordinary and isotactic polystyrene (Morawetz 1965) and to obtain composition distributions in copolymers.

Approach to Equilibrium

Archibald (1947) reexamined the boundary conditions for the ultracentrifuge equation and showed that measurement of c and dc/dr at the cell boundaries allows molecular weight to be determined at any stage in the equilibrium process. If measurements are made early enough, before the molecular species have time to redistribute in the cell, the weight-average molecular weight and second virial coefficient can be evaluated. In practice, measurements can be made about 10–60 min after the start of the experiment. At properly chosen times within this range, a graph of $(1/cr)(dc/dr)$ vs. r is linear near the ends of the cell, allowing simple extrapolation to the position of the meniscus or cell bottom. Redistribution of molecular species can be recognized and corrected for by extrapolating the apparent molecular weight to zero time. The method holds considerable promise for random coil polymers, but sedimentation equilibrium with short cells is often preferred.

Sedimentation transport

In the *sedimentation transport* or *sedimentation velocity* experiment, the ultracentrifuge is operated at high speed so that the solute is transported to the bottom of the cell. The rate of sedimentation is given by the *sedimentation constant*

$$s = \frac{1}{\omega^2 r}\frac{dr}{dt} = \frac{m(1 - \bar{v}\rho)}{f} \tag{3-37}$$

and related to particle mass through a *frictional coefficient f*. For rigid particles f is well defined (for solid spheres of diameter d, f is given by Stokes' law, $f = 3\pi\eta d$), but for random-coil polymers the evaluation of f is more difficult. It can be related to the diffusion coefficient D: at infinite dilution $D = RT/f$, whence,

$$\frac{D}{s} = \frac{RT}{M(1 - \bar{v}\rho)} \qquad (3\text{-}38)$$

However, even for a single solute species the equations have not been extended to take into account simultaneously the effects of diffusion in the ultracentrifuge cell, deviations from ideality, and compressibility of solvent and solute in the high centrifugal fields required. For polydisperse solutes, additional complications arise in assigning average molecular weights, since D and s must be averaged separately. In general, simple averages such as $\overline{M}_w$ and $\overline{M}_z$ are not obtained.

The diffusion constant may advantageously be measured in the ultracentrifuge, using a synthetic-boundary cell to create a boundary between solution and solvent. If the speed of rotation is low so that sedimentation is negligible, the spreading of the boundary results entirely from diffusion. Nevertheless, the diffusion constant is often not available, and it is desirable to seek another quantity related to f. If it is assumed that the frictional coefficient is the same for viscous flow as for free diffusion (and, indeed, also for sedimentation), the intrinsic viscosity is such a quantity. Mandelkern (1952a,b) writes

$$[\eta] = \frac{(f/K_f)^3}{M} \qquad (3\text{-}39)$$

whence

$$\frac{s}{[\eta]^{1/3}} = \frac{(1 - \bar{v}\rho)M^{2/3}}{K_f} \qquad (3\text{-}40)$$

This treatment is, of course, subject to the same limitations as the analysis involving s and D.

The most fruitful result of the sedimentation-transport experiment now appears to be examination of the distribution of sedimentation constants. By working in a Θ-solvent, the concentration dependence of s and D is reduced enough to avoid extrapolation to $c = 0$ [although Flory (1953) has shown that it is $d(s/D)/dc$ which vanishes at $T = \Theta$]. The observed distribution of s over the cell must still be extrapolated to infinite time to eliminate the effects of diffusion relative to sedimentation, and correction for compression must be made. The resulting distribution of s can be converted to a distribution of molecular weight, either through detailed knowledge of the dependence of s and D on M or from the relation

$$s = kM^{1/2} \qquad (3\text{-}41)$$

which is very nearly true at $T = \Theta$. The number of approximations and the complex treatment of the data required in this procedure suggest that its use be limited to qualitative observations. In this respect, however, the experiment can be extremely valuable, for it provides directly qualitative information about the nature of the distribution of species present.

Preparative separations The usefulness of the ultracentrifuge in the preparation of samples rather than in the production of analytical data should not be overlooked. Preparative ultracentrifuges have utility in fractionating polymer samples and in freeing them from easily sedimented impurities.

GENERAL REFERENCES

Svedberg 1940; Williams 1958, 1963; Schachman 1959; Baldwin 1960; Svensson 1961; Fujita 1962; Van Holde 1967; Adams 1968.

F. Polyelectrolytes

Polymers with ionizable groups along the chain, termed polyelectrolytes, normally exhibit properties in solution which are quite different from those with nonionizable structures. There are many examples of polyelectrolytes, including polyacids such as poly(acrylic acid) and hydrolyzed copolymers of maleic anhydride, polybases such as poly(vinyl amine) and poly(4-vinyl pyridine), polyphosphates, nucleic acids, and proteins.

When they are soluble in nonionizing solvents—for example, poly(acrylic acid) in dioxane—polyelectrolytes behave in completely normal fashion, but in aqueous solution they are ionized, with three major results. First, the mutual repulsion of their charges causes expansions of the chain far beyond those resulting from changes from good to poor solvents with ordinary polymers. The size of the polyelectrolyte random coil is, moreover, a function of the concentrations of polymer and added salt, if any, since both influence the degree of ionization.

Second, the ionization of the electrolyte groups leads to a variety of unusual effects in the presence of small amounts of added salt. The intensity of light scattering decreases because of the ordering of the molecules in solution, while the osmotic pressure and ultracentrifugation behavior are determined predominantly by the total charge on the molecule (Donnan effect).

Finally, the ionic charges attached to the chains create regions of high local charge density, affecting the activity coefficients and properties of small ions in these localities. Although the various effects cannot be separated completely, the results of chain expansion are of primary interest for the measurement of molecular weight and size.

Those properties depending on the size of the chain, such as viscosity and angular dependence of light scattering, are strongly affected by chain expansion. The viscosity may even increase markedly as polymer concentration *decreases* with consequent increase in the degree of ionization of the polymer. When very high chain extensions are reached (up to half of the fully extended chain length) the effect reverses, but it does not disappear at infinite dilution. On the other hand, the addition of low-molecular-weight electrolyte (salt) to the aqueous solution increases the ionic strength of the solution outside the polymer coil relative to that inside, and also reduces the thickness of the layer of "bound" counterions around the chain. Both effects cause the chain to contract, and when the concentration of added salt reaches, say, 0.1 molar, behavior is again normal. With some special precautions, molecular weights may be measured by light scattering and equilibrium ultracentrifugation, and intrinsic viscosity-molecular weight relations may be established, for polyelectrolytes in the presence of added salt.

Because of the preponderance of small ions, the colligative properties of polyelectrolytes in ionizing solvents measure counterion activities rather than molecular weight. In the presence of added salt, however, correct molecular weights of polyelectrolytes can be measured by membrane osmometry, since the small ions can equilibrate across the membrane. The second virial coefficient differs from that previously defined, since it is determined by both ionic and nonionic polymer-solvent interactions.

The transport and electrophoretic properties of polyelectrolytes are beyond the scope of this book.

GENERAL REFERENCES

Morawetz 1965, Chap. VII; Miller 1966, Chap. 12; Armstrong 1969.

BIBLIOGRAPHY

Adams 1968. E. T. Adams, Jr., "Molecular Weights and Molecular-Weight Distributions from Sedimentation-Equilibrium Experiments," pp. 84–142 in Donald McIntyre, ed., *Characterization of Macromolecular Structure,* National Academy of Sciences Publication No. 1573, Washington, D.C., 1968.

Allen 1959. P. W. Allen, *Techniques of Polymer Characterization,* Butterworths, London, 1959.

Aminco. American Instrument Co., Inc., 8030 Georgia Avenue, Silver Spring, Maryland, 20910.

Archibald 1947. W. J. Archibald, "A Demonstration of Some New Methods of Determining Molecular Weights from the Data of the Ultracentrifuge," *J. Phys. Coll. Chem.* **51,** 1204-1214 (1947).

Armstrong 1968. Jerold L. Armstrong. "Critical Evaluation of Commercially Available Hi-Speed Membrane Osmometers," pp. 51-55 in Donald McIntyre, ed., *Characterization of Macromolecular Structure,* National Academy of Sciences Publication No. 1573, Washington, D.C., 1968.

Armstrong 1969. R. W. Armstrong and U. P. Strauss, "Polyelectrolytes," pp. 781-861 in Herman F. Mark, Norman G. Gaylord, and Norbert M. Bikales, eds., *Encyclopedia of Polymer Science and Technology,* Vol. 10, Interscience Div., John Wiley and Sons, New York, 1969.

Badgley 1949. W. J. Badgley and H. Mark, "Osmometry and Viscometry of Polymer Solutions," pp. 75-112 in R. E. Burk and Oliver Grummit, eds., *High Molecular Weight Compounds* (*Frontiers in Chemistry* Vol. VI), Interscience Publishers, New York, 1949.

Baldwin 1960. R. L. Baldwin and K. E. van Holde, "Sedimentation of High Polymers," *Fortschr. Hochpolym.-Forsch* (*Advances in Polymer Science*) **1,** 451-511 (1960).

Baum 1961. F. J. Baum and F. W. Billmeyer, Jr., "An Automatic Photometer for Measuring the Angular Dissymmetry of Light Scattering," *J. Opt. Soc. Am.* **51,** 452-456 (1961).

Bausch and Lomb. Bausch and Lomb Co., Rochester, New York 14602.

Benoit 1968. Henri Benoit, "Use of Light Scattering and Hydrodynamic Methods for Determining the Overall Conformation of Helical Molecules" (in French), *J. Chem. Phys.* **65,** 23-30, (1968).

Bevington 1954a. J. C. Bevington, H. W. Melville, and R. P. Taylor, "The Termination Reaction in Radical Polymerization. Polymerizations of Methyl Methacrylate and Styrene at 25°," *J. Polymer Sci.* **12,** 449-459 (1954).

Bevington 1954b. J. C. Bevington, H. W. Melville, and R. P. Taylor, "The Termination Reaction in Radical Polymerizations. II. Polymerizations of Styrene at 60° and of Methyl Methacrylate at 0 and 60°, and the Copolymerization of These Monomers at 60°," *J. Polymer Sci.* **14,** 463-476 (1954).

Billmeyer 1955. F. W. Billmeyer, Jr., and C. B. de Than, "Dissymmetry of Molecular Light Scattering in Polymethyl Methacrylate," *J. Am. Chem. Soc.* **77,** 4763-4767 (1955).

Billmeyer 1964a. Fred W. Billmeyer, Jr., "Principles of Light Scattering," Chap. 56 in I. M. Kolthoff and Philip J. Elving, eds., with the assistance of Ernest B. Sandell, *Treatise on Analytical Chemistry,* Part I, Vol. 5, Interscience Div., John Wiley and Sons, New York, 1964.

Billmeyer 1964b. F. W. Billmeyer, Jr., and V. Kokle, "The Molecular Structure of Polyethylene. XV. Comparison of Number-Average Molecular Weights by Various Methods," *J. Am. Chem. Soc.* **86,** 3544-3546 (1964).

Billmeyer 1965. Fred W. Billmeyer, Jr., "Characterization of Molecular Weight Distributions in High Polymers," *J. Polymer Sci.* **C8,** 161-178 (1965).

Billmeyer 1966a. Fred W. Billmeyer, Jr., "Practical Applications of Polymer Solutions," *Polymer Eng. Sci.* **6,** 359-362 (1966).

Billmeyer 1966b. Fred W. Billmeyer, Jr., "Measuring the Weight of Giant Molecules," *Chemistry,* **39,** No. 3, 8-14 (1966).

Billmeyer 1969. Fred W. Billmeyer, Jr., "Recent Advances in Determining Polymer Molecular Weights and Sizes," *Applied Polymer Symposia* **10,** 1-6 (1969).

Bonnar 1958. R. U. Bonnar, M. Dimbat, and F. H. Stross, *Number-Average Molecular Weights*, Interscience Publishers, New York, 1958.

Brice 1950. B. A. Brice, M. Halwer, and R. Speiser, "Photoelectric Light-Scattering Photometer for Determining High Molecular Weights," *J. Opt. Soc. Am.* **40**, 768–778 (1950).

Brice 1951. B. A. Brice and M. Halwer, "A Differential Refractometer," *J. Opt. Soc. Am.* **41**, 1033–1037 (1951).

Bryant 1959. W. M. D. Bryant, F. W. Billmeyer, Jr., L. T. Muus, J. T. Atkins, and J. E. Eldridge, "The Molecular Structure of Polyethylene. X. Optical Studies of Cross-linked Networks," *J. Am. Chem. Soc.* **81**, 3219–3223 (1959).

Burge 1963. David E. Burge, "Osmotic Coefficients in Aqueous Solutions. Studies with the Vapor Pressure Osmometer," *J. Phys. Chem.* **67**, 2590–2593 (1963).

Bushuk 1958. W. Bushuk and H. Benoit, "Light-Scattering Studies of Copolymers. I. Effect of Heterogeneity of Chain Composition on the Molecular Weight," *Can. J. Chem.* **36**, 1616–1626 (1958).

Cabannes 1929. J. Cabannes and Y. Rocard, *La diffusion moléculaire de la lumière*, Les Presses Universitaires de France, Paris, 1929.

Carpenter 1966. Dewey K. Carpenter, "Light-Scattering Study of the Molecular Weight Distribution of Polypropylene," *J. Polymer Sci.* **A-2 4**, 923–942 (1966).

Carr 1950. C. I. Carr, Jr., and B. H. Zimm, "Absolute Intensity of Light Scattering from Pure Liquids and Solutions," *J. Chem. Phys.* **18**, 1616–1626 (1950).

Chiang 1964. R. Chiang "Characterization of High Polymers in Solutions–with Emphasis on Techniques at Elevated Temperatures," Chap. XII in Bacon Ke, ed., *Newer Methods of Polymer Characterization*, Interscience Div., John Wiley and Sons, New York, 1964.

Chylewski 1964. Ch. Chylewski and W. Simon, "The Vaporometric (Thermoelectric) Molecular Weight Determination, Part IV: Theoretical Treatment of the Measuring Apparatus" (in German), *Helv. Chim. Acta* **47**, 515–526 (1964).

Coll 1968. Hans Coll and F. H. Stross, "Determination of Molecular Weights by Equilibrium Osmotic-Pressure Measurements," pp. 10–27 in Donald McIntyre, ed., *Characterization of Macromolecular Structure*, National Academy of Sciences Publication No. 1573, Washington, D.C., 1968.

Conix 1958. André Conix, "On the Molecular Weight Determination of Poly(ethylene terephthatate)," *Makromol. Chem.* **26**, 226–235 (1958).

Cowie 1966. J. M. G. Cowie, "Estimation of Unperturbed Polymer Dimensions from Viscosity Measurements in Non-Ideal Solvents," *Polymer* **7**, 487–495 (1966).

Cragg 1946. L. H. Cragg, "The Terminology of Intrinsic Viscosity and Related Functions," *J. Colloid. Sci.* **1**, 261–269 (1946).

Daniels 1958. W. W. Daniels and R. E. Kitson, "Infrared Spectroscopy of Polyethylene Terephthalate," *J. Polymer Sci.* **33**, 161–170 (1958).

Debye 1944. P. Debye, "Light Scattering in Solutions," *J. Appl. Phys.* **15**, 338–342 (1946).

Debye 1947. P. Debye, "Molecular-Weight Determination by Light Scattering," *J. Phys. & Coll. Chem.* **51**, 18–32 (1947).

Debye 1948. P. Debye and A. M. Bueche, "Intrinsic Viscosity, Diffusion, and Sedimentation Rates of Polymers in Solutions," *J. Chem. Phys.* **16**, 573–579 (1948).

Dobry 1935. A. Dobry, "Apparatus for the Measurement of Very Small Osmotic Pressures in Colloidal Solutions" (in French), *J. chim. phys.* **32**, 46–49 (1935).

Dohner 1967. R. E. Dohner, A. H. Wachter, and W. Simon, "Apparatus for Molecular Weight Determination on Highly Dilute (10^{-4}M) Solutions by Means of Vapor Pressure Osmometry" (in German), *Helv. Chim. Acta* **50**, 2193–2200 (1967).

Einstein 1910. A. Einstein, "Theory of the Opalescence of Homogeneous Liquids and Liquid Mixtures in the Neighborhood of the Critical State" (in German), *Ann. Physik* **33**, 1275–1298 (1910).

Elias 1958. H.-G. Elias and T. A. Ritscher, "A Simple Two-Compartment Glass Osmometer with a Glass Membrane" (in German), *J. Polymer Sci.* **28**, 648–651 (1958).

Elias 1966. Hans-Georg Elias and Hanspeter Lys, "Virial Coefficients and Association of Polyethyleneglycols" (in German), *Makromol. Chem.* **92**, 1–24 (1966).

Elias 1967. Hans-Georg Elias and Rolf Bareiss, "Association of Macromolecules" (in German), *Chimia* **21**, 53–65 (1967).

Elias 1968. Hans-Georg Elias, "Dynamic Osmometry," pp. 28–50 in Donald McIntyre, ed., *Characterization of Macromolecular Structure*, National Academy of Sciences Publication No. 1573, Washington, D.C., 1968.

Eskin 1964. V. E. Eskin, "Light Scattering as a Method of Studying Polymers," *Soviet Phys. Uspekhi* **7**, 270–304 (1964).

Evans 1947. M. G. Evans, "Polymerization of Monomers in Aqueous Solution," *J. Chem. Soc.* **1947**, 266–274.

Ewart 1946. R. H. Ewart, "Significance of Viscosity Measurements on Dilute Solutions of High Polymers," pp. 197–251 in H. Mark and G. S. Whitby, eds., *Scientific Progress in the Field of Rubber and Synthetic Elastomers (Advances in Colloid Science, Vol. II)*, Interscience Publishers, New York, 1946.

Ezrin 1968. Myer Ezrin, "Determination of Molecular Weight by Ebulliometry," pp. 3–9 in Donald McIntyre, ed., *Characterization of Macromolecular Structure*, National Academy of Sciences Publication No. 1573, Washington, D.C., 1968.

F & M. F & M Scientific Div., Hewlett-Packard Corp., Avondale, Pa., 19311.

Flory 1943. Paul J. Flory, "Molecular Weights and Intrinsic Viscosities of Polyisobutylenes," *J. Am. Chem. Soc.* **65**, 372–382 (1943).

Flory 1949. Paul J. Flory, "The Configuration of Real Polymer Chains," *J. Chem. Phys.* **17**, 303–310 (1949).

Flory 1950. Paul J. Flory and Thomas G Fox, Jr., "Molecular Configuration and Thermodynamic Parameters from Intrinsic Viscosities," *J. Polymer Sci.* **5**, 745–747 (1950).

Flory 1951. P. J. Flory and T. G Fox, Jr., "Treatment of Intrinsic Viscosities," *J. Am. Chem. Soc.* **73**, 1904–1908 (1951).

Flory 1953. Paul J. Flory, *Principles of Polymer Chemistry*, Cornell University Press, Ithaca, N. Y., 1953.

Fujita 1962. Hiroshi Fujita, *Mathematical Theory of Sedimentation Analysis*, Academic Press, New York, 1962.

Fuoss 1943. Raymond M. Fuoss and Darwin J. Mead, "Osmotic Pressures of Polyvinyl Chloride Solutions by a Dynamic Method," *J. Phys. Chem.* **47**, 59–70 (1943).

Gans 1925. R. Gans, "Diagrams of Radiation from Ultra-Microscopic Particles" (in German), *Ann. Physik.* **76**, 29–38 (1925).

Gee 1944. Geoffrey Gee, "The Molecular Weights of Rubber and Related Materials. V. The Interpretation of Molecular Weight Measurements on High Polymers," *Trans. Faraday Soc.* **40**, 261–266 (1944).

Glover 1966. Clyde A. Glover, "Determination of Molecular Weights by Ebulliometry," pp. 1–67 in Charles N. Reilley and Fred W. McLafferty, eds., *Advances in Analytical Chemistry and Instrumentation*, Vol. 5, Interscience Div., John Wiley and Sons, New York, 1966.

Graessley 1968. William W. Graessley, "Detection and Measurement of Branching in Polymers," pp. 371–388 in Donald McIntyre, ed., *Characterization of Macromolecular*

Structure, National Academy of Sciences Publication No. 1573, Washington, D.C., 1968.

Hallikainen. Hallikainen Instruments, 750 National Court, Richmond, California, 94804.

Hellman 1962. Max Hellman and Leo A. Wall, " End-Group Analysis," Chap. V in Gordon M. Kline, ed., *Analytical Chemistry of Polymers*, Part III, Interscience Div., John Wiley and Sons, New York, 1962.

Hoffmann 1965. M. Hoffmann and M. Unbehend, "A New Possible Application of Osmometry " (in German), *Makromol. Chem.* **88**, 256–271 (1965).

Holleran 1968. Peter M. Holleran and Fred W. Billmeyer, Jr., " Rapid Osmometry with Diffusible Polymers," *J. Polymer Sci.* **B6**, 137–140 (1968).

Huggins 1942. Maurice L. Huggins, "The Viscosity of Dilute Solutions of Long-Chain Molecules. IV. Dependence on Concentration," *J. Am. Chem. Soc.* **64**, 2716–2718 (1942).

Immergut 1954. E. H. Immergut, S. Rollin, A. Salkind, and H. Mark, "New Types of Membranes for Osmotic Pressure Measurements," *J. Polymer Sci.* **12**, 439–443 (1954).

International Union 1952. International Union of Pure and Applied Chemistry, "Report on Nomenclature in the Field of Macromolecules," *J. Polymer Sci.* **8**, 255–277 (1952).

Joyce 1948. R. M. Joyce, W. E. Hanford, and J. Harmon, "Free Radical-Initiated Reaction of Ethylene with Carbon Tetrachloride," *J. Am. Chem. Soc.* **70**, 2529–2532 (1948).

Kerker 1964. Milton Kerker, Josip P. Kratohvil and Egon Matijević, "Calibration of Light-Scattering Instruments. II. The Volume Correction," *J. Polymer Sci.* **A2**, 303–311 (1964).

Kerker 1969. Milton Kerker, *The Scattering of Light and Other Electromagnetic Radiation*, Academic Press, New York, 1969.

Kirkwood 1948. John G. Kirkwood and Jacob Riseman, "The Intrinsic Viscosities and Diffusion Constants of Flexible Molecules in Solution," *J. Chem. Phys.* **16**, 565–573 (1948).

Kraemer 1938. Elmer O. Kraemer, " Molecular Weights of Cellulose and Cellulose Derivatives," *Ind. Eng. Chem.* **30**, 1200–1203 (1938).

Kratohvil 1962. J. P. Kratohvil, Gj. Deželić, M. Kerker, and E. Matijević, "Calibration of Light-Scattering Instruments: A Critical Survey," *J. Polymer Sci.* **57**, 59–78 (1962).

Kratohvil 1964. Josip P. Kratohvil, "Light Scattering," *Anal. Chem. Ann. Revs.* **36**, 458R–472R (1964).

Kratohvil 1965. Josip P. Kratohvil and Colin Smart, "Calibration of Light-Scattering Instruments. III. Absolute Angular Intensity Measurements on Mie Scatterers," *J. Colloid Sci.* **20**, 875–892 (1965).

Kratohvil 1966a. Josip P. Kratohvil, "Calibration of Light Scattering Instruments. IV. Corrections for Reflection Effects," *J. Colloid Interface Sci.* **21**, 498–512 (1966).

Kratohvil 1966b. J. P. Kratohvil, L. E. Oppenheimer, and M. Kerker, "Correlation of Turbidity and Activity Data. III. The System Tungstosilicic Acid-Sodium Chloride-Water," *J. Phys. Chem.* **70**, 2834–2839 (1966).

Kratohvil 1966c. Josip P. Kratohvil, "Light Scattering," *Anal. Chem. Ann. Revs.* **38**, 517R–526R (1966).

Krause 1961. Sonja Krause, "Light Scattering of Copolymers. I. The Composition Distribution of a Styrene-Methyl Methacrylate Block Copolymer," *J. Phys. Chem.* **65**, 1618–1622 (1961).

Krigbaum 1952. W. R. Krigbaum and P. J. Flory, "Treatment of Osmotic Pressure Data," *J. Polymer Sci.* **9**, 503–588 (1952).

Krigbaum 1953. W. R. Krigbaum and P. J. Flory, "Statistical Mechanics of Dilute Polymer Solutions. IV. Variation of the Osmotic Second Coefficient with Molecular Weight," *J. Am. Chem. Soc.* **75**, 1775–1784 (1953).

Krigbaum 1967. W. R. Krigbaum and R. J. Roe, "Measurement of Osmotic Pressure," Chap. 79 in I. M. Kolthoff and Philip J. Elving, eds., with the assistance of Ernest B. Sandell, *Treatise on Analytical Chemistry*, Part I, Vol. 7, Interscience Div., John Wiley and Sons, New York, 1967.

Kurata 1960. Michio Kurata, Walter H. Stockmayer, and Antonio Roig, "Excluded Volume Effect of Linear Polymer Molecules," *J. Chem. Phys.* **33**, 151–155 (1960).

Kurata 1963. M. Kurata and W. H. Stockmayer, "Intrinsic Viscosities and Unperturbed Dimensions of Long Chain Molecules," *Fortschr. Hochpolym. Forsch. (Advances in Polymer Science)* **3**, 196–312 (1963).

Kurata 1966. Michio Kurata, Masamichi Iwawa, and Kensuke Kamada, "Viscosity-Molecular Weight Relationships and Unperturbed Dimensions of Linear Chain Molecules," pp. IV-1–IV-72 in J. Brandrup and E. H. Immergut, eds., with the collaboration of H.-G. Elias, *Polymer Handbook*, Interscience Div., John Wiley and Sons, New York, 1966.

Livesey 1969. P. J. Livesey and F. W. Billmeyer, Jr., "Particle-Size Determination by Low-Angle Light Scattering: New Instrumentation and a Rapid Method of Interpreting Data," *J. Colloid Interface Sci.* **30**, 447–472 (1969).

Lyons 1967. John W. Lyons, "Measurement of Viscosity," Chap. 83 in I. M. Kolthoff and Philip J. Elving, eds., with the assistance of Ernest B. Sandell, *Treatise on Analytical Chemistry*, Part I, Vol. 7, Interscience Div., John Wiley and Sons, New York, 1967.

Mandelkern 1952a. Leo Mandelkern and Paul J. Flory, "The Frictional Coefficient for Flexible Chain Molecules in Dilute Solution," *J. Chem. Phys.* **20**, 212–214 (1952).

Mandelkern 1952b. L. Mandelkern, W. R. Krigbaum, H. A. Scheraga, and P. J. Flory, "Sedimentation Behavior of Flexible Chain Molecules: Polyisobutylene," *J. Chem. Phys.* **20**, 1392–1397 (1952).

Maron 1954. Samuel H. Maron and Richard L. H. Lou, "Calibration of Light-Scattering Photometers with Ludox," *J. Polymer Sci.* **14**, 29–36 (1954).

Mayo 1943. Frank R. Mayo, "Chain Transfer in the Polymerization of Styrene: The Reaction of Solvents with Free Radicals," *J. Am. Chem. Soc.* **65**, 2324–2329 (1943).

McIntyre 1964. D. McIntyre and F. Gornick, eds., *Light Scattering from Dilute Polymer Solutions*, Gordon and Breach, New York, 1964.

McIntyre 1968. Donald McIntyre, ed., *Characterization of Macromolecular Structure*, National Academy of Sciences Publication No. 1573, Washington, D.C., 1968.

Melabs. Melabs, Inc., Scientific Instruments Dept., 3300 Hillview Ave., Palo Alto, Calif., 94304. [Reiff osmometers are now manufactured by Wescan.]

Meyerhoff 1961. G. Meyerhoff, "The Viscometric Determination of Molecular Weight of Polymers" (in German), *Fortschr. Hochpolym-Forsch. (Advances in Polymer Science)* **3**, 59–105 (1961).

Mie 1908. G. Mie, "Optics of Turbid Media" (in German), *Ann. Physik.* **25**, 377–445 (1908).

Miller 1966. M. L. Miller, *The Structure of Polymers*, Reinhold Publishing Corp., New York, 1966.

Moore 1967. W. R. Moore, "Viscosities of Dilute Polymer Solutions," Chap. 1 in A. D. Jenkins, ed., *Progress in Polymer Science*, Vol. 1, Pergamon Press, New York, 1967.

Morawetz 1965. Herbert Morawetz, *Macromolecules in Solution*, Interscience Div., John Wiley and Sons, New York, 1965.

Muus 1957. L. T. Muus and F. W. Billmeyer, Jr., "The Molecular Structure of Polyethylene. VI. Molecular Weight from Dissymmetry of Scattered Light," *J. Am. Chem. Soc.* **79**, 5079–5082 (1957).

Newitt 1966. E. J. Newitt and V. Kokle, "Molecular Structure of Polyethylene. XIII. An Improved Cryoscopic Method for Determining Number-Average Molecular Weight of Polyethylene," *J. Polymer Sci. A-2,* **4**, 705–714 (1966).

Ogg 1945. C. L. Ogg, W. L. Porter, and C. O. Willits, "Determining the Hydroxyl Content of Certain Organic Compounds; Macro and Semimicro Methods," *Ind. Eng. Chem., Anal Ed.* **17**, 394–397 (1945).

Onyon 1959. P. F. Onyon, "Viscometry," Chap. 6 in P. W. Allen, ed., *Techniques of Polymer Characterization,* Butterworths, London, 1951.

Pasternak 1962. R. A. Pasternak, P. Brady, and H. C. Ehrmantraut, "Apparatus for the Rapid Determination of Molecular Weight," *Dechema Monograph* **44**, 205–207 (1962).

Patat 1956. F. Patat, "The Problem of the Membrane in Osmotic Measurements of High Polymers" (in German), *Z. Electrochem.* **60**, 208–218 (1956).

Patat 1959. F. Patat, "Membranes for Osmotic Measurements" (in German), *Makromol. Chem.* **34**, 120–138 (1959).

Peaker 1959. F. W. Peaker, "Light-Scattering Techniques," Chap. 5 in P. W. Allen, ed., *Techniques of Polymer Characterization,* Butterworths, London, 1959.

Pohl 1954. Herbert A. Pohl, "Determination of Carboxyl End Groups in a Polyester, Polyethylene Terephthalate," *Anal. Chem.* **26**, 1614–1616 (1954).

Price 1943. Charles C. Price and Bryce E. Tate, "The Polymerization of Styrene in the Presence of 3,4,5-Tribromobenzoyl Peroxide," *J. Am. Chem. Soc.* **65**, 517–520 (1943).

Price 1959. G. F. Price, "Techniques of End-Group Analysis," Chap. 7 in P. W. Allen, ed., *Techniques of Polymer Characterization,* Butterworths, London, 1959.

Rayleigh 1871a. Lord Rayleigh, "On the Light from the Sky, its Polarization and Color," *Phil. Mag.* (4) **41**, 107–120, 274–279 (1871).

Rayleigh 1871b. Lord Rayleigh, "On the Scattering of Light by Small Particles," *Phil. Mag.* (4) **41**, 447–454 (1871).

Rayleigh 1881. Lord Rayleigh, "On the Electromagnetic Theory of Light," *Phil. Mag.* (5) **12**, 81–101 (1881).

Rayleigh 1914. Lord Rayleigh, "On the Diffraction of Light by Spheres of Small Relative Index," *Proc. Royal Soc.* **A90**, 219–225 (1914).

Reiff 1959. Theodore R. Reiff and Marvin J. Yiengst, "Rapid Automatic Semimicro Colloid Osmometer," *J. Lab. Clin. Med.* **53**, 291–298 (1959).

Rolfson 1964. F. B. Rolfson and Hans Coll, "Automatic Osmometer for Determination of Number Average Molecular Weights of Polymers," *Anal. Chem.* **36**, 888–894 (1964).

Schachman 1959. Howard K. Schachman, *Ultracentrifugation in Biochemistry,* Academic Press, New York, 1959.

Schaefgen 1948. John R. Schaefgen and Paul J. Flory, "Synthesis of Multichain Polymers and Investigation of their Viscosities," *J. Am. Chem. Soc.* **70**, 2709–2718 (1948).

Scholte 1968a. Th. G. Scholte, "Molecular Weights and Molecular Weight Distribution of Polymers by Equilibrium Ultracentrifugation. Part I. Average Molecular Weights," *J. Polymer Sci. A-2,* **6**, 91–109 (1968).

Scholte 1968b. Th. G. Scholte, "Molecular Weights and Molecular Weight Distribution of Polymers by Equilibrium Ultracentrifugation. Part II. Molecular Weight Distribution," *J. Polymer Sci. A-2,* **6**, 111–127 (1968).

Scholte 1970. Th. G. Scholte, "Determination of the Molecular Weight Distribution of Polymers from Equilibriums in the Ultracentrifuge," *Eur. Polym. J.* **6**, 51–56 (1970).

Sharma 1967. R. K. Sharma and A. F. Sirianni, "Solution Properties of Polyvinylacetate by Thermoelectric Measurements," *Can. J. Chem.* **45,** 1069–1072 (1967).

Simon 1960. W. Simon and C. Tomlinson, "Thermoelectric Microdetermination of Molecular Weights, Part I: Survey Article" (in German), *Chimia* **14,** 301–308 (1860).

Simon 1966. W. Simon, J. T. Clerc, and R. E. Dohner, "Thermoelectric (Vaporometric) Determination of Molecular Weight in 0.001 Molar Solutions: A New Detector for Liquid Chromatography," *Microchem. J.* **10,** 495–508 (1966).

Smoluchowski 1908. M. Smoluchowski, "Molecular-kinetic Theory of Opalescence of Gases" (in German), *Ann. Physik* **25,** 205–226 (1908).

Smoluchowski 1912. M. Smoluchowski, "Opalescence of Gases in the Critical Condition," *Phil. Mag.* (6) **23,** 165–173 (1912).

Spinco. Beckman Instruments, Inc., Spinco Division, 1117 California Ave., Palo Alto, Calif. 94309.

Stabin 1954. J. V. Stabin and E. H. Immergut, "A High-Speed Glass Osmometer," *J. Polymer Sci.* **14,** 209–212 (1954).

Staudinger 1930. H. Staudinger and W. Heuer, "Highly Polymerized Compounds. XXXIII. A Relation Between the Viscosity and the Molecular Weight of Polystyrenes" (in German), *Ber.* **63B,** 222–234 (1930).

Steele 1963. R. E. Steele, W. E. Walker, D. E. Burge, and H. C. Ehrmantraut, paper presented at the Pittsburgh Conference on Analytical Chemistry, March, 1963.

Stockmayer 1952. Walter H. Stockmayer and Edward F. Casassa, "The Third Virial Coefficient in Polymer Solutions," *J. Chem. Phys.* **20,** 1560–1566 (1952).

Stockmayer 1963. W. H. Stockmayer and Marshall Fixman, "On the Estimation of Unperturbed Dimensions from Intrinsic Viscosities," *J. Polymer Sci.* **C1,** 137–141 (1963).

Svedberg 1940. The Svedberg and Kai O. Pedersen, *The Ultracentrifuge,* Clarendon Press, Oxford, 1940; Johnson Reprint Corp., New York, 1959.

Svensson 1961. Harry Svensson and Thomas E. Thompson, "Translational Diffusion Methods in Protein Chemistry," Chap. 3 in P. Alexander and R. J. Block, eds., *Analytical Methods of Protein Chemistry,* Vol. 3, Pergamon Press, New York, 1961.

Tomlinson 1963. C. Tomlinson, Ch. Chylewski, and W. Simon, "The Thermoelectric Microdetermination of Molecular Weight-III," *Tetrahedron* **19,** 949–960 (1963).

Van de Hulst 1957. H. C. van de Hulst, *Light Scattering by Small Particles,* John Wiley and Sons, New York, 1957.

Van Holde 1967. K. E. Van Holde, "Measurement of Sedimentation," Chap. 80 in I. M. Kolthoff and Philip J. Elving, eds., with the assistance of Ernest B. Sandell, *Treatise on Analytical Chemistry,* Part I, Vol. 7, Interscience Div., John Wiley and Sons, New York, 1967.

Van Oene 1968. H. Van Oene, "Measurement of the Viscosity of Dilute Polymer Solutions," pp. 353–367 in Donald McIntyre, ed., *Characterization of Macromolecular Structure,* National Academy of Sciences Publication No. 1573, Washington, D.C., 1968.

Vofsi 1957. D. Vofsi and A. Katchalsky, "Kinetics of Polymerization of Nitroethylene. II. Study of Molecular Weights (by Cryoscopic Method)," *J. Polymer Sci.* **26,** 127–139 (1957).

Wachter 1969. Alfred H. Wachter and Wilhelm Simon, "Molecular Weight Determination of Polystyrene Standards by Vapor Pressure Osmometry," *Anal. Chem.* **41,** 90–94 (1969).

Waltz 1947. J. E. Waltz and Guy B. Taylor, "Determination of the Molecular Weight of Nylon," *Anal. Chem.* **19,** 448–450 (1947).

Ward 1957. I. M. Ward, "The Measurement of Hydroxyl and Carbonyl End Groups in Polyethylene Terephthalate," *Trans. Faraday Soc.* **53**, 1406–1414 (1957).

Wegmann 1962. Dorothée Wegmann, C. Tomlinson, and W. Simon, "Thermoelectric Microdetermination of Molecular Weights, Part II: Routine Determination," *Microchem J. (Symp. Ser.)* **2**, 1069–1085 (1062).

Wescan. Wescan Instruments, Inc., P.O. Box 525, Cupertino, Calif., 95014.

Williams 1958. J. W. Williams, Kensal E. van Holde, Robert L. Baldwin, and Hiroshi Fujita, "The Theory of Sedimentation Analysis," *Chem. Revs.* **58**, 715–806 (1958).

Williams 1963. J. W. Williams, ed., *Ultracentrifugal Analysis in Theory and Experiment,* Academic Press, New York, 1963.

Wippler 1954. C. Wippler and G. Scheibling, "Description of an Apparatus for the Study of Light Scattering" (in French), *J. chim. phys.* **51**, 201–205 (1954).

Wood. C. N. Wood Manufacturing Company, Newtown, Pa., 18940. [Manufacture and repair. Sales through Phoenix Precision Instrument Division, VirTis Co., Route 208, Gardiner, N.Y., 12525.]

Zimm 1946. B. H. Zimm and I. Myerson, "A Convenient Small Osmometer," *J. Am. Chem. Soc.* **68**, 911–912 (1946).

Zimm 1948a. Bruno H. Zimm, "The Scattering of Light and the Radial Distribution Function of High Polymer Solutions," *J. Chem. Phys.* **16**, 1093–1099 (1948).

Zimm 1948b. Bruno H. Zimm, "Apparatus and Methods for Measurement and Interpretation of the Angular Variation of Light Scattering: Preliminary Results on Polystyrene Solutions," *J. Chem. Phys.* **16**, 1099–1116 (1948).

Zimm 1949. Bruno H. Zimm and Walter H. Stockmayer, "The Dimensions of Chain Molecules Containing Branches and Rings," *J. Chem. Phys.* **17**, 1301–1314 (1949).

Zimm 1962. Bruno H. Zimm and Donald M. Crothers, "Simplified Rotating Cylinder Viscometer for DNA," *Proc. Natl. Acad. Sci.* **48**, 905–911 (1962).

4

Analysis and Testing of Polymers

The purpose of this chapter is to assemble in one location information on the applicability to polymers of various standard methods of physical and chemical analysis. A background is thus established for the uninterrupted discussion in succeeding chapters of the results of such analyses and tests. In no case is the treatment in this chapter exhaustive; it is intended to provide, with minimum detail, a broad picture of the applicability of each method to polymer systems and of the type of information obtained.

A. Analytical Chemistry of Polymers

Chemical analysis

The chemical analysis of polymers is not basically different from the analysis of low-molecular-weight organic compounds, provided appropriate modification is made to ensure solubility or the availability of sites for reaction (e.g., insoluble specimens should be ground to expose a large surface area). The usual methods for functional group and elemental analysis are generally applicable, as are many other standard analytical techniques. Several books on the identification and analysis of plastics have appeared (Haslam 1965; Saunders 1966; Lever 1968; Wake 1969), as has an extensive review (Mitchell 1969) containing 1900 references to the subjects treated in Chapters 2E, 3, and 4. The analytical chemistry of the more common plastics is treated by Kline (1959). Chemical reactions of polymers (Chapter 12D) provide additional means of chemical analysis, as do their reactions of degradation (Chapter 12E).

105

This section, although far from comprehensive, describes two very powerful physical techniques for analyzing low-molecular-weight products of reactions (usually degradation) of polymers: mass spectrometry and gas chromatography.

Mass spectrometry (Wall 1962a)   In the most common applications of mass spectrometry to polymer systems, the polymer is allowed to react (as in thermal degradation, Chapter 12E) to form low-molecular-weight fragments which are condensed at liquid-air temperature. They are then volatilized, ionized, and separated according to mass and charge by the action of electric and magnetic fields in a typical mass spectrometer analysis. From the abundances of the various ionic species found, the structures of the low-molecular-weight species can be inferred.

Gas chromatography (Cassel 1962, Stevens 1969)   Gas chromatography is a method of separation in which gaseous or vaporized components are distributed between a moving gas phase and a fixed liquid phase or solid adsorbent. By a continuous succession of adsorption or elution steps, taking place at a specific rate for each component, separation is achieved. The components are detected by one of several methods as they emerge successively from the chromatographic column. From the detector signal, proportional to the instantaneous concentration of the dilute component in the gas stream, is obtained information about the number, nature, and amounts of the components present.

GENERAL REFERENCES

Kline 1959; Brauer 1965; Feinland 1965; Kenyon 1966; Ayrey 1969.

B. Infrared Spectroscopy

Emission or absorption spectra arise when molecules undergo transitions between quantum states corresponding to two different internal energies. The energy difference ΔE between the states is related to the frequency of the radiation emitted or absorbed by the quantum relation $\Delta E = h v$. Infrared frequencies in the wavelength range 1–50 μm* are associated with molecular vibration and vibration-rotation spectra.

A molecule containing N atoms has 3N normal vibration modes, including rotational and translational motions of the entire molecule. For highly symmetrical molecules with very few atoms, the entire infrared and Raman† spectrum can be correlated to and explained by the vibrational modes, but even for most low-molecular-weight substances N is too large for such analysis. Useful information can still be obtained, however, because some vibrational modes involve localized motions of small groups of atoms and give rise to absorption bands at frequencies characteristic of these groups and the type of motions they undergo.

In polymers, the infrared absorption spectrum is often surprisingly simple considering the large number of atoms involved. This simplicity results first from the fact that many of the normal vibrations have almost the same frequency and therefore appear in the spectrum as one absorption band; and second from the strict selection rules that prevent many of the vibrations from causing absorptions.

The approximate wavelengths of some infrared absorption bands arising from functional group and atomic vibrations found in polymers are shown in Fig. 4-1.

Experimental methods

Sources The usual sources of radiation in infrared spectrometers are electrically heated silicon carbide or rare earth oxide rods which emit infrared radiation as black bodies. The amount of energy thus emitted varies in a well-defined way with wavelength, usually reaching a maximum at about 2μm and falling below useful levels above 50 μm.

Dispersion methods Both diffraction gratings and prisms of inorganic salts are used to disperse the radiation in infrared spectrometers. Common prism materials are NaCl (useful below 15μm), KBr (useful below 25 μm), and CsBr (useful below 35 μm). For maximum dispersion, LiF is often used below 6 μm and CaF$_2$ below 9 μm. With reduced cost and improved quality and availability of gratings, however, the modern trend is strongly toward grating spectrometers.

---

* Infrared absorption wavelengths are often expressed in wave numbers (1/wavelength) in cm^{-1}: 1 μm = 10,000 cm^{-1}, 50 μm = 200 cm^{-1}, etc., where μm = micrometer, 10^{-6} meter, formerly (and still widely) called the micron.
† The Raman effect results when the frequency of visible light is altered in the scattering process, by the absorption or emission of energy produced by changes in molecular vibration and vibration-rotation quantum states.

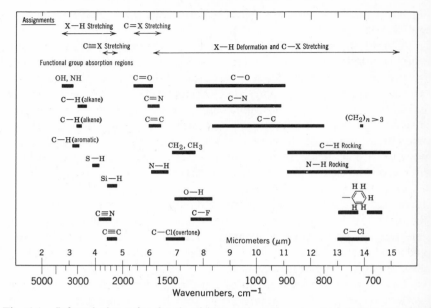

Fig. 4-1. Infrared absorption bands of interest in polymers arranged by approximate wavelength and frequency.

Detectors   The receivers of infrared energy are usually black bodies of very small mass coupled to sensitive heat detectors such as multiple-junction thermocouples. Their output may be read on sensitive galvanometers, but commercial instruments usually incorporate automatic recording features. Other types of detectors include bolometers (heat-sensitive resistance elements), pneumatic detectors (depending on expansion of a gas when heated), and PbS or PbTe photocells. The first three types of detectors are sensitive to all wavelengths at which their associated black bodies absorb energy. Lead sulfide photocells are sensitive below about 2 μm, and PbTe below about 4 μm. In their useful range, the photocells are approximately 100–1000 times as sensitive as other types of detectors.

Samples and sample cells   The infrared method can be applied to solid, liquid, or gaseous samples. Cells for operation at high or low temperatures are available, as are microscope attachments for studying small samples and single fibers.

To obtain transparency to infrared radiation, sample cells must be made with special window materials. The common prism materials and silver chloride are often used. With most of these materials, precautions against moisture and high humidity conditions must be observed. Infrared transmitting glasses, which are free of these difficulties, can be used at the shorter wavelengths.

One of the greatest experimental difficulties in work with substances which absorb heavily in the infrared, including many polymers, is obtaining sufficiently thin samples. Common methods of sample preparation include the following: compression molding, by far the most widely used preparation technique; dissolving the polymer in a solvent, such as carbon disulfide or tetrachloroethylene, whose spectrum is relatively free of intense absorption bands; preparing a thin film by microtoming or milling; casting a thin film from solution; and pressing a finely ground mixture of the sample with KBr to form a disc or wafer.

Application to polymers

Detection of chemical groups in typical spectra Figure 4-1 gives a few of the many chemical linkages or groups which can be detected in polymer spectra, together with the approximate wavelengths at which they occur. Among the infrared absorption bands of interest in common polymers are those in polyethylene corresponding to C—H stretching (3.4 μm), C—H bending of CH_2 groups (6.8 μm) and CH_3 groups (shoulder at 7.25 μm on an amorphous band at 7.30 μm), and CH_2 rocking of sequences of methylene groups in paraffin structures (13.9 μm). (The structure of polyethylene is discussed in more detail in Chapter 13.) Other absorption bands of interest are those due to C=C at 6.1 μm in natural rubber; carbonyl at 5.8 μm and ether at 8.9 μm in poly(methyl methacrylate); aromatic structures at 6.2, 6.7, 13.3, and 14.4 μm in polystyrene; C—Cl at 14.5 μm in poly(vinyl chloride); peptide groups at 3.0, 6.1, and 6.5 μm in nylon; and CF_2 at 8.2–8.3 μm in polytetrafluoroethylene.

Far infrared and Raman spectra It is desirable to supplement as much as possible the observations of infrared absorption usually obtained in the 2–15 μm wavelength range. For this purpose observations at longer wavelengths (say up to 200 μm) become important. Despite serious experimental difficulties, valuable information is available in this region: for some polymers, such as polytetrafluoroethylene, most of the absorption bands occur above 15 μm.

Since the requirements for activity of a vibration for causing absorption in the infrared and for causing Raman scattering are often different, information from Raman experiments in general supplements that obtainable from infrared absorption. Utilization of lasers as light sources has greatly decreased experimental difficulties in this technique, and research is quite active (Koenig 1970).

Dichroism   For any molecular vibration leading to infrared absorption, there is a periodic change in electric dipole moment. If the direction of this change is parallel to a component of the electric vector of the infrared radiation, absorption occurs; otherwise not. In oriented bulk polymers, the dipole-moment change can be confined to specified directions. The use of polarized infrared radiation then leads to absorption which is a function of the orientation of the plane of polarization. This phenomenon is called *dichroism* and is usually measured as the *dichroic ratio*, the ratio of the optical densities of an absorption band measured with radiation polarized parallel and perpendicular, respectively, to a specified direction in the sample. Dichroic ratios depend upon both the degree of orientation and the angle between the direction of the transition moment and the selected direction in the sample (for example, the axial direction in a fiber). They usually range between 0.1 and 1.0. (See also Chapter 5*F*).

Crystallinity   The infrared absorption spectra of the same polymer in the crystalline and amorphous states can differ for at least two reasons. First, specific intermolecular interactions may exist in the crystalline polymer which lead to sharpening or splitting of certain bands; and second, some specific conformations may exist in one but not the other phase, leading to bands which are characteristic exclusively of either crystalline or amorphous material. An example of the latter effect is poly(ethylene terephthalate), in which the —OCH_2CH_2O— portion of each repeat unit is restricted to the all-*trans* conformation in the crystal, but can exist in part in the *gauche* form in the melt. Several bands characteristic of each conformation have been identified (Ward 1957). In favorable cases, such as 6,6 nylon (Stark-weather 1956), per cent crystallinity can be determined in absolute terms from infrared absorption data.

Typical applications   The determination of the various types of unsaturation possible in polymer chains is of great importance, for example in the study of the structure of modern synthetic rubbers (Chapter 13). Table 4-1

TABLE 4-1.   Absorption wavelengths of olefinic groups (Cross 1950)

Group Containing C=C	Wavelength, μm
Vinyl, $R_1CH{=}CH_2$	10.1 and 11.0
trans-$R_1CH{=}CHR_2$	10.4
Vinylidene, $R_1R_2C{=}CH_2$	11.3
$R_1R_2C{=}CHR_3$	12.0
cis-$R_1CH{=}CHR_2$	14.2 (variable)

lists some of the important infrared absorption bands resulting from olefinic groups. In synthetic "natural" rubber, *cis*-1,4-polyisoprene, relatively small amounts of 1,2- and 3,4-addition can easily be detected, though it is more difficult to distinguish between the *cis* and *trans* configurations. Methods have been developed (Richardson 1953) to determine the relative amounts of the four isomers.

In addition to *cis*, *trans* configurational differences, those resulting from tacticity (Chapter 5*A*) can also be distinguished through infrared absorption as can those related to the α and β forms of polypeptides.

GENERAL REFERENCES

Cryon 1962; Zbinden 1964; Smith 1965; Hummel 1966; Kössler 1967; Miyazawa 1968; Elliott 1969.

C. X-Ray Diffraction Analysis

The *x-ray diffraction* method is a powerful tool for investigating orderly arrangements of atoms or molecules through the interaction of electromagnetic radiation to give interference effects with structures comparable in size to the wavelength of the radiation. If the structures are arranged in an orderly array or lattice, the interferences are sharpened so that the radiation is scattered or diffracted only under specific experimental conditions. Knowledge of these conditions gives information regarding the geometry of the scattering structures. The wavelengths of x-rays are comparable to interatomic distances in crystals; the information obtained from scattering at wide angles describes the spatial arrangements of the atoms. Low-angle x-ray scattering (Porod 1961) is useful in detecting larger periodicities, which may arise from lamellar crystallites (Chapter 5*C*) or from voids.

The results of the application of the x-ray method in determining the crystal structures of polymers are described in Chapter 5*B*. Collected crystallographic data for polymers have been tabulated (Miller 1966).

Experimental methods

X-rays are usually produced by bombarding a metal target with a beam of high-voltage electrons. This is done inside a vacuum tube, the x-rays passing out through a beryllium or mica window in the tube in a well-defined beam. Choice of the target metal and the applied voltage determines the wavelength or wavelengths of x-rays produced. Experiments in which nearly monochromatic x-rays are used are of the greatest interest.

The diffracted x-rays may be detected by their action on photographic films or plates, or by means of a radiation counter and electronic amplifier feeding a recorder. Each method has its advantages. Qualitative examination of the diffraction pattern and accurate measurement of angles and distances are best made from a photographic record, whereas for precise measurement of the intensity of the diffracted beam the counting technique is preferred.

X-rays of a given wavelength are diffracted only for certain specific orientations of the sample. If the sample is a single crystal, it must be placed in all possible orientations during the experiment, usually by rotating or oscillating the sample about one of its axes to achieve the desired orientations with respect to the x-ray beam. Alternatively, a sample made up of a powder of very small crystals may be used. If the minute powdered particles are randomly oriented, all orientations will be included within the sample. The powder method is more convenient but gives less information than the single-crystal method, since the latter allows orientations about one crystal axis at a time to be investigated.

Application to polymers

Since polymer single crystals as now prepared (Chapter 5C) are too small for x-ray diffraction experiments, the crystal structure of a polymer is usually determined from x-ray patterns of a fiber drawn from the polymer. Because of the alignment of the crystalline regions with the long axes of the molecules parallel to the fiber axis (Chapter 5F), the pattern is essentially identical to a rotation pattern from a single crystal. In such a pattern diffraction maxima occur in rows perpendicular to the fiber axis, called *layer lines* (seen, for example, in Fig. 5-26b).

Chain conformations   As the periodicity of molecular structure of a polymer is characterized by the existence of a *repeat unit*, so the periodicity of its crystal is characterized by a *repeat distance*. The repeat distance is directly determined by measuring the distance between the layer lines: the greater the repeat distance, the closer together are the layer lines. Determination of the repeat unit is considerably more difficult. Classically, it proceeds through derivation, from the positions of the diffracted x-ray beams, of the dimensions of the *unit cell* (usually the simplest geometric volume unit which by repetition builds up the three-dimensional crystalline array). The positions of the atoms in the unit cell are then derived from the relative intensities of the diffracted beams. The diffraction patterns of polymers, however, do not provide sufficient information to allow such analyses to be carried to completion. Additional structural information is utilized from other sources, such as normal bond lengths and angles and atomic arrangements along the chain suggested by the chemical structure of the sample. For example, the repeat distance of 2.55 A in the crystals of polyethylene is readily identified with a single repeat unit in the planar zigzag conformation.

Many polymer conformations can be described in terms of atoms regularly spaced along helices (Chapter 5B). Methods for analyzing the diffraction patterns from helical structures provide a highly versatile technique for the determination of the repeat unit in such polymers.

Chain packing The packing of the chains is described most completely in terms of the unit cell and its contents (shown for polyethylene in Fig. 5-3). The volume of the unit cell, and hence the volume occupied in the crystal by a single repeat unit, can be obtained from the repeat distance and the positions of the diffraction spots on the layer lines. This volume measures the density of the crystals, which is useful in determining the degree of crystallinity, as outlined in Chapter 5E.

The atomic arrangement within the unit cell is much more difficult to determine than the cell dimensions. Trial structures can often be deduced from these dimensions and a knowledge of the chain conformation, and can be tested in terms of calculated and observed diffraction intensities, using well-established crystallographic techniques.

Disorder in the crystal structure In the discussion so far it has been assumed that polymer crystals consist of perfect geometrical arrays of atoms. However, defects and distortions are present in crystals of all materials; and, together with grosser types of disorder, they play an important role in the crystal structures of polymers. X-ray diffraction data give a great deal of information about the qualitative aspects of disorder in the crystal structure, since disordering results in the broadening of the diffraction maxima. Since different classes of diffraction maxima are sensitive in different ways to particular types of disorder, a systematic study of the broadening yields detailed information on the types of disturbances in the crystals (Clark 1962).

Quantitative measurement of the disorder is difficult because broadening of the diffraction maxima is caused both by small crystallite size and by distortions within larger crystals. As discussed in Chapter 5D, the former explanation was traditionally adopted in interpreting the diffraction pattern from polymers. However, the distortion of perfect crystals into *paracrystals* (Hosemann 1962) is now considered important in many polymers.

Orientation Unless the crystalline regions in polymers are oriented, as by drawing a fiber, the diffraction maxima merge into rings, made up of the maxima from a large number of crystallites in many different orientations (Chapter 5F, especially Fig. 5-26a). If orientation is intermediate between these extremes, the rings split into arcs, and a quantitative evaluation of the crystallite orientation is made by measuring the angular spread and intensity of these arcs.

GENERAL REFERENCES

Posner 1962; Liebhafsky 1964; Alexander 1969.

D. Nuclear Magnetic and Electron Paramagnetic Resonance Spectroscopy

Although the techniques of nuclear magnetic resonance (NMR) and electron paramagnetic resonance (EPR) spectroscopy detect different phenomena and, therefore, yield quite different information, they are similar in basic principles and in experimental techniques. Nuclear magnetic resonance spectroscopy detects the motion of nuclei (such as protons) having nonzero spin, while EPR spectroscopy is useful for the detection of free radical species.

NMR spectroscopy

The major application of NMR spectroscopy in polymers is in the study of chain configurations and microstructure. Here NMR is probably the most powerful tool available, with the possible exception of x-ray diffraction for polymers where crystallinity is highly developed (Section 5C). This recent expansion of the utility of NMR has resulted from the development of techniques for observing so-called narrow-line or high-resolution spectra (Woodbrey 1968) in contrast to the broad-line spectra observed a few years ago. In broad-line work, NMR provides a useful method for studying molecular motion in solid polymers (Slichter 1970), equaling the utility of the older methods of dynamic mechanical testing (Chapter 6C) and dielectric studies (Section G). NMR has an advantage in that it allows study of the motion and positions of protons, which are not readily detected by most other means.

Experimental methods The NMR technique utilizes the property of spin (angular momentum and its associated magnetic moment) possessed by nuclei whose atomic number and mass number are not both even. Such nuclei include the isotopes of hydrogen, C^{13}, O^{17}, and F^{19}. Application of a strong magnetic field to material containing such nuclei splits the energy levels into two, representing states with spin parallel and antiparallel to the field. Transitions between the states lead to absorption or emission of an energy

$$E = hv_0 = 2\mu H_0 \tag{4-1}$$

where the frequency v_0 is in the microwave region for fields of strength H_0 of the order of 10,000 gauss, and μ is the magnetic moment of the nucleus. The energy change is observed as a resonance peak or line in experiments where either H or v_0 is varied with the other held constant.

With an assembly of nuclei, the field on any one is modified by the presence of the others:

$$hv_0 = 2\mu(H_0 + H_L) \tag{4-2}$$

where H_L is the local field with a strength of 5–10 gauss. A distribution of local fields usually exists so that the resonance line becomes broadened. The variables ordinarily observed are the line width and its *second moment*, the mean-square deviation of the field from the center of the line, H_0. Both the line width (for simple systems) and the second moment can be calculated from the structure of the molecule.

Much of the value of NMR studies arises from the effect on the resonance line of the development of molecular motion. As the nucleus moves, it experiences a variety of local fields H_L, so that the average field approaches H_0 more closely and the resonance envelope narrows. If molecular motion is further enhanced by working with a solution, rather than the pure compound, and by increasing the temperature, quite narrow lines can be observed. The positions of these lines on the scale of frequency or magnetic field depend on the local fields, which in turn result from the nature and location of the atomic groups in the vicinity of the protons. Deuterated solvents are often used, since they themselves contribute no resonances.

Since there is no fundamental scale unit or zero of reference on which to base these shifts, they are described in terms of two conventions: the size of the scale unit is one part per million relative change in field strength, and a fixed reference point is provided by the spectrum of a standard reference substance, tetramethyl silane. Two scales are in common use for measuring displacements in resonance, called *chemical shifts*, relative to this standard. On the δ scale, tetramethyl silane appears at zero; on the τ scale, at 10.00. Thus, $\delta = 10 - \tau$. Figure 4-2 shows the magnitude of the chemical shifts corresponding to proton-containing groups common in

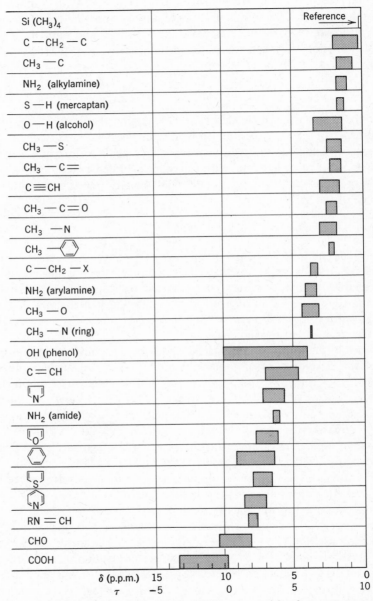

Fig. 4-2. Chart of proton chemical shifts in NMR spectra arising from groups commonly found in polymers (Bovey 1965).

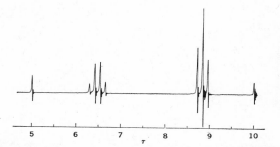

Fig. 4-3. NMR spectrum of a solution of ethyl *o*-formate, $HC(OCH_2CH_3)_3$, in carbon tetrachloride (Bovey 1965). Resonances for the CH, CH_2, and CH_3 protons appear at 5.00, 6.48, and 8.84τ. A small amount of tetramethylsilane added to the solution produced a reference resonance at 10.00τ. The observed chemical shifts result primarily from the varying proximity of the protons to the oxygen atom. The CH_2 and CH_3 protons are coupled to each other through the intervening bonds, resulting in the splitting of their resonances into n + 1 peaks where *n* is the number of equivalent neighboring protons. The CH proton is too distant to experience observable coupling.

polymers. A typical NMR spectrum for a simple compound is shown in Fig. 4-3.

NMR spectra can become quite complex as a result of the coupling described in the caption to Fig. 4-3. To aid in interpretation, two experimental modifications are becoming increasingly important. The first of these is the use of high magnetic field strengths, in the range 60,000 to 220,000 gauss compared to the few tens of thousand gauss used in early broad-line work. Superconducting magnets are used to obtain the highest field strengths. As Fig. 4-4 shows, the increased resolution at the higher field strength is a great aid in the unambiguous interpretation of the spectra.

The second recent development, double resonance or spin decoupling, effects great simplifications in the spectra. A second radio-frequency field is used which has the effect of removing the coupling and collapsing multiplet spectra to much simpler ones (Fig. 4-5). Mention should also be made of the technique of deuteration to provide model compounds with spectra simplified by elimination of proton resonances.

Applications of high-resolution NMR to polymers are discussed in Chapter 5*A*.

EPR spectroscopy

The major use of EPR spectroscopy lies in the detection of free radicals (Chapter 9*A*). These species are uniquely characterized by their magnetic moment, arising from the presence of an unpaired electron. Measurement of a magnetic property of a material containing free radicals, such as its

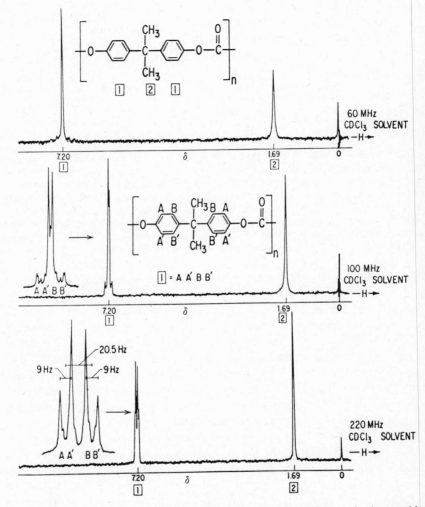

Fig. 4-4. High-resolution NMR spectra of bisphenol-A polycarbonate in deuterochloroform solution showing the effect of frequency on resolution (Sudol 1969).

magnetic susceptibility, gives the concentration of free radicals, but the method lacks sensitivity and cannot reveal the structure of the radicals. Electron paramagnetic resonance spectroscopy is essentially free from these limitations.

Experimental methods   As in NMR spectroscopy, the action of a strong magnetic field on a material containing free radicals removes the degeneracy

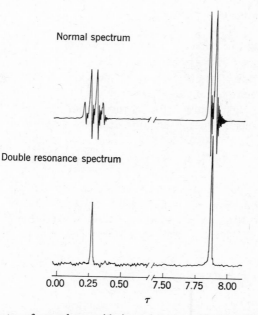

Normal spectrum

Double resonance spectrum

0.00   0.25   0.50   7.50   7.75   8.00

τ

Fig. 4-5.   NMR spectra of propylene oxide in carbon-tetrachloride solution. Top, conventional spectrum; bottom, double-resonance spectrum showing simplification resulting from collapse of the splitting due to coupling (Bovey 1965).

of their ground-state energy levels. For low radical concentrations, the new energy levels are given by two terms. The first is

$$E = h\nu_0 = g\beta\mu_0 H_0 \tag{4-3}$$

where g is a tensor relating the field direction and the symmetry directions in the radical, β is the magnetic moment of the electron spin, and μ_0 is the magnetic permeability of a vacuum. The second term represents coupling of the electron spin with the nuclear spins in the molecule. This coupling results in the splitting of the resonance line into a symmetrical group of lines whose positions and amplitudes depend on, and therefore give information about, the structure of the radical.

Applications to polymers   Only a few systematic investigations have been made of the EPR spectra of polymer systems, and much theoretical and experimental work remains to be done before their results are completely understood. The investigation of free radicals produced by high-energy irradiation (Chapter 12F) of polytetrafluoroethylene (Rexroad 1958, Lebedev 1960) serves as an example.

The EPR spectrum of irradiated polytetrafluoroethylene is interpreted as arising from radicals of the type

$$-(CF_2)_x \dot{C}F(CF_2)_y-$$

remaining trapped in the polymer after a C-F bond has been broken and the fluorine atom has diffused away. The spectrum shows fine structure due to coupling between the unpaired electron and neighboring F^{19} nuclei. The fine structure is lost below 270K as the motion of the F^{19} nuclei is slowed down. The spectrum does not change at higher temperatures, indicating that the radical is stable to at least 550K. Secondary products, resulting from the reaction of the primary radical with such substances as O_2 and NO, were also studied.

It may be noted that, as discussed in Chapters 1C and 12F, the strength of the C-F bond and the tendency of polytetrafluoroethylene to degrade rather than crosslink on irradiation would suggest the breaking of a C-C rather than a C-F bond as a likely source of radicals in this polymer.

GENERAL REFERENCES

Pake 1962; Wall 1962b; Bovey 1965, 1968; Bresler 1966.

E. Thermal Analysis

Perhaps no field of polymer analysis has expanded so rapidly in recent years as that of thermal analysis. Since the introduction of modern instruments in 1962 (Du Pont, Perkin-Elmer), popularity of a wide variety of thermal-analysis techniques has increased tremendously, and instrument sales are currently estimated to be growing 30% per year. In addition to the traditional calorimetric and differential thermal analysis, the field now includes equipment for thermogravimetric analysis, thermomechanical analysis, electrical thermal analysis, and effluent gas analysis. Not only do we study the enthalpy changes associated with heating, annealing, crystallizing, or otherwise thermally treating polymers, but we can now study a wide variety of responses of the system to temperature, including polymerization, degradation, or other chemical changes.

Differential scanning calorimetry

Experimental methods   In contrast to earlier use of a large, expensive adiabatic calorimeter for measurements of specific heat and enthalpies of transition, these measurements are now usually carried out on quite small samples in a *differential scanning calorimeter* (*DSC*). The term is applied to two different modes of analysis, of which the one more closely related to traditional calorimetry is described here. In this type of DSC (Perkin-Elmer), an average temperature circuit measures and controls the temperature of sample and reference holders to conform to a predetermined time-temperature program. This temperature is plotted on one axis of an *x-y* recorder. At the same time, a temperature-difference circuit compares the temperatures of the sample and reference holders, and proportions power to the heater in each holder so that the temperatures remain equal. When the sample undergoes a thermal transition, the power to the two heaters is adjusted to maintain their temperatures, and a signal proportional to the power difference is plotted on the second axis of the recorder. The area under the resulting curve is a direct measure of the heat of transition.

Although the DSC is less accurate than a good adiabatic calorimeter (1–2% vs. 0.1%), its accuracy is adequate for most uses and its advantages of speed and low cost make it the outstanding instrument of choice for most modern calorimetry.

Application to polymers As a typical result, specific heat-temperature curves obtained (by adiabatic calorimetry) (Smith 1956) on heating quenched (amorphous) specimens of poly(ethylene terephthalate) are shown in Fig. 4-5. Each curve rises linearly with temperature at low temperatures, then rises more steeply at the glass transition, 60–80°C. With the onset of mobility of the molecular chains above this transition, crystallization takes place, as indicated by the sharp drop in the specific heat curve. At still higher temperatures, 220–270°C, the crystals melt with a corresponding rise in the specific heat curve.

Differential thermal analysis

Experimental methods In differential thermal analysis (DTA) the sample and an inert reference substance, undergoing no thermal transition in the temperature range of interest, are heated at the same rate. The temperature difference between sample and reference is measured and plotted as a function of sample temperature. The temperature difference is finite only when heat is being evolved or absorbed because of exothermic or endothermic activity in the sample, or when the heat capacity of the sample is changing abruptly. Since the temperature difference is directly proportional to the heat capacity, the curves resemble specific heat curves, but are inverted because, by convention, heat evolution is registered as an upward peak and heat absorption as a downward peak.

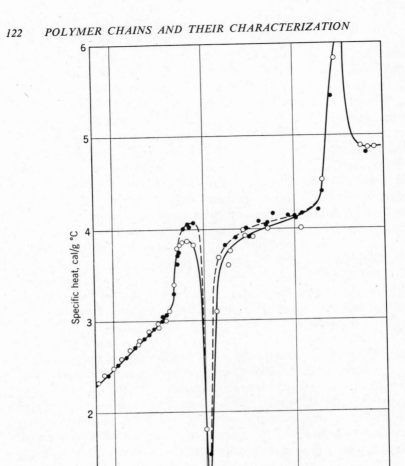

Fig. 4-6. Curves of specific heat as a function of increasing temperature for quenched (amorphous) poly(ethylene terephthalate) (Smith 1956).

Application to polymers   A typical DTA result (Ke 1960) is the differential thermal analysis curve for poly(ethylene terephthalate) shown in Fig. 4-7. Features of the curve may readily be identified by comparison with Fig. 4-6. The lower crystalline melting range in the specimen of Fig. 4-7 is attributed to impurities in the polymer.

Other thermal methods

Thermogravimetric analysis   In thermogravimetric analysis (TGA), a sensitive balance is used to follow the weight change of the sample as a function

| 1978 | | | JUNE | | | 1978 |
SUN	MON	TUE	WED	THU	FRI	SAT
				1	2	3
4	5	6	7	8	9	10
11	12	13	14	15	16	17
18	19	20	21	22	23	24
25	26	27	28	29	30	

15

**Thursday
June
1978**

| May | | | | 1978 |
S M	T W	T	F	S
1	2 3	4	5	6
7 8	9 10	11	12	13
14 15	16 17	18	19	20
21 22	23 24	25	26	27
28 29	30 31			

June   **1978**
S	M	T	W	T	F	S
				1	2	3
4	5	6	7	8	9	10
11	12	13	**14**	15	16	17
18	19	20	21	22	23	24
25	26	27	28	29	30	

Thursday, June 15

July   **1978**
S	M	T	W	T	F	S
						1
2	3	4	5	6	7	8
9	10	11	12	13	14	15
16	17	18	19	20	21	22
23	24	25	26	27	28	29
30	31					

199

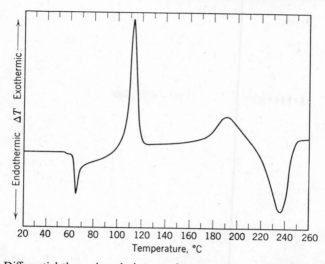

Fig. 4-7. Differential thermal analysis curve for amorphous poly(ethylene terephthalate) (Ke 1960); compare with Fig. 4-6.

of temperature. Typical applications include the assessment of thermal stability and decomposition temperature (Fig. 4-8), extent of cure in condensation polymers, composition and some information on sequence distribution in copolymers (Baer 1964*b*), and composition of filled polymers (Chiu 1966), among many others (Miller 1969).

Thermomechanical analysis   Thermomechanical analysis (TMA) measures the mechanical response of a polymer system as the temperature is changed. Typical measurements include dilatometry, penetration or heat deflection, torsion modulus, and stress-strain behavior.

GENERAL REFERENCES

Ke 1960, 1966; Kissinger 1962; Slade 1966–1970; Miller 1969; Wunderlich 1970.

F. Microscopy

While the techniques of light and electron microscopy are well described in the literature, their application to polymer analysis involves sufficient extension beyond the ordinary techniques to warrant brief comments.

Light microscopy   Reflected-light microscopy is valuable for examining the texture of solid opaque polymers. For materials which can be prepared as

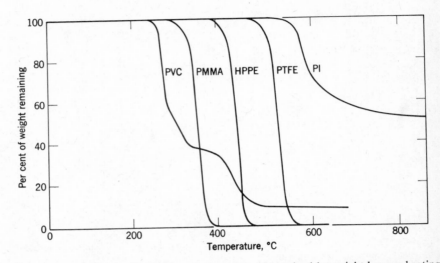

Fig. 4-8. Relative thermal stability of polymers as determined by weight loss on heating at 5°C/min in nitrogen in thermogravimetric analysis (Chiu 1966). Poly(vinyl chloride) (PVC) first loses HCl; later, the mixture of unsaturated carbon-carbon backbone and unchanged poly(vinyl chloride) partly chars and partly degrades to small fragments. Poly(methyl methacrylate) (PMMA), linear polyethylene (HPPE) and polytetrafluoroethylene (PTFE) degrade completely to volatile fragments (Chapter 12E), while a polyimide (PI) partially decomposes, forming a char above 800°C.

thin films (often, for example, by casting on the microscope slide), examination by transmitted light is usual, but little detail can be seen without some type of enhancement. Two common techniques are used. One is polarized-light microscopy, in which advantage is taken of the ability of crystalline materials to rotate the plane of polarized light. Thus, the structure of spherulites (Chapter 5D) is studied with the sample between crossed polarizers, and the crystalline melting point is taken as the temperature of disappearance of the last traces of crystallinity when using a hot-stage polarizing microscope. A second useful technique is phase-contrast microscopy, which allows observations of structural features involving differences in refractive index rather than absorption of light as in the conventional case. Interference microscopy, allowing measurement of thicknesses as low as a few angstrom units, has proved valuable in the study of polymer single crystals (Chapter 5C). Resolution, of course, is limited to object sizes about half the wavelength; practically, about 2000 A.

Electron microscopy and electron diffraction Resolution of smaller objects can be achieved in electron microscopy, though the theoretical resolving

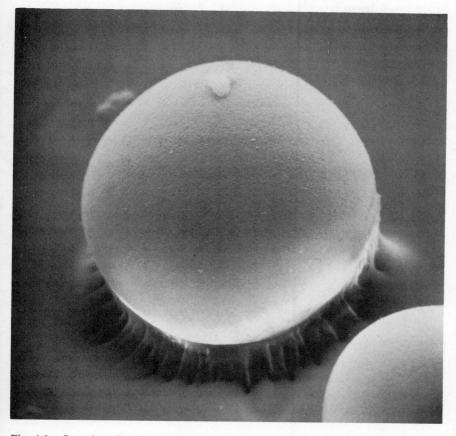

Fig. 4-9. Scanning electron micrograph of a porous poly(styrene-divinyl benzene) bead used in gel permeation chromatography (Chapter 2F) (Godwin 1969). At a magnification of approximately 1000, the porous nature of the bead surface can just be discerned.

power mentioned above is scarcely approached, the practical limit of resolution being a few angstrom units. Electron microscopy has been a powerful tool in the study of the morphology of crystalline polymers (Chapter 5C-D). The usual techniques of replication, heavy-metal shadowing, and solvent etching are widely used. The direct observation of thin specimens, such as polymer single crystals, is also possible and allows the observation of the electron-diffraction pattern of the same specimen area, invaluable for determining crystallographic directions and relating them to morphology. A severe problem, however, is damage of the specimen by the electron beam. Typically, polymer single crystals are severely damaged in times of a few seconds to a minute. This problem can be alleviated by maintaining the sample well below room temperature with a cold stage, by the use of accelerating voltages several times higher than the usual 50,000–100,000 V; or by the use of an image intensifier providing a television-tube image with far less beam current than that ordinarily required (Sauer 1969).

Scanning electron microscopy In 1965 equipment became commercially available for scanning electron microscopy, in which a fine beam of electrons is scanned across the surface of an opaque specimen to which a light-conducting film has been applied by evaporation. Secondary electrons, back-scattered electrons, or (in the *electron microprobe*) x-ray photons emitted when the beam hits the specimen are collected to provide a signal used to modulate the intensity of the electron beam in a television tube, scanning in synchronism with the microscope beam. Because the latter maintains its small size over large distances relative to the specimen, the resulting images have great depth of field and a remarkable three-dimensional appearance (conventional stereoscopic pairs of photographs can be made). Resolution is currently limited to the order of 100 A. Figure 4-9 shows, by way of example, a scanning electron micrograph of a crosslinked poly(styrene-divinyl benzene) bead used in gel permeation chromatography (Chapter 2*F*).

GENERAL REFERENCES

Birbeck 1961; Hartshorne 1964; Cocks 1965; Hirsch 1965; McCrone 1965; Reisner 1965; Fischer 1966; Geil 1966; Sjöstrand 1967; Thornton 1968; Kimoto 1969.

G. Physical Testing

The purpose of this section is to provide brief descriptions of the more important test methods for measuring the physical properties of polymers. The discussion is not intended to be comprehensive or detailed. Test methods related to rheology and viscoelasticity are discussed in Chapter 6.

The literature related to the physical testing of polymers is extensive. Several compilations are useful: the series edited by Schmitz (1965*a*, 1966, 1968) and Brown (1969); the extensive bibliography of Schmitz (1965*b*); and those volumes of methods of test and recommended practices dealing with plastics issued by the American Society for Testing and Materials (ASTM). There are also many pertinent articles in the *Encyclopedia of Polymer Science and Technology* (Mark 1964–1970). Beyond this listing, specific references have largely been omitted from this section.

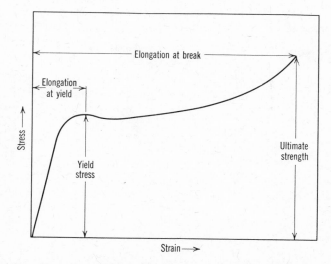

Fig. 4-10. Generalized tensile stress-strain curve for plastics (Winding 1961).

Mechanical properties

Stress-strain properties in tension One of the most informative mechanical experiments for any material is the determination of its *stress-strain curve in tension*. This is usually done by measuring continuously the force developed as the sample is elongated at constant rate of extension.

The generalized stress-strain curve for plastics shown in Fig. 4-10 serves to define several useful quantities, including *modulus* or *stiffness* (the slope of the curve), *yield stress*, and *strength* and *elongation at break*. This type of curve is typical of a plastic such as polyethylene. Figure 4-11 shows stress-strain curves typical of some other classes of polymeric materials. The properties of these polymer types are related to the characteristics of their stress-strain curves in Table 4-2.

Tensile properties are usually measured at rates of strain of 1–1000%/ min. At higher rates of strain—up to 10^6%/min—tensile strength and modulus usually increase several-fold, while elongation decreases. The interpretation of these results is complicated by large temperature rises in the test specimen.

In addition to tensile measurements, tests may also be performed in *shear, flexure, compression*, or *torsion*. For materials in film form, flexural tests are often used. These may include (for stiffer materials) measurement of *flexural modulus* or (for less stiff materials) *flexural* or *folding endurance* tests (see next paragraph).

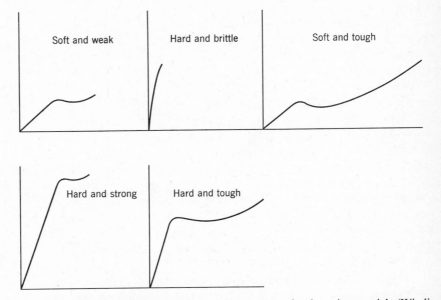

Fig. 4-11.   Tensile stress-strain curves for several types of polymeric materials (Winding 1961).

Fatigue tests   When subjected to cyclic mechanical stresses, most materials fail at a stress considerably lower than that required to cause rupture in a single stress cycle. This phenomenon is called *fatigue*. Various modes of fatigue testing in common use include alternating tensile and compressive stress and cyclic flexural stress. Results are reported as plots of stress vs. number of cycles to fail. Many materials show a fatigue endurance limit, or a maximum stress below which fatigue failure never takes place.

TABLE 4-2.   Characteristic features of stress-strain curves as related to polymer properties (Winding 1961).

Description of Polymer	Characteristics of Stress-Strain Curve			
	Modulus	Yield Stress	Ultimate Strength	Elongation at Break
Soft, weak	Low	Low	Low	Moderate
Soft, tough	Low	Low	Yield Stress	High
Hard, brittle	High	None	Moderate	Low
Hard, strong	High	High	High	Moderate
Hard, tough	High	High	High	High

Fatigue failure may arise from the absorption of energy in a material which is not perfectly elastic. This energy is manifested as heat, leading to a temperature rise, a lower modulus, and rapid failure. Energy absorption is accompanied and measured by a difference in phase between stress and strain in the cyclic test.

Impact tests In most cases, rupture in polymer samples may be divided into two classes: brittle and ductile. Brittle rupture occurs if the material behaves elastically up to the point of failure, i.e., does not yield or draw, whereas in ductile rupture the specimen is permanently distorted near the point of failure. Brittle failure is characterized by lack of distortion of the broken parts. Two aspects of brittle rupture are of interest: the temperature below which brittle failure occurs under a given set of experimental conditions, and measures of the toughness of materials at ambient temperature. The test methods used to obtain both types of information are highly empirical. Since the specimen is destroyed, multiple testing is usually employed to allow statistical evaluation of the results.

The *brittle point*, or temperature at the onset of brittleness, is usually determined by subjecting a specimen to impact in a standardized but empirical way. The temperature of the test is varied until that temperature is found where, statistically, half the specimens fail by brittle rupture. The brittle point is roughly related to the glass transition temperature (Chapter 6D).

Impact strength of plastics is commonly measured by tests in which a pendulum with a massive striking edge is allowed to hit the specimen. From the travel of the pendulum after breaking the specimen can be calculated the energy required to cause the break. The specimen is often notched in an effort to improve the reproducibility of the mode of failure.

Other forms of impact test include a large variety in which falling objects (balls, darts, etc.) strike the specimen. Usually, the height from which the missile must be dropped to cause failure in half the specimens is taken as a measure of the toughness of the material.

Tear resistance When plastics are used as films, particularly in packaging applications, their resistance to tearing is an important property. In one test of *tear strength* a specimen is torn apart at a cut made by a sharp blade. Energy is provided by a falling pendulum, and the work done is measured by the residual energy of the pendulum. Tear strength and tensile strength are closely related.

Hardness As usually conceived, *hardness* is a composite property combining concepts of resistance to *penetration*, *scratching*, *marring*, etc. Most hardness tests for plastics are based on resistance to penetration by an indentor pressed into the plastic under a constant load.

Abrasion Resistance   Abrasion resistance in plastics usually takes the form of a scratch test, in which the material is subjected to many scratches, usually from contact with an abrasive wheel or a stream of falling abrasive material. The degree of abrasion can be determined by loss of weight for severe damage, but is more usually measured by evidence of surface marring, such as loss of gloss or development of haze in transparent specimens.

Friction, hardness, and abrasion resistance are related closely to the viscoelastic properties of polymers (Chapter 6C).

Thermal properties

Softening temperature   In addition to the rheological tests discussed in Chapter 6A, a thermal property of great interest is the *softening temperature* of a plastic. Various ways of measuring this property include observation of the temperatures at which (*a*) an indentor under fixed load penetrates a specified distance into the material (*Vicat test*-Chapter 7D); (*b*) a bar, held in flexure under constant load, deforms a specified amount (*deflection temperature* or *heat distortion test*); (*c*) a polymer sample becomes molten and leaves a trail when moved across a hot metal surface with moderate pressure (*polymer melt or stick temperature test*); and (*d*) a polymer specimen fails in tension under its own weight (*zero strength temperature test*).

Flammability   The flammability of plastics is usually tested as the burning rate of a specified sample. The self-extinguishing tendency of the material on the removal of an external flame is also important.

Optical properties

Transmittance and reflectance   A major determinant of the appearance of a *transparent* material (one which does not scatter light) is its *transmittance*, the ratio of the intensities of light passing through and light incident on the specimen. Similarly, the appearance of an *opaque* material (one which may reflect light but does not transmit it) is characterized by its *reflectance*, the ratio of the intensities of the reflected and the incident light. A *translucent* substance is one which transmits part and reflects part of the light incident on it.

Transmittance and reflectance may be measured as a function of the wavelength of light in a spectrophotometer. When adjusted to correspond to visual perception, by weighting the effects at various wavelengths according to the power incident from a specified light source (often daylight) and the response of the human eye to light flux, these quantities are called *luminous transmittance* and *luminous reflectance*.

Color Color is the subjective sensation in the brain resulting from the perception of those aspects of the appearance of objects which result from the spectral composition of the light reaching the eye. Other aspects of appearance (gloss, haze, transparency—see below) are properly not part of the phenomenon of color. Since color is subjective, it cannot be described completely in physical terms, though to a first approximation color depends largely on the spectral-power distribution of a light source, the spectral reflectance of the illuminated object, and the spectral response curves of the eye. In terms of visually perceived quantities, the description of color requires specification of three variables, a common set of which are *hue*, the attribute which determines whether the color is red, green, blue, etc.; *lightness*, the attribute which permits a color to be classified equivalent to some member of a scale of grays ranging from white to black; and *saturation*, the attribute of any color possessing a hue which determines the degree of difference of the color from the gray of the same lightness. The instrumental "measurement" of color consists in determining sets of numbers, correlating approximately with visually perceived quantities for sufficiently restrictive conditions, which allow one to judge whether two colors are alike or different. Despite the many limitations involved, the ability to make such determinations quantitatively and objectively is of great commercial importance. The principles of color technology, including visual perception, the measurement of color and color difference, color matching, and the coloring of plastics and other materials, have been discussed by Billmeyer (1966).

Gloss Gloss is the geometrically selective reflectance of a surface responsible for its shiny or lustrous appearance. Surface reflectance is commonly at a maximum in or near the specular direction, i.e., the direction at which a mirror would reflect light. Photoelectric instruments are available for measuring gloss at a variety of angles of incidence and reflection.

Haze For transparent materials, haze is that percentage of transmitted light which in passing through the specimen deviates from the incident beam by forward scattering. In commercial hazemeters, only light deviating more than 2.5° from the transmitted beam direction is considered haze. The effect of haze is to impart a cloudy or milky appearance to the sample, but its transparency (defined in the next paragraph) need not be reduced.

Transparency For a transparent plastic material, transparency is defined as the state permitting perception of objects through or beyond the specimen.

A sample of low transparency may not exhibit haze, but objects seen through it will appear blurred or distorted. Transparency can be measured as that fraction of the normally incident light which is transmitted without deviation from the primary beam direction of more than 0.1° (Webber 1957).

Dielectric constant and loss factor   The dielectric constant of an (insulating) material is the ratio of the capacities of a parallel plate condenser measured with and without the dielectric material placed between the plates. The difference is, of course, due to the polarization of the dielectric. If the field applied to the condenser is time dependent (as in alternating current), the polarization is time dependent also. However, because of the resistance to motion of the atoms in the dielectric, there is a delay between changes in the field and changes in the polarization. This delay is often expressed as a phase difference or *loss angle*, δ. The *power factor* is then defined as sin δ and the *dissipation factor* as tan δ. The product of the dielectric constant and the power factor is called the *loss factor*. It is proportional to the energy absorbed per cycle by the dielectric from the field. Dielectric constant and loss factor are usually measured over a frequency range from 60 cycles/sec to thousands of megacycles/sec.

Resistivity   The resistance of most polymers to the flow of direct current is very high, and conductivity probably results from the presence of ionic impurities whose mobility is limited by the high viscosity of the medium. Both surface and volume resistivity are important properties for applications of polymers as insulating materials.

Dielectric strength   Insulators will not sustain an indefinitely high voltage: as the applied voltage is increased, a point is reached where a catastrophic decrease in resistance takes place, accompanied by a physical breakdown of the dielectric. As in mechanical fatigue experiments, application of lower voltages causes eventual breakdown: curves of time to fail vs. voltage can be plotted. Such curves show two distinct regions of failure. At short times failure is presumed to occur as a result of the inability of the electrons conducting the current to dissipate rapidly enough the energy they receive from the field. Breakdown at longer times appears to be due to corona attack, which is sensitive to the atmosphere surrounding the dielectric and to the presence of mechanical strains.

Arc resistance   The surfaces of some polymers may become carbonized and conduct current readily when exposed to an electrical discharge. This property is important for such applications as insulation for gasoline engine ignition systems.

Chemical properties

The following properties may not involve chemical treatment or attack exclusively, but are conveniently grouped together as different from those considered in previous sections.

Resistance to solvents The effect of solvents (broadly defined as liquids in general) on polymers may take several forms: *solubility* (Chapter 2A); *swelling*, including the absorption of water; *environmental stress cracking*, in which the specimen fails by breaking when exposed to mechanical stress in the presence of an organic liquid or an aqueous solution of a soap or other wetting agent; and *crazing*, in which a specimen fails by the development of a multitude of very small cracks in the presence of an organic liquid or its vapor, with or without the presence of mechanical stress.

Vapor permeability The *permeability* of a polymer to a gas or vapor is the product of the *solubility* of the gas or vapor in the polymer and its *diffusion coefficient*. Permeability is directly measured as the rate of transfer of vapor through unit thickness of the polymer in film form, per unit area and pressure difference across the film.

Weathering For convenience and reproducibility, the behavior of materials on long exposure to weather is often simulated by exposure to *artificial weathering* sources, such as filtered carbon-arc lamps in controlled atmospheres. Such tests are usually not appreciably accelerated over natural weathering, except that they may be applied continuously.

GENERAL REFERENCES

ASTM; Baer 1964*a*; Mark 1964–1970; Ritchie 1965; Schmitz 1965*a*, 1965*b*, 1966, 1968; Brown 1969.

BIBLIOGRAPHY

Alexander 1969. Leroy E. Alexander, *X-Ray Diffraction Methods in Polymer Science*, John Wiley and Sons, New York, 1969.

ASTM 1970. American Society for Testing and Materials, *1970 Book of ASTM Standards*, Parts 26 and 27, *Plastics*, American Society for Testing and Materials, Philadelphia, annually.

Ayrey 1969. G. Ayrey, "The Use of Isotopes in Polymer Analysis," *Fortschr. Hochpolym. Forsch.* (*Advances in Polymer Science*) **6,** 128–148 (1969).

Baer 1964a. Eric Baer, ed., *Engineering Design for Plastics*, Reinhold Publishing Corp., New York, 1964.

Baer 1964b. M. Baer, "Anionic Block Polymerization. II. Preparation and Properties of Block Copolymers," *J. Polymer Sci.* **A2,** 417–436 (1964).

Billmeyer 1966. Fred W. Billmeyer, Jr., and Max Saltzman, *Principles of Color Technology*, Interscience Div., John Wiley and Sons, New York, 1966.

Birbeck 1961. M. S. C. Birbeck, "Techniques for the Electron Microscopy of Proteins," Chap. 1 in *Analytical Methods of Protein Chemistry*, P. Alexander and R. J. Block, eds., Vol. 3, Pergamon Press, New York, 1961.

Bovey 1965. Frank A. Bovey, "Nuclear Magnetic Resonance," *Chem. Eng. News* **43,** No. 35, 98–121 (Aug. 30, 1965).

Bovey 1968. F. A. Bovey, "Nuclear Magnetic Resonance," pp. 356–396 in Herman F. Mark, Norman G. Gaylord, and Norbert M. Bikales, eds., *Encyclopedia of Polymer Science and Technology*, Vol. 9, Interscience Div., John Wiley and Sons, New York, 1968.

Brauer 1965. G. M. Brauer and G. M. Kline, "Chemical Analysis," pp. 632–665 in Herman F. Mark, Norman G. Gaylord, and Norbert M. Bikales, eds., *Encyclopedia of Polymer Science and Technology*, Vol. 3, Interscience Div., John Wiley and Sons, New York, 1965.

Bresler 1966. S. E. Bresler and E. N. Kazbekow, "Electron-Spin Resonance," pp. 669–692 in Herman F. Mark, Norman G. Gaylord, and Norbert M. Bikales, eds., *Encyclopedia of Polymer Science and Technology*, Vol. 5, Interscience Div., John Wiley and Sons, New York, 1966.

Brown 1969. W. E. Brown, ed., *Testing of Polymers*, Vol. IV, Interscience Div., John Wiley and Sons, New York, 1969.

Cassel 1962. James M. Cassel, "Chromatography," Chap. X in Gordon M. Kline, ed., *Analytical Chemistry of Polymers*, Part II, Interscience Div., John Wiley and Sons, New York, 1962.

Chiu 1966. Jen Chiu, "Applications of Thermogravimetry to the Study of High Polymers," *Appl. Poly. Symp.* **2,** 25–43 (1966).

Clark 1962. E. S. Clark and L. T. Muus, "Partial Disordering and Crystal Transitions in Polytetrafluoroethylene," *Z. Krist.* **117,** 119–127 (1962).

Cocks 1965. George C. Cocks, "Electron Microscopy," Chap. 73 in I. M. Kolthoff and Philip J. Elving, eds., with the assistance of Ernest B. Sandell, *Treatise on Analytical Chemistry*, Part I, Vol. 5, Interscience Div., John Wiley and Sons, New York, 1965.

Cross 1950. L. H. Cross, R. B. Richards, and H. A. Willis, "The Infrared Spectrum of Ethylene Polymers," *Disc. Faraday Soc.* **9,** 235–245 (1950).

Du Pont. E. I. du Pont de Nemours and Co., Inc., Wilmington, Delaware, 19898.

Elliott 1969. Arthur Elliott, *Infrared Spectra and Structure of Organic Long-Chain Polymers*, St. Martin's Press, New York, 1969.

Feinland 1965. R. Feinland, "Chromatography," pp. 731–762 in Herman F. Mark, Norman G. Gaylord, and Norbert M. Bikales, eds., *Encyclopedia of Polymer Science and Technology*, Vol. 3, Interscience Div., John Wiley and Sons, New York, 1965.

Fischer 1966. E. W. Fischer and H. Goddar, "Electron-Diffraction Analysis," pp. 641–661 in Herman F. Mark, Norman G. Gaylord, and Norbert M. Bikales, eds., *Encyclopedia of Polymer Science and Technology*, Vol. 5, Interscience Div., John Wiley and Sons, New York, 1966.

Geil 1966. Phillip H. Geil, "Electron Microscopy," pp. 662–669 in Herman F. Mark, Norman G. Gaylord, and Norbert M. Bikales, eds., *Encyclopedia of Polymer Science and Technology*, Vol. 5, Interscience Div., John Wiley and Sons, New York, 1966.

Godwin 1969. R. W. Godwin, Celanese Fibers Co., Charlotte, N.C., private communication, 1969.

Hartshorne 1964. N. H. Hartshorne and A. Stuart, *Practical Optical Crystallography*, 2nd ed., American Elsevier, New York, 1964.

Haslam 1965. John Haslam and Harry A. Willis, *Identification and Analysis of Plastics*, D. Van Nostrand Co., Princeton, N.J., 1965.

Hirsch 1965. P. B. Hirsch, A. Howie, R. B. Nicholson, D. W. Pashley, and M. J. Wheelan, *Electron Microscopy of Thin Crystals*, Plenum Press, New York, 1965.

Hosemann 1962. R. Hosemann and S. N. Bagchi, *Direct Analysis of Diffraction by Matter*, North-Holland Publishing Co., Amsterdam, 1962.

Hummel 1966. Dieter O. Hummel, *Infrared Spectra of Polymers: in the Medium and Long Wavelength Region*, John Wiley and Sons, New York, 1966.

Ke 1960. Bacon Ke, "Application of Differential Thermal Analysis to High Polymers," pp. 361–392 in John Mitchell, Jr., I. M. Kolthoff, E. S. Proskauer, and A. Weissberger, eds., *Organic Analysis*, Vol. IV, Interscience Publishers, New York, 1960.

Ke 1966. Bacon Ke, "Differential Thermal Analysis," pp. 37–65 in Herman F. Mark, Norman G. Gaylord, and Norbert M. Bikales, eds., *Encyclopedia of Polymer Science and Technology*, Vol. 5, Interscience Div., John Wiley and Sons, New York, 1966.

Kenyon 1966. A. S. Kenyon, "Effluent-Gas Analysis," Chap. 5 in Philip E. Slade, Jr., and Lloyd T. Jenkins, *Techniques and Methods of Polymer Evaluation*, Vol. 1, *Thermal Analysis*, Marcel Dekker, New York, 1966.

Kimoto 1969. S. Kimoto and J. C. Russ, "The Characteristics and Applications of the Scanning Electron Microscope," *Am. Scientist* **57**, No. 1, 112–133 (1969).

Kissinger 1962. H. E. Kissinger and S. B. Newman, "Differential Thermal Analysis," Chap. IV in Gordon M. Kline, ed., *Analytical Chemistry of Polymers*, Part II, Interscience Div., John Wiley and Sons, New York, 1962.

Kline 1959. Gordon M. Kline, ed., *Analytical Chemistry of Polymers*, Part I, Interscience Publishers, New York, 1959.

Koenig 1970. J. L. Koenig, "Raman Scattering in Synthetic Polymers," paper presented before the American Physical Society Division of High Polymer Physics, Dallas, Texas, March 23, 1970.

Kössler 1967. Ivo Kössler, "Infrared-Absorption Spectroscopy," pp. 620–642 in Herman F. Mark, Norman G. Gaylord, and Norbert M. Bikales, eds., *Encyclopedia of Polymer Science and Technology*, Vol. 7, Interscience Div., John Wiley and Sons, New York, 1967.

Lebedev 1960. Ya. S. Lebedev, Iu. D. Tsvetkov, and V. V. Voevodskii, "Paramagnetic Resonance Spectra of Fluoroalkyl and Nitrose-fluoroalkyl Radicals in Irradiated Teflon," *Opt. & Spectr.* **8**, 426–428 (1960).

Lever 1968. A. E. Lever and J. A. Rhys, *Properties and Testing of Plastics Materials*, 3rd ed., Chemical Rubber Co., New York, 1968.

Liebhafsky 1964. H. A. Liebhafsky, H. G. Pfeiffer, and E. H. Winslow, "X-Ray Methods: Absorption, Diffraction, and Emission," Chap. 60 in I. M. Kolthoff and Philip J. Elving, eds., with the assistance of Ernest B. Sandell, *Treatise on Analytical Chemistry*, Part I, Vol. 5, Interscience Div., John Wiley and Sons, New York, 1964.

Mark 1964–1970. Herman F. Mark, Norman G. Gaylord, and Norbert M. Bikales, eds., *Encyclopedia of Polymer Science and Technology*, Interscience Div., John Wiley and Sons, New York, 14 volumes + supp., 1964–1970.

McCrone 1965. Walter C. McCrone and Lucy B. McCrone, "Chemical Microscopy," Chap. 72 in I. M. Kolthoff and Philip J. Elving, eds., with the assistance of Ernest B. Sandell, *Treatise on Analytical Chemistry*, Part I, Vol. 6, Interscience Div., John Wiley and Sons, New York, 1965.

Miller 1966. Robert L. Miller, "Crystallographic Data for Various Polymers," pp. III-1–III-59 in J. Brandrup and E. H. Immergut, eds., with the collaboration of H.-G. Elias, *Polymer Handbook*, Interscience Div., John Wiley and Sons, New York, 1966.

Miller 1969. Gerald W. Miller, "The Thermal Characterization of Polymers," *Appl. Poly. Symp.* **10**, 35–72 (1969).

Mitchell 1969. John Mitchell, Jr., and Jen Chiu, "Analysis of High Polymers," *Anal. Chem. Ann. Revs.* **41**, 248R–298R (1969).

Miyazawa 1968. Tatsuo Miyazawa, "Vibrational Analyses of the Infrared Spectra of Stereoregular Polypropylene," Chap. 3 in A. D. Ketley, ed., *The Stereochemistry of Macromolecules*, Vol. 3, Marcel Dekker, New York, 1968.

Pake 1962. George E. Pake, *Paramagnetic Resonance*, W. A. Benjamin, New York, 1962.

Perkin-Elmer. The Perkin-Elmer Corporation, Main Avenue, Norwalk, Connecticut, 06852.

Porod 1961. G. Porod, "Application and Results of Low-Angle X-ray Scattering in Solid High Polymers" (in German), *Fortschr. Hochpolym.-Forsch. (Advances in Polymer Science)* **2**, 363–400 (1961).

Posner 1962. Aaron S. Posner, "X-Ray Diffraction," Chap. II in Gordon M. Kline, ed., *Analytical Chemistry of Polymers*, Part II, Interscience Div., John Wiley and Sons, New York, 1962.

Reisner 1965. John H. Reisner, "Electron Diffraction and Scattering Techniques," Chap. 68 in I. M. Kolthoff and Philip J. Elving, eds., with the assistance of Ernest B. Sandell, *Treatise on Analytical Chemistry*, Part I, Vol. 6, Interscience Div., John Wiley and Sons, New York, 1965.

Rexroad 1958. Harvey N. Rexroad and Walter Gordy, "Electron Spin Resonance Studies of Irradiated Teflon: Effects of Various Gases," *J. Chem. Phys.* **30**, 399–403 (1958).

Richardson 1953. W. S. Richardson and A. Sacher, "Infrared Examination of Various Polyisoprenes," *J. Polymer Sci.* **10**, 353–370 (1953).

Ritchie 1965. P. D. Ritchie, ed., *Physics of Plastics*, D. Van Nostrand Co., Princeton, N.J., 1965.

Sauer 1969. J. A. Sauer, "Morphology and Structure of Polymer Crystals," *Ann. N.Y. Acad. Sci.* **155**, 517–538 (1969).

Saunders 1966. K. J. Saunders, *The Identification of Plastics and Rubbers*, Chapman and Hall, Ltd., London, 1966.

Schmitz 1965a. John V. Schmitz, ed., *Testing of Polymers*, Vol. I, Interscience Div., John Wiley and Sons, New York, 1965.

Schmitz 1965b. J. V. Schmitz, *Bibliography on Polymer Testing, Processing and Applications*, Plastics Institute of America, Inc., New York, 1965.

Schmitz 1966. John V. Schmitz, ed., *Testing of Polymers*, Vol. II, Interscience Div., John Wiley and Sons, New York, 1966.

Schmitz 1968. John V. Schmitz, ed., *Testing of Polymers*, Vol. III, Interscience Div., John Wiley and Sons, New York, 1968.

Sjöstrand 1967. Fritiof S. Sjöstrand, *Electron Microscopy of Cells and Tissues*, Vol. I, *Instrumentation and Techniques*, Academic Press, New York, 1967.

Slade 1966–1970. Philip E. Slade, Jr., and Lloyd T. Jenkins, eds., *Techniques and Methods of Polymer Evaluation*, Vol. 1, *Thermal Analysis*, 1966; Vol. 2, *Thermal Characterization Techniques*, 1970; Marcel Dekker, New York.

Slichter 1970. W. P. Slichter, "Nuclear Magnetic Resonance Studies of Molecular Motion in Polymers," *J. Chem. Educ.* **47**, 193–198 (1970).

Smith 1956. Carl W. Smith and Malcolm Dole, "Specific Heat of Synthetic High Polymers. VII. Polyethylene Terephthalate," *J. Polymer Sci.* **20**, 37–56 (1956).

Smith 1965. Lee A. Smith, "Infrared Spectroscopy," Chap. 66 in I. M. Kolthoff and Philip J. Elving, eds., with the assistance of Ernest B. Sandell, *Treatise on Analytical Chemistry*, Part I, Vol. 6, Interscience Div., John Wiley and Sons, New York, 1965.

Starkweather 1956. Howard W. Starkweather and Robert E. Moynihan, "Density, Infrared Absorption, and Crystallinity in 66 and 610 Nylons," *J. Polymer Sci.* **22**, 363–368 (1956).

Stevens 1969. Malcolm P. Stevens, *Characterization and Analysis of Polymers by Gas Chromatography*, Marcel Dekker, New York, 1969.

Sudol 1969. Robert S. Sudol, "Determination of Polymer Structure by High-Resolution Nuclear Magnetic Resonance Spectroscopy," *Anal. Chim. Acta* **46**, 231–237 (1969).

Thornton 1968. P. R. Thornton, *Scanning Electron Microscopy; Applications to Materials and Device Science*, Chapman and Hall, London, 1968.

Tryon 1962. M. Tryon and E. Horowitz, "Infrared Spectrophotometry," Chap. VIII in Gordon M. Kline, ed., *Analytical Chemistry of Polymers*, Part II, Interscience Div., John Wiley and Sons, New York, 1962.

Wake 1969. William C. Wake, *The Analysis of Rubber and Rubber-Like Polymers*, 2nd ed., John Wiley and Sons, New York, 1969.

Wall 1962a. Leo A. Wall, "Mass Spectrometry," Chap. VI in Gordon M. Kline, ed., *Analytical Chemistry of Polymers*, Part II, Interscience Div., John Wiley and Sons, New York, 1962.

Wall 1962b. Leo A. Wall and Roland E. Florin, "Magnetic Resonance Spectroscopy," Chap. XII in Gordon M. Kline, ed., *Analytical Chemistry of Polymers*, Part II, Interscience Div., John Wiley and Sons, New York, 1962.

Ward 1957. I. M. Ward, "Configurational Changes in Polyethylene Terephthalate," *Chem. & Ind.* **1957**, 1102.

Webber 1957. Alfred C. Webber, "Method for the Measurement of Transparency of Sheet Materials," *J. Opt. Soc. Am.* **47**, 785–789 (1957).

Winding 1961. Charles C. Winding and Gordon D. Hiatt, *Polymeric Materials*, McGraw-Hill Book Co., New York, 1961.

Woodbrey 1968. James C. Woodbrey, "High-Resolution Nuclear Magnetic Resonance of Synthetic Polymers," Chap. 2 in A. D. Ketley, ed., *The Stereochemistry of Macromolecules*, Vol. 3, Marcel Dekker, New York, 1968.

Wunderlich 1970. Bernhard Wunderlich, "Differential Thermal Analysis," Chap. XVII in A. Weissberger, ed., *Physical Methods of Chemistry*, Part IV, Interscience Div., John Wiley and Sons, New York, in press, 1970.

Zbinden 1964. Rudolf Zbinden, *Infrared Spectroscopy of High Polymers*, Academic Press, New York, 1964.

II

Structure and Properties of

Bulk Polymers

5

Morphology and Order in Crystalline Polymers

A. Configurations of Polymer Chains

In this book, the word *configuration* is used to describe those arrangements of atoms which cannot be altered except by breaking and reforming primary chemical bonds. In contrast, as discussed in Chapter 2*B*, arrangements which can be altered by rotating groups of atoms around single bonds are called *conformations*. Examples involving conformations of polymer chains include *trans* vs. *gauche* arrangements of consecutive carbon-carbon single bonds and the helical arrangements found in some polymer crystal structures (Section *B*). Examples involving configurations include head-to-head, tail-to-tail, or head-to-tail arrangements in vinyl polymers (Chapter 9*A*) and the several *stereoregular* arrangements described in this section, including 1,2 and 1,4 addition, *cis* and *trans* isomers, and arrangements around asymmetric carbon atoms.

Configurations involving an asymmetric carbon atom

Staudinger (1932) recognized early that polymers of monosubstituted olefins should contain a series of asymmetric carbon atoms along the chain, and much speculation followed over the possibility of synthesizing such polymers in stereoregular forms. As discussed in Chapter 10*D*, such syntheses have been achieved by several routes. The regular structure of the resulting polymers, notably poly(α-olefins), was recognized by Natta (1955, 1959*a*), who devised the nomenclature now accepted (Huggins 1962) to describe stereoregular polymers of this type. As indicated in Fig. 5-1, where the polymer chain is depicted in the fully extended (all *trans*) planar zigzag conformation, the configuration resulting when all the substituent groups R on the vinyl

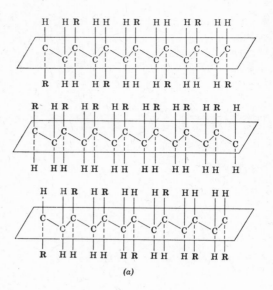

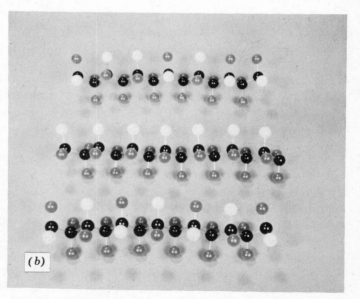

Fig. 5-1. (a) Diagrams and (b) models showing the stereoregular syndiotactic (top) and isotactic (center) and the irregular atactic (bottom) configurations in a vinyl polymer (Natta 1959b). The main carbon-carbon chain is depicted in the fully extended (all *trans*) planar zigzag conformation.

polymer lie above (or below) the plane of the main chain is called *isotactic*. If substituent groups lie alternatively above and below the plane, the configuration is called *syndiotactic*, whereas a random sequence of positions is said to lead to the *atactic* configuration.

The synthesis of isotactic and syndiotactic polymers has now become almost commonplace, and many examples of each type are known. In addition to poly(α-olefins), polystyrene and poly(methyl methacrylate) are examples of polymers which can be made in the isotactic configuration; while poly(methyl methacrylate) and polybutadiene in which 1,2 addition occurs exclusively, giving the structure

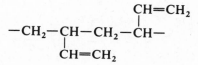

are examples of polymers which can be prepared in the syndiotactic configuration. It has been shown that some stereoregular polymers had been made earlier but were not recognized at the time, a notable example being poly(vinyl isobutyl ether) (Schildknecht 1948, 1949), which is isotactic with a structure similar to that of poly(5-methyl-hexene-1):

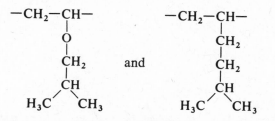

The assignment of an isotactic or syndiotactic structure can be made through determination of the crystal structure of the polymer. Several examples are treated in Section *B*. Independent confirmation of the structures has been obtained for poly(methyl methacrylate), using NMR spectroscopy (Chapter 4*D*). The method depends on differences in the environment of chain methylene groups which are part of isotactic or syndiotactic sequences. It can be applied also to polymers which are not predominantly stereoregular, yielding information about the probability σ that a monomer adds to the growing chain with the configuration established by the previous monomer: $\sigma = 0$, $\frac{1}{2}$, and 1 correspond to completely syndiotactic, atactic, and isotactic polymer, respectively. By this means it is found that free radical poly(methyl methacrylate) is predominantly syndiotactic ($\sigma \simeq 0.2$) but becomes increasingly atactic when polymerized at higher temperatures.

Configurations involving a carbon-carbon double bond

As discussed in Chapter 12C, polymers of 1,3 dienes, containing one residual double bond per repeat unit after polymerization, can contain sequences with several different configurations. For a monosubstituted butadiene such as isoprene, the following structures are possible:

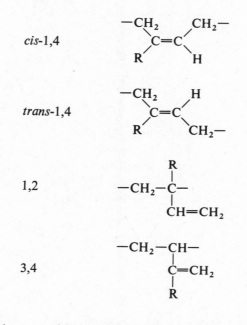

Stereoregular polymers with several of these configurations are known (Chapter 13E,F). Their structures are confirmed by x-ray diffraction and infrared spectroscopy.

Other stereoregular configurations

There is no doubt that stereoregularity plays an important role in the structures of proteins, nucleic acids, and other substances of biological importance. Deoxyribonucleic acid (DNA), e.g., has a highly stereoregular double helix structure, now famous (Watson 1953) (Fig. 5-2). Ladder polymers (Fig. 1-3) and other stereoregular structures in nonbiological polymers are also well known (Chapter 15).

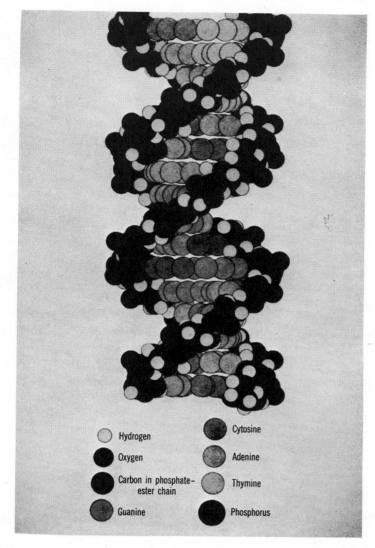

Fig. 5-2.   The famous double helix of DNA (Kornberg 1962, after Feughelman 1955).

Optically active polymers

Many polymers, both biological and synthetic, are capable of rotating the plane of polarization of light and are said to be optically active (Goodman 1967, Schulz 1968). While in simple low-molecular-weight compounds optical activity is associated with the presence of asymmetric carbon atoms; this is not universally true in polymers. Staudinger (1932) pointed out that every second chain carbon atom in vinyl polymers is formally asymmetric. Yet such polymers are not usually optically active even when isotactic or syndiotactic, because of intermolecular compensation (Pino 1965).

Optically active polymers can be prepared, however, in several ways (Schulz 1968): (1) polymerization of optically active monomers in such a way that their asymmetric centers are preserved; (2) introduction of activity into optically inactive polymers by reactions which place asymmetric centers on side chains; (3) stereospecific polymerization of optically inactive monomers with optically active catalysts in such a manner that the repeat units have one or more asymmetric carbon atoms of a given configuration; or (4) polymerization of racemic monomer mixtures to yield polymer from only one of the monomers (as by using a selective initiator) or from both polymerizing simultaneously and independently to yield a mixture of dextrorotatory and levorotatory polymer which can be separated.

An example of an optically active polymer is that made from L-propylene oxide (Price 1956, Furukawa 1967):

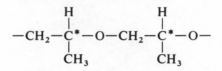

GENERAL REFERENCES

Meares 1965, Chap. 2; Miller 1966*a*, Chap. 8; Natta 1967; Tsuruta 1967; Aubrey 1968; Farina 1968; Bryan 1969.

B. Crystal Structures of Polymers

Structural requirements for crystallinity

Many polymers, including most fibers, are partially crystalline. The most direct evidence of this fact is provided by x-ray diffraction studies. The x-ray patterns of crystalline polymers show both sharp features associated with regions of three-dimensional order, and more diffuse features characteristic of molecularly disordered substances like liquids. The occurrence of both types of feature is evidence that ordered and disordered regions coexist in most crystalline polymers. Additional evidence to this effect comes from other polymer properties, such as densities, which are intermediate between those calculated for completely crystalline and amorphous species.

The close relation between regularity of molecular structure and crystallizability has long been recognized. Typical crystalline polymers are those whose molecules are chemically and geometrically regular in structure. Occasional irregularities, such as chain branching in polyethylene or copolymerization of a minor amount of isoprene with polyisobutylene as in butyl rubber, limit the extent of crystallization but do not prevent its occurrence. Typical noncrystalline polymers, on the other hand, include those in which irregularity of structure occurs: copolymers with significant amounts of two or more quite different monomer constituents, or atactic polymers.

Before the recognition of the importance of stereoregularity to crystallinity in polymers, it was thought that the bulkiness of the side groups in some atactic polymers, such as polystyrene and poly(methyl methacrylate), alone prevented their crystallization. The argument was based in part on the fact that poly(vinyl alcohol) can be made to crystallize, although it is derived by hydrolysis from poly(vinyl acetate), which has never been crystallized. The crystallinity of poly(vinyl alcohol) was once taken as evidence that it was isotactic. Therefore, poly(vinyl acetate) should have been similarly isotactic and would be expected to crystallize—perhaps not in a completely extended conformation—despite the bulkiness of the acetate groups. Poly(vinyl alcohol) is in fact atactic. Other irregular polymers, such as copolymers of ethylene and vinyl alcohol, ethylene and tetrafluoroethylene, and ethylene and carbon monoxide, are also crystalline. All these polymers have very similar structures. It is now clear that the CH_2, $CHOH$, CF_2, and $C{=}O$ groups are near enough to the same size to fit into similar crystal lattices despite the stereochemical irregularity of the polymers. On the other hand, the $CHCl$ group is apparently too large, since chlorinated polyethylene is noncrystalline. The absence of crystallinity in poly(vinyl acetate) and similar polymers is due to the combination of their atactic structure and the size of their substituent groups.

When the disordering features of atactic configuration and bulky side groups are combined with strong interchain forces, the polymer may have a structure which is intermediate in kind between crystalline and amorphous. An example is polyacrylonitrile, in which steric and intramolecular dipole repulsions lead to a stiff, irregularly twisted, backbone chain conformation. The stiff molecular chains pack like rigid rods in a lattice array; lateral order between chains is present, but longitudinal order along chains is absent.

Structures based on extended chains

There is little doubt that the fully extended planar zigzag is the conformation of minimum energy for an isolated section of a hydrocarbon chain, the energy of the *trans* conformation being some 800 cal/mole less than that of the *gauche*. It is to be expected, therefore, that fully extended chain conformations will be favored in polymer crystal structures unless substituents on the chains cause steric hindrance relieved by the assumption of other forms. Fully extended chains are found in the crystal structures of polyethylene, poly(vinyl alcohol), syndiotactic polymers including poly(vinyl chloride) and poly(1,2-butadiene), most polyamides, and cellulose.

Polyethylene   Except for end groups, the arrangement of chains in the crystal structure of polyethylene is essentially the same as that in crystals of linear paraffin hydrocarbons containing 20–40 carbon atoms. The unit cell of polyethylene is rectangular with dimensions $a = 7.41$ A, $b = 4.94$ A, and $c = 2.55$ A. The latter is the chain repeat distance, which is identical with the fully extended zigzag repeat distance. The arrangement of the chains with respect to one another is shown in Fig. 5-3. The positions of the hydrogen atoms are not detectable by the x-ray method (since the scattering power of an atom is proportional to its atomic weight) but have been calculated assuming tetrahedral bonds and a C—H distance of 1.10 A.

Poly(vinyl alcohol)   The crystal structure of poly(vinyl alcohol) is similar to that of polyethylene, as might be expected, since the CHOH group is small enough to fit into the polyethylene structure in place of a CH_2 group with little difficulty. The unit cell is monoclinic (i.e., distorted from rectangular

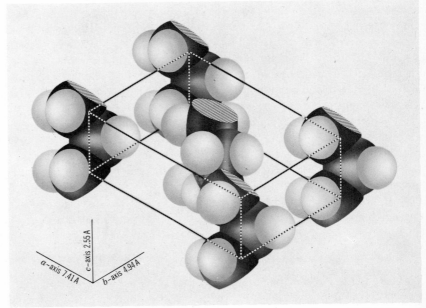

Fig. 5-3.   Arrangement of chains in the unit cell of polyethylene (Geil 1963).

by making the face perpendicular to the chain axis a nonrectangular parallelo-gram with one included angle designated β) with $a = 7.81$ A, $b = 2.52$ A (chain axis), $c = 5.51$ A; $\beta = 91°42'$. Pairs of chains are linked together by hydrogen bonds, and similarly linked into sheets, insofar as the stereochemical irregularity allows.

Poly(vinyl chloride)   Crystallinity in poly(vinyl chloride) is not well devel-oped; the polymer is primarily syndiotactic but with considerable irregularity. Its crystal structure appears to be typical of syndiotactic polymers, with a repeat distance corresponding to four chain carbon atoms. Poly(1,2-buta-diene) has a similar structure, but the carbon-carbon chains are slightly distorted from the *trans* conformation to relieve repulsion between side-chain and neighboring-chain atoms.

Polyamides   The structures of poly(hexamethylene adipamide) (66 nylon), poly(hexamethylene sebacamide) (610 nylon), and polycaprolactam (6 nylon) are made up of fully extended chains linked together by hydrogen bonds to form sheets which may be packed together in two different ways, giving two crystal modifications. Oxygen atoms of one molecule are always found opposite NH groups of a neighboring molecule, with the N—H---O distance of 2.8 A smaller than normal because of hydrogen bond formation. Other nylons (99, 106, 1010, and 11) contain chains slightly distorted from the planar zigzag form.

Cellulose   The crystal structure of cellulose is of historical interest, since it represents the first instance in which the concepts of a crystal unit cell and a chain structure were reconciled by the concept of the chain extending through successive unit cells. The basic feature of the crystal structure of cellulose is the repeating segment of two cellobiose units, joined in 1,4 linkages, with the —CH$_2$OH side chains in successive units occurring on opposite sides of the chain. Many forms and derivatives of cellulose have crystal structures in which these chains are packed in slightly different ways.

Distortions from fully extended chains

Polyesters   In most aliphatic polyesters and in poly(ethylene terephthalate) the polymer chains are shortened by rotation about the C—O bonds to allow close packing. As a result, the main chains are no longer planar. The group

in poly(ethylene terephthalate) is planar, however, as required by resonance.

Polyisoprenes and polychloroprene   Polymers of isoprene and chloroprene in which 1,4 addition predominates can occur in all-*cis* or all-*trans* configurations; natural rubber and gutta-percha are naturally occurring *cis*-1,4-polyisoprene and *trans*-1,4-polyisoprene, respectively. All these polymers have similar crystal structures except that the repeat distance of the *trans* polymers corresponds to one monomer unit and that of the *cis* polymers to two.

The repeat distance of *trans*-1,4-polyisoprene (gutta-percha) (4.72–4.77 A) is smaller than that corresponding to a fully extended chain (5.04 A).

The structure of natural rubber (*cis*-polyisoprene) is similar to that of gutta-percha except as required by the longer repeat distance due to the *cis* configuration. The chain is slightly out of plane at the single bonds and in plane around the double bond, the observed repeat distance of 8.1 A being less than the extended distance of 10.08 A.

The structure of polychloroprene is similar to that of gutta-percha except that the molecules are oriented differently with respect to one another in the crystal because of differences in polarity between the chlorine and methyl groups.

Polypeptides   Proteins, including the natural fibers wool and silk, are made up of α-amino acids joined into long polypeptide chains with the planar structure

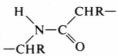

where R represents the amino acid chain. These polymer chains can exist in two quite different types of crystal structure.

One of these forms, the β-keratin structure, is made up of nearly extended polypeptide chains arranged into sheets through hydrogen bonding. The sheets are not quite planar and have been described as pleated or rippled. A number of possible structures have been proposed, differing in the detailed arrangement of the chains into sheets. One of these, the antiparallel-chain, pleated sheet structure, is now considered the most likely structure for silk fibroin.

The helical α-keratin structure of polypeptides is described on page 152.

Helical structures

Polymers with bulky substituents closely spaced along the chain often take on a helical conformation in the crystalline phase, since this allows the substituents to pack closely without appreciable distortion of chain bonds. Most isotactic polymers, as well as polymers of some 1,1-disubstituted ethylenes such as isobutylene, fall in this class. Other helical structures of interest are those of polytetrafluoroethylene and the α-keratin structure.

Isotactic polymers   Many isotactic polymers crystallize with a helical conformation in which alternate chain bonds take *trans* and *gauche* positions. For the *gauche* positions, the rotation is always in that direction which relieves steric hindrance by placing R and H groups in juxtaposition, generating either a left-hand or a right-hand helix.

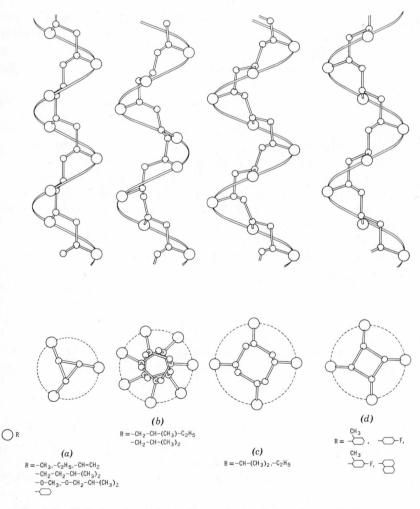

Fig. 5-4.   Helical conformations of isotactic vinyl polymers (Gaylord 1959).

If the side group is not too bulky, the helix has exactly three units per turn and the arrangement is similar to that in Fig. 5-4a. This arrangement is found in isotactic polypropylene, one form of poly(1-butene), polystyrene, and others. More bulky side groups require more space, resulting in the formation of looser helices as shown in Fig. 5-4b,c, and d. Isotactic poly(methyl methacrylate) forms a helix with five units in two turns, while polyisobutylene forms a helix with eight units in five turns.

Polypeptides   Helical structures are of great importance in biological materials. In addition to helical conformations proposed for DNA and other polymers of biological interest, the α-keratin structure assumed by many polypeptides is a helix, in which about 3.6–3.7 polypeptide residues occur per turn (Fig. 5-5). Adjacent turns are held together by intramolecular hydrogen bonding. Several variations of this structure have been proposed.

Order-disorder transitions in crystalline phases

Several instances are known where a polymer exists in two or more crystal modifications. In many cases one of these forms is significantly less ordered or less closely packed than the other. Thus, poly(1-butene) crystallizes in the four-fold helical structure of Fig. 5-4c when first cooled from the melt, but these crystals are transformed to the more closely packed threefold helical conformation (Fig. 5-4a) when the polymer is allowed to stand at temperatures between T_m and T_g or is drawn.

Polytetrafluoroethylene   This polymer exists in two helical conformations which may be described as twisted ribbons in which the fully extended planar form is distorted to have a 180° twist in thirteen CF_2 units in the more stable (low-temperature) form. Above 19°C this form is replaced by a slightly untwisted conformation with fifteen CF_2 units per half-twist. Above 19°C the x-ray diffraction pattern shows diffuse streaks which are interpreted as resulting from small angular displacements of molecular segments about their long axes. Above 30°C, more diffuseness occurs in the x-ray pattern, increasing as the temperature is raised to and above the melting point, 327°C. This additional diffuseness is attributed to random angular displacement of the molecules about their long axes.

In terms of molecular motion, it is considered that the molecules undergo slight torsional motion—i.e., twisting and untwisting of the helices—which causes the lattice to expand laterally, and at 19°C is strong enough to cause the untwisting from thirteen to fifteen CF_2 groups in each 180° twist. Between 19 and 30°C the molecules are "locked" into the threefold helix corresponding to the fifteen CF_2 repeat distance; their motion is restricted to small-angle

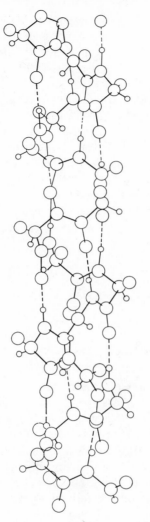

Fig. 5-5.   Helical structure of α-keratin proteins (Pauling 1951).

oscillations about their "rest positions." Above 30°C the molecules lose their preferred angular orientation, becoming randomly misaligned about their axes. Nearly cylindrical in shape, they pack like parallel rods. Above 30°C torsional motion increases, with the twisting and untwisting becoming more pronounced, until at T_m all crystalline order is lost.

Helix-coil transitions in polypeptides (Poland 1970)   The α-keratin helix structure which has been described as characteristic of polypeptides and many polymers of biological origin can undergo several phase transitions, including melting to random coils at T_m or in the presence of a solvent. A transition of particular interest results from the fact that the helix can exist as a stable conformation in solution, so that the helix-coil transition can occur without involving a solid phase.

The stability of the helix in solution results from intramolecular hydrogen bonding. Since disruption of the helix requires breaking about three such bonds per turn, some 5–10 kcal of enthalpy must be expended to realize the greater entropy of the random coil state. Once the transition is initiated, by a change in temperature or composition, the preferred conformation is perpetuated over many bonds. In the simplest case, the helical conformation occurs either not at all or in long sequences with infrequent interruption if not throughout the chain. The change is fairly abrupt and resembles a phase transition.

GENERAL REFERENCES

Geil 1963; Meares 1965, Chap. 4; Miller 1965, 1966b,c; Corradini 1968; Bryan 1969.

C. Morphology of Polymer Single Crystals

Although the formation of single crystals of polymers was observed during polymerization many years ago, it was long believed that such crystals could not be produced from polymer solutions because of molecular entanglement. In 1953 (Schlesinger 1953) and in several laboratories in 1957 (Fischer 1957, Keller 1957, Till 1957), the growth of such crystals was reported. The phenomenon has been reported for so many polymers, including gutta-percha, polyethylene, polypropylene and other poly(α-olefins), polyoxymethylene, polyamides, and celluose and its derivatives, that it appears to be quite general and universal (Geil 1963; Keller 1964).

Lamellae

All the structures described as polymer single crystals have the same general appearance, being composed of thin, flat platelets about 100 A thick and often many microns in lateral dimensions. They are usually thickened by the spiral growth of additional lamellae from screw dislocations. A typical crystal of 6 nylon is shown in Fig. 5-6. The size, shape, and regularity of the crystals depend on their growth conditions, such factors as solvent, temperature, and growth rate being important. The thickness of the lamellae depends on crystallization temperature and any subsequent annealing treatment.

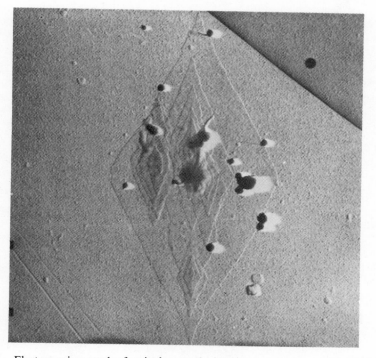

Fig. 5-6. Electron micrograph of a single crystal of 6 nylon grown by precipitation from dilute glycerine solution (Geil 1960). The lamellae are about 60 A thick.

Chain folding

Electron diffraction measurements, performed in almost all the researches cited above, indicate that the polymer chains are oriented normal or very nearly normal to the plane of the lamellae. Since the molecules in the polymers are at least 1000 A long and the lamellae are only about 100 A thick, the only plausible explanation is that the chains are folded (Keller 1957). This arrangement has been shown to be sterically possible. In polyethylene, for example, the molecules can fold in such a way that only about five chain carbon atoms are involved in the fold itself (Fig. 5-7).

Pyramidal structure

Many single crystals of essentially linear polyethylene show secondary structural features, including corrugations (Fig. 5-8) and pleats (Fig. 5-9). Both these features result from the fact that many crystals of polyethylene, and perhaps other polymers too, grow in the form of hollow pyramids. When solvent is removed during preparation of the crystals for microscopy, surface tension forces cause the pyramids to collapse.

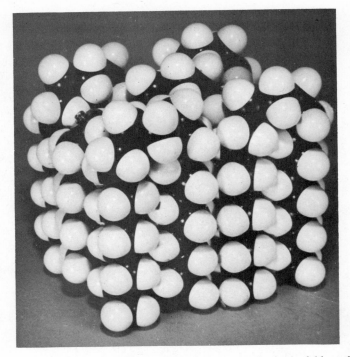

Fig. 5-7.   Model of the conformation of polyethylene molecules in the folds at the surface of a single crystal (Reneker 1960).

The hollow pyramidal structure is related to the packing of the folded chains. If successive planes of folded molecules are displaced from their neighbors by an integral (or, with rotation of 180°, by a half-integral) number of repeat distances, a pyramidal structure with a slope of 32° (or 17°) results. [It should be mentioned that the orientation of the planes along which folds occur differs in the four quadrants of the crystal, as suggested by the pattern of the corrugations in Fig. 5-8.]

More complex structure.

The growth of crystal structures from polymer solutions can result in more complex structures sometimes reminiscent of features found in polymers crystallized from the melt (Section *D*). Typical of these are sheaflike arrays (Fig. 5-10) corresponding to the nuclei and initial growth habit of spherulites. Each ribbon in the array is composed of a number of lathlike lamellae with the usual 50–100 A thickness. Twinned crystals, dendritic growths, clusters of hollow pyramids, spiral growths, steps, Bragg extinction lines, dislocation

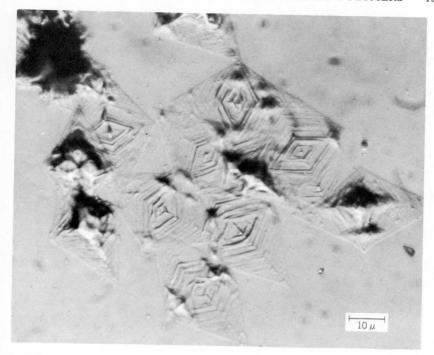

Fig. 5-8.   Optical micrograph showing corrugations in a single crystal of linear poly-ethylene grown from a solution in perchlorethylene (Reneker 1960).

networks, moiré patterns, and epitaxial growths have been described, among others.

Disorder and nature of the fold surfaces

The major unsolved problem (Keller 1969) in polymer single crystal studies has for some time centered on the nature of the fold surfaces. Although there is strong structural and kinetic evidence favoring regular folding with immediate adjacent reentry of the chain into the crystal, as depicted in Fig. 5-7, there is even more convincing evidence that considerable molecular disorder exists in polymer single crystals. The extent of this disorder appears too great to be compatible with intercrystalline defects (Section *D*) and to require some sort of amorphous region or layer at the fold surface. Such a region might involve loose or irregular folds or a more random "switch-board" model as suggested by Flory (1962), and as indicated schematically in Fig. 5-11. Studies of folding in model compounds consisting of large cyclic hydrocarbons have been instructive but not definitive (Newman 1967; Schonhorn 1969).

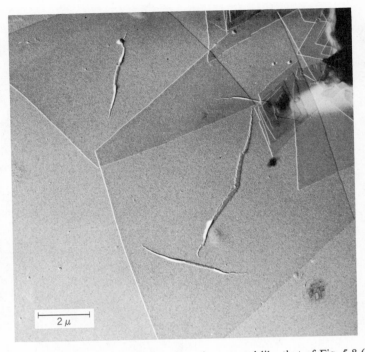

Fig. 5-9. Electron micrograph showing pleats in a crystal like that of Fig. 5-8 (Reneker 1960).

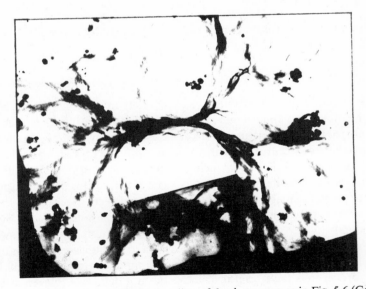

Fig. 5-10. Electron micrograph of spherulites of 6 nylon grown as in Fig. 5-6 (Geil 1960).

GENERAL REFERENCES

Geil 1963; Keller 1964, 1969; Blackadder 1967; Ingram 1968; Sauer 1969.

D. Structure of Polymers Crystallized from the Melt

The fringed micelle concept

Although x-ray diffraction studies show recognizable crystalline features in some high polymers, the Bragg reflections appear broad and diffuse compared to those from well-developed simple crystals. Diffraction theory indicates that this broadening can arise from either small crystallite size or the presence of lattice defects (Bunn 1961), but the diffraction patterns from

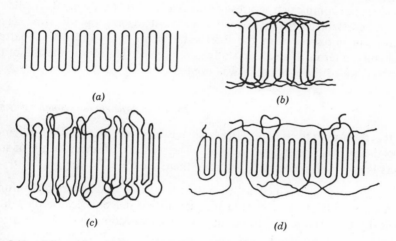

(a)

(b)

(c)

(d)

Fig. 5-11.   Schematic two-dimensional representations of models of the fold surface in polymer lamellae (Ingram 1968): (*a*), sharp folds; (*b*), "switchboard" model; (*c*), loose loops with adjacent reentry; (*d*), a combination of several features.

polymers are usually too weak to permit discrimination between these possibilities. Historically, the hypothesis of small crystallite size was selected as more probable.

With this assumption, rough estimates from the width of diffraction rings indicate that crystallite size rarely exceeds a few hundred angstroms. A substantial background of diffuse scattering suggests the coexistence of an appreciable amorphous fraction. It was suggested that the polymer chains are precisely aligned over distances corresponding to the dimensions of the crystallites, but include more disordered segments which do not crystallize, and hence are included in amorphous regions. Since the chains are very long, they are visualized as contributing segments to several crystalline and amorphous regions, leading to a *composite single-phase* structure known as the *fringed micelle* or *fringed crystallite* model (Fig. 5-12), the fringes representing transition material between the crystalline and amorphous phases.

The fringed micelle concept enjoyed popularity for many years, largely because it led to simple and appealing consequences such as the strong bonding of crystalline and amorphous regions into a composite structure of good mechanical properties, and the simple interpretation of degree of crystallinity in terms of the percentages of well-defined crystalline and amorphous regions. However, it tends to draw attention away from the details of fine structure and gives little insight into the structures of larger entities such as spherulites.

Almost simultaneously with the discovery of polymer single crystals with a lamellar structure involving chain folding (Section *C*), it was found that spherulites are complex structures of lamellae, and that individual lamellae in polymers crystallized from the melt are almost macroscopic in size. These features, incompatible with the interpretation of x-ray diffraction pattern broadening in terms of small crystallites, suggested that reexamination of the interpretation of the x-ray data was needed.

The defect structure of crystalline polymers

It is rewarding to recognize the existence of defects in crystalline polymers just as in any other crystalline material. The presence of several kinds of such defects is now well recognized. Among them are the following:

(1) Point defects, such as vacant lattice sites and interstitial atoms. Chain ends, which themselves must be considered defects because they differ chemically from the rest of the chain, are usually accompanied by vacancies.

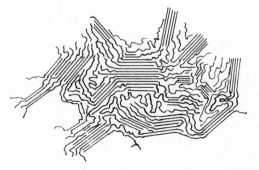

Fig. 5-12. Fringed micelle model of the crystalline-amorphous structure of polymers (Bryant 1947).

Interstitial atoms or groups may be foreign material or may be associated with the chain, for example as certain kinds of side chains.

(2) Dislocations, primarily screw dislocations and edge dislocations. Screw dislocations resulting in growth spirals have been observed many times in both polymer single crystals and bulk polymers. Edge dislocations have been discovered more recently through the study of moiré patterns.

(3) Two-dimensional imperfections, those of particular interest being the fold surfaces (Fig. 5-11).

(4) Chain-disorder defects, including folds, changes of alignment and the like.

(5) Amorphous defects, defined as disorders large enough to disrupt the lattice surrounding their own immediate area by forcing other atoms out of normal lattice positions.

The net effect of all these defects, in particular the fold surfaces and surrounding regions, is to provide both localized amorphous regions contributing to diffuse x-ray scattering, plus deformed or distorted lattices, termed paracrystalline (Hosemann 1962), which contribute in part to x-ray line broadening. In summary, the fringed-micelle concept has largely been supplanted, especially for highly crystalline polymers, by a defect-crystal concept accounting equally well for the observable facts, and utilizing in addition experimentally verified defect structures. Much still remains to be learned, however, about the structure of polymers with low or intermediate crystallinity; here the fringed-micelle concept may well still be the more appropriate one.

Extended-chain crystals

In addition to producing folded-chain structures, polymers can also crystallize from the melt in the form of extended-chain crystals (Fig. 5-13) (Geil 1964; Wunderlich 1968). This morphology is more easily obtained with low-molecular-weight polymers, but this depends on crystallization conditions including applied pressure and supercooling. The extended-chain morphology, in which the size of the crystal in the chain direction is essentially equal to the extended chain length, is now considered the equilibrium morphology for polyethylene and probably many other crystalline polymers.

Structure of spherulites and polymers crystallized from the melt

Microscopic features of spherulites   The most prominent structural organization in polymers on a scale larger than lamellae is the *spherulite*, (ideally) a spherical aggregate ranging from submicroscopic in size to millimeters in diameter in extreme cases. Spherulites are recognized (Fig. 5-14) by their characteristic appearance in the polarizing microscope, where they are seen as (ideally) circular birefringent areas possessing a dark Maltese cross pattern. The birefringence effects are associated with molecular orientation resulting from the characteristic lamellar morphology now recognized in spherulites.

Relation of spherulites to crystallites   Although there has not been general agreement on whether spherulites are formed by rearrangement of previously crystallized material or as products of primary crystallization, the latter interpretation is considered more probable. It has been shown by examination of polymers which are partially spherulitic that in the usual case the

Fig. 5-13.   Extended-chain crystals of linear polyethylene (Prime 1969). Fracture surface of a sample crystallized at 5000 atm pressure.

Fig. 5-14.   Spherulites in a siliconelike polymer, observed in the optical microscope between crossed polarizers (Price 1958). The large and small spherulites were grown by crystallization at different temperatures.

spherulites are crystalline whereas the intermediate nonspherulitic material is amorphous. Spherulites thus appear to represent the crystalline portions of the sample, growing at the expense of the noncrystalline melt. The point of initiation of spherulite growth, its nucleus, may be a foreign particle (heterogeneous nucleation) or may arise spontaneously in the melt (homogeneous nucleation).

Morphology of spherulites   Electron microscopic evidence indicates that spherulites have a lamellar structure for almost all polymers. Crystallization spreads by the growth of individual lamellae. Growth may begin from screw dislocations at the nucleus or at spiral growths originating at crystal defects farther out in the spherulite. When two spherulites meet during crystallization, lamellae from both extend across the boundary into any uncrystallized material available. This results in interleaving, which holds the material together. Evidence that the lamellar structure persists throughout the body of spherulites, and that lamellae thus are basic structural elements of most solid polymers, is furnished by electron microscopic examination of fracture surfaces.

Ringed spherulites   The morphology of spherulites in some polymers is complicated by the development of a ring structure (Fig. 5-15). In the electron

Fig. 5-15. Ringed spherulites of poly(trimethylene glutarate) observed in the optical microscope between crossed polarizers (Keller 1959).

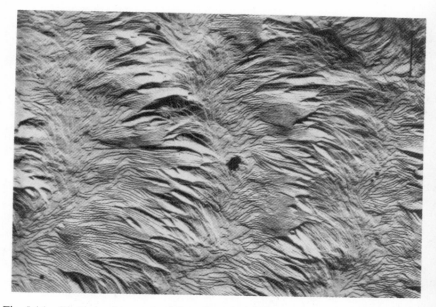

Fig. 5-16. Electron micrograph of a portion of a ringed spherulite in linear polyethylene. (Photograph by E. W. Fischer, from Geil 1963.)

microscope it is seen (Fig. 5-16) that the ring structure is associated with a periodic twisting of the lamellae; the cause of the twisting is unknown.

GENERAL REFERENCES

Geil 1963; Peterlin 1967; Ingram 1968; Wunderlich 1969.

E. Crystallization Processes

Crystallization kinetics

Experimental observations of the development of crystallinity in polymers are of two types. In one case a property varying with total amount of crystallinity, such as specific volume, is observed as a function of time at constant temperature. Alternatively the rate of formation and growth of spherulites is observed directly with the microscope. The two techniques lead to essentially the same results.

The development of crystallinity in polymers is not instantaneous. Curves of specific volume as a function of time at temperatures below the crystalline melting point (Fig. 5-17) show that crystallization sometimes cannot be considered complete for long periods. Since the time for complete crystallization is somewhat indefinite, it is customary to define the rate of crystallization at a given temperature as the inverse of the time needed to attain one-half the total volume change.

The rate so defined is a characteristic function of temperature (Fig. 5-18). As the temperature is lowered, the rate increases, goes through a maximum, and then decreases as the mobility of the molecules decreases and crystallization becomes diffusion controlled. At temperatures at which the rate is very low, the polymer may be supercooled and maintained in the amorphous state.

If data for specific volume as a function of time during crystallization, such as those for natural rubber in Fig. 5-17, are plotted against log time, all the curves have the same shape and can be superposed by a shift along the time axis. This behavior is also typical of the crystallization of low-molecular-weight substances, and curves of this type can be fitted with an equation due to Avrami:

$$\ln [(V_\infty - V_t)/(V_\infty - V_0)] = -(1/w_c)kt^n \tag{5-1}$$

where V_∞, V_t, and V_0 are specific volumes at the times indicated in the subscripts, w_c is the weight fraction of material crystallized, k is a constant

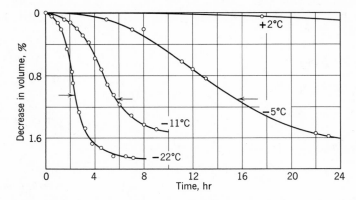

Fig. 5-17. Volume-time relations during the crystallization of natural rubber (Wood 1946). Arrows indicate points where crystallization is half complete.

describing the rate of crystallization, and n is an exponent varying with the type of nucleation and growth process.

Although this analysis provides a satisfying analogy between crystallization in polymers and that in low-molecular-weight substances, quantitative aspects of the agreement with theory are not completely satisfactory. The shape of the curves, as in Fig. 5-17, is not very sensitive to values of the exponent n in the expected range of 2-4, and it is not always possible to determine n or indeed to ascertain whether n is constant throughout the crystallization process. It is generally found that crystallization continues in polymer systems for much longer times than Eq. 5-1 predicts. This results from the fact that crystallization in polymers involves the steps of (primary) nucleation and relatively rapid spherulite growth, followed by a slow, kinetically difficult improvement in crystal perfection. "Amorphous" chains

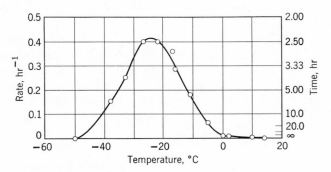

Fig. 5-18. Rate of crystallization of rubber as a function of temperature (Wood 1946).

remaining between the spherulite lamellae after the first stage also crystallize in the slow second step.

Origins of chain folding The causes of chain folding and, in particular, of the regularity with which the fold period is maintained, are still obscure. Both thermodynamic and kinetic theories of chain folding have been devised. In the thermodynamic theory, folds or other defects are shown to stabilize the crystals by their influence on torsional and vibrational modes of molecular motion. The kinetic theories relate the fold period to the stability of nuclei rather than fully grown crystals: nuclei with too short fold periods are energetically unstable and those with too long periods are less likely to occur. The fold period which persists is that which forms and grows fastest.

Molecular motion and chain folding The concept of the lamellar structure of polymers implies large-scale molecular motion in both the melt and the solid crystalline polymer. Either prefolding and some local orientation exist in the crystallizing polymer, or the molecules must undergo a considerable degree of motion during crystallization. Extensive melting and crystallization with an extremely rapid, several-fold increase in fold period occur during annealing of polymer single crystals and melt-crystallized lamellae; the extent to which the crystal morphology is disrupted, even at temperatures well below T_m, is illustrated in Fig. 5-19. The development of spherulites in polymers quenched from the melt and then held at temperatures between T_m and T_g is additional evidence of high chain mobility below T_m. The existence of mobile lattice defects provides a plausible mechanism for this mobility in crystalline material.

The disappearance of a polymer crystalline phase at the melting point is accompanied by changes in physical properties: the material becomes a (viscous) liquid, with discontinuous changes in density, refractive index, heat capacity, transparency, and other properties. Measurement of any of these properties may be used to detect the *crystalline melting point,** T_m. Since melting occurs over a temperature range, the melting point determined by a particular method may depend to some extent on the sensitivity of the method. For example, crystalline melting points may be determined by

---

* Note that, while T_m is often reported in °C, its value in K is required in equations such as those found in this section.

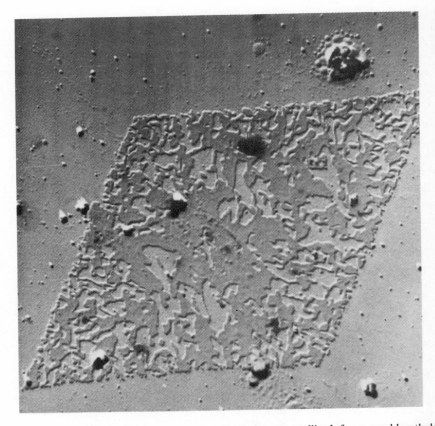

Fig. 5-19. A single crystal of linear polyethylene crystallized from perchlorethylene solution, then annealed for 30 min at 125°C, at least 10° below T_m (Statton 1960). The fold period increased from about 100 A to almost 200 A during annealing. Compare Fig. 5-9.

noting the temperature of disappearance of the last traces of crystallinity as evidenced by birefringence observed between crossed polarizers on a hot-stage microscope. This melting point is usually several degrees higher than those obtained from methods depending on gross changes in properties such as specific volume and specific heat. Because of retention of order above T_m and the existence of a melting range in polymers, properties sensitive to the amount of crystalline material present undergo change over a relatively wide range in temperature, as evidenced by experimental data for specific volume (Fig. 5-20) or specific heat (Fig. 5-21).

Today, almost all measurements of T_m are made by differential thermal analysis (Chapter 4E). Other methods which could be used to detect T_m are x-ray diffraction and infrared and NMR spectroscopy, each of these being suited to the detection of crystalline material as described in Chapter 4. As the experiments are usually performed, they are not adapted to detecting T_m but rather are carried out at constant temperature.

Thermodynamics of crystalline melting

For a polymer homogeneous in molecular weight, a statistical thermodynamic analysis based on a lattice model yields the relation

$$\frac{1}{T} - \frac{1}{T_m} = \frac{R}{\Delta H_m}\left(\frac{1}{xw_a} + \frac{1}{x - \zeta + 1}\right) \tag{5-2}$$

where w_a is the weight fraction of material that is amorphous at temperature T. The equilibrium melting point of the pure polymer at infinite chain length is $T_m = \Delta H_m/\Delta S_m$. The degree of polymerization is x, and ζ is a parameter characterizing the crystallite size. Equation 5-2 indicates that, even for a polymer of uniform chain length, melting occurs over a finite temperature range. The physical reason for this behavior is the higher surface free energy of small crystallites. Equation 5-2 also shows that the temperature interval for fusion decreases with increasing molecular weight,

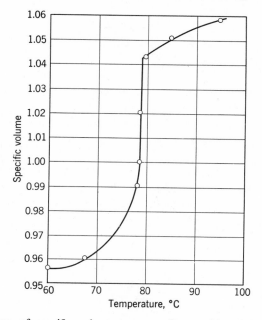

Fig. 5-20. Change of specific volume on crystalline melting in poly(decamethylene sebacate) (Evans 1950).

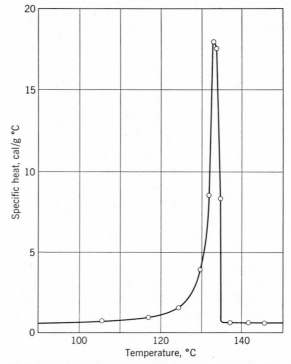

Fig. 5-21.   Specific heat as a function of temperature over the crystalline melting range in linear polyethylene (Wunderlich 1957).

and further analysis indicates that it broadens with increasing breadth o molecular weight distribution. On the other hand, as w_a approaches unity its derivative $\partial w_a/\partial T$ does not vanish; in consequence, the last traces o crystallinity disappear at a well-defined temperature, as confirmed by experi ment.

The extended-chain crystals described in Section D are now considerec to be the equilibrium state for high polymers, and thus the only state whose melting is properly described by Eq. 5-2. The more common folded-chair crystals are not in their most stable state, and as T_m is approached, they car undergo several possible changes in structure, exemplified by the reorganiza tion and recrystallization seen during annealing. The melting point ir systems of this kind occurs at the point where the free-energy change ΔG is zero, even though the system is not at equilibrium. The process is termec *zero entropy-production melting.* The melting point is a measure of the degree of imperfection of the metastable system, and is greatly dependent on the thermal history of the sample and in particular the heating rate (Jaffe 1967).

For polymer-diluent systems, use of the Flory-Huggins expression for the free energy of mixing (Chapter 2C) leads to the expression

$$\frac{1}{T_m} - \frac{1}{T_m{}^0} = \frac{R}{\Delta H_m} \frac{V_2}{V_1} (v_1 - \chi_1 v_1{}^2) \qquad (5\text{-}3)$$

where T_m and $T_m{}^0$ are the melting points of the polymer-diluent mixture and the pure polymer, respectively, V is molar volume and v volume fraction, and χ_1 is the polymer-diluent interaction parameter defined in Chapter 2C. Thus the melting point depression depends on the volume fraction of diluent added and its interaction with the polymer. For thermodynamically good solvents, large depressions (40–50°C) are sometimes observed in dilute solutions. Since the melting temperature is only slightly dependent on molecular weight, crystallization does not provide an efficient method of molecular weight fractionation.

Heats and entropies of fusion

Measurement of the heat and entropy of fusion yields important experimental data for assessing the effect of polymer structure on T_m, as discussed in Chapter 7C.

Heats of fusion are readily derived from several experimental sources, at present most often differential scanning calorimetry (Chapter 4E). The observed heat of fusion ΔH^* calculable from specific heat measurements is not ΔH_m, the heat of fusion per mole of repeating unit, but is related to it by the degree of crystallinity w_c: $\Delta H^* = w_c \Delta H_m$. It is ΔH_m which appears directly in Eq. 5-3, however, and in the related equation for the melting points of copolymers,

$$\frac{1}{T_m} - \frac{1}{T_m{}^0} = -\frac{R}{\Delta H_m} \ln n \qquad (5\text{-}4)$$

where n is the mole fraction of crystallizing units. Thus ΔH_m can be determined from melting point data for these two types of systems even if w_c is not known.

In some cases ΔH_m can be inferred from the heats of fusion of low-molecular-weight homologs. It is found, for example, that the increment per CH_2 group in the heat of fusion of n-paraffins becomes constant for chain length greater than six carbon atoms. With certain assumptions, this increment may be taken as the heat of fusion of crystalline polyethylene. The method can be extended to the calculation of the lattice or sublimation energy at 0K, a measure of the intermolecular forces in the crystal.

A comparison is made in Table 5-1 of the values of ΔH_m for polyethylene determined from specific heat measurements, the melting point depressions

TABLE 5-1.   Heat and entropy of fusion of crystalline polymers

Polymer	ΔH_m, cal/g	Reference	ΔS_m, cal/g K	$(\Delta S_m)_v$, cal/g K	Reference
Polyethylene			0.162	0.126	d
From specific heat	66.2	a			
From diluents	67.0	b			
From extrapolation	65.9	c			
Polyoxymethylene	59.6	d	0.131	0.079	d
Polytetrafluoroethylene	13.7	d	0.023	0.015	d
Natural rubber	15.3	e	0.051	0.025	f
Gutta-percha	45.3	f	0.130	0.075	f

a, Wunderlich 1957; b, Quinn 1958; c, Billmeyer 1957; d, Starkweather 1960; e, Robert 1955; f, Mandelkern 1956.

of polymer-diluent mixtures, and extrapolation of data on paraffins. Heat of fusion for several other polymers are included.

The total entropy change on fusion can be calculated from T_m and the heat of fusion, since $\Delta G_m = \Delta H_m - T\Delta S_m = 0$ at the melting point. The total change in entropy is made up of the change in entropy on fusion at constant volume $(\Delta S_m)_v$ plus the change in entropy due to expansion to the volume of the liquid phase. The latter can be evaluated by a simple thermodynamic relation to yield the equation

$$(\Delta S_m)_v = \Delta S_m - (V_L - V_S)\left(\frac{\partial P}{\partial T}\right)_{v.n} \qquad (5-5$$

where V_S and V_L are the volumes of the solid and liquid phases, respectively. Values of $(\Delta S_m)_v$ so calculated, as well as of ΔS_m, are included in Table 5-1.

Degree of crystallinity

The determination of the degree of crystallinity of a polymer is affected by the conceptual difficulties discussed in Section D. The interpretation of measurements of crystallinity depends on the model used in designing the experiments: each of the methods to be described involves a simplifying assumption, usually concerned with the degree to which defects and disordered regions in paracrystalline material are weighted in the determination of the amorphous fraction. Fortunately, independent techniques lead to unexpectedly good experimental agreement, implying that differences in their resolution, specifically the difference in the lateral extent of lattice order detected, are relatively minor (Richardson 1969).

The three major methods of determining crystallinity, discussed in the following paragraphs, are based on specific volume, x-ray diffraction, and infrared spectroscopy. Polyethylene is used throughout as an example, although each method is applicable, at least in principle, to other polymers as well.

Another well-established method, based on measurement of heat content as a function of temperature through the fusion range, is now easily carried out using differential scanning calorimetric measurements. Other methods are based on increased swelling or increased chemical reactivity of the amorphous regions. Nuclear magnetic resonance spectroscopy (Chapter 4D) has been correlated with crystallinity, but this technique places more emphasis on molecular motion than directly on differences between crystalline and amorphous character.

Specific volume   The crystallinity of a material is given in terms of the specific volumes of the specimen (V), the pure crystals (V_c), and the completely amorphous material (V_a), as

$$w_c = \frac{V_a - V}{V_a - V_c} \qquad (5\text{-}6)$$

This relation assumes additivity of the specific volumes, but this requirement is fulfilled for polymers of homogeneous structure, as far as is known. It is implicit that the sample be free of voids. Since specific volume can be determined to 1 or 2 parts in 10^4, the method can attain high precision. Its accuracy depends on the uncertainty in V_a and V_c, which are known to about 1 part in 10^3 for polyethylene. The crystalline specific volume is determined from x-ray unit cell dimensions, while V_a is usually obtained by extrapolation of the specific volumes of polymer melts.

X-ray diffraction   The x-ray method allows calculation of the relative amounts of crystalline and amorphous material in a sample if it is possible to resolve the contributions of the two types of structure to the x-ray diffraction pattern. A favorable situation is depicted in Fig. 5-22, where the scattering envelope of a polyethylene sample (outer line) has been resolved into contributions of two crystalline peaks (with indices 110 and 200) and a broad amorphous peak. The estimation of amount of crystallinity is usually based on comparison of the areas under the peaks, but other measures, such as peak height, may be used. With proper attention to experimental detail, this method provides one of the fundamental measures of crystallinity in polymers.

Infrared absorption   Suitable infrared absorption bands can be used as primary measures of crystallinity in polymers. The application requires

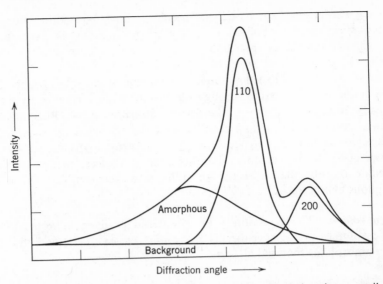

Fig. 5-22. Resolution of the x-ray scattering curve of a polyethylene into contributions from two crystalline peaks (indices 110 and 200), the amorphous peak, and background.

that the band bear a simple and unambiguous relationship to the crystalline or amorphous character of the polymer, and that absorption data can be obtained or inferred for the pure crystalline and amorphous polymers. These conditions are satisfied, e.g., for an infrared band at 7.67 μm in polyethylene identified with a methylene wagging motion in the amorphous regions, and a band at 10.6 μm in 610 nylon whose specific absorbance varies linearly with specific volume.

GENERAL REFERENCES

Mandelkern 1964; Sharples 1966; Price 1968, 1969; Wunderlich 1969.

F. Orientation and Drawing

When a polymer mass is crystallized in the absence of external forces, there is no preferred direction in the specimen along which the polymer chains lie. If such an unoriented crystalline polymer is subjected to an external stress, it undergoes a rearrangement of the crystalline material. Changes in the x-ray diffraction pattern suggest that the polymer chains align in the direction of the applied stress. At the same time the physical properties of the sample change markedly.

Cold drawing

When the orientation process is carried out below T_m but above the glass transition temperature T_g, as when an unoriented fiber is stretched rapidly, the sample does not become gradually thinner but suddenly becomes thinner at one point, in a process known as "necking down" (Fig. 5-23). As the stretching is continued, the thin or drawn section increases in length at the expense of the undrawn portion of the sample. The diameters of the drawn and undrawn portions remain about the same throughout the process. The *draw ratio*, or ratio of the length of the drawn fiber to that of the undrawn, is about 4 or 5 to 1 for a number of polymers, including branched

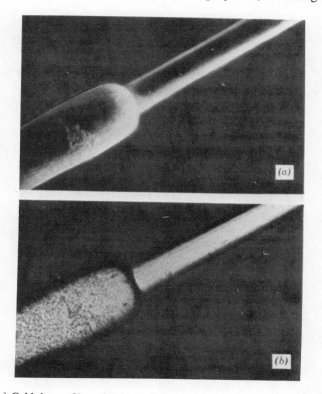

Fig. 5-23. (*a*) Cold drawn fiber of polyethylene, showing sections of drawn and undrawn material (Bunn 1953). (*b*) Same, observed between crossed polarizers, showing evidence of random orientation in the undrawn section and orientation parallel to the direction of stress in the drawn section.

polyethylene, polyesters, and polyamides, but is much higher (10 to 1 or more) in linear polyethylene.

In general, the degree of crystallinity in the specimen does not change greatly during drawing if crystallinity was previously well developed. If the undrawn polymer was amorphous or only partially crystallized, crystallinity is likely to increase during cold drawing.

Morphology changes during orientation

Observations in the electron microscope indicate that, when polymer single crystals are stressed, fibils 50–100 A in diameter are drawn across the break (Fig. 5-24). These fibrils must contain many molecules; smaller fibrils have not been resolved in the microscope. The fibrils appear to come from individual lamellae, and the sharp discontinuity between the drawn and

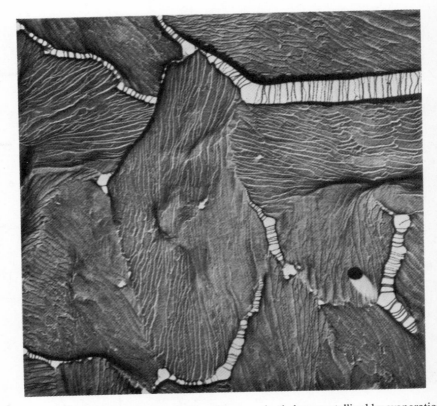

Fig. 5-24.   Electron micrograph of a film of linear polyethylene crystallized by evaporation of the solvent (Keller 1957). Fibrils drawn across the breaks are 50–100 A in diameter and appear to have a 100–300 A periodicity along their length.

the undrawn material suggests that the molecules are unfolding. In some cases the fibrils appear to have a periodicity of 100–400 A along their length; a similar period is observed by low-angle x-ray diffraction.

Examination of melt or solution cast specimens shows that the lamellae tend to rotate toward the draw direction and on further elongation break up into microlamellae and finally into submicroscopic units. Since the elongation at break in cold drawn specimens is less than could be obtained by complete unfolding, it has been suggested that the smallest units are microcrystallites in which the chains remain folded. Periodic spacing of these crystallites within the fibrils would account for their observed periodicity.

Spherulites tend to remain intact during the first stages of drawing, often elongating to markedly ellipsoidal shapes. Rupture of the sample usually occurs at spherulite boundaries.

The structure of fibers, whether formed by drawing or crystallized from an oriented melt, is still unsettled. X-ray evidence indicates that the molecules are aligned and shows that a long periodicity is present. Microfibrils are undoubtedly important structural elements. Among the models proposed are: fringed fibrils, a development of the fringed-micelle model (Section *D*) in which the crystalline regions are aligned in the direction of the fiber axis; a "string" model (Statton 1959), in which the microfibrils have alternating crystalline and amorphous regions, but there are fewer tie molecules between them than in the fringed-fibril concept; a "folded fibril" model in which the microfibrils consist of small stacked sections of lamellae; and a paracrystalline model in which the imperfect crystalline regions are thought to be much larger than required by the other models.

Perhaps the final stages in the formation of an oriented fiber by drawing can be imagined (Peterlin 1965, 1967) as shown in Fig. 5-25. A multilayer lamellar crystal is destroyed with tilting and slipping of the chains. At points of concentration of defects, blocks of folded chains break off and are incorporated along with unfolded chains into the new fiber structure. In consequence of these processes, the number of chain folds decreases and the number of tie molecules increases. The latter and the partial interpenetration of lamellae increase the mechanical stability, tensile strength, and stiffness of the fiber.

Degree of orientation

As in the measurement of degree of crystallinity (Section *E*), the determination of degree of orientation is complicated both by the complexity of the types of orientation which can occur and by the conceptual difficulties involved in the current picture of the structure of an oriented specimen. The interpretation of the measurements depends on the model and simplifying

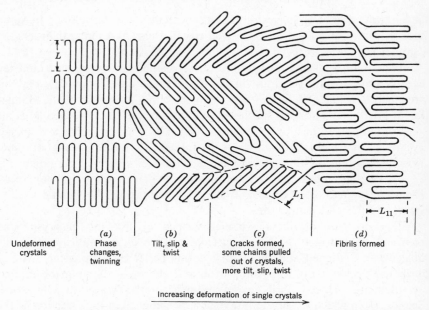

| | (a) | (b) | (c) | (d) |
| Undeformed crystals | Phase changes, twinning | Tilt, slip & twist | Cracks formed, some chains pulled out of crystals, more tilt, slip, twist | Fibrils formed |

Increasing deformation of single crystals

Fig. 5-25. Suggested model for fiber formation by chain tilting and slipping followed by breaking off of blocks of lamellae (Peterlin 1967).

assumptions in each method, but these differences are relatively minor, and the agreement among results by different methods is in general satisfactory.

X-ray diffraction As indicated in Chapter 4C, x-ray diffraction patterns of unoriented polymers resemble small-molecule powder photographs, characterized by rings rather than diffraction spots. As the specimen is oriented, these rings break into arcs, and at high degrees of orientation approach the relatively sharp patterns characteristic of paracrystals (Fig. 5-26). From the distribution of scattered intensity in the arcs or spots, the degree of orientation can be measured in terms of distribution functions. These functions describe the amount of material occurring in crystalline regions having angular coordinates describing their orientation which fall within certain limits. For practical purposes, only average values of the functions are important.

Birefringence Birefringence, or change in index of refraction with direction, is evidenced by the ability of a material to rotate the plane of polarized light. The birefringence of a crystalline polymer is made up of contributions from the crystalline and amorphous regions plus a contribution, *form birefringence*, resulting from the shape of the crystals or the presence of voids. For completely unoriented material, the contributions of both the

crystalline and the amorphous regions are zero. The increase in birefringence
on orientation is due primarily to the crystalline regions and is proportional
to the degree of crystallinity, the intrinsic birefringence of the crystals (which
can be calculated), and an orientation factor which can be identified with that
determined by the x-ray technique.

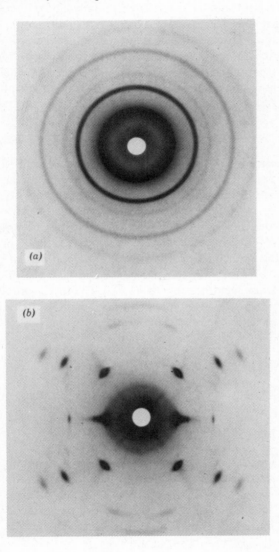

Fig. 5-26.   X-ray diffraction patterns for unoriented (*a*) and oriented (*b*) polyoxymethylene (courtesy of E. S. Clark).

Infrared dichroism   Infrared absorption is dependent upon change in dipole moment. Such a change is a vector quantity, i.e., one which is confined to definite directions for certain groups. In oriented samples the amount of absorption of plane polarized infrared radiation may vary markedly with change in direction of the plane of polarization. Thus the stretching vibrations of the C=O and N—H groups in polyamides involve changes in dipole moment which are almost exactly perpendicular to the chain axis; the corresponding absorption bands are weak for polarized radiation vibrating along the chain axis and strong for that vibrating perpendicular to the axis. When separate absorptions can be found for both the crystalline and the amorphous regions, the dichroism of these bands gives information about orientation in both regions of the specimen.

Orientation in noncrystalline polymers

The presence of crystalline material is not essential for the development of molecular orientation. Even polymers such as polystyrene, which do not crystallize, undergo considerable molecular orientation when subjected to stress at temperatures above T_g. This orientation (which is often seen in injection molded pieces) is stable below T_g and is accompanied by the development of birefringence and of enhanced strength properties in the direction of the applied stress. This type of orientation can be measured by birefringence, by a property change associated with the orientation, or by a change on subsequent heating above T_g. The force with which the specimen tends to retract on heating above T_g is a suitable and readily measured property characterizing orientation in amorphous polymers.

GENERAL REFERENCES

Meares 1965, Chapter 5; Peterlin 1967; Szabolcs 1967; Ingram 1968; Wilchinsky 1968.

BIBLIOGRAPHY

Aubrey 1968. Derek W. Aubrey, "Polymer Micro-Structure," Chap. 7 in Derek A. Smith, ed., *Addition Polymers: Formation and Characterization*, Plenum Press, New York, 1968.

Billmeyer 1957. Fred W. Billmeyer, Jr., "Lattice Energy of Crystalline Polyethylene," *J. Appl. Phys.* **28**, 1114–1118 (1957).

Blackadder 1967. D. A. Blackadder, "Ten Years of Polymer Single Crystals," *J. Macromol. Sci.-Rev. Macromol. Chem.* **C1** (2), 297–326 (1967).

Bryan 1969. W. P. Bryan, G. E. Hein, and M. F. Perutz, "Proteins," pp. 620–677 in Herman F. Mark, Norman G. Gaylord, and Norbert M. Bikales, eds., *Encyclopedia of Polymer Science and Technology*, Vol. 11, Interscience Div., John Wiley and Sons, New York, 1969.

Bryant 1947. W. M. D. Bryant, "Polythene Fine Structure," *J. Polymer Sci.* 2, 547–564 (1947).

Bunn 1953. C. W. Bunn, "Polymer Texture," Chap. 10 in Rowland Hill, ed., *Fibres from Synthetic Polymers*, Elsevier Publishing Co., New York, 1953.

Bunn 1961. Charles W. Bunn, *Chemical Crystallography*, 2nd ed., Oxford University Press, London, 1961.

Corradini 1968. P. Corradini, "Chain Conformation and Crystallinity," Chap. 1 in A. D. Ketley, ed., *The Stereochemistry of Macromolecules*, Vol. 3, Marcel Dekker, New York, 1968.

Evans 1950. Robert D. Evans, Harold R. Mighton, and Paul J. Flory, "Crystallization in High Polymers. V. Dependence of Melting Temperatures of Polyesters and Polyamides on Composition and Molecular Weight," *J. Am. Chem. Soc.* 72, 2018–2028 (1950).

Farina 1968. Mario Farina and Giancarlo Bressan, "Optically Active Stereoregular Polymers," Chap. 4 in A. D. Ketley, ed., *The Stereochemistry of Macromolecules*, Vol. 3, Marcel Dekker, New York, 1968.

Feughelman 1955. M. Feughelman, R. Langridge, W. E. Seeds, A. R. Stokes, H. R. Wilson, C. W. Hooper, M. H. F. Wilkins, R. K. Barclay, and L. D. Hamilton, "Molecular Structure of Deoxyribose Nucleic Acid and Nucleoprotein," *Nature* 175, 834–838 (1955).

Fischer 1957. E. W. Fischer, "Step and Spiral Crystal Growth of High Polymers" (in German), *Z. Naturforsch.* 12a, 753–754 (1957).

Flory 1962. P. J. Flory, "On the Morphology of the Crystalline State in Polymers," *J. Am. Chem. Soc.* 84, 2857–2867 (1962).

Furukawa 1967. J. Furukawa and T. Saegusa, "Stereoregular Epoxide Polyıers," pp. 175–195 in Herman F. Mark, Norman G. Gaylord, and Norbert M. Bikales, eds., *Encyclopedia of Polymer Science and Technology*, Vol. 6, Interscience Div., John Wiley and Sons, New York, 1967.

Gaylord 1959. Norman G. Gaylord and Herman F. Mark, *Linear and Stereoregular Addition Polymers*, Interscience Publishers, New York, 1959.

Geil 1960. P. H. Geil, "Nylon Single Crystals," *J. Polymer Sci.* 44, 449–458 (1960).

Geil 1963. Phillip H. Geil, *Polymer Single Crystals*, Interscience Div., John Wiley and Sons, New York, 1963.

Geil 1964. Phillip H. Geil, Franklin R. Anderson, Bernhard Wunderlich, and Tamio Arakawa, "Morphology of Polyethylene Crystallized from the Melt Under Pressure," *J. Polymer Sci.* A2, 3707–3720 (1964).

Goodman 1967. Murray Goodman, Akihiro Abe, and You-Ling Fan, "Optically Active Polymers," *Macromol. Revs.* 1, 1–33 (1967).

Hosemann 1962. R. Hosemann and S. N. Bagchi, *Direct Analysis of Diffraction by Matter*, North-Holland Publishing Co., Amsterdam, 1962.

Huggins 1962. M. L. Huggins, G. Natta, V. Desreux, and H. Mark, "Report on Nomenclature Dealing with Steric Regularity in High Polymers," *J. Polymer Sci.* 56, 152–161 (1962).

Ingram 1968. P. Ingram and A. Peterlin, "Morphology," pp. 204–274 in Herman F. Mark, Norman G. Gaylord, and Norbert M. Bikales, eds., *Encyclopedia of Polymer Science and Technology*, Vol. 9, Interscience Div., John Wiley and Sons, New York, 1968.

Jaffe 1967. M. Jaffe and B. Wunderlich, "Melting of Polyoxymethylene," *Kolloid Z. & Z. Polymere* 216–217, 203 216 (1967).

Keller 1957. A. Keller, "A Note on Single Crystals in Polymers: Evidence for a Folded Chain Configuration," *Phil. Mag.* [8] 2, 1171–1175 (1957).

Keller 1959. A. Keller, "Investigations on Banded Spherulites," *J. Polymer Sci.* **39**, 151–173 (1959).

Keller 1964. A. Keller, "Crystal Configurations and their Relevance to the Crystalline Texture and Crystalline Mechanisms in Polymers," *Kolloid Z. & Z. Polymere* **197**, 98–115 (1964).

Keller 1969. A. Keller, "Solution Grown Polymer Crystals—A Survey of Some Problematic Issues," *Kolloid Z. & Z. Polymere* **231**, 386–421 (1969).

Kornberg 1962. Arthur Kornberg, *Enzymatic Synthesis of DNA*, John Wiley and Sons, New York, 1962.

Mandelkern 1956. L. Mandelkern, F. A. Quinn, Jr., and D. E. Roberts, "Thermodynamics of Crystallization in High Polymers: Gutta-Percha," *J. Am. Chem. Soc.* **78**, 926–932 (1956).

Mandelkern 1964. Leo Mandelkern, *Crystallization of Polymers*, McGraw-Hill Book Co., New York, 1964.

Meares 1965. Patrick Meares, *Polymers: Structure and Properties*, D. Van Nostrand Co., Princeton, N.J., 1965.

Miller 1965. Robert L. Miller, "Crystalline and Spherulitic Properties," Chap. 12 in R. A. V. Raff and K. W. Doak, eds., *Crystalline Olefin Polymers*, Part I, Interscience Div., John Wiley and Sons, New York, 1965.

Miller 1966a. M. L. Miller, *The Structure of Polymers*, Reinhold Publishing Corp., New York, 1966.

Miller 1966b. Robert L. Miller, "Crystallinity," pp. 449–528 in Herman F. Mark, Norman G. Gaylord, and Norbert M. Bikales, eds., *Encyclopedia of Polymer Science and Technology*, Vol. 4, Interscience Div., John Wiley and Sons, New York, 1966.

Miller 1966c. Robert L. Miller, "Crystallographic Data for Various Polymer," pp. III-1–III-59 in J. Brandrup and E. H. Immergut, eds., with the collaboration of H.-G. Elias, *Polymer Handbook*, Interscience Div., John Wiley and Sons, New York, 1966.

Natta 1955. G. Natta, Piero Pino, Paolo Corradini, Ferdinando Danusso, Enrico Mantica, Giorgio Mazzanti, and Giovanni Moranglio, "Crystalline High Polymers of α-Olefins," *J. Am. Chem. Soc.* **77**, 1708–1710 (1955).

Natta 1959a. G. Natta and F. Danusso, "Nomenclature Relating to Polymers Having Sterically Ordered Structure," *J. Polymer Sci.* **34**, 3–11 (1959).

Natta 1959b. Giulio Natta and Paolo Corradini, "Conformation of Linear Chains and Their Mode of Packing in the Crystal State," *J. Polymer Sci.* **39**, 29–46 (1959).

Natta 1967. Giulio Natta and Ferdinando Danusso, eds., *Stereoregular Polymers: Stereoregular Polymers and Stereospecific Polymerization*, Pergamon Press, New York, 2 Vols., 1967.

Newman 1967. B. A. Newman and H. F. Kay, "Chain Folding in Polyethylene and Cyclic Paraffins," *J. Appl. Phys.* **38**, 4105–4109 (1967).

Pauling 1951. Linus Pauling, Robert B. Corey, and H. R. Branson, "The Structure of Proteins: Two Hydrogen-Bonded Helical Configurations of the Polypeptide Chain," *Proc. Natl. Acad. Sci.* **37**, 205–211 (1951).

Peterlin 1965. A. Peterlin, "Crystalline Character in Polymers," *J. Polymer Sci.* **C9**, 61–89 (1965).

Peterlin 1967. A. Peterlin, "The Role of Chain Folding in Fibers," pp. 283–340 in H. F. Mark, S. M. Atlas, and E. Cernia, eds., *Man-Made Fibers, Science and Technology*, Interscience Div., John Wiley and Sons, New York, 1967.

Pino 1965. Piero Pino, "Optically Active Addition Polymers," *Fortschr. Hochpolym. Forsch.* (*Advances in Polymer Science*) **4**, 393–456 (1965).

Poland 1970. Douglas Poland and Harold A. Scheraga, *Theory of Helix-Coil Transitions in Biopolymers,* Academic Press, New York, 1970.

Price 1956. Charles C. Price and Maseh Osgan, "The Polymerization of *l*-Propylene Oxide," *J. Am. Chem. Soc.* **78,** 4787–4792 (1956).

Price 1958. F. P. Price, p. 466 in R. H. Doremus, B. W. Roberts, and D. Turnbull, eds., *Growth and Perfection of Crystals,* John Wiley and Sons, New York, 1958.

Price 1968. Fraser P. Price, "Kinetics of Crystallization," pp. 63–83 in Herman F. Mark, Norman G. Gaylord, and Norbert M. Bikales, eds., *Encyclopedia of Polymer Science and Technology,* Vol. 8, Interscience Div., John Wiley and Sons, New York, 1968.

Price 1969. Fraser P. Price, "Nucleation in Polymer Crystallization," *Nucleation* **1969,** 405–488.

Prime 1969. R. Bruce Prime, Bernhard Wunderlich, and Louis Melillo, "Extended-Chain Crystals. V. Thermal Analysis and Electron Microscopy of the Melting Process in Polyethylene," *J. Polymer Sci. A-2* **7,** 2091–2097 (1969).

Quinn 1958. F. A. Quinn, Jr., and L. Mandelkern, "Thermodynamics of Crystallization in High Polymers: Poly-(ethylene)," *J. Am. Chem. Soc.* **80,** 3178–3182 (1958).

Reneker 1960. D. H. Reneker and P. H. Geil, "Morphology of Polymer Single Crystals," *J. Appl. Phys.* **31,** 1916–1925 (1960).

Richardson 1969. M. J. Richardson, "Crystallinity Determination in Polymers and a Quantitative Comparison for Polyethylene," *Brit. Polym. J.* **1,** 132–137 (1969).

Roberts 1955. Donald E. Roberts and Leo Mandelkern, "Thermodynamics of Crystallization in High Polymers: Natural Rubber," *J. Am. Chem. Soc.* **77,** 781–786 (1955).

Sauer 1969. J. A. Sauer, "Morphology and Structure of Polymer Crystals," *Ann. N.Y. Acad. Sci.* **155,** 517–538 (1969).

Schildknecht 1948. C. E. Schildknecht, S. T. Gross, H. R. Davidson, J. M. Lambert, and A. O. Zoss, "Polyvinyl Isobutyl Ethers—Properties and Structures," *Ind. Eng. Chem.* **40,** 2104–2115 (1948).

Schildknecht 1949. C. E. Schildknecht, A. O. Zoss, and Frederick Grosser, "Ionic Polymerization of Some Vinyl Compounds," *Ind. Eng. Chem.* **41,** 2891–2896 (1949).

Schlesinger 1953. Walter Schlesinger and H. M. Leeper, "Gutta. I. Single Crystals of Alpha-Gutta," *J. Polymer Sci.* **11,** 203–213 (1953).

Schonhorn 1969. Harold Schonhorn and J. P. Luongo, "Fold Structure of Polyethylene Single Crystals," *Macromolecules* **2,** 366–369 (1969).

Schulz 1968. Rolf C. Schulz, "Optically Active Polymers," pp. 507–524 in Herman F. Mark, Norman G. Gaylord, and Norbert M. Bikales, eds., *Encyclopedia of Polymer Science and Technology,* Vol. 9, Interscience Div., John Wiley and Sons, New York, 1968.

Sharples 1966. Allan Sharples, *Introduction to Polymer Crystallization,* St. Martin's Press, New York, 1966.

Starkweather 1960. Howard W. Starkweather, Jr., and Richard H. Boyd, "The Entropy of Melting of Some Linear Polymers," *J. Phys. Chem.* **64,** 410–414 (1960).

Statton 1959. W. O. Statton, "Polymer Texture: The Arrangement of Crystallites," *J. Polymer Sci.* **41,** 143–155 (1959).

Statton 1960. W. O. Statton and P. H. Geil, "Recrystallization of Polyethylene during Annealing," *J. Appl. Polymer Sci.* **3,** 357–361 (1960).

Staudinger 1932. Hermann Staudinger, *Die Hochmolekularen Organischen Verbindungen,* (*High Molecular Weight Organic Compounds*) (in German), Springer-Verlag, Berlin, 1932.

Szabolcs 1967. O. Szabolcs and I. Szabolcs, "Current Ideas on the Morphology of Synthetic Fibers," pp. 341–374 in H. F. Mark, S. M. Atlas, and E. Cernia, eds.,

Man-Made Fibers, Science and Technology, Vol. 1, Interscience Div., John Wiley and Sons, New York, 1967.

Till 1957. P. H. Till, Jr., "The Growth of Single Crystals of Linear Polyethylene," *J. Polymer Sci.* **24**, 301–306 (1957).

Tsuruta 1967. Teiji Tsuruta and Shohei Inoue, "Well-Ordered Polymers," *Science & Technology* No. 71, 66–74 (1967).

Watson 1953. J. D. Watson and F. H. C. Crick, "A Structure for Deoxyribose Nucleic Acid," *Nature* **171**, 737–738 (1953).

Wilchinsky 1968. Zigmond W. Wilchinsky, "Orientation," pp. 624–648 in Herman F. Mark, Norman G. Gaylord, and Norbert M. Bikales, eds., *Encyclopedia of Polymer Science and Technology*, Vol. 9, Interscience Div., John Wiley and Sons, New York, 1968.

Wood 1946. Lawrence A. Wood, "Crystallization Phenomena in Natural and Synthetic Rubbers," pp. 57–95 in H. Mark and G. S. Whitby, eds., *Scientific Progress in the Field of Rubber and Synthetic Elastomers* (Advances in Colloid Science, Vol. II), Interscience Publishers, New York, 1946.

Wunderlich 1957. Bernhard Wunderlich and Malcolm Dole, "Specific Heat of Synthetic High Polymers. VIII. Low Pressure Polyethylene," *J. Polymer Sci.* **24**, 201–213 (1957).

Wunderlich 1968. Bernhard Wunderlich and Louis Melillo, "Morphology and Growth of Extended Chain Crystals of Polyethylene," *Makromol. Chem.* **118**, 250–264 (1968).

Wunderlich 1969. Bernhard Wunderlich, *Crystalline High Polymers: Molecular Structure and Thermodynamics*, The American Chemical Society, Washington, D.C., 1969.

6

Rheology and the Mechanical Properties of Polymers

Rheology is, by definition, the science of deformation and flow of matter. The rheological behavior of polymers involves several widely different phenomena, which can be related to some extent to different molecular mechanisms. These phenomena and their associated major mechanisms are as follows:

a. Viscous flow, the irreversible bulk deformation of polymeric material, associated with irreversible slippage of molecular chains past one another.

b. Rubberlike elasticity, where the local freedom of motion associated with small-scale movement of chain segments is retained, but large-scale movement (flow) is prevented by the restraint of a diffuse network structure.

c. Viscoelasticity, where the deformation of the polymer specimen is reversible but time dependent, and is associated (as in rubber elasticity) with the distortion of polymer chains from their equilibrium conformations through activated segment motion involving rotation about chemical bonds.

d. Hookean elasticity, where the motion of chain segments is drastically restricted, and probably involves only bond stretching and bond angle deformation: the material behaves like a glass.

These four phenomena are discussed in Sections *A–D*, respectively. Together they form the basis for a description of the mechanical properties of amorphous polymers. The mechanical properties of semicrystalline polymers, however, depend intimately on the restraining nature of their crystalline regions and can be inferred only in part from the rheological behavior of amorphous polymers. The mechanical properties of crystalline polymers are therefore discussed separately in Section *E*.

A. Viscous Flow

Phenomena of viscous flow

If a force per unit area s causes a layer of liquid at a distance x from a fixed boundary wall to move with a velocity v, the viscosity η is defined as the ratio between the shear stress s and the velocity gradient $\partial v/\partial x$ or rate of shear $\dot{\gamma}$:

$$s = \eta \frac{\partial v}{\partial x} = \eta \dot{\gamma} \qquad (6\text{-}1)$$

If η is independent of the rate of shear, the liquid is said to be *Newtonian* or to exhibit ideal flow behavior (Fig. 6-1a). Two types of deviation from Newtonian flow are commonly observed in polymer solutions and melts (Bauer 1967). One is *shear thinning*, a reversible decrease in viscosity with increasing shear rate (Fig. 6-1b). Shear thinning results from the tendency of the applied force to disturb the long chains from their favored equilibrium conformation (Chapter 2B), causing elongation in the direction of shear. An opposite effect, *shear thickening*, in which viscosity increases with increasing shear rate, is rarely observed in polymers.

A second deviation from Newtonian flow is the exhibition of a *yield value*, a critical stress below which no flow occurs. Above the yield value, flow may be either Newtonian (as indicated in Fig. 6-1c) or non-Newtonian. For most polymer melts, only an apparent yield value is observed.

For low-molecular-weight liquids, the temperature dependence of viscosity is found to follow the simple exponential relationship

$$\eta = A e^{-(E/RT)} \qquad (6\text{-}2)$$

where E is an *activation energy for viscous flow*, and A is a constant.

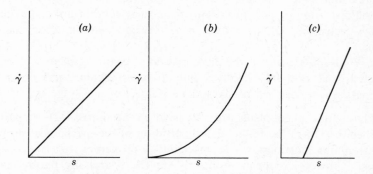

Fig. 6-1. Dependence of shear rate $\dot{\gamma}$ on shearing force s for three types of liquid viscosity: (a) Newtonian, (b) non-Newtonian, (c) yield stress followed by Newtonian flow.

TABLE 6-1. *Summary of methods for measuring viscosity*

Method	Approximate Useful Viscosity Range, Poises
Capillary pipette	$10^{-2}-10^{3}$
Falling sphere	$1-10^{5}$
Capillary extrusion	$1-10^{8}$
Parallel plate	$10^{4}-10^{9}$
Falling coaxial cylinder	$10^{5}-10^{11}$
Stress relaxation	$10^{3}-10^{10}$
Rotating cylinder	$1-10^{12}$
Tensile creep	$10^{5}->10^{12}$

These features of the flow of liquids can be explained in terms of several molecular theories. That of Eyring (Glasstone 1941) is based upon a lattice structure for the liquid, containing some unoccupied sites or holes. These sites move at random throughout the liquid as they are filled and created anew by molecules jumping from one site to another. Under an applied stress the probability of such jumps is higher in the direction which relieves the stress. If each jump is made by overcoming an energy barrier of height E, the theory leads to Eq. 6-2. The energy of activation E is expected to be related to the latent heat of vaporization of the liquid, since the removal of a molecule from the surroundings of its neighbors forms a part of both processes. Such a relation is indeed found and is taken as evidence that the particle which moves from site to site is probably a single molecule.

As molecular weight is increased in a homologous series of liquids up to the polymer range, the activation energy of flow E does not increase proportionally with the heat of vaporization but levels off at a value independent of molecular weight. This is taken to mean that in long chains the unit of flow is considerably smaller than the complete molecule. It is rather a segment of the molecule whose size is of the order of 5–50 carbon atoms. Viscous flow takes place by successive jumps of segments (with of course some degree of coordination) until the whole chain has shifted.

Flow measurement

Methods commonly used for measuring the viscosity of polymer solutions and melts are listed in Table 6-1. The most important of these methods involve rotational and capillary devices (Van Wazer 1963, Mendelson 1968).

Rotational viscometry   Rotational viscometers are available with several different geometries, including concentric cylinders, two cones of different angles, a cone and a plate, or combinations of these. Measurements with rotational devices become difficult to interpret at very high shear stresses owing to the generation of heat in the specimen because of dissipation of energy, and to the tendency of the specimen to migrate out of the region of high shear.

A simple rotational instrument used in the rubber industry is the *Mooney viscosimeter*. This empirical instrument measures the torque required to revolve a rotor at constant speed in a sample of the polymer at constant temperature. It is used to study changes in the flow characteristics of rubber during milling or mastication (Chapter 19). The *Brabender Plastograph* is a similar device.

Capillary viscometry   Capillary rheometers, usually made of metal and operated either by dead weight or by gas pressure, or at constant displacement rate, have advantages of good precision, ruggedness, and ease of operation. They may be built to cover the range of shear stresses found in commercial fabrication operations. However, they have the disadvantage that the shear stress in the capillary varies from zero at the center to a maximum at the wall.

An elementary capillary rheometer (extrusion plastometer) is used to determine the flow rate of polyethylene in terms of *melt index*, defined as the mass rate of flow of polymer through a specified capillary under controlled conditions of temperature and pressure.

Experimental results

Molecular weight and shear dependence   As discussed further in Chapter 7, the most important structural variable determining the flow properties of polymers is molecular weight or, alternatively, chain length, Z (the number of atoms in the chain). Although early data (Flory 1940) suggested that log η was proportional to $Z^{1/2}$, it has since been observed (Fox 1956) for essentially all polymers studied that, for values of Z above a critical value Z_c,

$$\log \eta = 3.4 \log \bar{Z}_w + k \tag{6-3}$$

where k is temperature dependent. This equation is valid only for shear stress sufficiently low (10^2–10^3 dynes/cm^2) that the viscosity is Newtonian. The weight-average chain length $\bar{Z}_w$ is usually assumed to be the appropriate average for the above conditions.

For chain lengths below Z_c, which is about 600 for many polymers, the viscosity is found to depend upon a power of $\bar{Z}_w$ (and hence $\bar{M}_w$) in the range 1.75–2.0. In this range, shear rate has little effect on viscosity. Typical experimental data for the two regions described above are shown in Fig. 6-2.

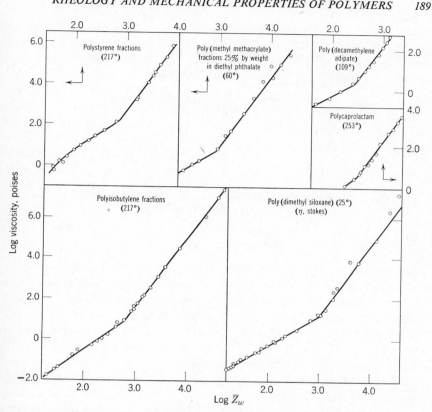

Fig. 6-2. Dependence of melt viscosity η on chain length $\bar{Z}_w$ for low shear rate, showing the regions below $Z_c \simeq 600$, where η is approximately proportional to $\bar{Z}_w^{1.75}$, and above Z_c, where $\eta \sim \bar{Z}_w^{3.4}$ (Fox 1956).

While the Newtonian melt viscosity is determined by $\overline{M}_w$ as described in Eq. 6-3, the dependence of viscosity on shear rate also depends upon the molecular-weight distribution. As depicted schematically in Fig. 6-3, the drop in melt viscosity below its Newtonian value begins at a lower shear rate and continues over a broader range of shear rates for polymers with broader distributions of molecular weight. At sufficiently high shear rates, the melt viscosity appears to depend primarily on $\overline{M}_n$ rather than $\overline{M}_w$. Qualitative information, at least, about the molecular-weight distribution can be obtained from melt viscosity-shear rate studies.

For branched polymers, the situation is complicated. The introduction of chain branching appears to have the same effect on the viscosity-shear rate curve as broadening of the molecular-weight distribution. Since the two effects go together (as in polyethylene, Chapter 13), measurement of more than one property is required to separate them.

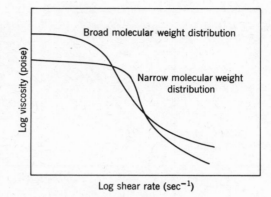

Fig. 6-3. Generalized melt viscosity-shear rate curves for polymers with broad and narrow distributions of molecular weight.

Flow instabilities   At shear stresses in the neighborhood of 2×10^6 dynes/cm² for many polymers, instabilities in the flow appear, with the result that the upper Newtonian region is rarely realized in bulk polymers. These instabilities are manifested as a striking and abrupt change in the shape of the polymer stream emerging from the capillary of the rheometer. At and above a critical stress, the shape of the emerging stream changes from that of a regular cylinder to a rough or distorted one.

Emergent streams of polymers extruded under unstable flow conditions have one of two rather different characteristic shapes: either a gross distortion or a fine-scale surface roughness without gross distortion. The gross distortion is sometimes called *melt fracture* (Tordella 1969).

Temperature dependence of viscosity   On close examination for polymer systems, the activation energy for viscous flow of Eq. 6-2 is found to be constant only for small ranges of temperature. For pure liquids, it was found many years ago that most of the change in viscosity with temperature is associated with the concurrent change in volume. This observation led to theories of viscosity based on the concept of free volume, whose application to polymers is discussed in Section *D*. A major result of these theories is the WLF equation (Eqs. 6-22 and 6-31), in which the temperature dependence of melt viscosity is expressed in terms of the glass transition temperature T_g (or another reference temperature) and universal constants. Since the terms describing the variation of melt viscosity with temperature and with molecular weight are independent, the WLF equation can be combined with Eq. 6-3 to yield the relation, for low shear rates,

$$\log \eta = 3.4 \log \bar{Z}_w - \frac{17.44(T - T_g)}{51.6 + T - T_g} + k' \qquad (6\text{-}4)$$

where k' is a constant depending only on polymer type. This equation holds over the temperature range from T_g to about $T_g + 100K$.

GENERAL REFERENCES

Eirich 1956–1969; Severs 1962; Van Wazer 1963; Meares 1965, Chap. 13; Reiner 1966; Mendelson 1968.

B. Kinetic Theory of Rubber Elasticity

Rubberlike elasticity is in many respects a unique phenomenon, involving properties markedly different from those of low-molecular-weight solids, liquids, or gases. The properties of typical elastomers are defined by the following requirements:

a. They must stretch rapidly and considerably under tension, reaching high elongations (500–1000%) with low damping, i.e., little loss of energy as heat.

b. They must exhibit high tensile strength and high modulus (stiffness) when fully stretched.

c. They must retract rapidly, exhibiting the phenomenon of *snap* or *rebound.*

d. They must recover their original dimensions fully on the release of stress, exhibiting the phenomena of *resilience* and *low permanent set.*

Although the thermodynamics associated with rubber elasticity was developed in the middle of the nineteenth century, the molecular requirements for the exhibition of rubbery behavior were not recognized until 1932. Theories of the mechanism relating these molecular structure requirements to the phenomena of rubber elasticity were developed soon after.

As discussed further in Chapter 7F, the molecular requirements of elastomers may be summarized as follows:

a. The material must be a high polymer.

b. It must be above its glass transition temperature T_g to obtain high local segment mobility.

c. It must be amorphous in its stable (unstressed) state for the same reason.

d. It must contain a network of crosslinks to restrain gross mobility of its chains.

A discussion of rubber elasticity in thermodynamic terms can be based on considerations of either free energy or work content. Since the constant-pressure experiment is more usual, free energy is used in the present treatment.

The ideal elastomer In discussions of the mechanical properties of a polymer, parameters related to its distortion must be included, as well as the variables defining the state of the system, such as pressure and temperature. In particular, when a sample of rubber is stretched, work is done on it and its free energy is changed. By restricting the development to stretching in one direction, for simplicity, the elastic work W_{el} equals $f\,dl$, where f is the retractive force and dl the change in length. If the change in free energy is G,

$$f = \left(\frac{\partial G}{\partial l}\right)_{T,p} = \left(\frac{\partial H}{\partial l}\right)_{T,p} - T\left(\frac{\partial S}{\partial l}\right)_{T,p} \tag{6-5}$$

In analogy with an ideal gas, where

$$\text{Ideal gas:} \qquad \left(\frac{\partial E}{\partial V}\right)_T = 0; \qquad p = T\left(\frac{\partial S}{\partial V}\right)_T \tag{6-6}$$

an *ideal elastomer* is defined by the condition that

$$\text{Ideal elastomer:} \left(\frac{\partial H}{\partial l}\right)_{T,p} = 0; \qquad f = -T\left(\frac{\partial S}{\partial l}\right)_{T,p} \tag{6-7}$$

The negative sign results since work is done on the specimen to increase its length.

Entropy elasticity Equation 6-7 shows that the retractive force in an ideal elastomer is due to its decrease in entropy on extension. The molecular origin of this *entropy elasticity* is the distortion of the polymer chains from their most probable conformations in the unstretched sample. As described in Chapter 2B, the distribution of these conformations is Gaussian, the probability of finding a chain end in a unit volume of the space coordinates x, y, z at a distance r from the other end being

$$W(x,\,y,\,z) = \left(\frac{b}{\sqrt{\pi}}\right)^3 e^{-b^2 r^2} \tag{6-8}$$

where $b^2 = \frac{3}{2}xl^2$, x being the number of links, with length l. Since the entropy of the system is proportional to the logarithm of the number of configurations it can have,

$$S = (\text{const.}) - kb^2r^2 \tag{6-9}$$

where k is Boltzmann's constant. It follows that, for a single polymer chain, the retractive force f' for an extension of magnitude dr is

$$f' = -T\frac{dS}{dr} = 2kTb^2r \tag{6-10}$$

It is customarily assumed that the retractive force f (Eq. 6-7) for a bulk polymer sample can be identified as the sum of the forces f' for all the chains in the specimen. This assumption, which is inaccurate in detail but justified in most cases, implies that individual chains contribute additively and without interaction to the elasticity of the macroscopic sample, and that the distribution of end-to-end distances undergoes transformation identically with the sample dimensions (affine transformation).

Temperature coefficient of retractive force   Equation 6-10 states that, unlike the Hookean restoring force of a spring, the retractive force of a rubber increases with increasing temperature. It was shown by Lord Kelvin in 1857 that

$$\left(\frac{\partial f}{\partial T}\right)_{p,l} = \frac{C_p}{T}\left(\frac{\partial T}{\partial l}\right)_{p,\text{ adiabatic}} \tag{6-11}$$

where C_p is the specific heat. Thus, as is readily checked, rubbers heat on adiabatic stretching. It follows also that

$$\left(\frac{\partial T}{\partial f}\right)_{p,\text{ adiabatic}} = -\frac{T}{C_p}\left(\frac{\partial l}{\partial T}\right)_{p,f} \tag{6-12}$$

which states that rubbers contract, rather than expand, on heating. These facts were demonstrated as long ago as 1806.

Deviations from ideal elastomeric behavior   Since stretching a rubber may involve overcoming energy barriers in real polymer chains, the internal energy of the sample may be expected to depend slightly on the elongation. Since

$$\left(\frac{\partial H}{\partial l}\right)_{T,p} = \left(\frac{\partial E}{\partial l}\right)_{T,p} + p\left(\frac{\partial V}{\partial l}\right)_{T,p} \tag{6-13}$$

and the second term on the right is negligible for normal pressures, this represents a departure from the ideal behavior defined in Eq. 6-7. As indicated in Fig. 6-4, the contribution of internal energy to the retractive force

$$f = \left(\frac{\partial E}{\partial l}\right)_{T,p} - T\left(\frac{\partial S}{\partial l}\right)_{T,p} \tag{6-14}$$

is small except for very low elongations. At much greater elongations, however, the energy term becomes very large negatively as crystallization takes place.

Stress-strain behavior of elastomers

The stress-strain curves of typical elastomers (Fig. 6-5) show marked deviations from the straight line required by Hooke's law. The relatively low slope of the curve (defined near the origin as the modulus) decreases to about one-third its original value over the first hundred per cent elongation, and later increases, often to quite high values at high elongations. To explain this behavior, it is convenient first to examine the stress-strain properties of a simple model, and then to consider the relation of the model to actual elastomers.

A simple model of an elastomeric network (Guth 1946) It is assumed that the actual tangled mass of polymer chains may be represented by an idealized

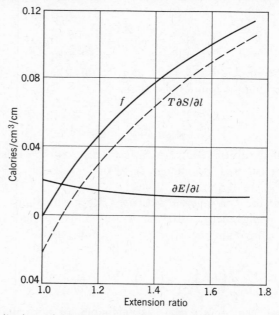

Fig. 6-4. Contributions of the energy and entropy terms to the retractive force in natural rubber (Treloar 1958).

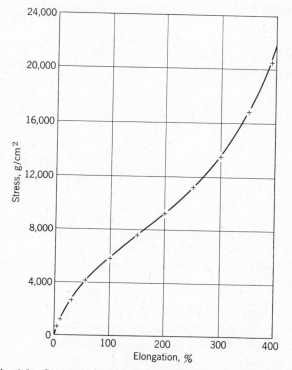

Fig. 6-5.   Stress-strain curve for a typical elastomer (Guth 1946).

network of flexible chains, irregular in detail but homogeneous and isotropic, extending throughout the sample. The network consists of m chains
of average length $(\overline{r^2})^{1/2}$ per unit volume directed along each of three perpendicular axes. To obtain the proper space-filling properties of the model
it is assumed that the space between the chains is filled with an incompressible fluid exerting a hydrostatic pressure p outward against the elastic tension of the chains.

If a cube of this material of unit dimensions is stretched by a force s along
the z-axis, it will be converted into a parallelepiped of reduced dimensions
α, β, and γ, where $\alpha = \beta$. Since it is incompressible, its volume is $\alpha\beta\gamma = 1$.
The condition for equilibrium on a face parallel to the stress is obtained by
equating the inward and outward forces. By the conventions of the model
the outward force is $p\alpha\gamma$. The inward force is that developed by the m
chains of the network seeking to find their equilibrium length and is given
by Eq. 6-10. Setting the outward and inward forces equal yields

$$p\alpha\gamma = 2mkTb^2\beta \qquad (6\text{-}15)$$

On an end face the stretching force must be considered also:

$$s + p\alpha\beta = 2mkTb^2\gamma \tag{6-16}$$

Rearranging and eliminating p,

$$s = 2mkTb^2\left(\gamma - \frac{1}{\gamma^2}\right) \tag{6-17}$$

The modulus* is given by

$$G = \frac{ds}{d\gamma} = 2mkTb^2\left(1 + \frac{2}{\gamma^3}\right) \tag{6-18}$$

As the strain increases from zero ($\gamma = 1$) to large values of γ, the slope decreases to one-third its initial value as required by the experimental facts. Equation 6-18 predicts the stress-strain curve of actual elastomers very well up to elongations of 300% or more (Fig. 6-6).

It can be shown that this oversimplified model yields results in accord with more general cases.

Behavior at high extensions   Figure 6-6 suggests that, at elongations greater than about 300%, the stress-strain curves of actual elastomers have a higher slope than that predicted by Eq. 6-18. This is far below the ultimate elongation of 1000% for a good elastomer. One reason for the failure of the theory is that the Gaussian distribution of chain lengths which holds at low elongations fails in the region of interest for elastomers, as the actual lengths approach those of fully extended chains. A better approximation than the Gaussian is available. With its use the theory predicts the actual stress-strain curves rather well for elastomers which do not crystallize on stretching. In the cases where crystallinity does develop, the slope still increases faster than predicted. As is true in general for the problem of the mechanical behavior of crystalline polymers (Section *E*), a quantitative theory of the effect of crystallinity on the modulus of elastomers has not been developed.

Network structure and elasticity

In any real polymer network, all the chains cannot respond to external stress. Some of them are attached to other chains at one end only and contribute nothing to the elasticity. These are designated *terminal sections* in contrast to the *principal sections* attached at both ends. Other types of flaws,

---

* Note that the symbol G is conventionally used to represent both the Gibbs free energy (Eq. 6-5) and, throughout the remainder of this chapter, the modulus of elasticity.

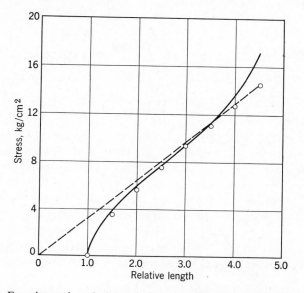

Fig. 6-6. Experimental and theoretical stress-strain curves (Guth 1946). Solid line, experimental; circles, computed from Eq. 6–18; dashed line, asymptote of Eq. 6-18 for high elongations.

in which one chain loops around another, can also be visualized. For a constant number of crosslinks per unit volume or mass of polymer, the number of terminal sections increases as the average molecular weight of the polymer decreases, and the number of chain ends correspondingly increases. At some critical molecular weight just equal to the average molecular weight of a primary segment, the network structure can no longer exist.

If the network is assumed to consist of N chains crosslinked together at $v/2$ points (i.e., with v mers chemically linked together in $v/2$ pairs), there are $2N$ terminal segments and $v + N$ segments in all, hence $v - N$ principal segments. If the number of crosslink points were just $N - 1$ ($\simeq N$), all the chains would be joined together, but with no closed loops an elastic network would not exist. Each additional crosslink point, however, adds one loop useful in generating elasticity. Thus the number of crosslink points effective in producing elasticity (the number of loops) is $(v_e/2) = (v/2) - N$, defining the number of effective crosslinked mers as $v_e = v - 2N = v(1 - 2N/v)$.

It can be shown (Flory 1953) that the entropy of the network is given, in analogy with Eq. 6-9, by

$$S = (\text{const.}) - k\left(\frac{v_e}{2}\right)r^2 \qquad (6\text{-}19)$$

If the volume of the network containing v_e effective crosslinked mers in V, it follows from a derivation similar to that on pages 195-196 that

$$s = \frac{kTv_e}{V}\left(\gamma - \frac{1}{\gamma^2}\right) = \frac{kTv}{V}\left(1 - \frac{2N}{v}\right)\left(\gamma - \frac{1}{\gamma^2}\right) \tag{6-20}$$

$$G = \frac{kTv_e}{V}\left(1 + \frac{2}{\gamma^3}\right) = \frac{kTv}{V}\left(1 - \frac{2N}{v}\right)\left(1 + \frac{2}{\gamma^3}\right) \tag{6-21}$$

The first term on the right shows that, for the usual case of $v \gg 2N$, the stress and modulus are proportional to the density of crosslinks v/V. The second term introduces the dependence on molecular weight: at $v = 2N$, a critical molecular weight is reached below which (higher N) the network cannot exist. This conclusion suggested a method (Flory 1953) for determining the critical molecular weight: a series of butyl rubber fractions was vulcanized to a constant density of crosslinks. It was found that network structures were obtained only for molecular weights above a limiting value. For higher molecular weights, G varied with M as predicted by Eq. 6-21 and shown in Fig. 6-7.

GENERAL REFERENCES

Flory 1953, Chap. XI; Meares 1965, Chap. 6, 8; Treloar 1958; Vincent 1965.

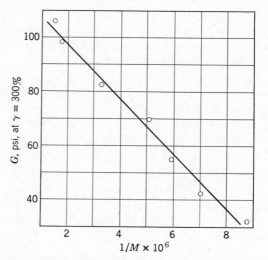

Fig. 6-7.   Dependence of the modulus of vulcanized butyl rubber on its molecular weight before vulcanization (Flory 1944).

C. Viscoelasticity

Sections *A* and *B* have dealt with the equilibrium response of linear and network polymer structures to external stress. In this section are considered the time-dependent mechanical properties of amorphous polymers. As in studies of rate phenomena, theories of dynamic mechanical phenomena in polymers are less thoroughly developed than those referring to the equilibrium states. They are also more dependent on the details of models. The present problem is, first, to find suitable models or mathematical functions with which to describe the several types of molecular motion postulated earlier to be associated with the processes of elasticity, viscoelasticity, and viscous flow in polymers; and, second, to utilize these models in relating the behavior of the polymers to their molecular structure.

The description of the viscoelastic response of amorphous polymers to small stresses is greatly simplified by the application of the following two general principles which are widely applicable to these systems.

The Boltzmann superposition principle   This principle states that strain is a linear function of stress, so that the total effect of applying several stresses is the sum of the effects of applying each one separately. Application of the superposition principle makes it possible to predict the mechanical response of an amorphous polymer to a wide range of loading conditions from a limited amount of experimental data. The principle applies to both static and time-dependent stresses.

Time-temperature equivalence   An increase in temperature accelerates molecular and segmental motion, bringing the system more rapidly to equilibrium or apparent equilibrium and accelerating all types of viscoelastic processes. A convenient way of formulating this effect of temperature is in terms of the ratio a_T of the time constant (*relaxation time*) of a particular response, τ, at temperature T to its value τ_0 at a convenient reference temperature T^0. For many cases, including most nonpolar amorphous polymers, a_T does not vary with τ, so that changes in temperature shift the distribution of relaxation times, representing all possible molecular responses of the system, to smaller or greater values of τ but do not otherwise alter it. Time and temperature affect viscoelasticity only through the product of a_T and actual time, and a_T is called a *shift factor*. Its application to experimental data is described in the following section.

Despite the marked dependence on molecular structure of the relation between a_T and absolute temperature, nearly general empirical relations have been derived by expressing the temperature for each material in term of its glass transition temperature T_g or some nearly equivalent reference temperature. Among the most successful of these relations is the William Landel-Ferry (WLF) equation (Williams 1955):

$$\log a_T = \frac{-17.44(T - T_g)}{51.6 + T - T_g} \tag{6-2:}$$

This equation holds over the temperature range from T_g to about $T_g + 100$ K The constants are related to the free volume as described in Section D.

Experimental metho

Stress relaxation   If elongation is stopped during the determination of th stress-strain curve of a polymer (Chapter 4G), the force or stress decrease with time as the specimen approaches equilibrium or quasi-equilibrium unde the imposed strain. The direct observation and measurement of this pheno menon constitutes the *stress relaxation* experiment. Usually the sample i deformed rapidly to a specified strain, and stress at this strain is observe for periods ranging from several minutes to several days or longer.

It has been shown both experimentally and theoretically that the stres relaxation behavior of rubbery polymers can be factored into independer functions of strain and time. At small strains, the stress-strain function i almost linear and can be represented by a time-dependent modulus o elasticity $G(t)$.

Typical stress relaxation data for a well-studied sample of polyisobu tylene are shown in Fig. 6-8, where $G(t)$ is plotted against time for experiment performed at several temperatures. By means of time-temperature equiva lence these data may be shifted to produce a "master curve," as shown i Fig. 6-9. The curve indicates that any specified modulus can be observe after a period that depends, through the factor a_T, on temperature. As tem perature is increased, a given modulus is observed at shorter times.

Master curves such as that of Fig. 6-9 show the different types of visco elastic behavior usually observed with amorphous materials. At small time the high modulus and low slope are characteristic of glassy behavior. The following region of rapidly decreasing modulus represents the glass transition described further in Section D. This is followed by a flat region ot rubber behavior produced either through entanglements among relatively lon molecules or through permanent crosslinks. The ultimate slope at long time represents the region of viscous flow in uncrosslinked polymers.

Creep   Creep is studied by subjecting a sample rapidly to a constant stres and observing the resulting time-dependent strain for relatively long period

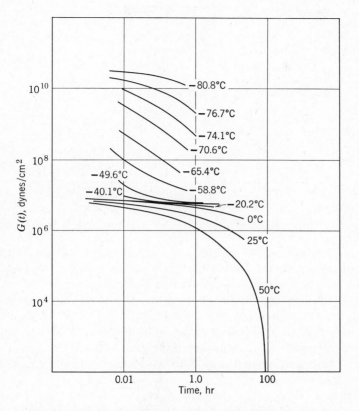

Fig. 6-8.   Stress relaxation of polyisobutylene (Tobolsky 1956).

of time, frequently for a week or more or even for a year or more. Creep and stress relaxation are complementary aspects of plastics behavior and in many cases may provide equivalent information for studies of both fundamental viscoelastic properties and performance in practical applications. Creep experiments are usually easier, more economical, and more feasible for long periods of time.

Dynamic methods The delayed reaction of a polymer to stress and strain also affects its dynamic properties. If a simple harmonic stress of angular frequency ω is applied to the sample, the strain lags behind the stress by a phase angle whose tangent measures the *internal friction* $\Delta E/E$, where ΔE is the energy dissipated in taking the sample through a stress cycle, and E is the energy stored in the sample when the strain is a maximum. The internal friction is a maximum when the dynamic modulus of the material (the ratio

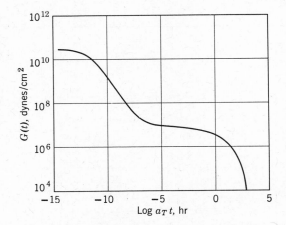

Fig. 6-9.   Stress relaxation master curve for polyisobutylene (Tobolsky 1956).

of the stress to that part of the strain which is in phase with the stress) is in a region of relaxation. In terms of frequency the relaxation behavior of the dynamic modulus and the internal friction is illustrated in Fig. 6-10.

At low frequencies (about 1 sec^{-1}) it is convenient to measure the internal friction of polymers by observing the free decay of torsional oscillations of a thin fiber or film loaded with a large moment of inertia (*torsion pendulum*). The internal friction is the logarithm of the ratio of the amplitudes of successive free oscillations, while the modulus is calculated from the frequency of the oscillations.

A natural extension of this technique is to force oscillation with an external driving force. In such techniques the specimen is driven in flexural, torsional, or longitudinal oscillations by an oscillating force of constant amplitude but variable frequency, and the displacement amplitude of the resulting oscillation is observed. These techniques are particularly useful in the kilocycle frequency range. At very high frequencies (above 1 megacycle), the dimensions of the sample become awkwardly small for mechanical techniques, and it is convenient to measure the attenuation of sound waves in the sample. These methods are discussed fully by Ferry (1970).

Models of viscoelastic behavior

Before attempting to devise a model to duplicate the viscoelastic behavior of an actual polymer, it is well to examine the response γ of ideal systems to a stress s. An ideal elastic element is represented by a spring which obeys Hooke's law, with a modulus of elasticity G. The elastic deformation is instantaneous and independent of time: $\gamma = (1/G)s$ (Fig. 6-11a). A completely viscous response is that of a Newtonian fluid, whose deformation is linear

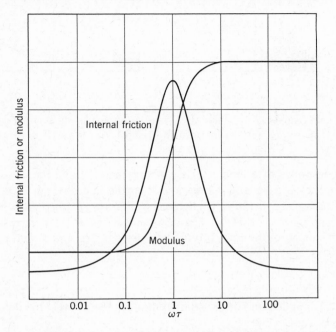

Fig. 6-10. Variation of dynamic modulus and internal friction with frequency.

with time while the stress is applied and is completely irrecoverable: $d\gamma/dt = (1/\eta)s$, where η is the viscosity of the fluid (Fig. 6-11b). A simple mechanical analogy of a Newtonian fluid is a dashpot.

The two elements of spring and dashpot can be combined in two ways. If they are placed in series, the resulting *Maxwell element* (Fig. 6-12a) exhibits flow plus elasticity on the application of stress (Fig. 6-11c): when the stress is applied, the spring elongates while the dashpot slowly yields. On the removal of the stress the spring recovers but the dashpot does not. The strain is given by the equation

$$\frac{d\gamma}{dt} = \frac{1}{\eta}s + \frac{1}{G}\frac{ds}{dt} \tag{6-23}$$

The relation of creep to stress relaxation may be seen by considering the experiment in which a strain is obtained and then held by fixing the ends of the system: $d\gamma/dt = 0$ in Eq. 6-23. The equation can then be solved:

$$s = s_0 e^{-(G/\eta)t} = s_0 e^{-t/\tau} \tag{6-24}$$

where the stress s relaxes from its initial value s_0 exponentially as a function of time. The time η/G after which the stress reaches $1/e$ of its initial value is the relaxation time τ.

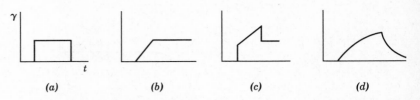

Fig. 6-11.   Strain-time relationships at constant stress for simple models (Alfrey 1948
(a) ideal elastic spring; (b) Newtonian fluid (dashpot); (c) Maxwell element; (d) Voig
element.

The response of a tangled mass of polymer chains to a stress is bette
represented by a parallel combination of spring and dashpot in a *Kelvin* c
Voigt element (Fig. 6-12b). This element shows a *retarded elastic* or *visc*
elastic response (Fig. 6-11d). The dashpot acts as a damping resistance t
the establishment of the equilibrium of the spring. The equation for the strai
is

$$\eta \frac{d\gamma}{dt} + G\gamma = s \tag{6-2}$$

If a stress is applied and after a time removed (Fig. 6-11d), the deformation
time curve is given by:

$$\gamma = \frac{s}{G}(1 - e^{-(G/\eta)t}) = \frac{s}{G}(1 - e^{-t/\tau}) \tag{6-2}$$

where τ is a *retardation time*. When the stress is removed the sample return
to its original shape along the exponential curve:

$$\gamma = \gamma_0 e^{-t/\tau} \tag{6-2}$$

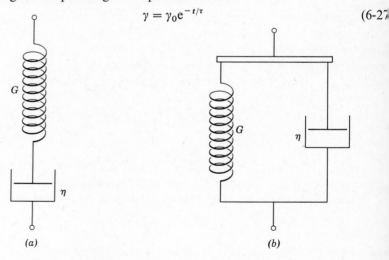

Fig. 6-12.   A Maxwell element (a), and a Voigt element (b).

eneral mechanical models for an amorphous polymer   The three components
hich make up the simplest behavior of an actual polymer sample in creep
an be represented by a mechanical model which combines a Maxwell and a
oigt element in series. If a stress is suddenly applied to this model, the strain
nanges with time as shown in Fig. 6-13; the corresponding behavior of the
nodel is shown in Fig. 6-14.

The model departs from the initial conditions (*a*) at time t_1 by an elastic
eformation s/G_1 (*b*). A viscoelastic response approaching s/G_2 as an equili-
rium value and a viscous flow at the rate s/η_3 follow (*c*). On the removal of
ne stress at time t_2, the elastic element relaxes immediately (*d*) and the
scoelastic one slowly (*e*), but the viscous flow is never recovered.

Alternatively, the creep and stress relaxation experiments can be
escribed with a generalized model consisting of a Maxwell and a Voigt
ement arranged in parallel. The two generalized models are entirely equiv-
ent.

Although these models exhibit the chief characteristics of the viscoelastic
ehavior of polymers, they are nevertheless very much oversimplified. The
ow of the polymer is probably not Newtonian, and its elastic response may
ot be Hookean. Moreover, the behavior of a real polymer cannot be
naracterized by a single relaxation time, but requires a spectrum of relaxation
mes to account for all phases of its behavior.

Treatment of experimental data

istribution of relaxation times   Equation 6-26 indicates that most of the
elaxation associated with a single element takes place within one cycle of
og time. In contrast, experiments indicate (as in Fig. 6-9) that relaxation
henomena in polymers extend over much wider ranges of time. Thus, the
ctual behavior of polymers can be described only in terms of a distribution
f model elements and an associated distribution of relaxation or retardation
mes:

$$\gamma(t) = s \int_{-\infty}^{\infty} \bar{J} \, (\log \tau)(1 - e^{-t/\tau}) \, d \log \tau \qquad (6\text{-}28)$$

here $J = 1/G$ is the *elastic compliance*, and $\bar{J}$ (log τ) the distribution of
etardation times.

In principle, knowledge of $\gamma(t)$ over the entire range of time allows
(log τ) to be evaluated explicitly. This is difficult in practice, however, and
is convenient for mathematical simplicity to adopt simple empirical forms
f the distribution function $\bar{J}$ (log τ) and evaluate their parameters from the
nore limited experimental data at hand. Alternatively, the experimental data
nay be fitted graphically by summing the theoretical curves corresponding
a small number (three or four) of discrete relaxation times. The usefulness

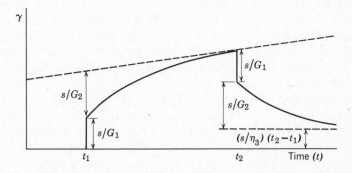

Fig. 6-13. Strain-time relationship for a generalized mechanical model for creep combinir elasticity, viscoelasticity, and flow (Alfrey 1948).

of the distribution of relaxation times arises from its identification wit certain molecular parameters of the specimen, as described below.

Molecular structure and viscoelastici

The phenomena of relaxation processes can be thought of in terms of th effect of thermal motion on the orientation of polymer molecules. When mechanical stress is applied to a polymer, introducing deformations of th chains, the entropy of the system decreases as less probable conformatior are taken up. The free energy correspondingly increases. If the sample kept in the deformed state, stress relaxation takes place as a result of th thermal motions of the chains, the molecular deformations are obliterate(

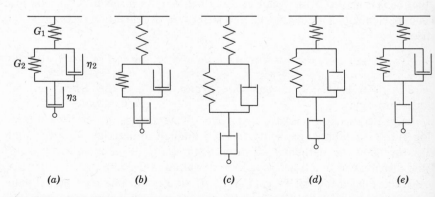

Fig. 6-14. Interpretation of strain-time curve of Fig. 6-13 in terms of the generalize mechanical model for creep.

ᴉd the excess free energy is dissipated as heat. The details of the stress ᴉlaxation process depend upon the multiplicity of ways in which the polymer olecules can regain their most probable conformations through thermal otion. These complex motions of a polymer molecule can be expressed as series of characteristic modes requiring various degrees of long-range ᴉooperation among the segments of the chain. Thus the first mode corre- ᴉonds to translation of the entire molecule, requiring maximum cooperation, ᴉe second corresponds to motion of the ends of the chain in opposite direc- ᴉons, requiring somewhat less cooperation, and so on. With each of these ᴉodes is associated a characteristic relaxation time; there are so many modes ᴉat over most of the time scale the discrete spectrum of relaxation times can ᴉ approximated by a continuous distribution.

Theories based on the above considerations, derived for dilute solutions ᴉ polymer molecules, have been combined (Ferry 1970) with the concepts of ᴉtanglement of long molecules successful in predicting the molecular weight ᴉpendence of viscous flow (Section A). Although exact numerical agreement not achieved, e.g., in calculating the dynamic modulus of polyisobutylene ᴉver a wide range of frequency and molecular weight, the qualitative situation ᴉems extremely satisfactory.

In the rubbery region, the maximum relaxation time is strongly dependent ᴉn molecular weight. In this region, the motions of the molecules are long ᴉnge in nature, involving motions of units of the order of the length of the ᴉolecule itself. This is the region where entanglements are important.

In the region around the glass transition, however, only vibrations of the ᴉarts of the molecule between entanglements are important. So long as the ᴉolecules are long enough for entanglements to exist, this portion of the ᴉrve is expected, and found, to be essentially independent of the molecular ᴉeight and to depend primarily upon the local structure of the polymer.

ᴉENERAL REFERENCES

ᴉfrey, 1948, 1956, 1967; Ferry 1958, 1970; Leaderman 1958; Tobolsky 1958, 1960; Van Wazer 1963; Hilton 1964; Passaglia 1964; Thorkildsen 1964; Meares 1965, Chaps. 9, 11; Vincent 1965; Reiner 1966; McCrum 1967; Payne 1968.

ᴉ. The Glassy State and the Glass Transition

As anticipated in the previous sections, and in particular in Fig. 6-9, all ᴉmorphous polymers assume at sufficiently low temperatures the characteris- ᴉcs of glasses, including hardness, stiffness, and brittleness. One property associated with the glassy state is a low volume coefficient of expansion. Thi low coefficient occurs as the result of a change in slope of the curve of volum vs. temperature at the point called the *glass transition temperature* T_g. Thi behavior is shown for natural rubber in Fig. 6-15. In the high-temperatur region, the slope of the curve (expansion coefficient) is characteristic of rubber; below T_g at about $-70°C$, it is characteristic of a glass.

Figure 6-15 illustrates another general phenomenon: the amorphou regions in partially crystalline polymers also assume a glassy state, T_g bein independent of degree of crystallinity to a first approximation. The magn tude of the phenomena associated with T_g decreases with decreasing amor phous content, however. As a result, T_g is sometimes difficult to detect i highly crystalline polymers. In terms of the lamellar model (Chapter 5*D* the glass transition is considered to involve defect regions within or at th boundaries of the lamellae.

In contrast to crystalline melting at a temperature T_m (about $+10°C$ i Fig. 6-15), there is not an abrupt change in volume at T_g, but only a change i the slope of the volume-temperature curve. In analogy to thermodynami first- and second-order transitions, T_g is sometimes referred to as a second order or apparent second-order transition. This nomenclature is considere poor, however, since it implies more thermodynamic significance than th nature of the transition warrants.

Measurement of T

The glass transition temperature can be detected in a variety of experi ments, which can be roughly classified into those dealing with bulk propertie of the polymer, and those measuring the nature and extent of molecula

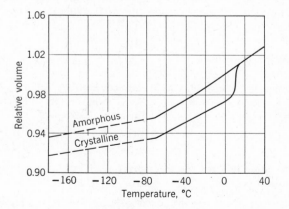

Fig. 6-15. Volume behavior of rubber near the glass transition temperature (Bekkedah 1934).

motion. The classification is to some extent arbitrary, since, as indicated below, T_g is in fact the temperature of onset of extensive molecular motion.

Bulk properties Perhaps the most common way of estimating T_g is by means of the volume expansion coefficient, as indicated above. Other bulk properties whose temperature coefficients undergo marked changes at T_g, and which can therefore be used for its determination, include heat content (Chapter 4*E*), refractive index, stiffness, and hardness.

Molecular motion Experiments which are sensitive to the onset of molecular motion in polymer chains may be used to detect the glass transition. Such methods include the measurement of internal friction (Section *C*), dielectric loss in polar polymers, and NMR spectroscopy (Chapter 4*D*).

Phenomena related to T_g The onset of brittleness, as measured in impact tests, and the softening of amorphous polymers, as measured in thermal tests of various sorts, take place at temperatures near T_g (see Chapter 4*G*).

Time effects near T_g If a polymer sample is cooled rapidly to a temperature just below T_g, its volume continues to decrease for many hours. In consequence, the value observed as T_g in a volume-temperature experiment depends on the time scale of the measurements. It is convenient to define T_g in terms of an arbitrary convenient (but not highly critical) time interval, such as 10 min to 1 hr.

For similar reasons, other tests for T_g give results somewhat dependent on the time scale of the experiments, with tests requiring shorter times yielding higher values for T_g. The brittle temperature as determined in an impact test is, e.g., normally somewhat higher than T_g as otherwise measured.

Molecular interpretation of T_g

In the glassy state, large-scale molecular motion does not take place. Rather, atoms and small groups of atoms move against the local restraints of secondary bond forces, much as atoms vibrate around their equilibrium positions in a crystal lattice, except that the glassy state does not have the regularity of the crystalline state. The glass transition corresponds to the onset of liquidlike motion of much longer segments of molecules, characteristic of the rubbery state. This motion requires more free volume than the short-range excursions of atoms in the glassy state. The rise in the relative free volume with increasing temperature above T_g leads to the higher observed volume expansion coefficient in this region.

A theoretical examination (Gibbs 1956, 1958; DiMarzio 1958, 1959) of the temperature dependence of the number of conformations available for the polymer molecule leads to the following line of reasoning. Since the fully extended chain is the conformation of minimum energy (Chapter 5B), it tends to be assumed more frequently as the temperature is lowered. As the molecules thus straighten out, the free volume decreases. In consequence, flow becomes more difficult. The glass transition (observed at infinite time)—or, alternatively, the onset of crystallization where possible—is taken as the point where the number of possible conformations of the amorphous phase decreases sharply toward one.

The fraction f of "free" volume may be defined as

$$\begin{aligned} f &= f_g + (T - T_g)\Delta\alpha & T \geq T_g \\ f &= f_g & T < T_g \end{aligned} \tag{6-29}$$

Thus f is constant at the value f_g for all temperatures below T_g. Here the volume expansion coefficient α is that resulting from the increase in amplitude of molecular vibrations with temperature. Above T_g new free volume is created as the result of an increase $\Delta\alpha$ in the expansion coefficient.

Williams, Landel, and Ferry (Williams 1955) proposed that log viscosity varies linearly with $1/f$ above T_g, so that

$$ln\left(\frac{\eta}{\eta_g}\right) = \frac{1}{f} - \frac{1}{f_g} \tag{6-30}$$

Substitution into Eq. 6-29 leads to

$$\log\left(\frac{\eta}{\eta_g}\right) = -\frac{a(T - T_g)}{b + T - T_g} \tag{6-31}$$

which is the WLF equation presented in Section C (Eq. 6-22), the numerical constants for a and b given there being determined by fitting literature data on the viscosity-temperature behavior of many glass-forming substances. The shift factor a_T is seen to be just the ratio of the viscosity at T relative to that at T_g. The latter is about 10^{13} poise for many substances.

Equation 6-31 also implies that both the viscosity of the polymer and the activation energy for viscous flow $\Delta E = 2.3R\ d(\log\eta)/d(1/T)$ should become infinite at $T = T_g - b = T_g - 51.6$. Thus by extrapolating downwards from behavior well above T_g one would predict that all molecular motion should become completely frozen at $T < T_g - 51.6$. What happens, of course, is that new mechanisms of deformation take over more or less sharply as this critical range is approached, in fact at T_g. That such a change is to be expected on statistical-mechanical grounds is the major result of the Gibbs-DiMarzio theory.

Molecular motion below T_g

The foregoing discussion is concerned with the transition involving the motion of long segments of the polymer chain. At lower temperatures, other transitions may occur, produced by the motion of short sections of the main chain or of side chains. Although some characteristics of the glassy state, such as brittleness, may occasionally occur only below one of these lower transitions, it is proposed that the transition of highest temperature be called T_g. Alternatively, the transitions may be denoted α, β, γ, etc., in order of descending temperature.

Transitions due to the motion of short segments of the main polymer chain occur most prominently in crystalline polymers such as polyethylene, polypropylene, and polytetrafluoroethylene. Such polymers also typically exhibit an α-transition.

The internal friction results for isotactic polypropylene (Fig. 6-16) are typical. The onset of brittleness in this polymer is accounted for by the β-transition occurring at 0°C at the torsion pendulum frequency of 1 cycle/sec but shifted to room temperature or higher at the frequencies prevalent in impact tests. In linear polyethylene, however, the β-transition appears as a shoulder on the large α-transition; the associated decrease in modulus is spread out over a wider temperature range because of the combined effects of the two phenomena. The onset of brittleness occurs in the temperature range of the γ-transition, at about -120°C in the torsion pendulum experiment.

Side-chain transitions occur in methacrylate polymers as a result of the relaxation of the carbomethoxy side chain at about 20°C (torsion pendulum) and the relaxation of the aliphatic ester group below -150°C.

GENERAL REFERENCES

Boyer 1963; Gordon 1965; Meares 1965, Chap. 10; Vincent 1965; Shen 1966; McCrum 1967; Kaelble 1969; Eisenberg 1970.

E. The Mechanical Properties of Crystalline Polymers

The models developed in the preceding sections represent well the rheological and mechanical properties of amorphous polymers. The viscoelastic properties of crystalline polymers are much more complex, however, and are not amenable to adequate theoretical explanation for three reasons.

First, an amorphous polymer is isotropic. This means that models suitable for describing shear stress, e.g., are adequate to describe tensile stress or other types. Since crystalline polymers are not isotropic, this universality does not hold and the range of application of any model is severely limited.

Second, the homogeneous nature of amorphous polymers ensures that an applied stress is distributed uniformly throughout the system, at least down to very small dimensions. In crystalline polymers the relatively large

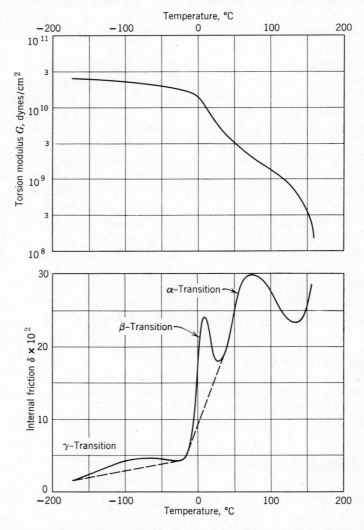

Fig. 6-16. Dependence on temperature of the internal friction and torsion modulus of isotactic polypropylene (Muus 1959).

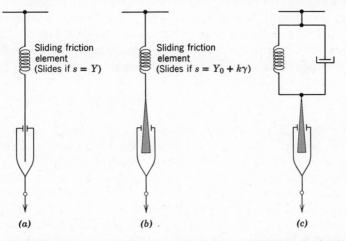

Fig. 6-17. Mechanical models incorporating yield points: (*a*), ideal elastic-plastic flow model; (*b*), same with strain hardening; (*c*), viscoelastic model with plastic flow and strain hardening (Alfrey 1967).

crystallites are bound together in such a way that large stress concentrations inevitably develop.

Finally, a crystalline polymer is a mixture of regions of different degrees of order ranging all the way from completely ordered crystallites to completely amorphous regions. As the stress on the sample changes, the amounts of these regions change continuously as the crystallites melt or grow. This change of composition with respect to ordering is the most difficult obstacle to overcome in formulating a theory of the mechanical behavior of crystalline polymers. Even in the simplest cases the necessity of having the mechanical model change continuously with the applied stress has led to serious difficulties.

In consequence, neither the Boltzmann superposition principle nor time-temperature equivalence apply to crystalline polymers. Without these simplifying principles, attempts to explain the viscoelastic response of crystalline polymers in terms of models become complex and are only qualitative. With these limitations, however, it is instructive to consider mechanical models containing elements which exhibit yield stresses, typical of the actual behavior of crystalline polymers. Several such models are shown in Fig. 6-17, and the stress-strain curves corresponding to them are depicted in Fig. 6-18.

Figures 6-17*a* and 6-18*a* contain a sliding friction element for which only elastic deformation occurs at stresses below the yield value *Y*, whereas plastic deformation occurs if this stress is exceeded. Figures 6-17*b* and 6-18*b* represent the case of *work hardening* or *strain hardening*, in which the yield stress increases with the plastic strain. These features are combined with viscoelastic rather than pure elastic response in Figs. 6-17*c* and 6-18*c*.

All three models exhibit permanent set following loading-unloading cycles which reach stresses where plastic deformation occurs. Viscoelastic plastics exhibiting work hardening display the behavior shown in Fig. 6-18*d* on successive cycles of loading and unloading.

Typical behavior

An orderly, if qualitative, discussion of the mechanical properties of crystalline polymers requires their classification into several categories, as indicated in Table 6-2. It is in the range of intermediate degree of crystallinity that the properties, unique to polymers, of most importance for mechanical and engineering applications are found. Further, these properties are found for the most part in the temperature range between T_g and T_m, and at temperatures not far below T_g. Well below the glass transition, molecular motion is essentially absent and the material behaves as a hard, glassy solid with the presence or absence of crystallinity making little difference: for example, the properties of atactic and isotactic polystyrene are quite similar at room temperature. Above T_m, of course, crystallinity plays no part in the properties of the amorphous viscoelastic melt.

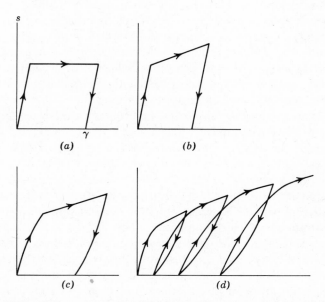

Fig. 6-18.   Stress-strain curves for the idealized materials of Fig. 6-17 (Alfrey 1967). Curve (*d*) shows the result of several stress-strain cycles on the model of Fig. 6-17(*c*).

TABLE 6-2. *Classification of crystalline polymers*

Predominant Properties in Temperature Range	Degree of Crystallinity		
	Low (5–10%)	Intermediate (20–60%)	High (70–90%)
Above T_g	Rubbery	Leathery, tough	Stiff, hard (brittle)
Below T_g	Glassy, brittle	Hornlike, tough	Stiff, hard brittle

Polymers with low crystallinity include plasticized poly(vinyl chloride) and elastic polyamides. These materials behave like lightly crosslinked amorphous polymers, their crystalline regions acting like crosslinks which are stable with respect to time but unstable with respect to temperature. The viscoelastic properties of these polymers are much like those of amorphous polymers except that the transition region between glassy and rubbery or liquid behavior is very much broadened on the temperature scale.

At very low extensions ($<1\%$), at temperatures well below T_m, and at not too long times, polymers with intermediate degrees of crystallinity (such as low-density polyethylene) behave much like those with very low crystallinity. Time-temperature equivalence is not applicable to such polymers. The transition regions of these polymers in modulus and the corresponding distributions of relaxation times are exceptionally broad, as indicated by the master curves of Fig. 6-19. (It should be pointed out that these curves were themselves derived by application of time-temperature equivalence and must be considered only as idealizations.)

At higher extensions, these polymers exhibit the phenomena of a yield stress and cold drawing, with the accompanying changes in crystalline morphology described in Chapter 5G. No adequate theory relating rheological and mechanical behavior in this region to structure has yet been formulated. Polymers of intermediate degree of crystallinity are characteristically leathery or horny in texture, and exhibit good impact resistance which in many cases is retained even below T_g. The exact structural features responsible for this toughness have not been well defined.

The major effects of a further increase in crystallinity to very high values include (a) a further increase in modulus, as the high modulus characteristic of the crystalline regions is approached; and (b) the onset of a tendency toward brittleness, in the sense of failure at low strains. Such polymers, of which high-density polyethylene is typical, can be cold drawn only with difficulty. Tensile failure usually occurs at or slightly beyond the yield stress, accompanied by distortion or deformation which appears to occur at slip boundaries or dislocations, reminiscent of the viscoelastic behavior of metals.

Crystallization on stressing   The application of a mechanical stress to a noncrystalline but crystallizable polymer can cause crystallinity to develop, either by raising T_m or by increasing the rate of crystallization. An example of the former effect is the crystallization of natural rubber on stretching described in Section *B*. At room temperature, unstretched natural rubber is above its crystalline melting point. As a tensile stress is applied, T_m is raised

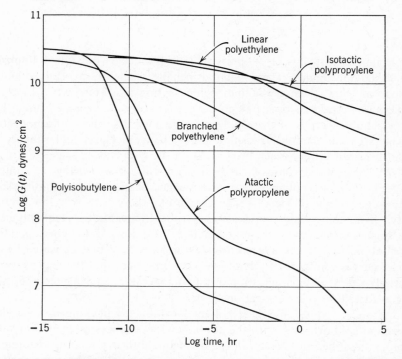

Fig. 6-19.   Stress relaxation master curves for several crystalline and amorphous polymers (Tobolsky 1960).

and the rubber crystallizes to an oriented structure. When the stress is released, T_m is reduced and the rubber melts as it retracts.

An example of the effect of stress on crystallization rate is seen in polymers such as poly(ethylene terephthalate) which can be quenched to a metastable amorphous state at temperatures well below T_m. Without an applied stress, the rate of nucleation and crystal growth is very low. When a tensile stress is applied, T_m is raised and the rate of crystallization is greatly increased so that crystallization takes place during stretching. These crystals do not melt when the stress is removed, of course.

GENERAL REFERENCES

Alfrey 1948, 1967; Hopkins 1960; Tobolsky 1960; Thorkildsen 1964; Vincent 1965; McCrum 1967; Stein 1969; Woodward 1969; Ferry 1970.

BIBLIOGRAPHY

Alfrey 1948. Turner Alfrey, Jr., *Mechanical Behavior of High Polymers*, Interscience Publishers, New York, 1948.

Alfrey 1956. Turner Alfrey, Jr., and E. F. Gurnee, "Dynamics of Viscoelastic Behavior," Chap. 11 in Frederick R. Eirich, ed., *Rheology—Theory and Applications*, Vol. 1, Academic Press, New York, 1956.

Alfrey 1967. Turner Alfrey, Jr., and Edward F. Gurnee, *Organic Polymers*, Prentice-Hall, Inc., Englewood Cliffs, N.J., 1967.

Bauer 1967. Walter H. Bauer and Edward A. Collins, "Thixotropy and Dilatancy," Chap. 8 in Frederick R. Eirich, ed., *Rheology—Theory and Applications*, Vol. 4, Academic Press, New York, 1967.

Bekkedahl 1934. Norman Bekkedahl, "Forms of Rubber as Indicated by Temperature-Volume Relationship," *J. Res. Natl. Bur. Stds.* **13**, 411–431 (1934).

Boyer 1963. Raymond F. Boyer, "The Relation of Transition Temperatures to Chemical Structure in High Polymers," *Rubber Chem. Tech.* **36**, 1303–1421 (1963).

DiMarzio 1958. E. A. DiMarzio and J. H. Gibbs, "Chain Stiffness and the Lattice Theory of Polymer Phases," *J. Chem. Phys.* **28**, 807–813 (1958).

DiMarzio 1959. E. A. DiMarzio and J. H. Gibbs, "Glass Temperature of Copolymers," *J. Polymer Sci.* **40**, 121–131 (1959).

Eirich 1956–1969. Frederick R. Eirich, ed., *Rheology—Theory and Applications*, Academic Press, New York, Vol. 1, 1956; Vol. 2, 1958; Vol. 3, 1960; Vol. 4, 1967; Vol. 5, 1969.

Eisenberg 1970. A. Eisenberg and M. Shen, "Recent Advances in Glass Transitions in Polymers," *Rubber Chem. Tech.* **43**, 156–170 (1970).

Ferry 1958. John D. Ferry, "Experimental Techniques for Rheological Measurements on Viscoelastic Bodies," Chap. 11 in Frederick R. Eirich, ed., *Rheology—Theory and Applications*, Academic Press, New York, Vol. 2, 1958.

Ferry 1970. John D. Ferry, *Viscoelastic Properties of Polymers*, 2nd ed., John Wiley and Sons, New York, 1970.

Flory 1940. Paul J. Flory, "Viscosities of Linear Polyesters. An Exact Relationship between Viscosity and Chain Length," *J. Am. Chem. Soc.* **62**, 1057–1070 (1940).

Flory 1944. Paul J. Flory, "Network Structure and the Elastic Properties of Vulcanized Rubber," *Chem. Revs.* **35**, 51–75 (1944).

Flory 1953. Paul J. Flory, *Principles of Polymer Chemistry*, Cornell University Press, Ithaca, N.Y., 1953.

Fox 1956. T. G Fox, Serge Gratch, and S. Loshaek, "Viscosity Relationships for Polymers in Bulk and in Concentrated Solution," Chap. 12 in Frederick R. Eirich, ed., *Rheology—Theory and Applications*, Academic Press, New York, Vol. 1, 1956.

Gibbs 1956. Julian H. Gibbs, "Nature of the Glass Transition in Polymers," *J. Chem. Phys.* **25**, 185–186 (1956).

Gibbs 1958. Julian H. Gibbs and Edmund A. DiMarzio, "Nature of the Glass Transition and the Glassy State," *J. Chem. Phys.* **28**, 373–383 (1958).

Glasstone 1941. Samuel Glasstone, Keith J. Laidler, and Henry Eyring, *The Theory of Rate Processes*, McGraw-Hill Book Co., New York, 1941.

Gordon 1965. M. Gordon, "Thermal Properties of High Polymers," Chap. 4 in P. D. Ritchie, ed., *Physics of Plastics*, D. Van Nostrand Co., Princeton, N.J., 1965.

Guth 1946. E. Guth, H. M. James, and H. Mark, "The Kinetic Theory of Rubber Elasticity," pp. 253–299 in H. Mark and G. S. Whitby, eds., *Scientific Progress in the Field of Rubber and Synthetic Elastomers (Advances in Colloid Science*, Vol. II), Interscience Publishers, New York, 1946.

Hilton 1964. Harry H. Hilton, "Viscoelastic Analysis," Chap. 4 in Eric Baer, ed., *Engineering Design for Plastics*, Reinhold Publishing Corp., New York, 1964.

Hopkins 1960. I. L. Hopkins and W. O. Baker, "The Deformation of Crystalline and Crosslinked Polymers," Chap. 10 in Frederick R. Eirich, ed., *Rheology—Theory and Applications*, Academic Press, New York, Vol. 3, 1960.

Kaelble 1969. D. H. Kaelble, "Free Volume and Polymer Rheology," Chap. 5 in Frederick R. Eirich, ed., *Rheology—Theory and Applications*, Academic Press, New York, Vol. 5, 1969.

Leaderman 1958. Herbert Leaderman, "Viscoelasticity Phenomena in Amorphous High Polymeric Systems," Chap. 1 in Frederick R. Eirich, ed., *Rheology—Theory and Applications*, Vol. 2, Academic Press, New York, 1958.

McCrum 1967. N. G. McCrum, B. Read, and G. Williams, *Anelastic and Dielectric Effects in Polymeric Solids*, John Wiley and Sons, New York, 1967.

Meares 1965. Patrick Meares, *Polymers: Structure and Bulk Properties*, D. Van Nostrand Co., Princeton, N.J., 1965.

Mendelson 1968. Robert A. Mendelson, "Melt Viscosity," pp. 587–620 in Herman F. Mark, Norman G. Gaylord, and Norbert M. Bikales, eds., *Encyclopedia of Polymer Science and Technology*, Vol. 8, Interscience Div., John Wiley and Sons, New York, 1968.

Muus 1959. Laurits T. Muus, N. Gerald McCrum, and Frank C. McGrew, "The Relationship of Physical Properties to Structure in Linear Polymers of Ethylene and Propylene," *SPE J.* **15**, 368–372 (1959).

Passaglia 1964. E. Passaglia and J. R. Knox, "Viscoelastic Behavior and Time-Temperature Relationships," Chap. 3 in Eric Baer, ed., *Engineering Design for Plastics*, Reinhold Publishing Corp., New York, 1964.

Payne 1968. A. R. Payne, "Physics and Physical Testing of Polymers," pp. 1–93 in J. C. Robb and F. W. Peaker, eds., *Progress in High Polymers*, Vol. 2, CRC Press, Cleveland, Ohio, 1968.

Reiner 1966. M. Reiner, "Deformation," pp. 620–647 in Herman F. Mark, Norman G. Gaylord, and Norbert M. Bikales, eds., *Encyclopedia of Polymer Science and Technology*, Vol. 4, Interscience Div., John Wiley and Sons, New York, 1966.

Severs 1962. Edward T. Severs, *Rheology of Polymers*, Reinhold Publishing Corp., New York, 1962.

Shen 1966. Mitchel C. Shen and Adi Eisenberg, "Glass Transitions in Polymers," Chap. 9 in H. Reiss, ed., *Progress in Solid State Chemistry*, Vol. 3, Pergamon Press, New York, 1966 [reprinted in *Rubber Chem. Tech.* **43**, 95–155 (1970)].

Stein 1969. Richard S. Stein, "Studies of the Deformation of Crystalline Polymers," Chap. 6 in Frederick R. Eirich, ed., *Rheology—Theory and Applications*, Vol. 5, Academic Press, New York, 1969.

Thorkildsen 1964. R. L. Thorkildsen, "Mechanical Behavior," Chap. 5 in Eric Baer, ed., *Engineering Design for Plastics*, Reinhold Publishing Corp., New York, 1964.

Tobolsky 1956. Arthur V. Tobolsky and Ephriam Catsiff, "Elastoviscous Properties of Polyisobutylene (and Other Amorphous Polymers) from Stress-Relaxation Studies. IX. A Summary of Results," *J. Polymer Sci.* **19**, 111–121 (1956).

Tobolsky 1958. Arthur V. Tobolsky, "Stress Relaxation Studies of the Viscoelastic Properties of Polymers," Chap. 2 in Frederick R. Eirich, ed., *Rheology—Theory and Applications*, Vol. 2, Academic Press, New York, 1958.

Tobolsky 1960. Arthur V. Tobolsky, *Properties and Structure of Polymers*, John Wiley and Sons, New York, 1960.

Tordella 1969. John P. Tordella, "Unstable Flow of Molten Polymers," Chap. 2 in Frederick R. Eirich, ed., *Rheology—Theory and Applications*, Vol. 5, Academic Press, New York, 1969.

Treloar 1958. L. R. G. Treloar, *The Physics of Rubber Elasticity*, 2nd ed., Clarendon Press, Oxford, 1958.

Van Wazer 1963. J. R. Van Wazer, J. W. Lyons, K. Y. Kim, and R. E. Colwell, *Viscosity and Flow Measurement, A Laboratory Handbook of Rheology*, Interscience Div., John Wiley and Sons, New York, 1963.

Vincent 1965. P. I. Vincent, "Mechanical Properties of Polymers: Deformation," Chap. 2 in P. D. Ritchie, ed., *Physics of Plastics*, D. Van Nostrand Co., Princeton, N.J., 1965.

Williams 1955. Malcolm L. Williams, Robert F. Landel, and John D. Ferry, "The Temperature Dependence of Relaxation Mechanisms in Amorphous Polymers and Other Glass-Forming Liquids," *J. Am. Chem. Soc.* **77**, 3701–3707 (1955).

Woodward 1969. A. E. Woodward, "Deformation and Dissipative Processes in High Polymeric Solids," Chap. 7 in Frederick R. Eirich, ed., *Rheology—Theory and Applications*, Vol. 5, Academic Press, New York, 1969.

7

Polymer Structure and Physical Properties

Although interrelations between the molecular structure of polymers and their properties are inferred throughout this book, they are emphasized in this chapter. First the structural features of polymers most directly responsible for determining their properties are classified, and in succeeding sections we describe how these structures influence various classes of physical properties. Finally, the property requirements associated with the uses of various polymers are reviewed.

Since the acceptance of the macromolecular hypothesis in the 1920's it has been recognized that the unique properties of polymers—e.g., the elasticity and abrasion resistance of rubbers, the strength and toughness of fibers, and the flexibility and clarity of films—must be attributed to their long-chain structure. In the more recent examination of structure-property relationships it has been advantageous to classify properties into those involving large and small deformations. The former class includes such properties as tensile strength and phenomena observed in the melt, while properties involving only small deformations include electrical and optical behavior, such mechanical properties as stiffness and yield point, and the glass and crystalline melting transitions.

Properties involving large deformations depend primarily on the long-chain nature of polymers and the gross configuration of their chains. Important factors for this group of properties include molecular weight and its distribution, chain branching and the related category of side-chain substitution, and crosslinking.

Physical properties associated with small deformations are influenced most by factors determining the manner in which chain atoms interact at small distances. The ability of polymers to crystallize, set by considerations of symmetry and steric effects, has major importance here, as do the flexibility of the chain bonds and the number, nature, and spacing of polar groups. To the extent that they influence the achievement of local order, gross configurational properties are also important. Similar considerations apply to amorphous polymers below the glass transition.

In crystalline polymers, the nature of the crystalline state introduces another set of variables influencing mechanical properties. These variables include nature of the crystal structure, degree of crystallinity, size and number of spherulites, and orientation. Some of these phenomena are in turn influenced by the conditions of fabrication of the polymer.

Finally, the properties of polymers can be varied importantly by the addition of other materials, such as plasticizers or reinforcing fillers. Properties involving both large and small deformations may be influenced in this way.

As described in Chapter 1, one of the most important determinants of polymer properties is the location in temperature of the major transitions, the glass transition and the crystalline melting point. It is appropriate therefore first to discuss the relations between molecular structure and these transition temperatures.

A. The Crystalline Melting Point

As has been pointed out in Chapter 5, crystalline melting in polymers is at least a pseudo-equilibrium process, and it is convenient here to describe it in thermodynamic terms, realizing that in a given situation an observed melting point T_m may not be precisely the equilibrium value. Melting takes place when the free energy of the process is zero:

$$\Delta G = \Delta H_m - T_m \Delta S_m = 0$$

Thus, T_m is set by the ratio $\Delta H_m / \Delta S_m$, and it is necessary to explore the effect of molecular structure upon both of these quantities to gain insight into the melting process.

The crystalline melting point is usually taken to be independent of molecular weight in the polymer range. It is assumed that ΔH_m and ΔS_m are made up of molecular weight-independent terms H_0 and S_0 plus increments H_1 and S_1 for each chain unit. Thus, for degree of polymerization x,

$$T_m = \frac{\Delta H_m}{\Delta S_m} = \frac{H_0 + xH_1}{S_0 + xS_1} \to \frac{H_1}{S_1} \quad \text{as} \quad x \to \infty \qquad (7\text{-}1)$$

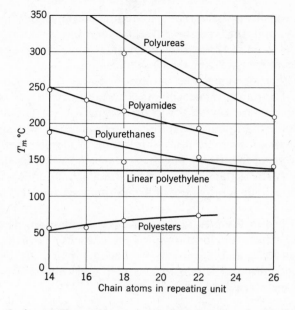

Fig. 7-1.   Trend of crystalline melting points in homologous series of aliphatic polymers (after Hill 1948).

Melting points of homologous series

The melting points of homologous series of various types of polymers are plotted in Fig. 7-1. The effect of the spacing of the polar groups on T_m is much as predicted by considering the polymers as derived from polyethylene by replacing methylenes with polar groups, but the over-all level of T_m for each series remains to be considered.

Polyesters   Although the low melting points of linear aliphatic polyesters were once attributed to unusual flexibility of the C—O chain bond (Bunn 1955), implying high entropies of fusion, both ΔS_m and ΔH_m are significantly lower in these polyesters than in polyethylene (Wunderlich 1958). Only recently have the reasons for these low values been elucidated in detail, as described later in this section.

Polyamides   For polyamides it is observed that ΔH_m is lower than for polyethylene; therefore, molar cohesion cannot account for high values of T_m. These result from low liquid-state entropies, leading to low values of

ΔS_m, which arise from partial retention of hydrogen bonding in the melt and from chain stiffening due to the tendency for resonance of the type

within the amide group.

Similar considerations undoubtedly apply to the melting points of polyurethanes and polyureas, but values of ΔH_m and ΔS_m are not available.

As the spacing between the polar groups is increased, the melting points approach that of polyethylene. When the homologous series are examined in more detail, the melting points are found to vary in a more complex way with the spacing of the polar groups than is suggested in Fig. 7-1. An alternation of T_m with spacing is typical (Fig. 7-2). It results from differences in the crystal structure, which alternates in type for chain repeat units with odd and even numbers of carbon atoms.

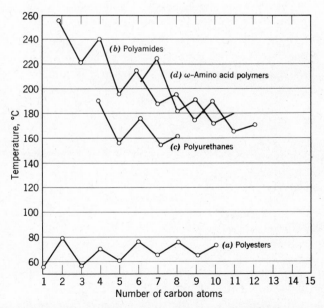

Fig. 7-2. Dependence of crystalline melting point on spacing of polar groups. Number of carbon atoms refers to (a) acid for polyesters made with decamethylene glycol; (b) diamine for polyamides made with sebacic acid; (c) diamine for polyurethanes made with tetramethylene glycol; and (d) ω-amino acid polymers (Bannerman 1956).

Effect of chain flexibility and other steric facto

Chain flexibility The flexibility of chain molecules arises from rotatio around saturated chain bonds. The potential energy barriers hindering thi rotation range from 1 to 5 kcal/mole, the same order of magnitude a molecular cohesion forces. It is not surprising therefore that the flexibilit of polymer chains is an important factor in determining their melting point: Thus polytetrafluoroethylene ($T_m = 327°C$) melts much higher than poly ethylene because of its low entropy of fusion (Starkweather 1960), whic results from the high stiffness of the polymer chains. Similarly, the hig melting point of isotactic polypropylene ($T_m = 165°C$) is attributed to lo entropy of fusion arising from stiffening of the chain in the melt because c the higher energy barrier for rotation about C—C bonds than in polyethylen In neither case can a high heat of fusion account for the high value of T_m since for both polytetrafluoroethylene and polypropylene ΔH_m is well belo that of polyethylene as indicated in Table 5-1 (Dole 1959).

The substitution of an inflexible group like the *p*-phenylene grou

for six chain CH_2 groups causes a marked rise in th

melting point of the polymer (Edgar 1952). Some examples of this change ar shown in Table 7-1. The rise is considerable when the *p*-phenylene group connected to CH_2 groups, but greater still when it is connected to carbony

since the group can resonate as a unit. Among othe

chain-stiffening groups are *p,p'*-diphenyl, 1,5- or 2,6-naphthyl, diketo

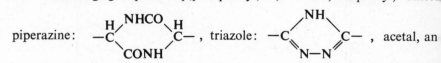

piperazine: , triazole: , acetal, an

thioketal.

Side-chain substitution In most cases, the substitution of nonpolar group for hydrogens of a polymer chain leads to a reduction in T_m or possibl complete loss of cystallinity. If the substitution is random, as in branche polyethylene, the primary effect is a reduction in the size and perfection o the crystalline regions, usually accompanied by a decrease in the degree o crystallinity. The crystalline melting point of polyethylene is lowered 20–25 on going from the linear to the branched material.

TABLE 7-1. *Effect of a p-phenylene group on the melting point of condensation polymers (Edgar 1952)*

Repeating Unit	T_m, °C
$-O(CH_2)_2OCO(CH_2)_6CO-$	45
$-O(CH_2)_2OCO\langle\bigcirc\rangle CO-$	265
$-NH(CH_2)_6NHCO(CH_2)_6CO-$	235
$-NH(CH_2)_6NHCO\langle\bigcirc\rangle CO-$	350*
$-O(CH_2)_8OCO(CH_2)_8CO-$	75†
$-OCH_2\langle\bigcirc\rangle CH_2OCOCH_2\langle\bigcirc\rangle CH_2CO-$	146
$-CH_2CH_2-$	135
$-CH_2\langle\bigcirc\rangle CH_2-$	380

* Decomposes
† Estimated.

Replacement of an amide hydrogen with an alkyl group has a much larger effect, since hydrogen bonding is destroyed. In general, N-methyl nylons melt at least 100° lower than their unsubstituted counterparts.

When an alkyl group is regularly substituted into a methylene chain, with the retention of stereoregularity so that crystallization is possible, two effects compete in setting T_m. As indicated for isotactic poly(α-olefins) in Table 7-2 (with polyethylene omitted because of its widely different crystal structure), an increase in the length of the side chain results in a looser crystal structure with an increasingly lower melting point. On the other hand, an increase in the bulkiness of the side chain increases T_m, since rotation in the side chain is hindered in the liquid state, with consequent decrease in ΔS_m.

Entropy and heat of fusion

Entropy of fusion The effect of structure on the entropy of fusion can now be calculated, in good agreement with experiment, for many polymers. This has led, for example, to a satisfactory explanation (Hobbs 1970) of the low entropy of fusion of the linear aliphatic polyesters whose melting points are depicted in Fig. 7-1. Following Starkweather (1960), the entropy of

TABLE 7-2.   *Effect of side-chain struc-*
ture on the crystalline
melting point of isotactic
poly(a-olefins) (Campbell
1959, Bawn 1960)

Side Chain	T_m, °C
—CH$_3$	165
—CH$_2$CH$_3$	125
—CH$_2$CH$_2$CH$_3$	75
—CH$_2$CH$_2$CH$_2$CH$_3$	−55
—CH$_2$CHCH$_2$CH$_3$	196

$$\underset{\text{CH}_3}{|}$$

—CH$_2$CCH$_2$CH$_3$	350

$$\underset{\text{CH}_3}{|}$$

fusion is considered to consist of independent contributions from volume change on melting and the change in the number of conformations which the chain can assume, from one in the crystal to a number determined by the type of chain bonds in the melt. The latter number was calculated by an enumeration scheme similar to that used by Smith (1965, 1966). This number was then combined with knowledge of the potential barrier to rotation to compute the desired conformational entropy.

$$\overset{\text{O}}{\overset{\|}{}}$$

For the linear aliphatic polyesters, the C—O bond is considered frozen in the *trans* conformation because of an extremely high potential barrier to rotation (17 kcal/mole) resulting from partial resonance with the carbonyl oxygen. The CH$_2$—CH$_2$ bond is like those in polyethylene, with a barrier of

$$\overset{\text{O}}{\overset{\|}{}}$$

about 3 kcal/mole, while the CH$_2$—C and O—CH$_2$ bonds are significantly more flexible, with barriers no higher than about 1 kcal/mole. In the enumeration, it was assumed that the three latter bonds could exist in the *trans*, *gauche*, or *gauche'* conformations, and conformations resulting in the equivalent of "pentane interference" were eliminated. Finally, correction was made for a contribution to the entropy of the crystal resulting from a distortion of the chain away from the all-*trans* conformation. (It is probable that variations in this distortion between repeat units with odd and even numbers of carbon atoms account for the alternation of melting points shown in

Fig. 7-2.) It was concluded that, in comparison to polyethylene, the stiffening effect of the rigid ester bonds outweighs to a small degree the increased flexibility of the ester chains.

Heat of fusion The low melting points of the linear aliphatic polyesters must thus be attributed to their low heats of fusion. These are less easily explained in detail, but a very approximate analysis (Hobbs 1970) suggests that the cohesive forces to be overcome in fusion result almost entirely from methylene interactions (dispersion forces) between neighboring chains. Since there are fewer methylene groups per unit chain length in the polyesters, their heats of fusion are lower.

Dipole interactions from the ester carbonyl groups appear not to contribute to the heats of fusion of these polymers, supporting the inference that the dipole bonds in the crystal are almost entirely reformed in the melt. A similar effect was postulated for the polyamides by Dole (1959).

Effect of copolymerization

When copolymers are made from monomers which form crystalline homopolymers, degree of crystallinity and crystalline melting point decrease as the second constituent is added to either homopolymer. The melting point depends on the mole fraction n of the crystallizing constituent by the relation of Eq. 5-4 (Flory 1949):

$$\frac{1}{T_m} - \frac{1}{T_m{}^0} = -\frac{R}{\Delta H_m} \ln n \tag{7-2}$$

where $T_m{}^0$ is the melting point of the homopolymer and ΔH_m is its heat of fusion. A typical case is the copolymer of hexamethylene terephthalamide and hexamethylene sebacamide (Fig. 7-3).

If, however, the comonomers are isomorphous, i.e., capable of replacing each other in the crystals, the melting point may vary smoothly over the composition range. An example is the copolymer of hexamethylene terephthalamide and hexamethylene adipamide, also shown in Fig. 7-3. Other variations may occur, including the formation of an alternating copolymer with a crystal structure and melting point far different from those of either homopolymer. Block and graft copolymers may exhibit two crystalline melting points, one for each type of chain segment.

GENERAL REFERENCES

Faucher 1965; Alfrey 1967; Sweeny 1969.

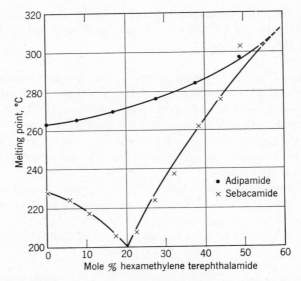

Fig. 7-3. Melting points of copolymers of hexamethylene adipamide and terephthalamide, and of hexamethylene sebacamide and terephthalamide (Edgar 1952).

B. The Glass Transition

Relation between T_m and T_g

With few exceptions, polymer structure affects the glass transition T_g and the crystalline melting point T_m similarly. This is not unexpected, since similar considerations of cohesive energy and molecular packing apply to the amorphous and crystalline or paracrystalline regions, respectively, in accounting for the temperature levels at which the transitions occur. In consequence, T_m and T_g are rather simply related for many polymers (Beaman 1952; Boyer, 1954, 1963): depending on symmetry, T_g K is approximately one-half to two-thirds of T_m K (Fig. 7-4). There are, however, some exceptions to this general rule. The most important appear (Pearce 1969) to be associated with crystallizability or hydrogen bonding in the amorphous regions of polymers. Within series of polymers where crystallizability is changed by control of tacticity (Reding 1962) or structure (Zimmerman 1968), ease of crystallization is associated with a low ratio of T_m to T_g, and difficulty with a high ratio. In polyamides, where T_m varies with ester group spacing as indicated in Section *A*, T_g is nearly constant at a level postulated (Woodward 1960; Komoto 1967) to be set by the energy required to break hydrogen bonds in the amorphous regions. Copolymers also have ratios of T_m to T_g differing from those shown in Fig. 7-4, as described below.

Effects of molecular weight and diluents

In the polymer range, T_g is more dependent on molecular weight than is T_m (Section A), the relation having the form

$$T_g = T_g{}^\infty - \frac{k}{\overline{M}_n} \qquad (7\text{-}3)$$

derived from temperature-volume considerations (Fox 1955), where $T_g{}^\infty$ is the glass transition temperature at infinite molecular weight, and k is about 2×10^5 for polystyrene (Fox 1950) and poly(methyl methacrylate) (Beevers 1960), and 3.5×10^{-5} for atactic poly(α-methyl styrene) (Cowie 1968).

From the considerations of Chapter 5, it is probable that chain ends, which lead to the terms H_0 and S_0 for low-molecular-weight substances, are usually associated with defects in the crystalline regions of polymers. Hence, they are unlikely to make a significant contribution to T_m, which is the melting point of the most perfect crystalline regions. In amorphous polymers, however, the effect of chain ends on free volume (Chapter 6D) should retain importance in proportion to the concentration of ends; thus it is not surprising that T_g is found to vary with $\overline{M}_n$ in the polymer range.

While the effect of molecular-weight distribution on T_g is accounted for by the appearance of $\overline{M}_n$ in Eq. 7-3, the effect of low-molecular-weight diluents is worth noting. The primary example of this effect is plasticization, widely used to improve the flexibility of certain polymers and allow them to remain flexible well below T_g of the unplasticized resin (Chapters 14, 17). Based on the weight of added constituent, plasticization is considerably more effective than copolymerization in lowering T_g.

Effect of chemical structure

The effects of the nature of the chain repeat units on T_g are closely related to intermolecular forces, chain stiffness, and symmetry. Probably the most important factor among these is hindrance to free rotation along the polymer chain resulting from the presence of stiff bonds or bulky side groups: compare polybutadiene, $T_g = -85°C$; styrene-butadiene copolymer (25/75), -55; polystyrene, $+100$; poly(α-methyl styrene, $+150$; polyacenaphthalene, $+285$. The effect of intermolecular forces is quantified by the cohesive energy density or solubility parameter (Chapter 2A): compare polypropylene, $\delta = 7.8$, $T_g = -20°C$; polyacrylonitrile, $\delta = 15.4$, $T_g = 90°C$. (On the

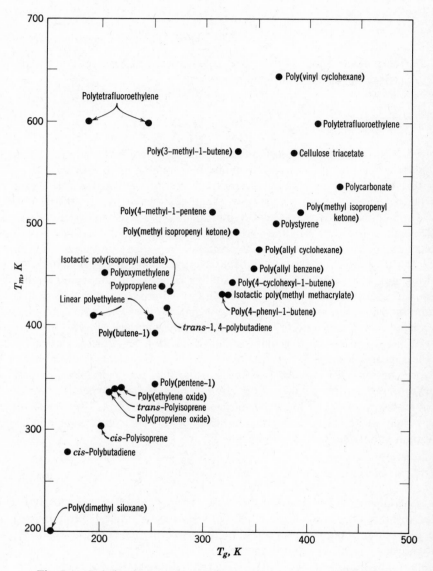

Fig. 7-4. Relation between T_m and T_g for various polymers (Boyer 1963).

other hand, the solubility parameters of polystyrene and polyacenaphthalene are not very different.) The effect of symmetry of the repeat unit is illustrated in Fig. 7-4: compare poly(vinyl chloride), $T_g = 85°C$; poly(vinylidene chloride), -17.

Effect of chain topology

Copolymerization   The glass transition temperatures of random copolymers usually fall between those of the corresponding homopolymers, T_g for the copolymer often being a weighted average given by

$$a_1 c_1 (T_g - T_{g1}) + a_2 c_2 (T_g - T_{g2}) = 0 \qquad (7\text{-}4)$$

where T_{g1} and T_{g2} refer to the homopolymers, c_1 and c_2 are the weight fractions of monomers 1 and 2 in the copolymer, and a_1 and a_2 depend on monomer type (Wood 1958). There are numerous deviations, both positive and negative, from this straight-line relationship, however.

The contrast between this behavior and the common depression of T_m by copolymerization is not surprising, since the changes at T_g do not require fitting a structure into a crystal lattice, and in consequence structural irregularity does not affect T_g as it does T_m.

While the above considerations for random copolymers lead to low ratios of T_m to T_g, it is sometimes found that block and graft copolymers may have long enough homogeneous chain segments to exhibit the properties of both homopolymers, rather than intermediate values. Thus, a block copolymer in which one homopolymer has high, and the other low, softening and brittle temperatures may exhibit both a high softening point and a low brittleness temperature. Hence it may have values of T_m/T_g higher than those found for other polymer types.

Branching and crosslinking   The effects of chain branching and crosslinking on T_g can be explained in terms of free volume. The higher concentration of chain ends in a branched polymer increases the free volume and lowers T_g, whereas crosslinking lowers free volume and raises T_g. Roughly, the latter change may be accounted for in terms of the average molecular weight of the segment between crosslinks by an equation like 7-3; more complete treatments are cited in Nielsen 1969.

GENERAL REFERENCES

Faucher 1965; Gordon 1965; Meares 1965; Alfrey 1967.

C. Properties Involving Large Deformations

Melt viscosity   As discussed in Chapter 6*A*, the viscosity of a polymer mel is a strong function of weight-average (or more properly viscosity-average molecular weight. Melt viscosity is also influenced by chain branching: i polyethylene and in silicone polymers melt viscosity decreases with increas ing degree of long-chain branching at constant weight-average molecula weight, whereas in poly(vinyl acetate) melt viscosity increases under the sam circumstances. The reason for this difference appears to lie in branch length: in experiments with poly(vinyl acetate) branched by graft polymeriza tion (Long 1964), melt viscosity decreased at constant molecular weigh when the added branches were shorter than the critical chain length Z_c a which the melt viscosity power law changes (Chapter 6*A*). Only whe branches longer than Z_c were added did the viscosity increase.

Crosslinking has a pronounced effect on melt viscosity in that the latte becomes essentially infinite along with $\overline{M}_w$ at the onset of gelation. How ever, small, tightly crosslinked network particles may behave like rigi spheres and have little effect on melt viscosity.

The addition of low-molecular-weight species, as in plasticization, re duces melt viscosity by lowering average molecular weight. Bulky sid groups may have a similar effect.

For many polymers an upper limit to $\overline{M}_w$ is set by fabrication require ments on the melt viscosity. At the same time, a lower limit on $\overline{M}_n$ may b set by requirements involving tensile strength, brittleness, or other mechan ical properties. In such cases, which include polypropylene and probabl other polyolefins, the best balance of properties is achieved when the distri bution of molecular weights is made as narrow as possible.

However, the anticipated gain in ease of fabrication on decreasing $\overline{M}_w$ at constant $\overline{M}_n$ may not be fully realized, since fabricability depends upon melt viscosity at high shear stress. In contrast to low shear (Newtonian) viscosity, which depends upon $\overline{M}_w$, high shear viscosity depends on a molecular weight average between $\overline{M}_w$ and $\overline{M}_n$ (Rudd 1960).

Other melt properties   For polymers with very high melt viscosity such as polytetrafluoroethylene, the tensile strength of the melt becomes a property of importance. Like melt viscosity, it increases with increasing molecular weight. The viscoelastic or elastic properties of polymer melts decrease in magnitude with increasing molecular weight and with increasing chain branching.

Tensile strength and related properties

Many polymer properties, including tensile strength, can be described by an equation of the type

$$\text{property} = a - \frac{b}{\overline{M}_n} \tag{7-5}$$

This is the type of relation predicted for properties depending on the number of ends of polymer chains. Many such properties, including density and refractive index, attain constant values at molecular weights well below the polymer range. Tensile strength, however, varies significantly with molecular weight in the range of interest for polymers, although the variation may in fact be with an average between $\overline{M}_n$ and $\overline{M}_w$ (McCormick 1959). It has been shown (Flory 1945) that dependence on $\overline{M}_n$ implies that the tensile strength TS of a mixture of components with tensile strengths $(TS)_i$ is the weight average

$$\overline{TS} = \sum_i w_i (TS)_i \tag{7-6}$$

If a polymer exhibits a yield point and then undergoes extensive elongation before tensile failure, its ultimate tensile strength increases with increasing molecular weight. Typical data for branched polyethylene are shown in Fig. 7-5. Important structure variables in this polymer were found to be degree of crystallinity and level of molecular weight; in Fig. 7-5 crystallinity is replaced by its equivalent, density, and molecular weight by the logarithm of melt viscosity.

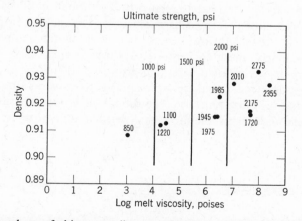

Fig. 7-5. Dependence of ultimate tensile strength of polyethylene on molecular structure variables (Sperati 1953).

Although Fig. 7-5 shows that tensile strength is independent of degree of crystallinity for branched polyethylene, this independence does not carry over to linear polyethylenes, with densities in the range 0.94–0.96: the latter polymers have higher tensile strengths than branched polyethylenes of the same molecular weight (see also Chapter 6E). The difference is attributed to changes in the crystal morphology of the high-density polymers during tensile elongation (Chapter 5F).

Morphology is indeed important in determining the mechanical properties of crystalline polymers. Both tensile strength and the mechanism of failure are influenced by such factors as spherulite size and structure and the nature of interlamellar ties. Polymers with smaller, finer-textured spherulites tend to fail at high elongations after drawing, while those with large, coarse spherulites often fail by brittle fracture between spherulites at low elongations (Collier 1969).

Toughness in rubber-modified glassy plastics

One of the major developments leading to plastics with outstanding toughness has been the production of rubber-modified glassy plastics such as the ABS resins (Chapter 14), made preferably by polymerizing a continuous glassy matrix in the presence of small rubber particles. It is known that for optimum toughness the two-phase structure is essential, only a small amount (5–15%) of rubber is needed, the optimum size of the rubber particles is 1–10 μm, and the rubber-matrix interface should be well grafted.

The mechanism (Bucknall 1967) by which toughness is developed in these materials, in contrast to the brittle failure characteristic of the unmodified glassy polymer, is intimately related to the formation of crazes, regions of low-density material formed as a precursor to cracking in glassy polymers (Wolock 1964). At low strains, the stress in the sample is largely borne by the matrix, and is concentrated at the rubber particles. As straining continues crazes are initiated here, and grow with the absorption of energy as the matrix deforms. Ultimately the applied stress is distributed between the rubber and the crazed matrix. The rubber, now under tension, strengthens the crazed matter and fracture is thus delayed in favor of craze initiation elsewhere.

GENERAL REFERENCES

Faucher 1965; Moore 1965; Vincent 1965; Alfrey 1967; Kargin 1968; Payne 1968.

D. Properties Involving Small Deformations

This section is concerned with structural determinants of a variety of properties involving small local deformations of polymers in contrast to the gross deformations discussed in Section C. Among these are such mechanical properties as stiffness, yield stress, elongation, and impact strength. Related to these properties are hardness, abrasion resistance, and flexural fatigue life, among others.

A second group of properties discussed includes solubility and related phenomena, such as swelling, cloud points, sorption of liquids, permeability to gases, and compatibility of plasticizers. The final class under discussion includes the effects of electromagnetic radiation, in such optical properties as refractive index and transparency and such electrical properties as dielectric constant, dielectric loss, and dielectric strength.

Effect of crystallinity

Mechanical properties   The properties of crystalline polymers are emphasized in this section for two reasons: it is these polymers which are most widely utilized because of their mechanical properties, and structural features related to crystallinity may have profound effects on these properties.

The degree of crystallinity alone is effective in determining the stiffness and yield point for most crystalline plastics. As indicated for branched polyethylene in Fig. 7-6, these properties are independent of molecular weight. As a result, they can be expressed as single-valued functions of degree of crystallinity. As crystallinity decreases, both stiffness and yield stress decrease. As a result of the latter change, the chance of brittle failure is reduced.

As the degree of crystallinity decreases with temperature during the approach to T_m (Chapter 5F), stiffness and yield stress decrease correspondingly. These factors often set limits on the temperature at which a plastic is useful for mechanical purposes.

A major determinant of the behavior of a polymer on impact is the relation between the yield stress and the tensile strength in brittle failure, which may be designated the brittle strength. If the yield stress or strength is lower than the brittle strength, plastic flow begins (at the yield point in a tensile experiment) and the polymer is tough. If the brittle strength is lower, brittle failure takes place on impact.

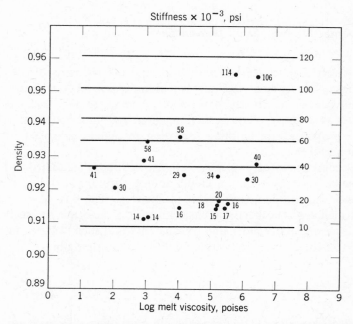

Fig. 7-6.   Dependence of stiffness of branched polyethylene on molecular structure vari
ables (Sperati 1953). (See discussion for Fig. 7-5.)

Solubility and related properties   As pointed out in Chapter 2, crystallinity
decreases the solubility of polymers markedly, since the process of solution
involves overcoming the heat and entropy factors associated with crystallization as well as those of the intermolecular interactions in the amorphous
regions.   Properties related to solubility, such as the cloud point of dilute
solutions, are often functions of crystallinity relatively independent of
molecular weight.

The solubility of liquids and gases in polymers is also strongly dependent
on crystallinity, since solubility is usually confined to the amorphous regions.
Permeability, the product of solubility and diffusivity, behaves similarly.

Plasticization is closely related to solubility, and the selection of an
efficient and compatible plasticizer (see Chapter 17) involves considerations
similar to those for the selection of solvents. Plasticization usually results
in loss of crystallinity; however, if crystallinity is well developed it may
not be possible to find a plasticizer sufficiently compatible with (soluble in)
the polymer to have a significant effect on its properties.

Electrical and optical properties   The primary effect of crystallinity on the
electrical and optical properties is associated with the changes in dielectric

onstant and refractive index arising from the difference in density between he crystalline and amorphous regions. In the case of visible light, this ifference leads to scattering, which may be large if the regions responsible crystallites or lamellae, and spherulites) are significant in size compared o the wavelength of the light. Thus crystalline plastics usually appear anslucent or opaque, their transparency increasing with decreasing pherulite size.

Effect of molecular weight

olubility Where the presence of a crystalline phase is not involved, olubility and related phenomena are inverse functions of molecular weight. his fact is reflected in the equations for the thermodynamic properties of olymer solutions, and forms the basis of fractionation methods, as discussed in Chapter 2.

lectrical and optical properties The interactions of electromagnetic radiation ith polymers involve, at the most, the cooperative movement of small roups of atoms. The molecular weight dependence of these properties, beying relations of the type of Eq. 7-1, vanishes at molecular weights far elow the polymer range. Except as molecular weight influences some more irect structural determinant of these properties, they are independent of his variable.

Combined effects of crystallinity and molecular weight

Mechanical properties A number of mechanical properties, including ardness, flexural fatigue resistance or flex life, softening temperature, longation at tensile break (where plastic flow occurs), and sometimes mpact strength, are influenced by both degree of crystallinity and molecular veight. Typical examples are the softening temperature of branched olyethylene as measured in the Vicat test (Chapter 4G), which increases vith increasing molecular weight and increasing crystallinity (Fig. 7-7), and he flex life of polytetrafluoroethylene, which increases with increasing molecular weight and decreasing crystallinity (Fig. 7-8; "standard specific ravity" is a measure of molecular weight, as discussed in Chapter 14). Flex life is sometimes associated with brittleness, samples having low esistance to flexing being brittle. The chance of brittle failure is decreased by aising molecular weight, which increases brittle strength, and by reducing rystallinity as indicated previously. For amorphous polymers, impact trength is found to depend on the weight-average molecular weight.

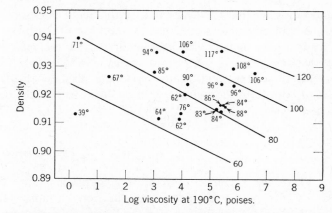

Fig. 7-7. Dependence of softening temperature, as measured by the Vicat test, of branched polyethylene on molecular structure parameters (Sperati 1953). (See discussion for Fig 7-5.)

Solubility and related properties In general, the introduction of polar groups into polymers tends to decrease solubility, since strong polymer polymer bonds usually develop. The situation is complicated, however, by factors such as the arrangement and bulkiness of the groups, which in turn influence crystallinity.

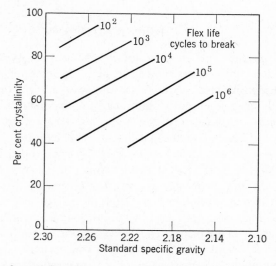

Fig. 7-8. Lines of equal flex life on a crystallinity-molecular weight "map" for poly tetrafluoroethylene (Thomas 1956).

The solubility behavior of cellulose and its derivatives provides examples ' the effect of polarity and substitution on solubility. Cellulose itself is soluble because of strong hydrogen bonding and stiff chains which prevent 'drating molecules from penetrating its crystalline regions. Substitution ' less polar groups for the hydroxyls on cellulose leads first to solubility alkalis, which only swell the parent polymer, and then to water solubility hen 0.5–1 hydroxyl per glucose unit is replaced. Crystallinity has usually sappeared at this stage. At higher degrees of substitution, water solubility replaced by alcohol solubility, and then by solubility in organic solvents hose type depends on the nature of the substituent group. When all the bstitutents are alike, complete substitution leads to crystallization and ore limited solubility.

The permeability of polymers to gases and liquids decreases with creasing polarity, since a more polar polymer has a higher activation energy r diffusion. Even methyl groups contribute sufficiently to high cohesion d high activation energy in rubbers to impart low gas permeability. Thus, tyl, nitrile, and chloroprene rubbers all have lower permeability than tural or butadiene-based rubbers.

ectrical properties The electrical properties of polymers depend more ı an unbalance or asymmetry of dipoles than on the presence of polar oups *per se.* Thus, both polyethylene and polytetrafluorethylene have w dielectric constant and dielectric loss (dissipation factor), but these antities are much larger for polymers containing both hydrogen and ıorine or chlorine. Similarly, the dielectric loss of poly(2,5-dichlorosty-ne) is much less than that of poly(3,4-dichlorostyrene) because of cancella-ɔn of the dipoles of the *para* chlorines.

Effect of copolymerization

echanical properties The addition of a comonomer to a crystalline ɔlymer usually causes a marked loss in crystallinity, unless the second onomer crystallizes isomorphous with the first (see Section A). Crystallinity pically decreases very rapidly, accompanied by reductions in stiffness, ırdness, and softening point, as relatively small amounts (10–20 mole per nt) of the second monomer are added. In many cases, a rigid, fiber-forming ɔlymer is converted to a highly elastic, rubbery product by such minor odification.

The dependence of mechanical properties on copolymer composition in stems which do not crystallize results primarily from changes in inter-olecular forces as measured by cohesive energy. Higher cohesive energy sults in higher stiffness and hardness and generally improved mechanical 'operties.

The additional variable introduced by block or graft copolymerizati can be used to alter the properties of such copolymers by changing the method of preparation. A block or graft copolymer consisting of long-cha segments of widely differing polarity can exist in solution with one or t other type of segment extended, the second relatively contracted, dependi on solvent type. When isolated from solution, the copolymer has properti resembling those of the homopolymer corresponding to the extended se ments (Merrett 1957). Thus, natural rubber with chains of poly(meth methacrylate) grafted to it was hard and stiff with a nontacky surface wh isolated from a solution where the rubber chains were collapsed and t poly(methyl methacrylate) chains were extended. When prepared under t opposite conditions, it was limp, flabby, and self-adherent like rubber. third form with intermediate properties was isolated from solvents in whi both chain segments were relatively extended.

Solubility   In essentially random copolymers of monomers whose hom polymers are noncrystalline, a property such as solubility varies more or le regularly from that of one homopolymer to that of the other as the relati proportions of the components are varied. The solubility of copolymers this type is frequently low in solvents for either homopolymer, but high mixtures of these solvents.

The solubility of graft and block copolymers is often unusually hig especially if the two components have widely different polarities. Bloc copolymers of polystyrene with poly(vinyl alcohol) or poly(acrylic acic e.g., are soluble in benzene, acetone, and water. In water the hydrophil blocks are solubilized and extended, holding the tightly coiled hydrocarbo segments in solution much as a detergent solubilizes a hydrocarbon by micel formation. In benzene the opposite situation occurs, but in an intermedia solvent such as acetone both blocks are relatively extended, as indicated b higher solution viscosity in intermediate solvents. These polymers act a efficient detergents and emulsifying and compatibilizing agents, but the virtually universal solubility makes it difficult to isolate or purify them.

Effect of plasticization, reinforcement, or crosslinki

Plasticization   The addition of a plasticizer usually reduces stiffness, har ness, and brittleness, and has a similar effect on other mechanical propertie since interchain forces are effectively reduced. These changes are accom panied by a reduction in T_g, as noted in Section B. As indicated in Chapt 17, plasticization is usually restricted to amorphous polymers or polyme with a low degree of crystallinity because of the limited compatibility plasticizers with highly crystalline polymers.

Reinforcement and crosslinking Whether carried out by chemical cross-linking in an unmodified amorphous polymer system or by the addition of a reinforcing filler such as carbon black in rubber (a process involving chemical bonding between polymer and filler), the addition of crosslinks leads to stiffer, stronger, tougher products, usually (in the case of rubbers) with enhanced tear and abrasion resistance as well. However, extensive crosslinking in a crystalline polymer may cause loss of crystallinity, with attendant deterioration of the mechanical properties depending on this factor. When this occurs, the initial trend of properties may be toward either enhancement or deterioration, depending on the degree of crystallinity of the unmodified polymer and the method of formation and location (crystalline or amorphous regions) of the crosslinks.

GENERAL REFERENCES

Faucher 1965; Hulse 1965; Kargin 1968.

E. Property Requirements and Polymer Utilization

The variables necessary to define the mechanical and physical properties of polymers have now been discussed (Fig. 7-9). The increase of T_m with molecular weight, leveling off as polymer molecular weights are reached, the related approximate behavior of T_g, and the continual increase of viscosity with molecular weight serve to define, in terms of the variables molecular

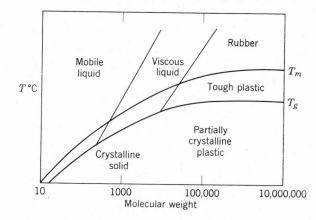

Fig. 7-9. Approximate relations among molecular weight, T_g, T_m, and polymer properties.

weight and temperature, regions in which the properties of typical plastics, rubbers, viscous liquids, etc., may be found.

Combinations of properties unique to polymers are evidenced in each of the major uses, including elastomers, fibers, and plastics, to which macromolecules are put. In this section the property requirements of these end uses are described and related to polymer structure.

Elastomers

It is a matter of experience that all substances exhibiting a high degree of rubberlike elasticity contain long-chain structures. The restoring force leading to elastic behavior results directly from the decrease in entropy associated with the distortion of a chain macromolecule from its most probable conformation. Two additional property requirements are imposed by the condition that there be sufficient freedom of molecular motion to allow the distortions to take place rapidly: first, the polymer must at its use temperature be above T_g; and, second, it must be amorphous, at least in the undistorted state.

In contrast to the high local mobility of chain segments implied by these requirements, the gross mobility of chains in elastomers must be low. The motions of chains past one another must be restricted in order that the material can regain its original shape when the stress is released. This restriction of gross mobility is usually obtained by the introduction of a network of primary bond crosslinks in the material. (It cannot be obtained through secondary bond forces; these must be kept low in order to gain local segment mobility.) The crosslinks must be relatively few and widely separated, however, so that stretching to large extensions can take place without rupture of primary bonds.

The requirement of low cohesive energy limits the family of elastomers to polymers which are largely hydrocarbon (or fluorocarbon or silicone) in nature, with polar groups distributed at random and in not too great number, so that crystallinity is absent and T_g is sufficiently low, ideally in the range -50 to $-80°C$. The requirement of sites for crosslinking often leads to the choice of a diene as a monomer or comonomer for an elastomer.

In contrast to its equilibrium properties, a stretched elastomer should have the high tensile strength and modulus usually associated with crystalline plastics. Thus, rubbers in which crystallinity can develop on stretching, such as natural rubber and its stereoregular synthetic counterparts, usually have more desirable properties than those with less regular structures. However, a reinforcing filler can sometimes impart to a rubber properties similar to those obtained on the development of crystallinity. Thus, reinforced styrene-butadiene rubber has properties nearly equivalent to those of reinforced natural rubber, whereas its properties when crosslinked but not reinforced are much poorer.

Fibers

In contrast to elastomers, the requirements of high tensile strength and modulus characteristic of fibers are almost always obtained by utilizing the combination of molecular symmetry and high cohesive energy associated with a high degree of crystallinity. Usually the fiber is oriented to provide optimum properties in the direction of the fiber axis.

The end use requirements of fibers, particularly those involving textiles, lead to restrictions on several properties. The crystalline melting point, T_m, must be above a certain minimum, say 200°C, if the resulting fabric is to be subjected to ironing. On the other hand, spinning the polymer into a fiber requires either that T_m be below, say, 300°C and well below the temperature of decomposition of the polymer, or that the polymer be soluble in a solvent from which it can be spun. Other requirements limit solubility, e.g., in solvents useful for dry cleaning.

The requirement of orientation in the fiber usually implies that T_g be not too high (since orientation by cold drawing and ironing are typically carried out at or above this temperature) or too low (since orientation and related characteristics, such as crease retention after ironing, must be maintained at room temperature).

Thus the selection of a polymer for use as a fiber involves a number of compromises, usually met by choosing a linear polymer with high symmetry and high intermolecular forces resulting from the presence of polar groups, high enough in molecular weight so that tensile strength and related properties are fully developed. Branching in the polymer chain is in general detrimental to fiber properties because branch points disrupt the crystalline lattice, lower the crystalline melting point, and decrease stiffness. Crosslinking, on the other hand, offers the possibility of obtaining strong interchain bonding. If crosslinks are formed after the polymer is spun into fiber, and are relatively few in number, improvement in fiber properties may result. Thus poly(vinyl alcohol), polyurethanes, and protein fibers may be crosslinked with formaldehyde to give higher melting point, lower solubility and moisture regain, and improved hand, while wool, a natural protein fiber, is crosslinked with cystine links.

General-purpose and specialty plastics

The wide range of end uses of plastics requires a variety of property combinations; correspondingly, a wider variety of structures is important. In general, the properties of plastics are intermediate between those of fibers and elastomers, with much overlapping on either end. Thus, typical plastics may have cohesive energies higher than those of elastomers but lower than those of fibers. However, a polymer useful as a fiber when oriented in this form may also be useful as a plastic, where orientation is not readily achieved in the massive pieces used; an example is nylon.

Optical applications   The requirement of good optical properties, especially in massive pieces, imposes severe limitations on the structure of polymers. In general, crystallinity must be absent, and most amorphous polymers exhibit softness and brittleness which exclude them from many applications. In thin films, structural requirements are not as severe, since crystallinity can often be tolerated if spherulite growth is inhibited and the material can be processed to give sufficiently smooth surfaces. Thus, the normally incompatible requirements of clarity and toughness can both be met.

Electrical applications   To achieve low dielectric loss over a wide frequency range, the structure of a polymer must be selected on the basis of low polarity. All other requirements are of considerably less importance. It follows that polyethylene and polytetrafluoroethylene are the best materials for low-loss applications, particularly at high frequencies. At low frequencies, however, other plastics, such as poly(vinyl chloride), are useful.

Mechanical applications   Perhaps the most important property requirement for the use of polymers in mechanical applications is toughness. This property is usually achieved by selection of a polymer with a moderate, but not too high, degree of crystallinity (see the discussion of the mechanical properties of crystalline polymers, Chapter 6E). Often, a delicate balance of structural features is needed to achieve the desired combination of properties. Composite structures (as in glass-reinforced or rubber-modified plastics) often provide unique property combinations.

Economic aspects   Though the preceding discussion could be extended considerably, it should be pointed out that a major factor in the selection of the appropriate plastic for a given use is economic. In many (some observers would say far too many) cases, a mass market is achieved by sacrificing properties for price. The plastics manufacturer must, therefore, consider what structures give the optimum combination of melt and solid-state properties. He would like, e.g., to lower molecular weight to achieve low melt viscosity and more rapid fabrication; but he must maintain levels of molecular weight consistent with the development of good mechanical properties. The selection of raw materials is also important: the rise in prominence of olefin polymers is closely related to the low cost of the monomers. The cost of polymerization itself appears less important in most cases, but it is clear that monomers which are expensive to polymerize have less opportunity to yield large-volume plastics.

A "profile" of the important properties of common general-purpose and specialty plastics is given in Fig. 7-10.

Engineering plastics

In contrast to general-purpose and specialty plastics, the term engineering polymers or plastics is applied to those materials which command a premium price, usually associated with relatively low production volume, because of their outstanding balance of properties which allows them to compete successfully with other materials (metals, ceramics) in engineering

	Poly(methyl methacrylate)	Modacrylic	Cellulose acetate	Cellulose acetate-butyrate	Cellulose propionate	Ethyl cellulose	Chlorinate polyether	Epoxy (cast)	FEP fluorocarbon	Ionomer	Melamine-formaldehyde (cellulose filled)	Phenol-formaldehyde (cellulose filled)	Polybutylene	Polyethylene (low density)	Polyphenylene oxide (modified)	Polystyrene	Poly(vinyl chloride)	Poly(vinyl chloride) (copolymer)	Poly(vinylidene chloride)	Poly(vinylidene fluoride)	SAN	Silicone	Urea-formaldehyde (cellulose filled)
Price	0	0	0	0	0	0	0	0	−	0	0	+	0	+	0	+	+	+	0	−	+	−	0
Processability	+	+	+	+	+	+	+	0	+	+	0	0	0	+	+	+	0	0	+	+	0	−	+
Tensile strength	+	0	0	0	0	0	0	+	−	−	0	0	−	−	0	+	0	−	0	0	+	−	+
Stiffness	+	+	+	0	0	0	−	0	−	−	+	+	−	−	0	+	+	−	−	−	+	−	+
Impact strength	−	0	0	0	+	0	−	−	+	+	−	−	+	+	−	−	+	+	−	0	−	−	−
Hardness	0	+	+	+	+	+	0	+	−	−	+	+	−	−	+	+	−	−	+	−	+	−	+
Useful temperature range	0	−	0	0	0	−	0	+	0	−	0	0	0	0	0	−	−	−	−	0	0	0	−
Resistance to chemicals	0	0	−	−	−	−	−	0	+	0	0	−	+	+	+	−	+	+	+	+	−	0	−
Resistance to weather	+	+	0	0	0	0	0	0	+	0	0	0	−	−	−	−	0	0	0	+	−	+	0
Resistance to water	0	0	−	−	−	−	−	+	−	+	0	0	0	+	+	0	0	0	0	0	0	0	0
Flammability	−	−	0	−	−	−	0	−	+	−	0	−	−	−	−	0	−	0	0	0	0	−	0

Fig. 7-10. A "profile" of the properties of some general-purpose and specialty plastics (Billmeyer 1968). Key: +, outstanding in the property indicated, among the best performers available; 0, acceptable performance in this property, still suitable in most cases; −, not recommended if this property is important to the intended use.

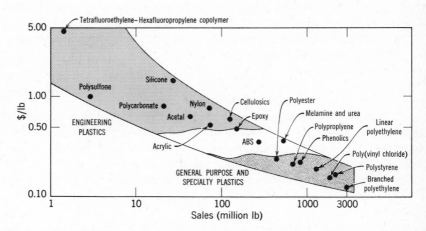

Fig. 7-11. Relationship between prices and production volumes of major engineering and general-purpose plastics (Gutoff 1969).

applications. They are strong, stiff, tough, abrasion-resistant materials capable of withstanding wide ranges of temperatures, and resistant to attack by weather, chemicals, and other hostile conditions. The value they contribute to the end product justifies their higher price per pound. Figure 7-11 compares engineering with general-purpose plastics for price and volume.

	ABS	Acetal	Polyetra-fluoroethylene	Polychlorotri-fluoroethylene	Nylon	Phenoxy	Polycarbonate	Polyimide	Poly(phenylene oxide)	Polyethylene (high density)	Polypropylene	Polysulfone
Price	0	0	−	−	−	−	0	−	−	+	+	−
Processability	0	+	−	+	+	0	0	−	0	+	+	+
Tensile strength	0		−	0	0	0	0	+	+	−	0	+
Stiffness	0		−	0	0	0	0	+	0	−	0	0
Impact strength	0	−	0	0	−	+	+	−	−	+	−	0
Hardness	0	+	−	0	0	+	+	+	+	−	0	+
Useful temperature range	−	0	+	0	0	−	0	+	0	0	0	0
Resistance to chemicals	0	0	+	+	0	0	0	+	0	+	+	+
Resistance to weather	0	0	+	+	−	0	0	+	0	−	−	0
Resistance to water	0	0	+	+	−	0	0	0	+	+	+	0
Flammability	−	−	+	+	0	0	0	+	0	+	+	0

Fig. 7-12. "Profile" of the properties of some engineering plastics (Billmeyer 1968). Key same as for Fig. 7-10.

The outstanding properties of engineering plastics come primarily from their crystalline nature and strong intermolecular forces. Most of them have quite high melting points, insuring retention of good physical properties to high temperatures, and good toughness over wide temperature ranges.

The necessary high melting points can be obtained in several ways, as discussed elsewhere. These include combining high degree of crystallinity with stiff polymer chains, or finding structural features substituting for crystallinity in imparting rigidity to the total material structure. This can be done in two ways: crosslinking and utilizing composite structures with an extremely rigid material, such as glass fibers.

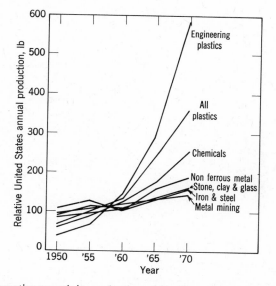

Fig. 7-13.   Comparative growth in production of plastics and other commodities over the last decade (Gutoff 1969).

So far, these alternatives have been utilized one at a time in almost all cases. That is, glass fibers are usually used to reinforce glassy plastics such as polyester-styrene copolymers; crosslinked polymers are not crystallizable, etc. We are now beginning to see instances where two or more of these property-enhancing approaches are combined, with outstanding results—for example, the development of ladder polymers combining the features of crosslinking with crystallizability, and the use of high-performance inorganic fibers (such as boron) to reinforce crystalline plastics. Clearly, much effort is being expended towards progress in these directions.

A "profile" of the important properties of engineering plastics is given in Fig. 7-12.

The role of plastics among materials

In the last decade, the role of plastics in the over-all materials market has changed so drastically that brief comment should not be omitted. Production of all plastics has increased about 3.5-fold in that period (Fig. 7-13),

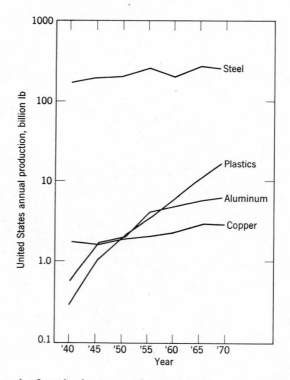

Fig. 7-14.   Growth of production on a weight basis (billions of pounds) of plastics and some metals over the last three decades (Gutoff 1969).

compared to a 2.5-fold increase for chemicals and smaller values for metals and ceramics. Engineering plastics have increased proportionately more, about 6-fold.

This means that, on a weight basis, the annual production of plastics now exceeds that of copper and aluminum (Fig. 7-14), but falls far short of the production of steel. On a volume basis, however (Fig. 7-15), even modest extrapolations suggest that the production of plastics by around 1980 (estimated at about 40 billion pounds or 700 million cubic feet) will exceed even

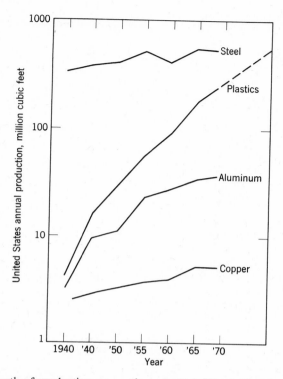

Fig. 7-15. Growth of production on a volume basis (millions of cubic feet) of plastics and some metals over the last three decades, with extrapolation to 1980 (Gutoff 1969).

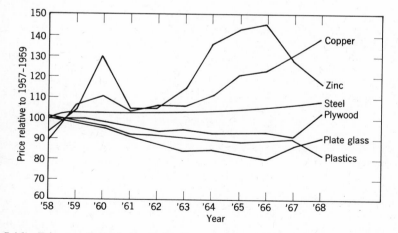

Fig. 7-16. Price trends in plastics and other materials over the last decade (Gutoff 1969).

that of steel. At the same time, the trend in plastics prices over the last decade has been consistently downward (Fig. 7-16) in contrast to the trends for most other materials. As a result, the plastics business has become far more competitive over the last few years.

We cannot help concluding that the burgeoning expansion in the use of polymeric materials we have seen in the last decade will continue for the foreseeable future.

GENERAL REFERENCES

Graf 1964; Riley 1968; Platzer 1969.

BIBLIOGRAPHY

Alfrey 1967. Turner Alfrey, Jr., and Edward F. Gurnee, *Organic Polymers*, Prentice-Hall, Inc., Englewood Cliffs, N.J., 1967.

Bannerman 1956. D. G. Bannerman and E. E. Magat, "Polyamides and Polyesters," Chap. VII in C. E. Schildknecht, ed., *Polymer Processes*, Interscience Div., John Wiley and Sons, New York, 1956.

Bawn 1960. C. E. H. Bawn, "High Polymers and Molecular Architecture," *Chem. & Ind.* **1960,** 388–391.

Beaman 1952. Ralph G. Beaman, "Relation between (Apparent) Second-Order Transition Temperature and Melting Point," *J. Polymer Sci.* **9,** 470–472 (1952).

Beevers 1960. R. B. Beevers and E. F. T. White, "Physical Properties of Vinyl Polymers. Part 1.—Dependence of the Glass-Transition Temperature of Polymethylmethacrylate on Molecular Weight," *Trans. Faraday Soc.* **56,** 744–752 (1960).

Billmeyer 1968. Fred W. Billmeyer, Jr., and Renée Ford, "The Anatomy of Plastics," *Science and Technol. No.* **73,** 22–37, 81–82 (1968).

Boyer 1954. R. F. Boyer, "Relationship of First-to-Second-Order Transition Temperatures for Crystalline High Polymers," *J. Appl. Phys.* **25,** 825–829 (1954).

Boyer 1963. Raymond F. Boyer, "The Relation of Transition Temperatures to Chemical Structure in High Polymers," *Rubber Chem. Tech.* **36,** 1303–1421 (1963).

Bucknall 1967. C. B. Bucknall, "The Relationship between the Structure and Mechanical Properties of Rubber-Modified Thermoplastics, Part One," *Brit. Plastics* **40** (11), 118–122 (1967); "Part Two," *ibid.* (12), 84–86.

Bunn 1955. C. W. Bunn, "The Melting Points of Chain Polymers," *J. Polymer Sci.* **16,** 323–343 (1955).

Campbell 1959. Tod W. Campbell and A. C. Haven, Jr., "The Relationship Between Structure and Properties of Crystalline, High-Melting Polyhydrocarbons," *J. Appl. Polymer Sci.* **1,** 73–83 (1959).

Collier 1969. John R. Collier, "Polymer Crystallization History and Resultant Properties," *Rubber Chem. Tech.* **42,** 769–779 (1969).

Cowie 1968. J. M. G. Cowie and P. M. Toporowski, "The Dependence of Glass Temperature on Molecular Weight for Poly(α-methyl styrene)," *Europ. Polymer J.* **4,** 621–625 (1968).

Dole 1959. M. Dole and B. Wunderlich, "Melting Points and Heats of Fusion of Polymers and Copolymers," *Makromol. Chem.* **34**, 29–49 (1959).

Edgar 1952. Owen B. Edgar and Rowland Hill, "The *p*-Phenylene Linkage in Linear High Polymers: Some Structure-Property Relationships," *J. Polymer Sci.* **8**, 1–22 (1952).

Faucher 1965. J. A. Faucher and F. P. Reding, "Relationship between Structure and Fundamental Properties," Chap. 13 in R. A. V. Raff and K. W. Doak, eds., *Crystalline Olefin Polymers*, Part I, Interscience Div., John Wiley and Sons, New York, 1965.

Flory 1945. Paul J. Flory, "Tensile Strength in Relation to Molecular Weights of High Polymers," *J. Am. Chem. Soc.* **67**, 2048–2050 (1945).

Flory 1949. Paul J. Flory, "Thermodynamics of Crystallization in High Polymers. IV. A Theory of Crystalline States and Fusion in Polymers, Copolymers, and Their Mixtures with Diluents," *J. Chem. Phys.* **17**, 223–240 (1949).

Fox 1950. Thomas G Fox, Jr., and Paul J. Flory, "Second-Order Transition Temperatures and Related Properties of Polystyrene. I. Influence of Molecular Weight," *J. Appl. Phys.* **21**, 581–591 (1950).

Fox 1955. T. G Fox and S. Loshaek, "Influence of Molecular Weight and Degree of Crosslinking on the Specific Volume and Glass Temperature of Polymers," *J. Polymer Sci.* **15**, 371–390 (1955).

Gordon 1965. M. Gordon, "Thermal Properties of High Polymers," Chap. 4 in P. D. Ritchie, ed., *Physics of Plastics*, D. Van Nostrand Co., Princeton, N.J., 1965.

Graf 1965. George L. Graf, Jr., "Applied Economics," Chap. 19 in Eric Baer, ed., *Engineering Design for Plastics*, Reinhold Publishing Corp., New York, 1964.

Gutoff 1969. Reuben Gutoff, "Engineering Polymers: Risks and Rewards in the Decade Ahead," talk before the Commercial Chemical Development Association, New York, March 1969.

Hill 1948. R. Hill and E. E. Walker, "Polymer Constitution and Fiber Properties," *J. Polymer Sci.* **3**, 609–630 (1948).

Hobbs 1970. Stanley Y. Hobbs and Fred W. Billmeyer, Jr., "Heats and Entropies of Linear Aliphatic Polyesters. II. Molecular Origins," *J. Polymer Sci.* A-2, **8**, 1395–1409 (1970).

Hulse 1965. G. Hulse, "Other Mechanical Properties of High Polymers," Chap. 3 in P. D. Ritchie, ed., *Physics of Plastics*, D. Van Nostrand Co., Princeton, N.J., 1965.

Kargin 1968. V. A. Kargin and G. L. Slonimsky, "Mechanical Properties," pp. 445–516 in Herman F. Mark, Norman G. Gaylord and Norbert M. Bikales, eds., *Encyclopedia of Polymer Science and Technology*, Vol. 8, Interscience Div., John Wiley and Sons, New York, 1968.

Komoto 1967. H. Komoto, "Physico-Chemical Studies of Polyamides. I. Polyamides having Long Methylene Chain Units," *Revs. Phys. Chem. Japan* **37**, 105–111 (1967).

Long 1964. V. C. Long, G. C. Berry, and L. M. Hobbs, "Solution and Bulk Properties of Branched Poly(vinyl Acetates). IV. Melt Viscosity," *Polymer* **5**, 517–524 (1964).

McCormick 1959. Herbert W. McCormick, Frank M. Brower and Leo Kin, "The Effect of Molecular Weight Distribution on the Physical Properties of Polystyrene," *J. Polymer Sci.* **39**, 87–100 (1959).

Meares 1965. Patrick Meares, *Polymers: Structure and Properties*, D. Van Nostrand Co., Princeton, N.J., 1965.

Merrett 1957. F. M. Merrett, "Graft Copolymers with Preset Molecular Configurations," *J. Polymer Sci.* **24**, 467–477 (1957).

Moore 1965. L. D. Moore, Jr., and R. M. Scholkin, Jr., "Rheological Properties," Chap. 10 in R. A. V. Raff and K. W. Doak, eds., *Crystalline Olefin Polymers*, Part I, Interscience Div., John Wiley and Sons, New York, 1965.

Nielsen 1969. Lawrence E. Nielsen, "Cross-Linking—Effect on Physical Properties of Polymers," *J. Macromol. Sci.—Revs. Macromol. Chem.* **C3**, 69–103 (1969).

Payne 1968. A. R. Payne, "Physics and Physical Testing of Polymers," pp. 1–93 in J. C. Robb and F. W. Peaker, eds., *Progress in High Polymers*, Vol. 2. CRC Press, Cleveland, Ohio, 1968.

Pearce 1969. Eli M. Pearce, "Polymer Synthesis: Philosophy and Approaches," *Trans. N.Y. Acad. Sci.*, **31**, 629–636 (1969).

Platzer 1969. Norbert Platzer, "Progress in Polymer Engineering," *Ind. Eng. Chem.* **61** (5), 10–29 (1969).

Reding 1962. F. P. Reding, E. R. Walter, and F. J. Welch, "Glass Transition and Melting Point of Poly(Vinyl Chloride)," *J. Polymer Sci.* **56**, 225–231 (1962).

Riley 1968. Malcolm W. Riley, "Materials, Selection," pp. 419–440 in Herman F. Mark, Norman G. Gaylord, and Norbert M. Bikales, eds., *Encyclopedia of Polymer Science and Technology*, Vol. 8, Interscience Div., John Wiley and Sons, New York, 1968.

Rudd 1960. John F. Rudd, "The Effect of Molecular Weight Distribution on the Rheological Properties of Polystyrene," *J. Polymer Sci.* **44**, 459–474 (1960).

Smith 1965. Richard P. Smith, "Polymethylene Chains and Rings on a Diamond Lattice with Atom Overlap Excluded," *J. Chem. Phys.* **42**, 1162–1166 (1965).

Smith 1966. Richard P. Smith, "Configurational Entropy of Polyethylene and Other Linear Polymers," *J. Polymer Sci. A-2* **4**, 869–880 (1966).

Sperati 1953. C. A. Sperati, W. A. Franta, and H. W. Starkweather, Jr., "The Molecular Structure of Polyethylene. V. The Effect of Chain Branching and Molecular Weight on Physical Properties," *J. Am. Chem. Soc.* **75**, 6127–6133 (1953).

Starkweather 1960. Howard W. Starkweather, Jr., and Richard H. Boyd, "The Entropy of Melting of Some Linear Polymers," *J. Phys. Chem.* **64**, 410–414 (1960).

Sweeny 1969. W. Sweeny and J. Zimmerman, "Polyamides," pp. 483–597 in Herman F. Mark, Norman G. Gaylord, and Norbert M. Bikales, eds., *Encyclopedia of Polymer Science and Technology*, Vol. 10, Interscience Div., John Wiley and Sons, New York, 1969.

Thomas 1956. P. E. Thomas, J. F. Lontz, C. A. Sperati, and J. L. McPherson, "Effects of Fabrication on the Properties of Teflon Resin," *SPE J.* **12** (6), 89–96 (1956).

Vincent 1965. P. I. Vincent, "Mechanical Properties of Polymers: Deformation," Chap. 2 in P. D. Ritchie, ed., *Physics of Plastics*, D. Van Nostrand Co., Princeton, N.J., 1965.

Wolock 1964. Irvin Wolock and Sanford B. Newman, "Fracture Topology," Chap. IIC in Bernard Rosen, ed., *Fracture Processes in Polymeric Solids—Phenomena and Theory*, Interscience Div., John Wiley and Sons, New York, 1964.

Wood 1958. Lawrence A. Wood, "Glass Transition Temperatures of Copolymers," *J. Polymer Sci.* **28**, 319–330 (1958).

Woodward 1960. A. E. Woodward, T. M. Crissman, and J. A. Sauer, "Investigations of the Dynamic Mechanical Properties of Some Polyamides," *J. Polymer Sci.* **44**, 23–34 (1960).

Wunderlich 1958. Bernhard Wunderlich and Malcolm Dole, "Specific Heat of Synthetic High Polymers. IX. Poly(ethylene Sebacate)," *J. Polymer Sci.* **32**, 125–130 (1958).

Zimmerman 1968. Joseph Zimmerman, "Melt Blend of Polyamides," U.S. Patent 3,393,252 (to E. I. du Pont de Nemours and Co.), 16 July 1968.

III

Polymerization

8

Step-Reaction (Condensation) Polymerization

A. Classification of Polymers and Polymerization Mechanisms

In 1929 W. H. Carothers suggested a classification of polymers into two groups, *condensation* and *addition* polymers. Condensation polymers are those in which the molecular formula of the repeat unit of the polymer chain lacks certain atoms present in the monomer from which it is formed (or to which it can be degraded). For example, a polyester is formed by typical condensation reactions between bifunctional monomers, with the elimination of water:

$$x\text{HO—R—OH} + x\text{HOCO—R'—COOH} \longrightarrow$$
$$\text{HO[—R—OCO—R'—COO—]}_x\text{H} + (2x - 1)\text{H}_2\text{O}$$

Addition polymers (Chapter 9) are those in which this loss of a small molecule does not take place. The most important group of addition polymers includes those derived from unsaturated vinyl monomers:

$$\underset{\displaystyle X}{\text{CH}_2\text{=CH}} \longrightarrow \underset{\displaystyle X}{\text{—CH}_2\text{—CH—}}\underset{\displaystyle X}{\text{CH}_2\text{—CH—}}, \text{ etc.}$$

Carothers' original distinction between addition and condensation polymers was amended by Flory, who placed emphasis on the *mechanisms* by which the two types of polymer are formed. Condensation polymers are usually formed by the stepwise intermolecular condensation of reactive groups; addition polymers ordinarily result from chain reactions involving some sort of active center. Different chain structures may result from these mechanisms: the structural units of condensation polymers are usually joined by interunit functional groups, whereas most addition polymers do not have such functional groups in their backbone chain. However, this distinction is not always observed, leading to the breakdown of the classification as applied to polymerization products, shown in the following examples:

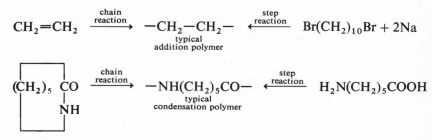

Polymerization mechanisms can be classified in two ways. If a stoichiometric basis is used, condensation polymerization may be said to involve the loss of a small molecule at each reaction step, whereas in addition polymerization this feature is absent. This scheme, however, classes the formation of polyurethanes with vinyl chain polymerization, which is kinetically quite different, rather than with the formation of polyesters, polyamides, etc. proceeding by entirely similar processes.

The classification adopted in this book is based on the reaction mechanism, dividing polymerization into *step reactions*, commonly producing step reaction or condensation polymers, and *chain reactions*, commonly producing chain-reaction or addition polymers (Mark 1950). Polymerizations are classi-

TABLE 8-1. Distinguishing features of chain- and step-polymerization mechanisms

Chain Polymerization	Step Polymerization
Only growth reaction adds repeating units one at a time to the chain.	Any two molecular species present can react.
Monomer concentration decreases steadily throughout reaction.	Monomer disappears early in reaction at DP* 10, less than 1 % monomer remains.
High polymer is formed at once; polymer molecular weight changes little throughout reaction.	Polymer molecular weight rises steadily throughout reaction.
Long reaction times give high yields but affect molecular weight little.	Long reaction times are essential to obtain high molecular weights.
Reaction mixture contains only monomer, high polymer, and about 10^{-8} part of growing chains.	At any stage all molecular species are present in a calculable distribution.

* Degree of polymerization.

fied without regard to loss of a small molecule (e.g., polyurethanes are formed by step-reaction polymerization) or type of interunit linkage (e.g., phenol-formaldehyde resins result from stepwise polymerization even though they lack interunit functional groups). Wherever precise differentiation on the basis of mechanism is required, the terms *step reaction* and *chain reaction* are used; in deference to well-established tradition, the common terms *condensation* and *addition* are permissible where no confusion can result.

Some of the consequences of the differences between the mechanisms of chain and stepwise polymerization are shown in Table 8-1.

GENERAL REFERENCES

Mark 1940; Flory 1953; Lenz 1967.

B. Chemistry of Stepwise Polymerization

Types of condensation polymers

Table 8-2 lists some representative products of stepwise polymerization. Most of them are, stoichiometrically, condensation polymers. Proteins and cellulose are included on the basis that they can be degraded hydrolytically to monomers differing from their repeating units by the addition of the elements of a molecule of water.

The type of product formed in a condensation reaction is determined by the *functionality* of the monomers, i.e., by the average number of reactive functional groups per monomer molecule. Monofunctional monomers give only low-molecular-weight products. Bifunctional monomers give linear polymers, as illustrated earlier in this chapter. Polyfunctional monomers, with more than two functional groups per molecule, give branched or cross-linked (three-dimensional) polymers. The properties of the linear and the three-dimensional polymers differ widely.

The chemistry of step-reaction polymerization is discussed in this section, following the classification scheme of Lenz (1967). Further information on the organic chemistry of the major types of condensation polymers having commercial utility is found in Part IV. Recent work has been reviewed (Lenz 1969, 1970).

Carbonyl addition-elimination mechanism

The most important reaction which has been used for the preparation of condensation polymers is that of addition and elimination at the carbonyl

TABLE 8-2. *Typical step-reaction polymers*

Type	Interunit Linkage	Examples
Polyester	$-\overset{\displaystyle O}{\overset{\|}{C}}-O-$	$HO(CH_2)_xCOOH \longrightarrow HO[-(CH_2)_xCOO-]_yH + H_2O$ $HO(CH_2)_xOH + HOOC(CH_2)_{x'}COOH \longrightarrow HO[-CH_2)_x\overset{O}{\overset{\|}{O}}C(CH_2)_{x'}\overset{O}{\overset{\|}{C}}O-]_yH + H_2O$ $\begin{array}{l} CH_2OH \\ | \\ CHOH + HOOC(CH_2)_xCOOH \longrightarrow \text{three-dimensional network} + H_2O \\ | \\ CH_2OH \end{array}$
Polyanhydride	$-\overset{\displaystyle O}{\overset{\|}{C}}-O-\overset{\displaystyle O}{\overset{\|}{C}}-$	$HOOC(CH_2)_xCOOH \longrightarrow HO[-CO(CH_2)_xCOO-]H + H_2O$
Polyacetal	$-O-\overset{\displaystyle H}{\underset{\displaystyle R}{\overset{\|}{\underset{\|}{C}}}}-O-$	$HO(CH_2)_xOH + CH_2(OR)_2 \longrightarrow HO[-(CH_2)_xOCH_2O-]_y(CH_2)_xOH + ROH$
Polyamide	$-\overset{\displaystyle O}{\overset{\|}{C}}-NH-$	$NH_2(CH_2)_xCOOH \longrightarrow H[-NH(CH_2)_xCO-]_yOH + H_2O$ $NH_2(CH_2)_xNH_2 + HOOC(CH_2)_{x'}COOH \longrightarrow H[-NH(CH_2)_xNHCO(CH_2)_{x'}CO-]_yOH + H_2O$
Polyurethane	$-O-\overset{\displaystyle O}{\overset{\|}{C}}-NH-$	$HO(CH_2)_xOH + OCN(CH_2)_{x'}CNO \longrightarrow [-O(CH_2)_xOCONH(CH_2)_{x'}NHCO-]_y$
Polyurea	$-NH-\overset{\displaystyle O}{\overset{\|}{C}}-NH-$	$NH_2(CH_2)_xNH_2 + OCN(CH_2)_xCNO \longrightarrow [-NH(CH_2)_xNHCONH(CH_2)_{x'}NHCO-]_y$
Silk fibroin	$-\overset{\displaystyle O}{\overset{\|}{C}}-NH-$	$NH_2CH_2COOH + NH_2CHRCOOH \longrightarrow H[-NHCH_2CONHCHRCO-]_yOH + H_2O$

Cellulose —C—O—C— $C_6H_{12}O_6 \longleftarrow -[C_6H_{10}O_4]-O-[C_6H_{10}O_4]- + H_2O$

Phenol-aldehyde —CH$_2$—
  —CH$_2$OCH$_2$—

$+ CH_2O \longrightarrow$ H$_2$O + three-dimensional network with —CH$_2$— and —CH$_2$OCH$_2$— bridges between positions on rings o- and p- to hydroxyls

Urea-aldehyde —NH—CHR—NH—
  —CHR—N—CHR—
  CHR—

$NH_2CONH_2 + CH_2O$ (1 to 1 ratio) $\longrightarrow$ —NHCONH—CH$_2$—NHCONH—CH$_2$— + H$_2$O

With excess CH$_2$O, three-dimensional network + H$_2$O

Polysulfide —S—
  —S—S—
  $\overset{S}{\underset{S}{\overset{\|}{\underset{\|}{-S-S-}}}}$

$Cl(CH_2)_xCl + Na_2S \longrightarrow Cl[-(CH_2)_xS-]_yNa + NaCl$

$HS(CH_2)_xSH + \text{oxidation} \longrightarrow HS[-(CH_2)_xSS-]_y(CH_2)_xSH + H_2O \text{ or } H_2S$

$Cl(CH_2)_xCl + Na_2S_4 \longrightarrow Cl[-(CH_2)_xS-\overset{S}{\underset{S}{\overset{\|}{\underset{\|}{-}}}}-S-]_yNa + NaCl$

Polysiloxane

$\longrightarrow$ three-dimensional network + H$_2$O

double bond of carboxylic acids and their derivatives. The generalized reaction is

$$
\underset{\substack{\|\\ \text{O}}}{R-C-X} + Y: \longrightarrow [R-\overset{\text{O:}}{\underset{\underset{X}{|}}{C}}-Y] \longrightarrow \underset{\substack{\|\\ \text{O}}}{R-C-Y} + X:
$$

where R and R' (below) may be alkyl or aryl groups, X may be OH, OR',

$$
\text{O}
$$
$$
\|
$$
NH$_2$, NHR', OCR', or Cl; and Y may be R'O$^-$, R'OH, R'NH$_2$, or R'COO$^-$. The species in the bracket is considered to be a metastable intermediate, which can either return to the original state by eliminating Y or proceed to the final state by eliminating X. The following paragraphs provide some typical examples of this reaction.

Direct reaction   The direct reaction of a dibasic acid and a glycol to form a polyester, or a dibasic acid and a diamine to form a polyamide, works well and is widely used in practice. In esterification, a strong acid or acidic salt often serves as a catalyst. The reaction may be carried out by heating the reactants together and removing water, usually applying vacuum in the later stages.

An important modification of the direct reaction is the use of a salt, as in the preparation of poly(hexamethylene adipamide) (66 nylon) by heating the hexamethylene diamine salt of adipic acid above the melting point in an inert atmosphere. The stringent requirement of stoichiometric equivalence to obtain high molecular weight is easily met by purifying the salt by recrystallization.

Interchange   The reaction between a glycol and an ester,

$$x\text{HO—R—OH} + x\text{R''OCO—R'—COOR''} \longrightarrow$$
$$\text{R''O(—CO—R'—COO—R—O)}x\text{H} + (2x-1)\text{R''—OH}$$

is often used to produce polyesters, especially where the dibasic acid has low solubility. Frequently the methyl ester is used, as in the production of poly(ethylene terephthalate) from ethylene glycol and dimethyl terephthalate. The reaction between a carboxyl and an ester link is much slower, but other interchange reactions, such as amine-amide, amine-ester, and acetal-alcohol, are well known.

Acid chloride or anhydride   Either of these species can be reacted with a glycol or an amine to give a polymer. The anhydride reaction is widely used to form an alkyd resin from phthalic anhydride and a glycol:

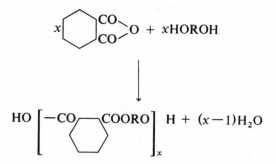

The condensation in bulk of an acid chloride with a glycol is not useful because of side reactions leading to low-molecular-weight products, but the reaction of an acid chloride with a diamine is a valuable means of preparing polyamides.

Interfacial condensation The reaction of an acid halide with a glycol or a diamine proceeds rapidly to high-molecular-weight polymer if carried out at the interface between two liquid phases, each containing one of the reactants (Morgan 1959). Very-high-molecular-weight polymer can be formed. Typically, an aqueous phase containing the diamine or glycol and an acid acceptor is layered at room temperature over an organic phase containing the acid chloride. The polymer formed at the interface can be pulled off as a continuous film or filament. The method has been applied to the formation of polyamides, polyurethanes, polyureas, polysulfonamides, and polyphenyl esters. It is particularly useful for preparing polymers which are unstable at the higher temperatures usual in step-reaction polymerization.

Ring versus chain formation In addition to polymer formation, bifunctional monomers may react intramolecularly to produce a cyclic product. Thus, hydroxy acids may give either lactones or polymers on heating,

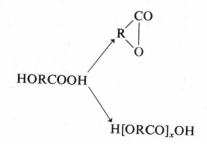

amino acids may give lactams or linear polyamides, etc.

The chief factor governing the type of product is the size of the ring which can be formed. If the ring contains less than five atoms or more than seven, the product is usually linear polymer. If a ring of five atoms can form, it will do so to the exclusion of linear chains. If six- or seven-membered rings can form, either type of product can result. Larger rings can sometimes be formed under special conditions.

The lack of formation of rings with less than five atoms is explained by the strain imposed by the valence angles of the ring atoms. Five-membered rings are virtually strain free, and all larger rings can be strain free if non-planar forms are possible. As the ring size increases, the statistical probability of forming rings becomes smaller. The formation of five-membered rings to the complete exclusion of linear polymer is not thoroughly understood.

The formation of high polymer from cyclic monomers of the type just discussed is often, kinetically, an anionic chain polymerization and is discussed in Chapter 10. This class of reaction includes the commercially important production of nylon from caprolactam.

Other mechanisms

Carbonyl addition-substitution reactions   The reaction of aldehydes with alcohols, involving first addition, then substitution at the carbonyl group is of great practical and historical importance in stepwise polymerization. The general reaction, leading to acetal formation, is

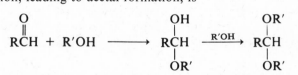

In addition to polyacetals, important polymers formed in this way include those of formaldehyde and phenol, urea, or melamine (Chapter 16).

Nucleophilic substitution reactions   These reactions are important from the standpoint of commercial organic polymers primarily because of their use in the polymerization of epoxides. The most common epoxide monomer is epichlorohydrin, which reacts with a nucleophile N: as follows:

$$\text{N: + } H_2C\!\!-\!\!CHCH_2Cl \longrightarrow NCH_2CHCH_2Cl$$
$$\diagdown\!\diagup\qquad\qquad\qquad\qquad |$$
$$O\qquad\qquad\qquad\qquad\quad O\!:^-$$

Typically, the nucleophile is a bifunctional hydroxy compound such as bisphenol A, and the reaction proceeds as described in Chapter 16.

A second nucleophilic substitution reaction of historical interest in polymer chemistry is that used for the production of the early polysulfide rubbers from aliphatic dichlorides and sodium sulfide:

$$x\text{ClCH}_2\text{CH}_2\text{Cl} + x\text{Na}_2\text{S}_n \longrightarrow -[\text{CH}_2\text{CH}_2\text{S}_n-]_x + 2x\text{NaCl}$$

This type of reaction can be used for the production of semiorganic and inorganic polymers, and is probably also the basis for the formation of natural polysaccharides and polynucleotides by polymerization in living organisms catalyzed by enzymes.

Double-bond addition reactions   Although addition reactions at double bonds are often associated with polymerization by chain mechanisms (Chapters 9, 10), this is not necessarily the case, and many important stepwise polymerizations are based on this reaction. Of major interest among these is the ionic addition of diols to diisocyanates in the production of polyurethanes (Chapter 15):

The free-radical addition of dithiols to unconjugated diolefins also proceeds by a stepwise mechanism.

Free-radical coupling   These reactions lead to several polymerization schemes of interest, including the preparation of arylene ether polymers, polymers containing acetylene units, and arylene alkylidene polymers. If [Ox] is an oxidizing agent, the first of these can be written

Aromatic electrophilic-substitution reactions   Reactions of this type, using standard Friedel-Crafts catalysts, produce polymers by step-growth mechanisms. A typical example is the production of poly(p-phenylene):

GENERAL REFERENCES

Mark 1940; Marvel 1959; Morgan 1965; Lenz 1967, Chaps. 4–8; Ravve 1967, Part IV; Sorenson 1968.

C. Kinetics and Statistics of Linear Stepwise Polymerization

Reactivity and molecular size

The similarity between monofunctional and polyfunctional step reactions is illustrated in the chemical equations in this chapter. Other similarities exist in the influence of temperature and catalysts on the rates of the reactions. It is tempting to substitute the simple concept of reactions between functional groups for that of the myriad separate reactions of each molecular species with all the others. With this substitution, the kinetics of stepwise polymerization is essentially identical with that of simple condensation; without it, analysis of the kinetics would be almost hopelessly difficult.

If the simplification is to be made, the rate of the reaction of a group must be independent of the size of the molecule to which it is attached. This assumption is amply justified by experimental evidence. The rate constants of condensation reactions in a homologous series reach asymptotic values independent of chain length quite rapidly and show no tendency to drop off with increase in molecular size. The diluting effect of the large chain must, of course, be taken into account by adjusting to constant molar composition of reactive groups. In addition, the rate constants for monofunctional and bifunctional reagents are identical for sufficiently long chains separating the reactive groups of the bifunctional compound.

In explanation of these results, it may be recalled that chemical reactions in condensed phases usually take place during a period of many collisions between reactive groups before they diffuse apart. Although a polymer chain as a whole diffuses very slowly, the mobility of the terminal functional group on the chain is much greater than that of the entire chain. Such a group diffuses readily over a considerable region through rearrangements in the conformations of nearby chain segments. Its collision rate with its neighbors is at least comparable to that prevailing in liquids. A somewhat lower diffusion rate not only prolongs the time before two groups diffuse into the same neighborhood, but proportionally prolongs the time during which the groups are close together and colliding. Stepwise polymerizations take place at rates such that only about one collision in 10^{13} leads to chemical reaction; within the length of time necessary for this number of collisions there is ample opportunity for diffusion to maintain the concentration of reactive pairs essentially at equilibrium.

In very dilute solution, the reactivity of a functional group may be reduced because the segments of its own molecule shield it from functional groups on other molecules. In concentrated systems, however, the molecules are extensively intertwined and show no preference for units of their own hain over those of other molecules. Here it is sufficient to dispose of the shielding problem by taking the diluent effect of nonreactive segments into account by computing the concentrations of functional groups per unit volume.

Kinetics of stepwise polymerization

With the concept of functional group reactivity independent of molecular weight, the kinetics of stepwise polymerization becomes quite simple. The formation of a polyester from a glycol and a dibasic acid may be taken as an example. It is well known that this reaction is catalyzed by acids. In the absence of added strong acid, a second molecule of the acid being esterified acts as a catalyst. The reaction is followed by measuring the rate of disappearance of carboxyl groups:*

$$-\frac{d[COOH]}{dt} = k[COOH]^2[OH] \tag{8-1}$$

If the concentrations c of carboxyl and hydroxyl groups are equal,

$$\frac{-dc}{dt} = kc^3$$

$$2kt = \frac{1}{c^2} - \text{const.} \tag{8-2}$$

and

It is convenient to introduce the *extent of reaction*, p, defined as the fraction of the functional groups that has reacted at time t. Then

$$c = c_0(1 - p) \tag{8-3}$$

and

$$2c_0^2kt = \frac{1}{(1 - p)^2} + \text{const.} \tag{8-4}$$

or a plot of $1/(1 - p)^2$ should be linear in time. Typical experimental data (Fig. 8-1) bear this out over a wide range of reaction times.

Stepwise polymerization typically involves equilibrium reactions of the type

$$A + B \underset{k_r}{\overset{k_f}{\rightleftharpoons}} C + D$$

where the rates of the forward and reverse reactions are $k_f[A][B]$ and $k_r[C][D]$, respectively. At equilibrium these rates are equal, whence $K = k_r/k_f = [A][B]/[C][D]$. If the system is far from equilibrium, as in the initial stages of polymerization, the reverse reaction is negligibly slow, and changes in the concentrations of the reactants may be considered to result from the forward reaction alone, as in Eq. 8-1 and those following.

If only bifunctional reactants are present and no side reactions occur, the number of unreacted carboxyl groups equals the total number of molecules in the system N. If acid or glycol groups separately (not in pairs) are defined as structural units, the initial number of carboxyls present is equal to the total number of structural units present, N_0. The number-average degree of polymerization $\bar{x}_n$ is simply

$$\bar{x}_n = \frac{N_0}{N} = \frac{c_0}{c} = \frac{1}{1-p} \tag{8-5}$$

The scale of DP in Fig. 8-1 shows that uncatalyzed esterifications require quite long times to reach high degrees of polymerization. Greater success is achieved by adding to the system a small amount of catalyst, whose concentration is constant throughout the reaction. In this case the concentration of the catalyst may be included in the rate constant:

$$- \frac{d[\text{COOH}]}{dt} = k'[\text{COOH}][\text{OH}]$$

$$- \frac{dc}{dt} = k'c^2 \tag{8-6}$$

$$c_0 k' t = \frac{1}{1-p} + \text{const.}$$

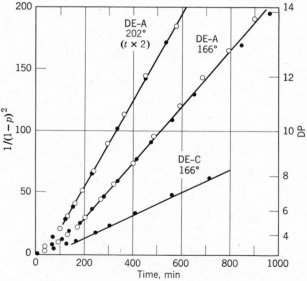

Fig. 8-1. Reactions of diethylene glycol (DE) with adipic acid (A) and caproic acid (C) (Flory 1946, 1949). Time values at 202°C have been multiplied by 2.

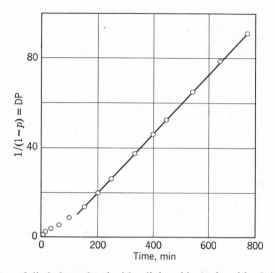

g. 8-2.   Reaction of diethylene glycol with adipic acid, catalyzed by 0.4 mole percent of toluenesulfonic acid (Flory 1946, 1949).

ere $\bar{x}_n$ increases linearly with reaction time (Fig. 8-2). Similar reactions ave been shown to be linear up to at least $\bar{x}_n = 90$, corresponding to molecular weights of 10,000.

Statistics of linear step-reaction polymerization

Rate equations similar to Eq. 8-6 may be written for each molecular ecies in the reaction mixture. From these it is possible to derive the distriution of molecular weights in a step-reaction polymer. The same relation , however, more readily derived from statistical considerations.

The statistical analysis of stepwise polymerization yields results which re equivalent to and in some cases more easily obtained than those of kinetic onsiderations. The method assumes independence of reaction rate and nolecular size. The extent of reaction p is defined as the probability that a nctional group has reacted at time t. This definition is entirely equivalent the previous definition of p as the fraction of the functional groups which as reacted. It follows that the probability of finding a functional group nreacted is $1 - p$.

It is now necessary to find the probability that a given molecule selected t random is an x-mer, i.e., contains x repeating units. Such a molecule conins $x - 1$ reacted functional groups (e.g., carboxyls) and, on the end, one nreacted group of this type. The probability of finding a single reacted carboxyl group in the molecule is p, and that of finding $x - 1$ of them in the same molecule is p^{x-1}. The presence of one unreacted group has a probability of $1 - p$. Hence the probability of finding the complete molecule i $p^{x-1}(1 - p)$. This is also equal to the fraction of all the molecules which are x-mers. If there are N molecules in all, the total number of x-mers is

$$N_x = Np^{x-1}(1 - p) \qquad (8-7)$$

If the total number of units present is N_0, $N = N_0(1 - p)$, and

$$N_x = N_0(1 - p)^2 p^{x-1} \qquad (8-8)$$

This is the number-distribution function for a linear stepwise polymerization at extent of reaction p.

The weight fraction w_x of x-mers is given by

$$w_x = \frac{x N_x}{N_0}$$
$$= x(1 - p)^2 p^{x-1} \qquad (8-9)$$

This is the weight-distribution function for a linear stepwise polymerization at extent of reaction p. Equations 8-8 and 8-9 are illustrated in Figs. 8-3 and 8-4, respectively. On a number basis, monomers are more plentiful than any other molecular species at all stages of the reaction. On a weight basis however, the proportion of low-molecular-weight material is very small and

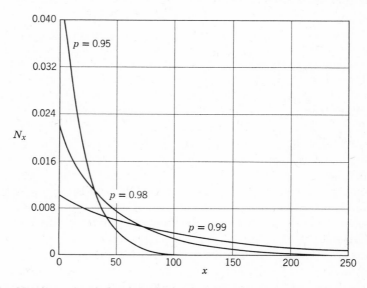

Fig. 8-3. Number or mole-fraction distribution of chain molecules in a linear step-reaction polymer for several extents of reaction p (Flory 1946, 1949).

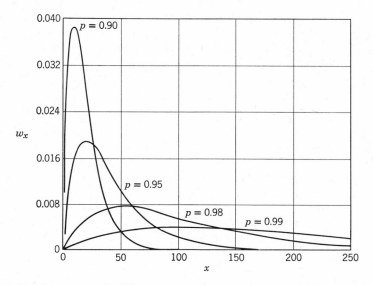

w_x

$p = 0.90$

$p = 0.95$

$p = 0.98$

$p = 0.99$

x

Fig. 8-4. Weight-fraction distribution of chain molecules in a linear step-reaction polymer for several extents of reaction p (Flory 1946, 1949).

decreases as the average molecular weight increases. The maximum in the weight-distribution curve occurs near the number-average molecular weight.

Various average degrees of polymerization may be derived from the weight- or number-distribution functions according to the definitions of the corresponding molecular weight averages in Chapter 3. The number-average degree of polymerization $\bar{x}_n$ calculated in this way is that given in Eq. 8-5. The weight-average degree of polymerization is

$$\bar{x}_w = \frac{1 + p}{1 - p} \tag{8-10}$$

The breadth of the molecular-weight distribution curve, $\bar{x}_w/\bar{x}_n = 1 + p$; at large extents of reaction $\bar{x}_w/\bar{x}_n \to 2.0$. This has been confirmed experimentally for nylon (Howard 1961).

Molecular weight control

Although initial molecular weight control in a stepwise polymerization can be achieved by stopping the reaction (e.g., by cooling) at the desired point, the stability of the product to subsequent heating may not be adequate with respect to changes in molecular weight.

The easiest way to avoid this situation is to adjust the composition of the reaction mixture slightly away from stoichiometric equivalence, by adding either a slight excess of one bifunctional reactant or (as in molecular weight stabilization of nylons by acetic acid) a small amount of a monofunctional reagent. Eventually the functional group deficient in amount is completely used up, and all chain ends consist of the group present in excess. If only bifunctional reactants are present, the two types of groups being designated A and B and initially present in numbers $N_A < N_B$ such that the ratio $r = N_A/N_B$, the total number of monomers present is $(N_A + N_B)/2 = N_A(1 + 1/r)/2$. At extent of reaction p (defined for A groups; for B groups, extent of reaction $= rp$), the total number of chain ends is $N_A(1 - p) + N_B(1 - rp) = N_A[1 - p + (1 - rp)/r]$. Since this is twice the number of molecules present,

$$\bar{x}_n = \frac{N_A(1 + 1/r)/2}{N_A[1 - p + (1 - rp)/r]/2} = \frac{1 + r}{1 + r - 2rp} \tag{8-11}$$

As $p \to 1$,

$$\bar{x}_n = \frac{1 + r}{1 - r} \tag{8-12}$$

Thus if 1 mole per cent of stabilizing groups is added,

$$\bar{x}_n = \frac{1 + (100/101)}{1 - (100/101)} = 201.$$

This illustrates the precision needed in maintaining stoichiometric balance in order to obtain high degrees of polymerization. Loss of one ingredient, side reactions, or the presence of monofunctional impurities may limit severely the degree of polymerization which can be achieved.

The analysis can be applied to the case of an added monofunctional reagent, retaining Eqs. 8-11 and 8-12 unchanged if r is appropriately defined in terms of the numbers of functional groups present.

Interchange reaction

In interchange reactions, which may occur freely at elevated temperature as in polymer melts, the molecular weights of the reacting molecules can change, since, e.g., two molecules of average length may react to give one longer and one shorter than average. However, the total number of molecules and hence $\bar{x}_n$ do not change.

It can be shown that, if free interchange takes place, the final molecular weight distribution is always the most probable distribution defined by Eqs. 8-8 and 8-9. The change can be demonstrated by mixing two polymers of different $\bar{x}_n$ and following a weight-average property (such as melt viscosity as a function of time at elevated temperature.

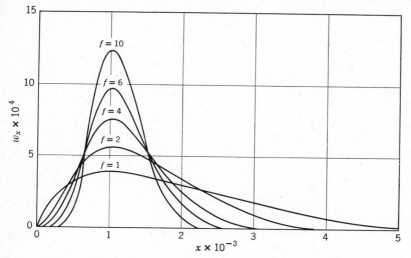

Fig. 8-5. Molecular weight distributions for multichain polymers with functionality f as shown (Schulz 1939).

Multichain polymer

Another type of step-reaction polymer of interest is produced by polymerizing a bifunctional monomer of type A—B with a small amount of monomer of functionality f of the type R—A$_f$. If the reaction is carried nearly to completion, the resulting polymer consists of f chains growing out from a central unit R. Network structures cannot occur, since no units of the type B—B are present. The case $f = 1$ corresponds to the most probable distribution of Eq. 8-9. The polymer formed when $f = 2$ is also linear; those with higher functionalities are branched.

Since the length of each of the branches is statistically determined independent of all the others, the probability of having a molecule with several branches all longer or shorter than average becomes smaller as the number of branches increases. The molecular weight distribution thus becomes narrower with increasing functionality, as shown in Fig. 8-5 (Schulz 1939). The distribution breadth is given by

$$\frac{\bar{x}_w}{\bar{x}_n} = 1 + \frac{1}{f} \qquad (8\text{-}13)$$

For $f = 2$, $\bar{x}_w/\bar{x}_n = 1.5$.

GENERAL REFERENCES

Flory 1946, 1949, 1953; Howard 1961; Lenz 1967; Odian 1970.

D. Polyfunctional Step-Reaction Polymerization

Three-dimensional step-reaction polymers are produced from the pol merization of reactants with more than two functional groups per molecul The structures of these polymers are more complex than those of linear ste reaction polymers. Three-dimensional polymerization is complicated exper mentally by the occurrence of *gelation*, or the formation of essentially infinite large polymer networks in the reaction mixture. The sudden onset of gelatic marks the division of the mixture into two parts: the *gel*, which is insolub in all nondegrading solvents; and the *sol*, which remains soluble and can l extracted from the gel. As the polymerization proceeds beyond the *gel po* the amount of gel increases at the expense of sol, and the mixture rapid transforms from a viscous liquid to an elastic material of infinite viscosit An important feature of the onset of gelation is the low number-avera molecular weight of the mixture at the gel point, where its weight-avera molecular weight becomes infinite.

In the statistical consideration of three-dimensional step polymerizatio it is assumed that all functional groups are equally reactive, independent molecular weight or viscosity. This assumption is not correct, e.g., for gl cerol, where the secondary hydroxyl group is known to be less reactive the the two primary hydroxyls; but this complication does not affect the gener conclusions of the theory.

It is also assumed that all the reactions occur between functional grou on different molecules. This is known to be somewhat in error, for it can l shown that the number of interunit links formed is always somewhat great than the corresponding decrease in the number of molecules present. Tl error again affects only the numerical agreement between theory and exper ment and not the over-all considerations of the theory.

Gelatic

Prediction of the gel point   In order to calculate the point in the reaction which gelation takes place, a *branching coefficient* α is defined as the prob bility that a given functional group on a *branch unit* (i.e., a unit of functionali greater than 2) is connected to another branch unit.

The value of α at which gelation becomes possible can be deduced follows. Consider the case where bifunctional A—A and B—B units a present, as well as polyfunctional units A_f with functionality f. The stru tures resulting in this system consist of chain segments of the type

$$A_{f-1}—A[B—BA—A]_i B—BA—A_{f-1}$$

where i may have any value, plus other segments which lack branch units on one or both ends. The criterion for gel formation is that at least one of the $f - 1$ segments radiating from the end of a segment of the type shown is in turn connected to another branch unit. The probability of this occurring is just $1/(f - 1)$; hence the critical value of α for gelation is

$$\alpha_c = \frac{1}{f - 1} \tag{8-14}$$

Here f is the functionality of the branch units; if more than one type of branch unit is present, an average f over all types of branch units may be used in Eq. 8-14.

To relate α to the extent of reaction, consider the probability of obtaining a segment of the type shown. If the extents of reaction for A and B groups are p_A and p_B, and the ratio of A groups on branch units to all A groups in the mixture is ρ, the probability that a B group has reacted with a branch unit is $p_B\rho$; with a bifunctional A, $p_B(1 - \rho)$. The probability that a segment of the type shown is obtained is given by

$$p_A[p_B(1 - \rho)p_A]^i p_B\rho$$

Summing over all values of i gives

$$\alpha = \frac{p_A p_B \rho}{1 - p_A p_B(1 - \rho)} \tag{8-15}$$

Either p_A or p_B can be eliminated from this expression by defining $r = N_A/N_B$, whence $p_B = rp_A$. Then

$$\alpha = \frac{rp_A^2\rho}{1 - rp_A^2(1 - \rho)} = \frac{p_B^2\rho}{r - p_B^2(1 - \rho)} \tag{8-16}$$

Simpler relations can be derived for several special cases. When equal numbers of A and B groups are present, $r = 1$ and $p_A = p_B = p$:

$$(r = 1) \quad \alpha = \frac{p^2\rho}{1 - p^2(1 - \rho)} \tag{8-17}$$

When there are no A—A units, $\rho = 1$ and

$$(\rho = 1) \quad \alpha = rp_A^2 = \frac{p_B^2}{r} \tag{8-18}$$

If both conditions apply,

$$(r = \rho = 1) \quad \alpha = p^2 \tag{8-19}$$

Finally, with only branch units present the probability that a functional group on a branch unit leads to another branch unit is just the probability that it has reacted:

$$\text{(branch units only)} \quad \alpha = p \qquad (8\text{-}20)$$

The equations hold for all branch unit functionalities with the definitions given for r and ρ.

Experimental observations of the gel point   The gel point can be observed precisely as the time when the polymerizing mixture suddenly loses fluidity, e.g., when bubbles no longer rise in it. If the extent of reaction has been followed as a function of time, say by removing aliquots of the solution and titrating for the number of functional groups present, the value of p at the gel point can be determined.

In a series of experiments with glycerol and dibasic acids, gelation occurred at $p = 0.765$. Here Eq. 8-19 was appropriate; hence $\alpha_c = p^2 = 0.58$. Correction for the lower reactivity of the secondary hydroxyl of glycerol would lower this value slightly. From Eq. 8-14 the theoretical value of α_c is 0.50.

In several other cases investigated by Flory, observed values of α_c were always slightly higher than those theoretically required. This discrepancy is attributed to the reaction of some functional groups to form intramolecular links, which do not contribute to network structures. The reactions must, therefore, be carried slightly farther to reach the critical point.

Molecular weight distributions in three-dimensional step-reaction polymers

The distribution functions for three-dimensional polymers are derived with somewhat more difficulty than those for the linear case, Eqs. 8-8 and 8-9. They depend upon the functionality and relative amounts of all the units involved. Only one example is discussed, for simplicity that of the reaction of equivalent quantities of two trifunctional monomers, all three of the functional groups of each monomer being equally reactive. In this case the weight-distribution function is

$$w_x = \left[\frac{(fx - x)!f}{(x - 1)!(fx - 2x + 2)!} \right] p^{x-1}(1 - p)^{fx - 2x + 2} \qquad (8\text{-}21)$$

This equation is analogous to Eq. 8-9. The first factor (in brackets) arises because of the numerous geometric isomers of a polyfunctional x-mer. (In the linear case this factor is just x.) The second term is the probability of

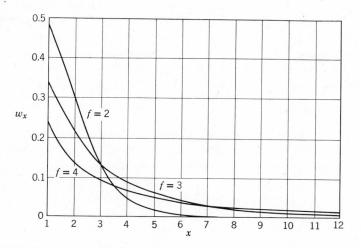

Fig. 8-6. Molecular size distribution in stepwise polymerizations for monomers with functionality f, at $p = 0.3$ (Mark 1950).

finding $x - 1$ links in an x-mer, and the last term is the probability of finding $(f - 2)x + 2 = fx - 2x + 2$ unreacted links or ends.

Equation 8-21 is illustrated in Figs. 8-6, 8-7, and 8-8. In comparison with the linear case, the weight distributions of branched step-reaction polymers of increasing functionality are progressively broader at equivalent extents of reaction (Fig. 8-6). Figure 8-7 shows that the distributions broaden out with increasing extent of reaction p (or, alternatively, α, in the trifunctional case).

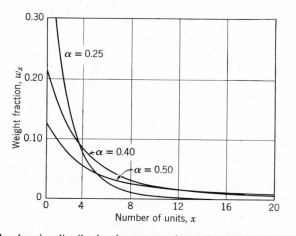

Fig. 8-7. Molecular size distribution in step-reaction polymer formed from trifunctional units at various stages in the reaction, denoted by $\alpha = p$ (Flory 1946, 1949).

In Fig. 8-8 the weight fraction of the various molecular species is plotted against $\alpha = p$. In contrast to the linear case (Figs. 8-7 and 8-8 vs. Fig. 8-4) the weight fraction of monomer is always greater than the amount of any one of the other species, the weight fractions of higher species being successively lower. The extent of reaction at which the weight fraction of any species reaches its maximum shifts continuously to higher values for higher molecular weights. In no case, however, does the maximum occur beyond the gel point ($\alpha = \frac{1}{2}$).

Up to the gel point the sum of the weight fractions of all the species present must equal unity; beyond this point the sum of the weight fractions of all finite species drops below unity as the weight fraction of gel increases. At $\alpha = 1$ only gel is present.

The number-average degree of polymerization is given by

$$\bar{x}_n = \frac{1}{1 - fp/2} \tag{8-22}$$

and the weight-average degree by

$$\bar{x}_w = x w_x = \frac{1 + p}{1 - (f - 1)p} \tag{8-23}$$

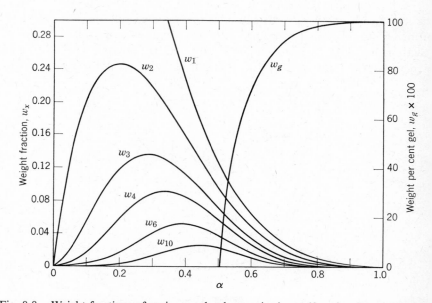

Fig. 8-8. Weight fractions of various molecular species in a trifunctional stepwise polymerization as a function of $a = p$ (Flory 1946, 1949). The weight fractions of the finite species are calculated from Eq. 8-21, that of the gel from Eq. 8-24.

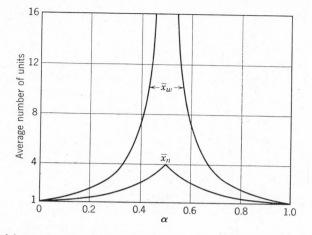

Fig. 8-9. Weight- and number-average degree of polymerization as a function of α for a trifunctional step-reaction polymer (Flory, 1946, 1949).

At the gel point the weight-average degree of polymerization becomes infinite. As may be seen in Fig. 8-9, where both averages are plotted against $\alpha = p$, $\bar{x}_n$ attains only a value of $1/(1 - \frac{3}{4}) = 4$ at this point. The very large values of $\bar{x}_w/\bar{x}_n$ near the gel point illustrate the extreme breadth of the distributions. (The portions of the curves in Fig. 8-9 above the gel point refer only to the sol fraction.)

The increase in distribution breadth with increasing degree of branching is illustrated in Fig. 8-10. These curves represent the calculated distributions for bifunctional step reactions in which the branching factor α was varied by varying the small amount of added trifunctional units. The curves were calculated for $\bar{x}_n = 50$ in each case. The curves are drawn so that the total area under each curve extended to infinite x is the same.

Equation 8-21 is valid for finite species beyond the gel point. If it is summed over all values of x and subtracted from unity, the weight fraction of gel results. For trifunctional units this is

$$w_g = 1 - \frac{(1-p)^3}{p^3} \tag{8-24}$$

The distribution functions for $\alpha > \frac{1}{2}$ are identical with those at the corresponding lower value $1 - \alpha$ except that they are reduced by a factor $w_s = 1 - w_g$, the total weight fraction of sol present at that value of α. Thus the complexity of the distribution of the sol fraction above the gel point is presumed to become less until, at $\alpha = 1$, only monomer is left in a vanishingly small amount. The reduction of $\bar{x}_w/\bar{x}_n$ shown for values of α increasing above $\frac{1}{2}$ in Fig. 8-9 is another way of expressing the same phenomenon.

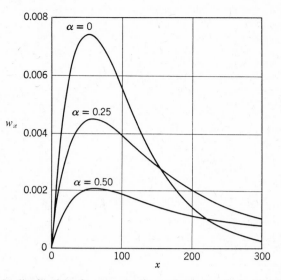

Fig. 8-10. Weight distributions for step-reaction polymers made to $\bar{x}_n = 50$ with various degrees of branching α induced by the addition of trifunctional units: $\alpha = 0$ indicates no branching, and $\alpha = 0.50$ is the critical point for gelation (Flory 1946, 1949).

Although the discussion was confined to the case of equivalent quantities of two trifunctional monomers, and Figs. 8-6 to 8-10 refer specifically to this case, Eqs. 8-21 to 8-23 are more broadly applicable if p is replaced with α, which is then determined in the case of interest by r, p, and ρ, as indicated in Eq. 8-16.

GENERAL REFERENCES

Flory 1946, 1949, 1953; Howard 1961; Lenz 1967; Odian 1970.

BIBLIOGRAPHY

Carothers 1929. W. H. Carothers, "An Introduction to the General Theory of Condensation Polymers," *J. Am. Chem. Soc.* **51**, 2548–2559 (1929).

Flory 1946. Paul J. Flory, "Fundamental Principles of Condensation Polymerization," *Chem. Revs.* **39**, 137–197 (1946).

Flory 1949. Paul J. Flory, "Condensation Polymerization and Constitution of Condensation Polymers," pp. 211–283 in R. E. Burk and Oliver Grummitt, eds., *High Molecular Weight Organic Compounds* (*Frontiers in Chemistry*, Vol. VI), Interscience Publishers, New York, 1949.

Flory 1953. Paul J. Flory, *Principles of Polymer Chemistry*, Cornell University Press, Ithaca, N.Y., 1953.

Howard 1961. G. J. Howard, "The Molecular Weight Distribution of Condensation Polymers," pp. 185–231 in J. C. Robb and F. W. Peaker, eds., *Progress in High Polymers*, Vol. 1, Academic Press, New York, 1961.

Lenz 1967. Robert W. Lenz, *Organic Chemistry of Synthetic High Polymers*, Interscience Div., John Wiley and Sons, New York, 1967.

Lenz 1969. Robert W. Lenz, "Applied Reaction Kinetics: Polymerization Reaction Kinetics," *Ind. Eng. Chem.* **61** (3), 67–75 (1969).

Lenz 1970. Robert W. Lenz, "Applied Polymerization Reaction Kinetics," *Ind. Eng. Chem.* **62** (2), 54–61 (1970).

Mark 1940. H. Mark and G. Stafford Whitby, eds., *Collected Papers of Wallace Hume Carothers on High Polymeric Substances*, Interscience Publishers, New York, 1940.

Mark 1950. H. Mark and A. V. Tobolsky, *Physical Chemistry of High Polymeric Systems*, Interscience Publishers, New York, 1950.

Marvel 1959. C. S. Marvel, *An Introduction to the Organic Chemistry of High Polymers*, John Wiley and Sons, New York, 1959.

Morgon 1959. Paul W. Morgan and Stephanie L. Kwolek, "The Nylon Rope Trick," *J. Chem. Educ.* **36**, 182–184 (1959).

Morgan 1965. Paul W. Morgan, *Condensation Polymers: by Interfacial and Solution Methods*, Interscience Div., John Wiley and Sons, New York, 1965.

Odian 1970. George Odian, *Principles of Polymerization*, McGraw-Hill Book Company, New York, 1970.

Ravve 1967. A. Ravve, *Organic Chemistry of Macromolecules*, Marcel Dekker, New York, 1967.

Schulz 1939. G. V. Schulz, "The Kinetics of Chain Polymerization. V. The Effect of Various Reaction Species on the Polymolecularity" (in German), *Z. physik. Chem.* **B43**, 25–46 (1939).

Sorenson 1968. Wayne R. Sorenson and Tod W. Campbell, *Preparative Methods of Polymer Chemistry*, 2nd ed., Interscience Div., John Wiley and Sons, New York, 1968.

9

Radical Chain (Addition) Polymerization

A. Chemistry of Vinyl Polymerization

The polymerization of unsaturated monomers typically involves a chain reaction. It can be initiated by methods typical for simple gas-phase chain reactions, including the action of ultraviolet light. It is susceptible to retardation and inhibition. In a typical chain polymerization, one act of initiation may lead to the polymerization of thousands of monomer molecules.

The characteristics of chain polymerization listed in Table 8-1 suggest that the active center responsible for the growth of the chain is associated with a single polymer molecule through the addition of many monomer units. Thus polymer molecules are formed from the beginning, and almost no species intermediate between monomer and high-molecular-weight polymer are found. Of several postulated types of active center, three have been found experimentally: cation, anion, and free radical. Free radical polymerization is discussed in this chapter, and the related cases of ionic and coordination polymerization are described in Chapter 10.

The concept of vinyl polymerization as a chain mechanism is not new, dating back to Staudinger's work in 1920. However, an alternative mechanism of a stepwise reaction involving hydrogen transfer was seriously considered as late as 1936. The controversy was largely settled by Flory's analysis of the kinetics of vinyl polymerization in 1937. He showed conclusively that radical polymerization proceeds by and requires the steps of initiation, propagation, and termination typical of chain reactions in low-molecular-weight species.

Vinyl monomers

Electronic structures of organic compounds As noted in Chapter 1*C*, most organic compounds involve covalent or shared-electron bonds. Single covalent bonds have bond electrons which are distributed within a considerable olume of space between the two nuclei and symmetrically about the line onnecting them. In a double bond, the presence of the second pair of lectrons destroys the axial symmetry, modifying the volume within which he electrons are found to a ribbonlike band between the nuclei. As a result here is a high energy barrier preventing free rotation, and *cis-trans* isomers •ecome possible. The fact that the second electron pair is held only about 0% as firmly as the first leads to high polarizability and chemical reactivity n unsaturated compounds.

The positions and reactivities of the electrons in unsaturated molecules re subject to several geometrical and electrical influences. *Mesomerism* and *esonance* are used to explain certain facts, such as the equivalence of the eactivity of all the hydrogen atoms in benzene. The presence of strongly lectropositive or electronegative groups in a molecule may cause an *inductive hift* of electrons. In chloroacetic acid, e.g., the shift of electrons toward the hlorine atom effectively increases the strength of the acid:

$$Cl \longleftarrow CH_2 \longleftarrow \overset{\overset{\displaystyle O}{\parallel}}{C} \longleftarrow O \longleftarrow H$$

\ further effect is purely *electrostatic* in nature: if, e.g., an atom or group of .toms is substituted for a hydrogen atom adjacent to a double bond, the lectrostatic force exerted by the electrons of the substituent on those of the louble bond affects the reactivity of the molecule.

The end result of these influences is that the reaction of the double bond vith a free radical proceeds well for compounds of the type $CH_2{=}CHX$ and ?$H_2{=}CXY$, called *vinyl monomers*. (Monomers in which fluorine is substi-uted for hydrogen may be included in this class.) The polymerization of nonomers with more than one double bond is considered in Chapter 12.

Not all vinyl monomers yield high polymer as a result of radical poly-nerization. Aliphatic hydrocarbons other than ethylene polymerize only to •ils; 1,2-disubstituted ethylenes not at all. Among compounds of the type ?$H_2{=}CXY$, those in which both groups are larger than CH_3 polymerize lowly if at all.

Mechanism of vinyl radical polymerization

;eneration of free radicals   Many organic reactions take place through inter-nediates having an odd number of electrons and, consequently, an unpaired lectron. Such intermediates are known as *free radicals*. They can be ,enerated in a number of ways, including thermal decomposition of organic •eroxides or hydroperoxides (Mageli 1968) or azo or diazo compounds (Zand 965). Other modes of generation of free radicals, occasionally used to nitiate polymerization, include photolytic decomposition of covalently •onded compounds (photoinitiation, Section *B*); dissociation of covalent

bonds by high-energy radiation (Chapter 12*F*); oxidation-reduction reaction (redox initiation, Chapter 12*B*); and electrochemical initiation (Friedlande 1966; Funt 1967).

Two reactions commonly used to produce radicals for polymerizatio are the thermal or photochemical decomposition of benzoyl peroxide:

$$(C_6H_5COO)_2 \longrightarrow 2C_6H_5COO\cdot \longrightarrow 2C_6H_5\cdot + 2CO_2$$

and of azobisisobutyronitrile:

$$(CH_3)_2\underset{\underset{CN}{|}}{C}N{=}N\underset{\underset{CN}{|}}{C}(CH_3)_2 \longrightarrow 2(CH_3)_2\underset{\underset{CN}{|}}{C}\cdot + N_2$$

The stability of radicals varies widely. Primary radicals are less stabl and more reactive than secondary radicals, which are in turn less stha tertiary ones. (Some tertiary radicals, such as the triphenylmethyl, can b isolated in the solid state without decomposition.) The phenyl radical is mor reactive than the benzyl radical, the allyl radical is quite unreactive, and so on

Initiation   When free radicals are generated in the presence of a vinyl mono mer, the radical adds to the double bond with the regeneration of anothe radical. If the radical formed by decomposition of the *initiator* I is designate R·,

$$I \longrightarrow 2R\cdot$$

$$R\cdot + CH_2{=}CHX \longrightarrow RCH_2\underset{\underset{X}{|}}{\overset{\overset{H}{|}}{C}}\cdot$$

The regeneration of the radical is characteristic of chain reactions.

Evidence for the radical mechanism of addition polymerization come not only from the capability of radicals to accelerate vinyl polymerization but also from the demonstration that the polymers so formed contain frag ments of the radicals. The presence of heavy atoms, such as bromine o iodine, or radioactive atoms, in the initiator has been shown many times to lead to polymers from which these atoms cannot be removed.

The efficiency with which radicals initiate chains can be estimated by comparing the amount of initiator decomposed with the number of polyme chains formed. The decomposition of the initiator can usually be followe by analytical methods. The most direct method of finding the initiator effi ciency then depends upon analyzing the polymer for initiator fragments This is not difficult in cases where the initiator leaves a reactive end grouf on the polymer or is radioactively tagged. In other polymerizations it is more useful to determine the number of polymer molecules formed from the number-average molecular weight of the polymer. Still another alternative sometimes of doubtful validity, is to react the chain radical stoichiometrically with an inhibitor (see page 285). Most initiators in typical vinyl polymerizations have efficiencies between 0.6 and 1.0; i.e., between 60 and 100% of all the radicals formed ultimately initiate polymer chains. The major cause of low efficiency is recombination of the radical pairs before they move apart (cage effect) (Noyes 1965).

Propagation The chain radical formed in the initiation step is capable of adding successive monomers to propagate the chain:

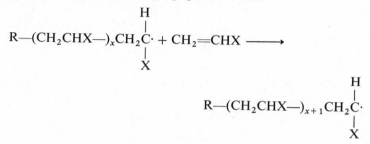

Termination Propagation would continue until the supply of monomer was exhausted were it not for the strong tendency of radicals to react in pairs to form a paired-electron covalent bond with loss of radical activity. This tendency is compensated for in radical polymerization by the small concentration of radical species compared to monomers.

The termination step can take place in two ways:

Combination or *coupling*:

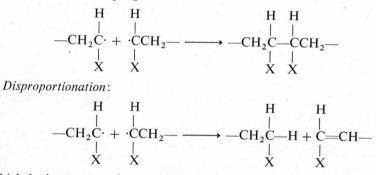

Disproportionation:

in which hydrogen transfer results in the formation of two molecules with one saturated and one unsaturated end group. Each type of termination is known. Studies (Bevington 1954*a,b*) of the number of initiator fragments per molecule show that polystyrene terminates predominantly by combination, whereas poly(methyl methacrylate) terminates entirely by disproportionation at polymerization temperatures above 60°C, and partly by each mechanism at lower temperatures.

Although the three steps of initiation, propagation, and termination are both necessary and sufficient for chain polymerization, other steps can take place during polymerization. As these often involve the reaction between a radical and a molecule, they are conveniently so classified.

Chain transfer It was recognized by Flory (1937) that the reactivity of a radical could be transferred to another species, which would usually be capable of continuing the chain reaction. The reaction involves the transfer of an atom between the radical and the molecule. If the molecule is saturated, like a solvent or other additive, the atom must be transferred to the radical:

$$\underset{\underset{X}{|}}{\overset{\overset{H}{|}}{-CH_2C\cdot}} + CCl_4 \longrightarrow \underset{\underset{X}{|}}{\overset{\overset{H}{|}}{-CH_2CCl}} + \cdot CCl_3$$

If the molecule is unsaturated, like a monomer, the atom transferred (usually hydrogen) can go in either direction:

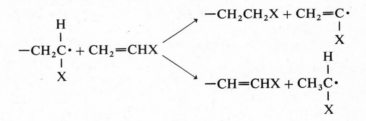

The major effect of chain transfer to a saturated small molecule (solvent, initiator, or deliberately added *chain transfer agent*) is the formation of additional polymer molecules for each radical chain initiated. Transfer to polymer, and transfer to monomer with subsequent polymerization of the double bond, lead to the formation of branched molecules. The latter reaction has a pronounced effect on molecular weight distribution (Section *E*) and is important in the production of graft copolymers (Chapter 11*D*).

The efficiency of compounds as chain transfer agents varies widely with molecular structure. Aromatic hydrocarbons are rather unreactive unless they have benzylic hydrogens. Aliphatic hydrocarbons become more reactive when substituted with halogens. Carbon tetrachloride and carbon tetrabromide are quite reactive. The reactivity of various polymer radicals to transfer varies widely. Transfer reactions offer a valuable means for comparing radical reactivities.

Inhibition and retardation A *retarder* is defined as a substance which can react with a radical to form products incapable of adding monomer. If the retarder is very effective, no polymer may be formed; this condition is sometimes called *inhibition* and the substance an *inhibitor*. The distinction is merely one of degree. The two phenomena are illustrated in Fig. 9-1, where the rate of polymerization is displayed in idealized fashion. The action of a retarder is twofold: it both reduces the concentration of radicals and shortens their average lifetime and thus the length of the polymer chains.

In the simplest case, the retarder may be a free radical, such as triphenylmethyl or diphenylpicrylhydrazyl, which is too unreactive to initiate a polymer chain. The mechanism of retardation is simply the combination or disproportionation of radicals. If the retarder is a molecule, the chemistry of retardation is more complex, but the product of the reaction must be a radical too unreactive to initiate a chain.

Inhibitors are useful in determining initiation rates, since their reaction with radicals is so rapid that the decomposition of inhibitor is independent of its concentration but gives directly the rate of generation of radicals. As a result the length of the *induction period* before polymerization starts is directly proportional to the number of inhibitor molecules initially present. This number then represents the number of radicals produced during the time of the induction period. For this analysis to be valid it must be known that one molecule of inhibitor is exactly equivalent to one radical. There are a number of reasons why this may not be so. Molecular inhibitors generate free radicals in their action; if these free radicals disappear by disproportionating, an inhibitor molecule reappears, so that eventually one inhibitor molecule stops two radicals. More severe discrepancies have been shown to exist: benzoquinone can undergo transfer with styrene chains,

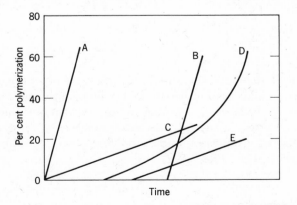

Fig. 9-1. Idealized time-conversion curves for typical inhibition and retardation effects (Goldfinger 1967). Curve A, normal polymerization; B, simple inhibition; C, simple retardation; D, nonideal inhibition; E, inhibition followed by retardation.

reducing molecular weight but not rate. With less reactive monomers (e.g., vinyl acetate) benzoquinone presumably acts only as an inhibitor. Triphenylmethyl initiates chains as well as inhibiting them (the latter much more rapidly). Diphenylpicrylhydrazyl was long presumed to act only as an inhibitor, but this is not certain.

Configuration of monomer units in vinyl polymer chains

The addition of a vinyl monomer to a free radical can take place in either of two ways:

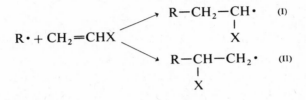

the reaction leading to the more stable product being favored. Since the unpaired electron can participate in resonance with the substituent X in structure I but not in II, reaction I is favored. Steric factors also favor reaction I.

The occurrence of reaction I exclusively would lead to a *head-to-tail* configuration in which the substituents occur on alternate carbon atoms:

$$-CH_2-CH-CH_2-CH-CH_2-CH-, \text{ etc.}$$
$$\quad\quad\;\; | \quad\quad\quad\;\; | \quad\quad\quad\;\; |$$
$$\quad\quad X \quad\quad\quad X \quad\quad\quad X$$

Alternative possibilities are a *head-to-head, tail-to-tail* configuration:

$$-CH_2-CH-CH-CH_2-CH_2-CH-CH-CH_2-, \text{ etc.}$$
$$\quad\quad\;\; | \quad | \quad\quad\quad\quad\quad\;\; | \quad |$$
$$\quad\quad X \quad X \quad\quad\quad\quad\quad X \quad X$$

or a random structure containing both arrangements. The possibility of obtaining a regular head-to-head, tail-to-tail configuration exclusively is remote, and it appears that only an occasional monomer unit enters the chain in a reverse manner to provide a single head-to-head, tail-to-tail linkage. General aspects of radical polymerization have been reviewed (Eastmond 1967, Lenz 1969, 1970).

GENERAL REFERENCES

Marvel 1959; Bevington 1961; Ham 1967; Lenz 1967; Ravve 1967.

B. Laboratory Methods in Vinyl Polymerization

This section is limited to the discussion of small-scale laboratory polymerizations carried to low conversion with pure monomer and minor amounts of initiator and possibly solvent, retarder, or transfer agent. Such experiments are typical of those used to obtain kinetic data. Further discussion of polymerization methods is found in Chapters 10–12 and Part IV.

As developed in the following sections, the experimental data of interest include the over-all rate of polymerization, the rate constants for initiation, propagation, and termination (and transfer where present), and the molecular weight of the polymer. The rate constants for propagation and termination cannot be separated in the simple experiments described here and are discussed in Section D; measurement of molecular weight is discussed in Chapter 3.

Rate of polymerization

Chemical methods or isolation and weighing of polymer can be used to measure the rate of polymerization, but it is customary to work in an evacuated and sealed system to avoid extraneous effects due to the presence of oxygen. Polymerization rate is then followed through changes in a physical property of the system, such as density, refractive index, or ultraviolet or infrared absorption. Measurement of density is usually the most sensitive and convenient technique. The reaction is carried out in a *dilatometer*, a vessel equipped with a capillary tube in which the liquid level can be measured precisely. The decrease in volume on polymerization is relatively large, being 21 % for methyl methacrylate and 27 % for vinyl acetate. In practice a few hundredths per cent polymerization can be detected. The experiments are usually carried out in the region below 5 % polymer so as to avoid deviations from constant polymerization rate due to depletion of reactants, etc. The dilatometer is placed in a constant-temperature bath to ensure isothermal conditions throughout the experiment.

Types of initiation

In the usual case, initiation is effected by thermal decomposition of an initiator as described in Section A. Direct thermal initiation through decomposition of monomer is known for some cases, notably styrene (Section C).

Thermal decomposition of an initiator has a disadvantage in that the rate of generation of free radicals cannot be controlled rapidly because of the heat capacity of the system. *Photoinitiated polymerization*, on the other hand, can be controlled with high precision, since the generation of radicals can be made to vary instantaneously by controlling the intensity of the initiating light (Oster 1969).

Light of short enough wavelength (i.e., high enough energy per quantum) can initiate polymerization directly. It is customary, however, to use a *photochemical initiator* such as benzoin or azobisisobutyronitrile, which is decomposed into free radicals by ultraviolet light in the 3600 A region, where direct initiation through decomposition of monomer does not occur. In photopolymerization these initiators are used at temperatures low enough so that they do not undergo appreciable thermal decomposition. Convenient sources of 3600 A light are available in the form of mercury-arc lamps or fluorescent lamps with special phosphors.

For photopolymerization the dilatometer is made in such a form as to serve as an absorption cell for the ultraviolet light. The distribution of free radicals throughout the reaction vessel must be uniform; this means that only a small fraction of the ultraviolet light may be absorbed. Also, the photoinitiator must not undergo any reaction other than to act as a source of free radicals which add rapidly to the monomer.

Rate of initiation

The number of polymer chains initiated per unit time can be calculated from the number of free radicals produced in the system, if the initiator efficiency is known. In photopolymerization, the number of free radicals produced is twice the number of quanta absorbed, since one quantum of light decomposes a single initiator molecule into two free radicals. The number of quanta absorbed in the reaction vessel can be measured in the following way. The fraction of the incident light absorbed in the reaction vessel, with monomer and initiator present exactly as in the polymerization experiment, is determined. The vessel is then filled with a solution, such as uranyl oxalate, which absorbs ultraviolet light by means of a known chemical reaction. The fraction of the light absorbed is again measured, and after exposure for a known time the number of quanta of light which have reacted with uranyl oxalate is determined by chemical analysis of the solution. The quanta absorbed under polymerization conditions can be calculated in a straightforward manner.

The initiation rate obtained in this way can be checked by computing the number of polymer molecules formed per unit time from the over-all rate of polymerization and $\bar{x}_n$, provided that chain transfer is absent and that the mechanism of termination is known. Since this is usually not the case, an independent check on the rate of initiation is desirable. This may be obtained by the use of an inhibitor or retarder for the polymerization.

GENERAL REFERENCES

Ravve 1967; Sorenson 1968.

C. Steady-State Kinetics of Vinyl Radical Polymerization

Conversion of monomer to polymer

The chemical equations of Section *A* for initiation, propagation, and termination of vinyl radical polymerization can be generalized in the following mathematical scheme.

Initiation in the presence of an initiator I may be considered in two steps: first, the rate-determining decomposition of the initiator into free radicals, R·,

$$I \xrightarrow{k_d} 2R· \tag{9-1}$$

and, second, the addition of a monomer unit to form a chain radical $M_1·$,

$$R· + M \xrightarrow{k_a} M_1· \tag{9-2}$$

where the k's in these and subsequent equations are *rate constants*, with subscripts designating the reactions to which they refer.

The successive steps in propagation,

$$M_1· + M \xrightarrow{k_p} M_2·$$
$$M_2· + M \xrightarrow{k_p} M_3· \tag{9-3}$$

or in general

$$M_x· + M \xrightarrow{k_p} M_{x+1}·$$

are assumed all to have the same rate constant k_p, since radical reactivity is presumed to be independent of chain length.

The termination step involves combination

$$M_x· + M_y· \xrightarrow{k_{tc}} M_{x+y} \tag{9-4}$$

or disproportionation

$$M_x· + M_y· \xrightarrow{k_{td}} M_x + M_y \tag{9-5}$$

Except where it is necessary to distinguish between the two mechanisms the termination rate constant is denoted k_t.

The rates of the three steps may be written in terms of the concentrations (in brackets) of the species involved and the rate constants. The rate of initiation is

$$v_i = \left(\frac{d[M·]}{dt}\right)_i = 2fk_d[I] \tag{9-6}$$

where the factor f represents the fraction of the radicals formed by Eq. 9- which is successful in initiating chains by Eq. 9-2. The rate of terminatio is

$$v_t = -\left(\frac{d[M\cdot]}{dt}\right)_t = 2k_t[M\cdot]^2 \tag{9-7}$$

For many cases of interest the concentration of free radicals [M·] become essentially constant very early in the reaction, as radicals are formed an destroyed at identical rates. In this *steady-state* condition $v_i = v_t$ an Eqs. 9-6 and 9-7 may be equated to solve for [M·]:

$$[M\cdot] = \left(\frac{fk_d[I]}{k_t}\right)^{1/2} \tag{9-8}$$

The rate of propagation is essentially the same as the over-all rate of dis appearance of monomer, since the number of monomers used in Eq. 9- must be small compared to that used in Eq. 9-3 if polymer is obtained Then

$$v_p = -\frac{d[M]}{dt} = k_p[M][M\cdot] \tag{9-9}$$

or, substituting from Eq. 9-8,

$$v_p = k_p\left(\frac{fk_d[I]}{k_t}\right)^{1/2}[M] \tag{9-10}$$

Thus the over-all rate of polymerization should, in the early stages o the reaction, be proportional to the square root of the initiator concentratio and, if f is independent of [M], to the first power of the monome concentration. This is true if the initiator efficiency is high. With ver; low efficiency, f may be proportional to [M], making v_p proportional t [M]$^{3/2}$.

The proportionality of the over-all rate to the square root of initiato concentration has been confirmed experimentally in a large number o cases. Typical experimental data are shown in Fig. 9-2. The straight line are drawn with the theoretical slope of $\frac{1}{2}$. The deviation for polystyren at low initiator concentration (line c) is due to the presence of some therma initiation (see page 292).

Evaluation of rate constants If the rate constant for initiator decomposi tion and the initiator efficiency are known, the ratio of rate constants k_p^2/k can be evaluated from the over-all polymerization rate: from Eqs. 9-6 an 9-10,

$$\frac{k_p^2}{k_t} = \frac{2v_p^2}{v_i[M]^2} \tag{9-11}$$

hese rate constants can be separated only by the nonsteady-state methods
escribed in Section *D*.   Measurements of $k_p{}^2/k_t$ as a function of temperature
ad to useful information about the thermochemistry of polymerization
ection *E*).

ver-all rate as a function of conversion   If the initiator concentration does
ot vary much during the course of polymerization and the initiator effi-
ency is independent of monomer concentration, polymerization proceeds
y first-order kinetics, i.e., the polymerization rate is proportional to mon-
mer concentration.   In some systems, such as the benzoyl peroxide-initiated
olymerization of styrene, the reaction is accurately first order up to quite
gh conversions (Fig. 9-3).

The polymerization of certain monomers undiluted or in concentrated
olution is accompanied by a marked deviation from first-order kinetics
the direction of an increase in reaction rate and molecular weight termed
utoacceleration or the *gel effect* (Norrish 1939, Schulz 1947, Trommsdorff
)48).   The effect is particularly pronounced with methyl methacrylate,
ethyl acrylate, or acrylic acid (Fig. 9-4).   It is independent of initiator and
due to a decrease in the rate at which the polymer molecules diffuse through
e viscous medium, thus lowering the ability of two long-chain radicals to
ome together and terminate.   Although termination is in fact diffusion

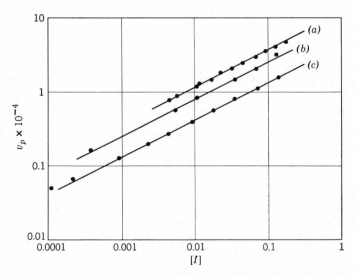

ig. 9-2.   Dependence of rate of polymerization v_p on concentration of initiator $[I]$ for:
) methyl methacrylate with azobisisobutyronitrile; (*b*) methyl methacrylate with benzoyl
eroxide; (*c*) styrene with benzoyl peroxide (Mayo 1959).

controlled for most liquid-phase polymerizations, even at low conversio
the dependence of diffusion rate on the viscosity of the medium leads
the gel effect at high polymer concentrations, with high-molecular-weig
polymer, or in the presence of inert solutes increasing the viscosity of th
medium (North 1968).

The decrease in termination rate leads to an increase in over-all poly
merization rate and in molecular weight, since the lifetime of the growin
chains is increased. At quite high conversions (70–90%), but long afte
propagation has become diffusion controlled, the rate of polymerizatio
drops to a very low value.

Kinetics of thermal polymerization   Certain monomers, including styren
which have been carefully purified to remove all traces of possible initiator
still undergo polymerization at elevated temperatures. In these cases th
initiation is presumably due to thermal decomposition of the monomer t
radicals. This decomposition takes place by a mechanism as yet unknowr
but the existence of diradicals which grow chains simultaneously fror
both ends can be ruled out. The rate of thermal initiation is usually propor
tional to monomer concentration, making the over-all polymerization rat
proportional to $[M]^2$. Initiation rates are many times smaller than those fc
polymerization in the presence of an initiator.

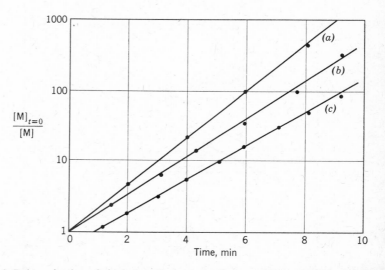

Fig. 9-3. Polymerization of vinyl phenylbutyrate in dioxane with benzoyl peroxide initiatc
at 60°C plotted as a first-order reaction (Marvel 1940). At $t = 0$, monomer concentratio
was 2.4 g/100 ml for curve (a); 6.0 g/100 ml for (b); and 7.3 g/100 ml for (c).

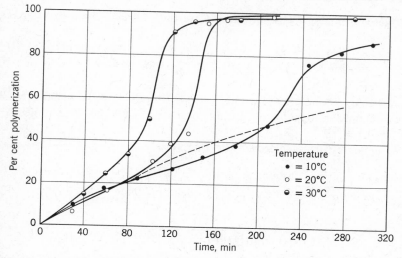

Fig. 9-4. Autoacceleration of polymerization rate in poly(methyl methacrylate) (Naylor 1953).

Degree of polymerization and chain transfer

Kinetic chain length   The *kinetic chain length* v is defined as the number of monomer units consumed per active center. It is therefore given by $v_p/v_i = v_p/v_t$:

$$v = \frac{k_p}{2k_t} \frac{[M]}{[M\cdot]}$$ (9-12)

Eliminating the radical concentration by means of Eq. 9-9,

$$v = \frac{k_p^2}{2k_t} \frac{[M]^2}{v_p}$$ (9-13)

For initiated polymerization, use of Eq. 9-8 leads to

$$v = \frac{k_p}{2(fk_d k_t)^{1/2}} \frac{[M]}{[I]^{1/2}}$$ (9-14)

If no reactions take place other than those already discussed, the kinetic chain length should be related to $\bar{x}_n$, the degree of polymerization: for termination by combination, $\bar{x}_n = 2v$, and for disproportionation, $\bar{x}_n = v$. This is found to be precisely true for some systems, but for others wide deviations are noted in the direction of more polymer molecules than active centers. These deviations are the result of chain transfer reactions:

$$M_x\cdot + XP \longrightarrow M_x X + P\cdot$$
$$P\cdot + M \longrightarrow PM\cdot, \text{ etc.}$$

where P may be monomer, initiator, solvent, or other added chain transfer agent. (Transfer to polymer is omitted since no new polymer molecule is produced.)

The degree of polymerization is therefore

$$\bar{x}_n = \frac{\text{rate of growth}}{\sum \text{rates of all reactions leading to dead polymer}}$$

$$= \frac{v_p}{fk_d[\text{I}] + k_{tr,\text{M}}[\text{M}][\text{M}\cdot] + k_{tr,\text{S}}[\text{S}][\text{M}\cdot] + k_{tr,\text{I}}[\text{I}][\text{M}\cdot]} \tag{9-15}$$

where the terms in the denominator represent termination by combination and transfer to monomer, solvent, and initiator, respectively. If termination is by disproportionation, the first term becomes $2fk_d[\text{I}]$. If transfer constants are defined as

$$C_\text{M} = \frac{k_{tr,\text{M}}}{k_p}, \qquad C_\text{S} = \frac{k_{tr,\text{S}}}{k_p}, \qquad C_\text{I} = \frac{k_{tr,\text{I}}}{k_p}, \tag{9-16}$$

then (assuming termination by combination)

$$\frac{1}{\bar{x}_n} = \frac{k_t}{k_p^2} \frac{v_p}{[\text{M}]^2} + C_\text{M} + C_\text{S} \frac{[\text{S}]}{[\text{M}]} + C_\text{I} \frac{k_t}{k_p^2 fk_d} \frac{v_p^2}{[\text{M}]^3} \tag{9-17}$$

The above analysis assumes that the radical formed in the transfer process is approximately as reactive as the original chain radical. Otherwise retardation or inhibition results.

Transfer to solvent In the presence of a solvent, and by properly choosing conditions to keep other types of chain transfer to a minimum, Eq. 9-17 reduces to

$$\frac{1}{\bar{x}_n} = \left(\frac{1}{\bar{x}_n}\right)_0 + C_\text{S} \frac{[\text{S}]}{[\text{M}]} \tag{9-18}$$

where $(1/\bar{x}_n)_0$ combines the polymerization and transfer to monomer terms. The linear dependence of $1/\bar{x}_n$ on $[\text{S}]/[\text{M}]$ is illustrated in Fig. 9-5. In experiments of this type the ratio $v_p/[\text{M}]^2$ must be held constant throughout i $(\bar{x}_n)_0$ is identified with the observed degree of polymerization in the absenc of solvent.

Transfer to monomer and initiator   If only the terms corresponding to polymerization, transfer to monomer, and transfer to initiator are kept in Eq. 9-17, the resulting equation for $\bar{x}_n$ is quadratic in v_p. This behavior has been observed, e.g., in the benzoyl peroxide-initiated polymerization of styrene. The contribution of transfer to monomer is constant and independent of polymerization rate; that of transfer to initiator increases rapidly with increasing rate, since high rate requires high initiator concentration (Fig. 9-6).

Control of molecular weight by transfer   Chain transfer agents with transfer constants near unity are quite useful in depressing molecular weight in polymerization reactions. This is often of great commercial importance, e.g., in the polymerization of dienes to synthetic rubbers, where chain length must be controlled for ease of processing.

The choice of transfer constant near unity ensures that the transfer agent, or *regulator*, is consumed at the same rate as the monomer so that the ratio [S]/[M] remains constant throughout the reaction. Too large quantities are needed of chain transfer agents with constants much lower than unity; and agents with transfer constants greater than about five are used up too early in the polymerization. Aliphatic mercaptans are suitable transfer agents for several common monomers.

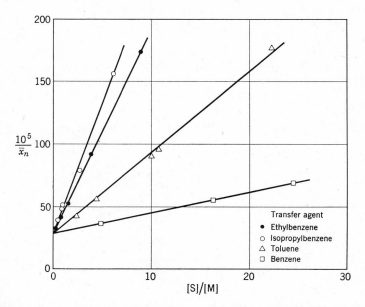

Fig. 9-5. Effect of chain transfer to solvent on the degree of polymerization of polystyrene (Gregg 1947).

Table 9-1 lists transfer constants for several solvents and radicals. T chemistry and kinetics of chain transfer have been reviewed (Palit 1965).

Inhibition and retardar

The kinetics of chain polymerization in the presence of added inhibit or retarder can be described (Goldfinger 1967; Odian 1970) by adding to t scheme of Eqs. 9-1–9-5 the reaction

$$M_x \cdot + Z \xrightarrow{k_z} M_x + Z \cdot \qquad (9\text{-}1$$

where Z is the inhibitor. It is assumed that the radical Z· does not initia polymerization, and is terminated without regenerating Z. Application the steady-state assumption leads, in direct analogy with Eq. 9-17, to t relation (again for termination by combination)

$$\frac{1}{\bar{x}_n} = \frac{k_t}{k_p^2} \frac{v_p}{[M]^2} + C_Z \frac{[Z]}{[M]} \qquad (9\text{-}2$$

where $C_Z = k_Z/k_p$.

Table 9-2 lists inhibition constants for some common monomers a inhibitors.

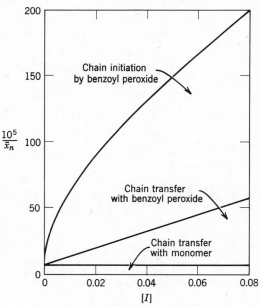

Fig. 9-6. Contribution of various sources of termination in the polymerization of styre with benzoyl peroxide at 60°C (Mayo 1951).

TABLE 9-1. *Chain transfer constants for various solvents and radicals at 60°C except as noted (Young 1966)*

| | $C_S \times 10^4$ for | | |
Solvent	Styrene	Methyl Methacrylate	Vinyl Acetate
Benzene	0.023	0.040	1.2
Cyclohexane	0.031	0.10 (80°C)	7.0
Toluene	0.125	0.20	22
Chloroform	0.5	1.77	150
Ethylbenzene	0.67	1.35 (80°C)	5.5
Triethylamine	7.1	8.3	370
Tetrachloroethane	18 (80°C)	0.155	107
Carbon tetrachloride	90	2.40	9600
Carbon tetrabromide	22,000	2700	28,700 (70°C)

GENERAL REFERENCES

Flory 1953; Walling 1957; Bamford 1958; Lenz 1967; Ravve 1967; Odian 1970.

D. Absolute Reaction Rates

It was shown in Section C that the rate constants for propagation and termination occur together as the ratio $k_p{}^2/k_t$ in both the equations for over-all polymerization rate and those for the kinetic chain length. Thus these rate constants cannot be evaluated separately through steady-state measurements. Recourse must be had to transient phenomena, such as

TABLE 9-2. *Inhibitor constants for various inhibitors and monomers at 50°C except as noted (Ulbricht 1966b)*

| | C_Z for | | | |
Inhibitor	Styrene	Methyl Acrylate	Methyl Methacrylate	Vinyl Acetate
Nitrobenzene	0.326	0.00464	—	11.2
Trinitrobenzene	64.2	0.204	—	404
p-Benzoquinone	518	—	5.5 (44°C)	—
DPPH	—	—	2000 (44°C)	—
Oxygen	14,600	—	33,000	—

the rate of polymerization before the free radical concentration has reached its steady-state value. The duration of this transient region depends on the length of time a free radical exists from its formation in the initiation step to its demise in termination.

The sector method   If a photopolymerization is begun by turning on light of intensity I, the concentration of free radicals grows gradually to its steady state value over a time interval which is a function of the lifetime of the radicals. When the light is turned off, the radical concentration decays in a similar manner. At the steady state the rate of polymerization is proportional to the square root of the light intensity: $v_p \simeq I^{1/2}$.

If the light is flashed on and off quite slowly, so that the steady-state concentration is reached early in each flash, the rate of polymerization is proportional to the fraction of the time the light is on. If, e.g., the light and dark periods are equal, $v_p \simeq \frac{1}{2} I^{1/2}$.

On the other hand, if the flashes are so rapid that the radical concentration changes very little during a single flash, the effect is the same as that produced by reducing the intensity by a constant factor but leaving on all the time: if the light and dark periods are equal, $v_p \simeq (I/2)^{1/2}$. The rates for long and short flashes differ, in this case, by $\sqrt{2}$.

The rate of flashing at which the transition occurs can be related to the mean lifetime τ_s of the free radicals. This is defined as

$$\tau_s = \frac{\text{number of radicals}}{\text{number disappearing per unit time}}$$

$$= \frac{[M\cdot]}{2k_t[M\cdot]^2} = \frac{1}{2k_t[M\cdot]} \tag{9-21}$$

or, if the radical concentration is substituted from Eq. 9-9,

$$\tau_s = \frac{k_p}{2k_t} \frac{[M]}{v_p} \tag{9-22}$$

Thus τ_s, [M], and the over-all polymerization rate v_p suffice to calculate k_p/k_t. With the ratio $k_p{}^2/k_t$ obtained from other data, the individual rate constants may be evaluated.

The relation between τ_s and the flashing rate is complex; in practice the experimental curves of polymerization rate (or radical concentration) against flashing rate are compared to the theoretical relation (Fig. 9-7). In the figure, the ratio t/τ_s, where t is the time light is on, is plotted against $\bar{v}_p/(v_p)_s$ where $\bar{v}_p$ is the rate with flashing illumination and $(v_p)_s$ is the rate with steady illumination, for several values of the ratio r of dark to light intervals.

In practice the light is usually flashed by placing a motor-driven *rotating sector* between the light source and the reaction vessel.

In the case of vinyl acetate, the radical lifetime is of the order of 4 sec, depending, of course, upon polymerization rate, according to Eq. 9-22. The rate constants show that the growing chain adds about 10^4 monomers per second; the kinetic chain length is the product of this figure and τ_s, or about 4×10^4. The value of $\bar{x}_n$ is about ten times less than this because of chain transfer to monomer; thus about ten transfers must occur during the life of a growing chain. This represents a relatively large amount of transfer to monomer.

The kinetic scheme for *emulsion polymerization*, described in Chapter 12, allows separation of the individual rate constants, since the growth of a single chain effectively takes place in an isolated system.

In a *viscosity method* (Bamford 1948; Onyon 1955), now little used, the polymerization is carried out in a viscometer, measurement of molecular weight by viscosity being substituted for measurement of rate. The change in viscosity after photoinitiation is stopped provides data obtained under transient conditions.

Rate of initiation

Dead-end polymerization   (Tobolsky 1960; Böhme 1966)   This term refers to polymerization with a limited amount of initiator. As the initiator is depleted, the reaction stops short of completion at the point where the half-life of the propagating chain is comparable to that of the initiator.

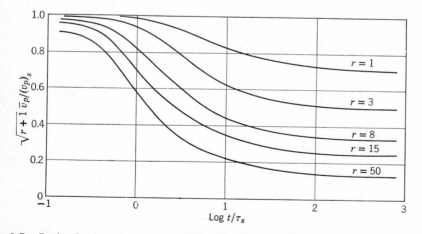

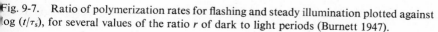

Fig. 9-7.   Ratio of polymerization rates for flashing and steady illumination plotted against log (t/τ_s), for several values of the ratio r of dark to light periods (Burnett 1947).

The kinetics of the situation is treated by integrating the equation fc depletion of initiator,

$$-\frac{d[I]}{dt} = k_d[I] \tag{9-2}$$

substituting the result, $[I] = [I]_0 e^{-k_d t}$, into Eq. 9-10, and again integratin to yield

$$-\ln\frac{[M]}{[M]_0} = 2k_p \left(\frac{f[I]_0}{k_t k_d}\right)^{1/2}(1 - e^{-k_d t/2}) \tag{9-2}$$

At long reaction times, $[M]$ reaches a limiting value $[M]_\infty$, and it can b shown that

$$\frac{\ln[M]_\infty - \ln[M]}{\ln[M]_\infty - \ln[M]_0} = e^{-k_d t/2} \tag{9-2}$$

A plot of the logarithm of the left-hand side of Eq. 9-25 vs. time allow evaluation of k_d and, via Eq. 9-10, f. Figure 9-8 shows typical data.

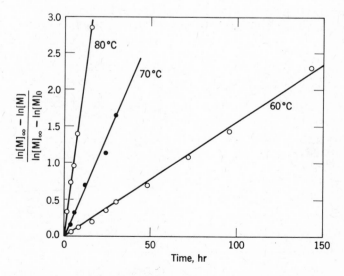

Fig. 9-8. Determination of k_d (slope of the lines) through dead-end polymerization o isoprene at the indicated temperatures with azobisisobutyronitrile initiator (Gobran 1960)

Values of the propagation and termination constants for several mono-
ɪers are tabulated in Table 9-3. The rate constants vary over wide range.
ince they depend upon the reactivity of both the radical and the monomer,
 is difficult to draw conclusions about the effect of structure on reactivity
om these data alone. Copolymerization experiments (Chapter 11) are
ɪore useful in this respect.

ABLE 9-3. *Rate constants for propagation and termination at 50°C. (Ulbricht 1966a)*

Monomer	k_p, liters/mole sec	k_t, liters/mole sec $\times$ 10^{-7}
tyrene	209	115
ɪethyl methacrylate	410	24
ɪethyl acrylate	1000	3.5
inyl acetate	2640	117
inyl chloride	11000	2100

ENERAL REFERENCES

lory 1953; Walling 1957; Bamford 1958; Odian 1970.

ʼ. Molecular Weight and Its Distribution

The distribution of molecular weights in vinyl polymers can readily be
alculated if the reaction is restricted to low conversion, where polymer
ɪade throughout the reaction has the same average molecular weight. If
ːrmination is by disproportionation (or transfer), let p be defined as the
robability that a growing chain radical will propagate rather than terminate:
$= v_p/(v_p + v_t)$.

The probability of formation of an x-mer, as a result of $x - 1$ propaga-
ɪons and a termination, is $p^{x-1}(1 - p)$. The situation is in complete ana-
ɔgy with that for linear stepwise polymerization, and the expressions for
$_n$, N_x, w_x, and $\bar{x}_w$ given in Eqs. 8-5 and 8-8 to 8-10 apply. If high polymer
 formed, i.e., $p \rightarrow 1$, $\bar{x}_w/\bar{x}_n = 2$.

Termination by combination is somewhat more complex, but it can be
hown that, in analogy to the case of multichain polymer (Eq. 8-13) with
$= 2$, $\bar{x}_w/\bar{x}_n = 1.5$ if transfer is absent.

The measurement of $\bar{x}_w/\bar{x}_n$ would distinguish between the possibilitie of termination by combination and by disproportionation. However, thi measurement is not easily carried out with the desired precision; a prefer able alternative procedure involves measuring the number of initiato fragments per molecule of polymer.

For values of p near 1, the most probable distribution of Eq. 8-9 reduce to

$$w_x = xp^x(\ln p)^2$$

$$\bar{x}_n = \frac{-1}{\ln p} \qquad (9\text{-}26)$$

$$\bar{x}_w = \frac{-2}{\ln p}$$

It can be shown that both this distribution, which describes low-conversio chain polymers terminating by disproportionation, and that describin termination by combination, are special cases of a more general distribu tion function (Zimm 1948):

$$w_x = \left(\frac{y^{z+1}}{z!}\right)x^z e^{-xy}$$

$$\bar{x}_n = \frac{z}{y} \qquad (9\text{-}27$$

$$\bar{x}_w = \frac{z+1}{y}$$

For $z = 1$, Eq. 9-27 reduces to 9-26 with $y = -\ln p$. For $z = 2$, it describe the case of termination by combination. Higher values of z correspond t narrower distributions such as might be found in fractionated samples In contrast to the most probable distribution, the number distributio corresponding to Eq. 9-27 does not indicate a large number of very smal molecules. Hence, it is preferred for describing polymers from which th small molecules have been removed, e.g., by precipitation of the polyme from solution.

As polymerization proceeds, the degree of polymerization of the poly mer being formed changes according to Eq. 9-14. If the rate of initiation i constant as is usual, $\bar{x}_n$ decreases throughout the reaction as [M] decreases approaching zero at complete conversion. At the same time, $\bar{x}_w/\bar{x}_n$ increase with increasing conversion, approaching infinity as a limit. Typical value of $\bar{x}_w/\bar{x}_n$ for high-conversion vinyl polymers seldom exceed 5, however except in special cases, such as autoacceleration in the polymerization o methyl methacrylate, where values of $\bar{x}_w/\bar{x}_n$ as high as 10 have been found The above distribution functions are not well suited to the description o these broader distributions. Among empirical functions suggested for these cases is the following form, which has a high-molecular-weight "tail" when the exponent ρ is positive and less than 1. Here Γ is the gamma function:

$$w(x) = \left[\rho/\Gamma\left(\frac{(z+1)}{\rho}\right)\right] y^{(z+1)/\rho} x^z e^{-yx^\rho}$$

$$\bar{x}_n = \Gamma\left(\frac{z+1}{\rho}\right)/y^{1/\rho} \tag{9-28}$$

$$\bar{x}_w = \Gamma\left(\frac{z+2}{\rho}\right)\Big/y^{1/\rho}\Gamma\left(\frac{z+1}{\rho}\right)$$

Where chain transfer to polymer is involved, extremely broad molecular weight distributions can result. The abnormal breadth arises from the fact that each transfer reaction adds a branch point to the molecule; after the branch has grown, the molecule has increased in molecular weight. But the probability of branching is proportional (approximately) to weight-average molecular weight. Hence, branched molecules tend to become more highly branched and still larger. The result is that the distribution of molecular weights has a long, high-molecular-weight "tail."

Among the empirical distribution functions suitable for describing fractions from which low-molecular-weight species have been removed is the so-called logarithmic normal distribution (Lansing 1935):

$$w(x) = \frac{e^{-\sigma^2/2}}{\sigma\sqrt{\pi}} \frac{1}{x_0} e^{-(\ln x - \ln x_0)^2/2\sigma^2}$$

$$\bar{x}_n = x_0 e^{\sigma^2/2} \tag{9-29}$$

$$\bar{x}_w = x_0 e^{3\sigma^2/2}$$

In this distribution the logarithm of the molecular weight follows a normal distribution function. The ratio $\bar{x}_w/\bar{x}_n$ is identical with $\bar{x}_z/\bar{x}_w$.

Other empirical equations have been proposed for cases of unusually broad distributions by Wesslau (1956), Tung (1956), and Gordon (1961; Roe 1961).

Polymers with extremely narrow molecular-weight distribution are produced by certain types of anionic polymerization ("living" polymers, Chapter 10C). Their molecular-weight distribution is described by the Poisson distribution (Flory 1940):

$$w(x) = \frac{\mu^{x+1} e^{-\mu}}{(x-1)!}$$

$$\bar{x}_n = \mu \tag{9-30}$$

$$\bar{x}_w = \mu + 1$$

GENERAL REFERENCES

Flory 1953; Walling 1957; Bamford 1958; Eastmond 1967; Odian 1970.

F. Thermochemistry of Chain Polymerization

Thermochemical and thermodynamic data give valuable insight int the mechanisms of vinyl polymerization reactions. By the usual Arrheniu treatment the temperature dependence of the rates of the various steps i radical chain polymerization can be separated into an energy of activatio representing the amount of energy which the reactant molecules must hav to be able to react on collision, and a frequency factor which allows est mation of the fraction of the collisions which lead to reaction. Heats c polymerization are readily measured and may also be calculated from bon dissociation energies. Differences between the two values give an indica tion of the amount of steric hindrance in the polymer. Free energies c polymerization reactions can be calculated for some simple cases and giv information on the thermodynamic equilibria involved in polymerization.

Activation energies and collision facto

Over-all polymerization   The ratio of rate constants $k_p{}^2/k_t$ or $k_p/\sqrt{k_t}$ obtained from the over-all polymerization rate and the rate of initiatio (Eq. 9-11). Expressing a rate constant by an Arrhenius type equation give e.g.,

$$k_p = A_p e^{-E_p/RT} \tag{9-31}$$

where A_p is the *collision frequency factor* for propagation and E_p the *energ* of activation for propagation. Application to the rate constant ratio give

$$\frac{k_p}{\sqrt{k_t}} = \left(\frac{A_p}{\sqrt{A_t}}\right) e^{-(E_p - E_t/2)/RT} \tag{9-32}$$

By plotting log $(k_p/\sqrt{k_t})$ against $1/T$, $E_p - E_t/2$ and $A_p/\sqrt{A_t}$ can be evalu ated. For many polymerization reactions, $E_p - E_t/2$ is about 5–6 kcal mole. This is equivalent to an increase in $k_p/\sqrt{k_t}$ of about 30–35% fo every 10°C near room temperature. Equation 9-32 refers only to the rati of rate constants $k_p/\sqrt{k_t}$ and does not include the effect of temperature o the rate of initiation. In photopolymerization, where the rate of initiation i independent of temperature, Eq. 9-32 represents the complete activation energy.

In an initiated polymerization, v_p is proportional to $k_p \sqrt{k_d/k_t}$, and the observed activation energy is $E_d/2 + (E_p - E_t/2)$. E_d, the activation energy for spontaneous decomposition of the initiator, is about 30 kcal/mole for benzoyl peroxide or azobisisobutyronitrile. The activation energy for initiated polymerization is therefore slightly greater than 20 kcal/mole. This corresponds to a two- to threefold increase in rate for a 10°C temperature change.

The activation energy for thermal polymerization is of the same order of magnitude, yet this reaction is much slower than normal because of the extremely low collision factor for thermal initiation. The normal value of A for a bimolecular reaction is 10^{11}–10^{13}; for the thermal initiation of polystyrene it is about 10^4–10^6.

Degree of polymerization If transfer reactions are negligible, the temperature coefficient of molecular weight depends on the initiation process. If an initiator is used, it follows from Eq. 9-14, by virtue of the differential form d ln $k/dT = E/RT^2$ of the Arrhenius equation, that

$$\frac{d \ln \bar{x}_n}{dT} = \frac{E_p - \dfrac{E_t}{2} - \dfrac{E_d}{2}}{RT^2} \tag{9-33}$$

Since $E_p - E_t/2$ is 5–6 kcal and E_d is about 30 kcal, this quantity is negative and the molecular weight decreases with increasing temperature. The same is true in thermal polymerization.

In photopolymerization in the absence of transfer, d ln $\bar{x}_n/dT = (E_p - E_t/2)/RT^2$, which is positive. This is the only case in which molecular weight increases with temperature.

If transfer reactions are controlling, it is found that molecular weight decreases with increasing temperature. By plotting log C_S or $-\log \bar{x}_n$ against $1/T$, the quantities $E_{tr,S} - E_p$ and $A_{tr,S}/A_p$ may be evaluated. $E_{tr,S}$ usually exceeds E_p by 5–15 kcal/mole, higher activation energies going with poorer transfer agents. Since the frequency factor for transfer usually exceeds that for propagation, the higher activation energy is responsible for the slower rate of transfer compared to monomer addition.

Absolute rates The activation energies and frequency factors for propagation and termination can be separated if the individual rate constants are known. For most monomers E_p is near 7 kcal/mole, and E_t varies from 3 to 5 kcal/mole. The frequency factors vary rather widely, suggesting that steric effects are more important than activation energies in setting the values of the rate constants. For example, A_p is much lower for methyl methacrylate with two substituents on the same ethylenic carbon atom than it is for less hindered monomers. All the values of A_p are considerably lower than normal, because, it is likely, of low probability of the transition leading to unpairing of the double bond electrons in the monomer.

Polymerization-depolymerization equilibrium   Some polymers decompose by stepwise loss of monomer units in a reaction which is essentially the reverse of polymerization (see Chapter 12). The depolymerization reaction can be incorporated into the kinetic scheme for chain polymerization. At ordinary temperatures the scheme is not changed because the rate constant for depolymerization is small. The activation energy of depolymerization is quite high (10–26 kcal/mole) compared to that of propagation, however, and at some elevated temperature the rates of polymerization and depolymerization become equal. This temperature is the *ceiling temperature* above which polymer free radicals, in the presence of monomer at 1 atm pressure, depolymerize rather than grow. For polystyrene, the ceiling temperature is about 300°C.

Heats and free energies of polymerization

Heats of polymerization   These may be obtained from the heats of combustion of monomer and polymer or directly by a calorimetric experiment. They may also be calculated from bond dissociation energies. When the two results are compared, it is seen that the observed heat of polymerization is equal to or somewhat lower than the calculated heat. For vinyl compounds with two substituents on the same carbon atom the difference is 3–9 kcal/mole. This is attributed to steric repulsion between the substituent groups in the polymer. For monosubstituted vinyl derivatives the difference is much lower; only for styrene is it definitely greater than 1 kcal/mole.

Free energies of polymerization   Free energy is a more precise indication than enthalpy of the thermodynamic plausibility of a chemical reaction. The heats of formation of free radicals and of their over-all reactions to polymer are of major importance in establishing the corresponding free energies, but the contribution of the entropy term in

$$\Delta G = \Delta H - T \Delta S$$

is also significant. Most chain polymerizations of alkenes have both negative heats and negative entropies of polymerization (Table 9-4), the latter falling in a narrow range of -25 to -30 cal/mole K. The free energies of polymerization are negative at usual temperatures. Experimental methods for determining these quantities were reviewed by Joshi (1967).

TABLE 9-4. Heats and entropies of polymerization at 25°C (Joshi 1967)

Monomer	$-\Delta H$, kcal/mole	$-\Delta S$, cal/mole K
Tetrafluoroethylene	37.2	26.8
Ethylene	22.7	24.0
Vinyl acetate	21.0	26.2
Propylene	20.5	27.8
Isoprene	17.8	24.2
Butadiene	17.4	20.5
Styrene	16.7	25.0
Methyl methacrylate	13.5	28.0
Isobutylene	12.3	28.8

The calculation of the heats and entropies of reaction of free radicals has been carried out for hydrocarbons and fluorocarbons and applied to the synthesis equations for polyethylene and polytetrafluoroethylene (Bryant 1951, 1962). One result of interest is that the values of heat and free energy for the addition of monomer to the radical become less negative as the length of the chain radical increases in the first few propagation steps. After the growing free radical reaches a length of about eight carbon atoms, all successive steps have the same heat and free energy of reaction, the same as those for the over-all polymerization step.

GENERAL REFERENCES

Flory 1953; Walling 1957; Joshi 1967; Sawada 1969; Odian 1970.

BIBLIOGRAPHY

Bamford 1948. C. H. Bamford and M. J. S. Dewar, "Studies in Polymerization. I. A Method for Determining the Velocity Constants in Polymerization Reactions and its Application to Styrene," Proc. Roy. Soc. (London) A192, 309–328 (1948).

Bamford 1958. C. H. Bamford, W. G. Barb, A. D. Jenkins, and P. F. Onyon, The Kinetics of Vinyl Polymerization by Radical Mechanisms, Butterworths, London, 1958.

Bevington 1954a. J. C. Bevington, H. W. Melville, and R. P. Taylor, "The Termination Reaction in Radical Polymerization. Polymerizations of Methyl Methacrylate and Styrene at 25°," J. Polymer Sci. 12, 449–459 (1954).

Bevington 1954b. J. C. Bevington, H. W. Melville, and R. P. Taylor, "The Termination Reaction in Radical Polymerizations. II. Polymerization of Styrene at 60° and of Methyl Methacrylate at 0 and 60°, and the Copolymerization of These Monomers at 60°," J. Polymer Sci. 14, 463–476 (1954).

Bevington 1961. J. C. Bevington, *Radical Polymerization*, Academic Press, New York, 1961.

Böhme 1966. R. D. Böhme and A. V. Tobolsky, "Dead-End Polymerization," pp. 594–605 in Herman F. Mark, Norman G. Gaylord, and Norbert M. Bikales, eds., *Encyclopedia of Polymer Science and Technology*, Vol. 4, Interscience Div., John Wiley and Sons, New York, 1966.

Bryant 1951. W. M. D. Bryant, "Free Energies of Formation of Hydrocarbon Free Radicals. I. Application to the Mechanism of Polythene Synthesis," *J. Polymer Sci.* **6,** 359–370 (1951).

Bryant 1962. W. M. D. Bryant, "Free Energies of Formation of Fluorocarbons and their Radicals. Thermodynamics of Formation and Depolymerization of Polytetrafluoroethylene," *J. Polymer Sci.* **56,** 277–296 (1962).

Burnett 1947. G. M. Burnett and H. W. Melville, "Determination of the Velocity Coefficients for Polymerization Processes. I. The Direct Photopolymerization of Vinyl Acetate," *Proc. Roy. Soc. (London)* **A189,** 456–480 (1947).

Eastmond 1967. G. C. Eastmond, "Free-Radical Polymerization," pp. 361–431 in Herman F. Mark, Norman G. Gaylord, and Norbert M. Bikales, eds., *Encyclopedia of Polymer Science and Technology*, Vol. 7, Interscience Div., John Wiley and Sons, New York, 1967.

Flory 1937. Paul J. Flory, "Mechanism of Vinyl Polymerizations," *J. Am. Chem. Soc.* **59,** 241–253 (1937).

Flory 1940. Paul J. Flory, "Molecular Size Distribution in Ethylene Oxide Polymers," *J. Am. Chem. Soc.* **62,** 1561–1565 (1940).

Flory 1953. Paul J. Flory, *Principles of Polymer Chemistry*, Cornell University Press, Ithaca, N.Y., 1953.

Friedlander 1966. Henry Z. Friedlander, "Electrochemical Initiation," pp. 629–641 in Herman F. Mark, Norman G. Gaylord, and Norbert M. Bikales, eds., *Encyclopedia of Polymer Science and Technology*, Vol. 5, Interscience Div., John Wiley and Sons, New York, 1966.

Funt 1967. B. L. Funt, "Electrolytically Controlled Polymerizations," *Macromol. Revs.* **1,** 35–56 (1967).

Gobran 1960. R. H. Gobran, M. B. Berenbaum, and A. V. Tobolsky, "Bulk Polymerization of Isoprene: Kinetic Constants from the Dead-End Theory," *J. Polymer Sci.* **46,** 431–440 (1960).

Goldfinger 1967. George Goldfinger, William Yee, and Richard D. Gilbert, "Inhibition and Retardation," pp. 644–664 in Herman F. Mark, Norman G. Gaylord, and Norbert M. Bikales, eds., *Encyclopedia of Polymer Science and Technology*, Vol. 7, Interscience Div., John Wiley and Sons, New York, 1967.

Gordon 1961. Manfred Gordon and Ryong-Joon Roe, "Surface-Chemical Mechanism of Heterogeneous Polymerization and Derivation of Tung's and Wesslau's Molecular Weight Distribution," *Polymer* **2,** 41–59 (1961).

Gregg 1947. R. A. Gregg and F. R. Mayo, "Chain Transfer in the Polymerisation of Styrene. III. The Reactivities of Hydrocarbons toward the Styrene Radical," *Disc. Faraday Soc.* **2,** 328–337 (1947).

Ham 1967. George E. Ham, "General Aspects of Free-Radical Polymerization," Chap. 1 in George E. Ham, ed., *Vinyl Polymerization*, Vol. 1, Part I, Marcel Dekker, New York, 1967.

Joshi 1967. R. M. Joshi and B. J. Zwolinski, "Heats of Polymerization and their Structural and Mechanistic Implications," Chap. 8 in George E. Ham, ed., *Vinyl Polymerization*, Vol. 1, Part I, Marcel Dekker, New York, 1967.

Lansing 1935. W. D. Lansing and E. O. Kraemer, "Molecular Weight Analysis of Mixtures by Sedimentation Equilibrium in the Svedberg Ultracentrifuge," *J. Am. Chem. Soc.* **57**, 1369–1377 (1935).

Lenz 1967. Robert W. Lenz, *Organic Chemistry of Synthetic High Polymers,* Interscience Div., John Wiley and Sons, New York, 1967.

Lenz 1969. Robert W. Lenz, "Applied Reaction Kinetics: Polymerization Reaction Kinetics," *Ind. Eng. Chem.* **61** (3), 67–75 (1969).

Lenz 1970. Robert W. Lenz, "Applied Polymerization Reaction Kinetics," *Ind. Eng. Chem.* **62**, (2), 54–61 (1970).

Mageli 1968. O. L. Mageli and J. R. Kolcynski, "Peroxy Compounds," pp. 814–841 in Herman F. Mark, Norman G. Gaylord, and Norbert M. Bikales, eds., *Encyclopedia of Polymer Science and Technology,* Vol. 9, Interscience Div., John Wiley and Sons, New York, 1968.

Marvel 1940. C. S. Marvel, Joseph Dec and Harold G. Cooke, Jr., "Optically Active Polymers from Active Vinyl Esters. A Convenient Method of Studying the Kinetics of Polymerization," *J. Am. Chem. Soc.* **62**, 3499–3504 (1940).

Marvel 1959. Carl S. Marvel, *An Introduction to the Organic Chemistry of High Polymers,* John Wiley and Sons, New York, 1959.

Mayo 1951. Frank R. Mayo, R. A. Gregg, and Max S. Matheson, "Chain Transfer in the Polymerization of Styrene. VI. Chain Transfer with Styrene and Benzoyl Peroxide; the Efficiency of Initiation and the Mechanism of Chain Termination," *J. Am. Chem. Soc.* **73**, 1691–1700 (1951).

Mayo 1959. Frank R. Mayo, "Contributions of Vinyl Polymerization to Organic Chemistry," *J. Chem. Educ.* **36**, 157–160 (1959).

Naylor 1953. M. A. Naylor and F. W. Billmeyer, Jr., "A New Apparatus for Rate Studies Applied to the Photopolymerization of Methyl Methacrylate," *J. Am. Chem. Soc.* **75**, 2181–2185 (1953).

Norrish 1939. R. G. W. Norrish and E. F. Brookman, "The Mechanism of Polymerization Reactions. I. The Polymerization of Styrene and Methyl Methacrylate," *Proc. Roy. Soc. (London)* **A171**, 147–171 (1939).

North 1968. A. M. North, "Diffusion Control of Homogeneous Free-Radical Reactions," pp. 95–135 in J. C. Robb and F. W. Peaker, eds., *Progress in High Polymers,* Vol. 2, CRC Press, Cleveland, Ohio, 1968.

Noyes 1965. Richard M. Noyes, "Cage Effect," pp. 796–801 in Herman F. Mark, Norman G. Gaylord, and Norbert M. Bikales, eds., *Encyclopedia of Polymer Science and Technology,* Vol. 2, Interscience Div., John Wiley and Sons, New York, 1965.

Odian 1970. George Odian, *Principles of Polymerization,* McGraw-Hill Book Company, New York, 1970.

Onyon 1955. P. F. Onyon, "The Polymerization of 4-Vinyl Pyridine," *Trans. Faraday Soc.* **51**, 400–412 (1955).

Oster 1969. Gerald Oster, "Photopolymerization and Photocrosslinking," pp. 145–156 in Herman F. Mark, Norman G. Gaylord, and Norbert M. Bikales, eds., *Encyclopedia of Polymer Science and Technology,* Vol. 10, Interscience Div., John Wiley and Sons, New York, 1969.

Palit 1965. Santi R. Palit, Satya R. Chatterjee, and Asish R. Mukherjee, "Chain Transfer," pp. 575–610 in Herman F. Mark, Norman G. Gaylord, and Norbert M. Bikales, eds., *Encyclopedia of Polymer Science and Technology,* Vol. 3, Interscience Div., John Wiley and Sons, New York, 1965.

Ravve 1967. A. Ravve, *Organic Chemistry of Macromolecules,* Marcel Dekker, New York, 1967.

Roe 1961. Ryong-Joon Roe, "A Test between Rival Theories for Molecular Weight

Distributions in Heterogeneous Polymerization: The Effect of Added Terminating Agent," *Polymer* **2**, 60–73 (1961).

Sawada 1969. Hideo Sawada, "Thermodynamics of Polymerization," *J. Macromol. Sci.— Revs. Macromol. Chem.* **3**, 313–395 (1969).

Schulz 1947. G. V. Schulz and G. Harborth, "On the Mechanism of the Explosive Course of the Polymerization of Methyl Methacrylate" (in German), *Makromol. Chem.* **1** 106–139 (1947).

Sorenson 1968. Wayne R. Sorenson and Tod W. Campbell, *Preparative Methods of Polymer Chemistry*, 2nd ed., Interscience Div., John Wiley and Sons, New York, 1968.

Staudinger 1920. H. Staudinger, "Polymerization" (in German), *Ber.* **53B**, 1073–108 (1920).

Tobolsky 1960. A. V. Tobolsky, C. E. Rodgers, and R. D. Brickman, "Dead-End Radical Polymerization. II," *J. Am. Chem. Soc.* **82**, 1277–1280 (1960).

Trommsdorff 1948. Ernest Trommsdorff, Herbert Köhle, and Paul Lagally, "On the Polymerization of Methyl Methacrylate" (in German), *Makromol. Chem.* **1**, 169–198 (1948).

Tung 1956. L. H. Tung, "Fractionation of Polyethylene," *J. Polymer Sci.* **20**, 495–506 (1956).

Ulbricht 1966a. J. Ulbricht, "Propagation and Termination Constants in Free Radical Polymerization," pp. II-57–II-69 in J. Brandrup and E. H. Immergut, eds., with the collaboration of H.-G. Elias, *Polymer Handbook*, Interscience Div., John Wiley and Sons, New York, 1966.

Ulbricht 1966b. J. Ulbricht, "Inhibitors and Inhibition Constants in Free Radical Polymerization," pp. II-71– II-75 in J. Brandrup and E. H. Immergut, eds., with the collaboration of H.-G. Elias, *Polymer Handbook*, Interscience Div., John Wiley and Sons, New York, 1966.

Walling 1957. Cheves Walling, *Free Radicals in Solution*, John Wiley and Sons, New York, 1957.

Wesslau 1956. Hermann Wesslau, "The Molecular Weight Distributions of Some Low Pressure Polyethylenes" (in German), *Makromol. Chem.* **20**, 111–142 (1956).

Young 1966. L. J. Young, G. Brandrup, and J. Brandrup, "Transfer Constants to Monomer, Polymer, Catalyst and Solvent in Free Radical Polymerization," pp. II-77–II-139 in J. Brandrup and E. H. Immergut, eds., with the collaboration of H.-G. Elias, *Polymer Handbook*, Interscience Div., John Wiley and Sons, New York, 1966.

Zand 1965. Robert Zand, "Azo Catalysts," pp. 278–295 in Herman F. Mark, Norman G. Gaylord, and Norbert M. Bikales, eds., *Encyclopedia of Polymer Science and Technology*, Vol. 2, Interscience Div., John Wiley and Sons, New York, 1965.

Zimm 1948. Bruno H. Zimm, "Apparatus and Methods for Measurement and Interpretation of the Angular Variation of Light Scattering: Preliminary Results on Polystyrene Solutions," *J. Chem. Phys.* **16**, 1099–1116 (1948).

IO

Ionic and Coordination Chain (Addition) Polymerization

A. Chemistry of Nonradical Chain Polymerization

Chain-reaction polymerization is known to occur by several mechanisms other than those involving free radicals discussed in Chapter 9. Prominent among these are reactions in which the chain carriers are *carbonium ions* (*cationic polymerization*, Section *B*) or *carbanions* (*anionic polymerization*, Section *C*). In addition, polymerization may take place by mechanisms, not yet confirmed in detail, involving coordination compounds among the monomer, the growing chain, and a catalyst, usually a solid (*coordination polymerization*, Section *D*). Still other chain polymerization reactions are known and are discussed in Section *E*.

The mechanisms of these polymerizations are less thoroughly understood than that of radical polymerization, for several reasons. Reaction systems are often heterogeneous, involving inorganic catalysts* and organic monomers. Unusually large effects may be produced by a third component (*cocatalyst*) present in very low concentration. Polymerization often leads to very-high-molecular-weight polymer at an extremely high rate, increasing enormously the difficulty in obtaining kinetic data or even reproducible results.

As indicated in Table 10-1, the type of monomer polymerizing best by each mechanism is related to the polarity of the monomer and the acid-base strength of the ion formed. Monomers with electron-donating groups attached to the double-bonded carbons form stable carbonium ions and

---

* The word *catalyst* is preferred to *initiator* in reference to ionic polymerization, since the agent often exerts its influence throughout the growth of the chain, rather than at the instant of initiation only.

TABLE 10-1. *Types of chain polymerization suitable for common monomers (Lenz 1967)*

Polymerization Mechanism*

Monomer Type	Radical	Cationic	Anionic	Coordination
Ethylene	+	+	−	+
Propylene and α-olefins	−	−	−	+
Isobutylene	−	+	−	−
Dienes	+	−	+	+
Styrene	+	+	+	+
Vinyl chloride	+	−	−	+
Vinylidene chloride	+	−	+	−
Vinyl fluoride	+	−	−	−
Tetrafluoroethylene	+	−	−	+
Vinyl ethers	−	+	−	+
Vinyl esters	+	−	−	−
Acrylic and methacrylic esters	+	−	+	+
Acrylonitrile	+	−	+	+

* + = high polymer formed; − = no reaction or oligomers only.

polymerize best with cationic catalysts. Conversely, monomers with electron-withdrawing substituents form stable anions and require anionic catalysts. In what might be considered an intermediate case, free radical polymerization is favored by moderate electron withdrawal from the double bond plus conjugation in the monomer. Many monomers can polymerize by more than one mechanism, including coordination polymerization.

Initiation of ionic polymerization usually involves the transfer of an ion or an electron to or from the monomer, with the formation of an ion pair. It is thought that the counterion of this pair stays close to the growing chain end throughout its lifetime, particularly in media of low dielectric constant. In contrast to radical polymerization, termination never involves the reaction between two growing chains, but usually involves the unimolecular reaction of a chain with its counterion or a transfer reaction leaving a species too weak to propagate. An agent which retards radical polymerization, such as oxygen, often has little influence on ionic polymerization; but impurities that neutralize the catalysts prevent the reaction, and many aromatic, heterocyclic, olefinic, or acetylenic compounds retard or terminate ionic polymerization.

Recent aspects of ionic and coordination polymerization have been reviewed (Lenz 1969, 1970).

GENERAL REFERENCES

Lenz 1967, Odian 1970.

B. Cationic Polymerization

Typical catalysts for cationic polymerization are Lewis acids and Friedel-Crafts catalysts such as $AlCl_3$, $AlBr_3$, BF_3, $SnCl_4$, H_2SO_4, and other strong acids. All these are strong electron acceptors. Most of them, possibly excepting the strong protonic acids, require a cocatalyst to initiate polymerization, usually a Lewis base or other donor of a proton, which is presumed to be the effective initiator.

High rate of polymerization at low temperatures is a characteristic of ionic polymerizations. It is often difficult to establish uniform reaction conditions before the reactants are consumed. The polymerization of isobutylene by $AlCl_3$ or BF_3 takes place within a few seconds at $-100°C$, producing polymer of molecular weight up to several million. Both rate and molecular weight are much lower at room temperature.

Mechanism

The most satisfactory theory of cationic polymerization involves the carbonium ion as the chain carrier. For example, in the polymerization of isobutylene with boron trifluoride catalyst, the first step is the reaction of the catalyst and a cocatalyst, for example water, to form a *catalyst-cocatalyst complex*

$$BF_3 + H_2O \rightleftharpoons H^+(BF_3OH)^-$$

which donates a proton to an isobutylene molecule to give a carbonium ion,

$$H^+ + (CH_3)_2C{=}CH_2 \longrightarrow (CH_3)_3C^+$$

This ion then reacts with monomer with the reformation of a carbonium ion at the end of each step. The "head-to-tail" addition of monomer to ion is the only one possible for energetic reasons. Since the reaction is in general carried out in a hydrocarbon medium of low dielectric constant, the anion cannot be separated from the growing cationic end to any appreciable distance; rather, they form an ion pair.

The termination reaction can take place by the rearrangement of the ion pair to yield a polymer molecule with terminal unsaturation, plus the original complex,

$$CH_3[(CH_3)_2CCH_2]_xC(CH_3)_2{}^+(BF_3OH)^-$$

$$\downarrow$$

$$H^+(BF_3OH)^- + CH_3[(CH_3)_2CCH_2]_xC\overset{\displaystyle CH_2}{\underset{\displaystyle CH_3}{\diagup}}$$

Here the catalyst-cocatalyst complex is regenerated, and many kinetic chains can be produced from each catalyst-cocatalyst species. An alternative mechanism, in which the catalyst or cocatalyst combines with the growing chain, is unlikely in this example but can take place when a covalent bond is formed, as in the polymerization of styrene with trifluoroacetic acid:

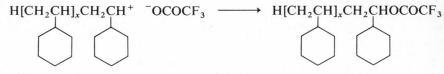

Chain transfer to monomer can also take place:

$$CH_3[(CH_3)_2CCH_2]_xC(CH_3)_2{}^+(BF_3OH)^- + (CH_3)_2C{=}CH_2$$

$$CH_3[(CH_3)_2CCH_2]_xC\overset{\diagup CH_2}{\underset{\diagdown CH_3}{}} + (CH_3)_3C^+(BF_3OH)^-$$

Chain transfer to polymer is known and leads to branched polymers.

The efficiency of the catalyst is dependent on the acid strength of the catalyst-cocatalyst complex. Also, the efficiency of the catalyst as a terminator should be related to the base strength of the complex anion. Thus, the more active a molecule is as a catalyst, the less active it is as a terminator. Although the requirements of the isobutylene-BF_3 system are fully satisfied by the mechanism just given, generalization of the scheme should be applied with caution. The experimental data on other systems, e.g., do not conclusively prove the necessity of a cocatalyst.

Kinetics

Although many cationic polymerizations proceed so rapidly that it is difficult to establish the steady state, the following kinetic scheme appears to be valid. Let the catalyst be designated by A and the cocatalyst by RH. Initiation, propagation, termination, and transfer may be represented thus:

$$A + RH \overset{K}{\rightleftharpoons} H^+\,{}^-AR$$
$$H^+\,{}^-AR + M \overset{k_i}{\longrightarrow} HM^+\,{}^-AR$$
$$HM_x{}^+\,{}^-AR + M \overset{k_p}{\longrightarrow} HM_{x+1}^+\,{}^-AR \tag{10-1}$$
$$HM_x{}^+\,{}^-AR \overset{k_t}{\longrightarrow} M_x + H^+\,{}^-AR$$
$$HM_x{}^+\,{}^-AR + M \overset{k_{tr}}{\longrightarrow} M_x + HM^+\,{}^-AR$$

The rate of initiation is

$$v_i = K k_i[A][RH][M] \tag{10-2}$$

where [A] is the catalyst concentration. (If the formation of $H^+ \ ^-AR$ is the rate-determining step, v_i may be independent of [M] and the kinetic scheme should be appropriately modified.)

Since, in strong contrast to radical polymerization, termination is first order,

$$v_t = k_t[M^+] \tag{10-3}$$

where $[M^+]$ is written as an abbreviation for $[HM^+ \ ^-AR]$. By the steady-state assumption (which appears to hold in most cases despite uncertainty as to whether such a state is actually achieved),

$$[M^+] = \frac{K k_i}{k_t} [A][RH][M] \tag{10-4}$$

The over-all polymerization rate is

$$v_p = k_p[M][M^+] = K \frac{k_i k_p}{k_t} [A][RH][M]^2 \tag{10-5}$$

If termination predominates over transfer,

$$\bar{x}_n = \frac{v_p}{v_t} = \frac{k_p[M^+][M]}{k_t[M^+]} = \frac{k_p}{k_t} [M] \tag{10-6}$$

whereas if transfer predominates,

$$\bar{x}_n = \frac{v_p}{v_{tr}} = \frac{k_p[M^+][M]}{k_{tr}[M^+][M]} = \frac{k_p}{k_{tr}} \tag{10-7}$$

Since v_p is proportional to $k_i k_p/k_t$ and, if termination predominates, $\bar{x}_n$ is proportional to k_p/k_t, the over-all rate of polymerization increases with decreasing temperature if $E_t > E_i + E_p$, and the molecular weight increases as the temperature of reaction decreases if $E_t > E_p$. Since the propagation step involves the approach of an ion to a neutral molecule in a medium of low dielectric constant, no activation energy is necessary. The termination step, on the other hand, requires the rearrangement of an ion pair and thus involves an appreciable activation energy. These conditions are exactly the reverse of those found in free radical polymerization.

Although some investigators suggest that polymerization also occurs by carbanion chains initiated by the anion of the initiator complex, the concept must be taken with great reservation. The charge separation necessary for independent growth of oppositely charged chains is much too great to occur in media of low dielectric constant.

The dielectric constant of the medium does have a significant effect on the nature of the ion pair, which can range between the extremes of complete covalency and completely free ions. It is often convenient to consider two types of propagating species, an ion pair and a free ion, in equilibrium. The nature of the solvent and the counterion can alter the course of the polymerization greatly by shifting this equilibrium. Large increases in v_p and $\bar{x}_n$ are usually observed as the solvating power of the medium is increased: the concentration of the free ion, which propagates more rapidly, is increased at the same time. Similarly, the larger and the less tightly bound the counterion, the more rapid is the propagation.

Plesch (1968) has concluded that many so-called cationic polymerizations do not proceed through ionic, but rather covalently bonded, intermediates; these are referred to as *pseudocationic polymerizations*.

GENERAL REFERENCES

Plesch 1954, 1963; Eastham 1960, 1965; Lenz 1967; Ravve 1967; Odian 1970.

C. Anionic Polymerization

Anionic polymerization was carried out on a commercial scale for many years before the nature of the polymerization was recognized, in the production of the buna-type synthetic rubbers in Germany and Russia by the polymerization of butadiene with sodium or potassium as the catalyst. The first anionic chain reaction to be so identified was the polymerization of methacrylonitrile by sodium in liquid ammonia at $-75°C$ (Beaman 1948).

In addition to the monomer types listed in Table 10-1, monomers that polymerize by ring scission (Section *F*) may do so by anionic mechanisms. Typical catalysts for anionic polymerization include the alkali metals, alkali metal amides, alkoxides, alkyls, aryls, hydroxides, and cyanides.

Mechanism

The conventional method of initiation of ionic chains involves the addition of a negative ion to the monomer, with the opening of a bond or ring, and growth at one end:

$$NH_2^- + CH_2{=}CHX \longrightarrow H_2NCH_2CHX^-$$

$$CH_3O^- + \underset{\underset{O}{\diagdown\diagup}}{CH_2CH_2} \longrightarrow CH_3OCH_2CH_2O^-$$

(In the equations of this section the counterions are omitted; however, their role is essentially the same as in cationic polymerization.) Simultaneous growth from more than one center can be obtained from polyvalent ions such as those derived from

$$C\left[\langle\bigcirc\rangle Li\right]_4 \quad \text{or} \quad N(CH_2CH_2O^-Na^+)_3.$$

The more basic the ion, the better it serves to initiate chains. Thus, OH^- will not initiate the anionic polymerization of styrene, NH_2^- initiates fairly well, but $\langle\bigcirc\rangle CH_2^-$ is powerful. Similarly, more acid monomers require less basic ions, with the acidity of the monomer depending on the strength of the $X\text{-}M^-$ bond formed in initiation and the stability of the resulting ion.

Initiation can also occur by the transfer of an electron to a monomer of high electron affinity. If D or D^- is an electron donor,

$$D + M \longrightarrow D^+ + M^-$$

or

$$D^- + M \longrightarrow D + M^-$$

Presumably, M^- is less reactive than a true carbanion or a free radical, but the addition of a monomer to M^- gives a species which contains one radical end and one anion end:

$$CH_2{=}\bar{C}HX + CH_2{=}CHX \longrightarrow \underset{\text{anion end}}{:\bar{C}H_2CHXCH_2}\underset{\text{radical end}}{\dot{C}HX}$$

This species can add monomer from the two ends by different mechanisms. Two radical ends may dimerize, however, leaving a divalent anion to propagate.

Propagation in anionic polymerization may be conventional or may be more complex, as in the elimination of CO_2 from N-carboxyanhydrides. In contrast to radical polymerization, the β (unsubstituted) carbon atom at the end of the growing chain is the site of addition of the next monomer.

As in cationic polymerization, termination is always unimolecular, usually by transfer. The recombination of a chain with its counterion or the transfer of a hydrogen to give terminal unsaturation, frequent in cationic polymerization, is unlikely in anionic mechanisms, as may be recognized by considering the small likelihood of transferring H^- when the counterion is Na^+. Thus, in anionic polymerization, termination usually involves transfer, and the kinetic chain is broken only if the new species is too weak to propagate.

This leads to the unique situation in which, by careful purification to eliminate all species to which transfer can occur, the termination step is effectively eliminated and the growing chains remain active indefinitely. This case is described below under the heading "living" polymers.

Kinetics

The kinetics of anionic polymerization may be illustrated by the polymerization of styrene with potassium amide in liquid ammonia (Higginson 1952, Wooding 1952):

$$KNH_2 \xrightarrow{K} K^+ + NH_2^-$$
$$NH_2^- + M \xrightarrow{k_i} NH_2 M^-$$
$$NH_2 M_x^- + M \xrightarrow{k_p} NH_2 M_{x+1}^-$$
$$NH_2 M_x^- + NH_3 \xrightarrow{k_{tr}} NH_2 M_x H + NH_2^-$$

(10-8)

The usual kinetic analysis leads for high degree of polymerization to

$$v_p = \frac{K k_p k_i}{k_{tr}} \frac{[NH_2^-]}{[NH_3]} [M]^2$$

(10-9)

and

$$\bar{x}_n = \frac{k_p}{k_{tr}} \frac{[M]}{[NH_3]}$$

(10-10)

"Living" polymers

Since the termination step usually involves transfer to some species not essential to the reaction, anionic polymerization with carefully purified reagents may lead to systems in which termination is lacking. The resulting species, called "living" polymers (Henderson 1968; Szwarc 1968b), can be prepared, for example, by polymerizing styrene with sodium naphthalene. Kinetic analysis shows that the polymer can have an extremely narrow distribution of molecular weight and for all practical purposes be essentially monodisperse; if initiation is rapid compared to propagation, the molecular-weight distribution is the Poisson function (Chapter 9E) for which $\bar{x}_w/\bar{x}_n \simeq 1 + 1/\bar{x}_n$. The polymer can be "killed" by addition of a terminating agent, e.g., water, at the end of the reaction.

The "living polymer" technique provides an unique opportunity for the preparation of block copolymers (Chapter 11D) of precisely defined composition (Henderson 1968; Fetters 1969).

GENERAL REFERENCES

Gaylord 1959; Mulvaney 1961; Robinson 1964; Overberger 1965; Lenz 1967; Morton 1967, 1969; Ravve 1967; Szwarc 1968a; Odian 1970.

D. Coordination Polymerization

Since 1954, a major development in polymer chemistry has been a new technique leading to polymers with unusual structures. From the fact that these structures may involve stereospecificity, the technique is sometimes called *stereospecific* or *stereoregular polymerization*, but the term *coordination polymerization* is used here to suggest the essential feature of a coordination complex or, at the very least, a directing force which governs the orientation with which a monomer approaches the growing chain end. Whether stereospecificity is involved depends on the nature of the monomer and the catalyst system. The mechanism of coordination polymerization may be anionic, cationic, or free radical. Quite often, coordination polymerization is carried out using a catalyst in the form of a slurry of small solid particles in an inert medium (a *fluid-bed* process) or a supported solid catalyst. Heterogeneous polymerization is not essential for coordination polymerization and the development of stereospecificity, however, for sufficiently polar monomers.

Perhaps the earliest record of a coordination polymerization was the cationic polymerization of vinyl isobutyl ether by Schildknecht (1947). Patent applications made by workers in the Du Pont Company in the period 1949–1954 describe what have since become known as coordination catalysts for the polymerization of ethylene. But the field of coordination polymerization actually came into existence with the work of Ziegler (1955) and Natta (1955) who developed new polymerization catalysts with unique stereoregulating powers. Their Nobel Prize addresses (Ziegler 1964; Natta 1965) form excellent introductions to the field.

Ziegler-Natta catalysts

The Ziegler-Natta or, often, Ziegler catalysts have the remarkable property of polymerizing a wide variety of monomers to linear and stereoregular polymers. Ethylene is polymerized to a highly linear chain, in contrast to the products of radical polymerization (Chapter 13). Polypropylene may be made in either the isotactic or syndiotactic form, but higher α-olefins yield only isotactic polymers. Dienes such as butadiene and isoprene may be polymerized to products which are almost exclusively *cis*-1,4; *trans*-1,4;

isotactic 1,2; or syndiotactic 1,2; depending on choice of catalyst and conditions. Many other examples exist.

Ziegler catalysts are complexes formed by the interaction of alkyls of metals of groups I-III in the periodic table with halides and other derivatives of transition metals of groups IV-VIII. A typical coordination catalyst is a complex between an aluminum alkyl and a titanium halide having the structure postulated in Fig. 10-1. Other forms are monometallic, with a typical structure being

where □ indicates an empty octahedral titanium orbital.

Mechanisms   The mechanism now considered most likely for the bimetallic Ziegler catalyst systems is represented in Fig. 10-2. The monomer forms a coordination complex with the catalyst, and is then inserted into the polarized Ti-C bond at the growing end of the polymer chain. One mechanism proposed for propagation with the monometallic species is that depicted in Fig. 10-3. Here the monomer is coordinated at the vacant orbital (*b*), and then inserted into the chain (*c*) with the regeneration of the vacant orbital with a different orientation (*d*). Continued addition would lead to a syndiotactic polymer; to obtain isotactic product the chain would have to migrate back to its original site (*e*). An alternate mechanism not requiring this migration has

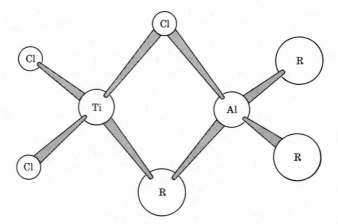

Fig. 10-1. Structure postulated for the metal halide-metal alkyl complex of a typical coordination catalyst.

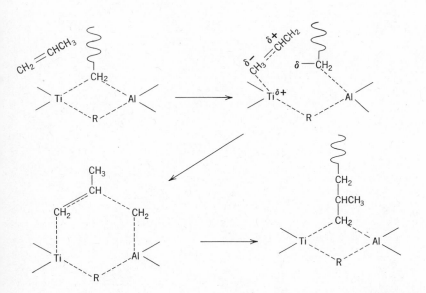

ig. 10-2.   Bimetallic mechanism for polymerization with a Ziegler-Natta catalyst (Odian)70, after Patat 1958 and Natta 1960).

een proposed (Boor 1967). Many other mechanisms have been presented t the literature, but virtually none has been substantiated, even in its major atures, much less in detail.

Only heterogeneous Ziegler-Natta catalysts yield isotactic polymers from onpolar monomers such as α-olefins. The nature of the catalyst surface ppears to be important in imparting stereoregularity in these cases. Propaga- on probably takes place at active sites on the crystal surface of the transition- 1etal compound, TiCl$_3$ for example. The sites are formed by the reaction f the Group I-III soluble metalorganic compound, for example Al(C$_2$H$_5$)$_3$ r Al(C$_2$H$_5$)$_2$Cl, with the crystal surface, yielding species of the type shown 1 Fig. 10-1. The chemical and crystal structures of the catalysts determine 1e orientation of the monomer as it is added to the growing chain. The riving force for isotactic propagation thus results from interaction between 1e monomer and the ligands of the transition metal at the active site (Boor 967). Several other models have been proposed, but none appears to be as idely applicable.

Kinetics   Since the kinetics of the homogeneous Ziegler-Natta catalyst sys- ems is quite similar to that of the ionic systems described in Sections *B* and

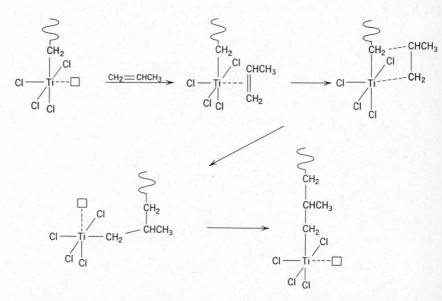

Fig. 10-3. Monometallic mechanism for coordination polymerization (Odian 1970, after Cossee 1964).

C, attention is limited here to the far more complex case of the heterogeneous polymerizations. Denoting the surface of the transition-metal compound by a heavy bar, ▬▬▬▬▬, the metal alkyl by AR, and the monomer by M, the following reactions are pertinent:

Adsorption of AR from solution to form the active site, and of M from solution:

$$\text{▬▬▬▬} + AR \xrightarrow{K_1} \text{▬▬▬▬}—AR$$
$$\text{▬▬▬▬} + M \xrightarrow{K_2} M—\text{▬▬▬▬}$$

(10-11)

Initiation:

$$M—\text{▬▬▬▬}—AR \xrightarrow{k_i} \text{▬▬▬▬}—A—M—R$$

(10-12)

Propagation:

$$M—\text{▬▬▬▬}—A—M_x—R \xrightarrow{k_p} \text{▬▬▬▬}—A—M_{x+1}—R$$

(10-13)

Transfer with monomer:

$$M—\text{▬▬▬▬}—A—M_x—R \xrightarrow{k_{tr}} \text{▬▬▬▬}—A—M—R + M_x$$

(10-14)

where M_x is polymer. Other transfer reactions are possible. Termination of the kinetic chain can occur by several reactions, an example being that with monomer to form an inactive site:

$$M\text{——}\blacksquare\text{——}A\text{—}M_x\text{—}R \xrightarrow{k_t} \text{—}\blacksquare\text{————}A\text{—}M\text{—} + M_xR \tag{10-15}$$

Rate equations can be derived if the adsorption law is known. If it is the usual Langmuir isotherm (Ashmore 1963), in which the equilibria of Eq. 10-11 are attained and maintained rapidly and in competition, the fractions of the available surface covered by AR and M, θ_A and θ_M, are given by

$$\theta_A = \frac{K_1[AR]}{1 + K_1[AR] + K_2[M]} \tag{10-16}$$

$$\theta_M = \frac{K_2[M]}{1 + K_1[AR] + K_2[M]} \tag{10-17}$$

The over-all rate is given by

$$\frac{-d[M]}{dt} = (k_p + k_{tr})\theta_M[C^*] \tag{10-18}$$

where $[C^*]$ is the concentration of growing chains. Assuming that $[C^*]$ reaches the steady state, the equations for initiation and termination can be equated:

$$k_i\theta_A\theta_M = k_t[C^*]\theta_M \tag{10-19}$$

from which by elimination of $[C^*]$ we obtain

$$-\frac{d[M]}{dt} = \frac{(k_p + k_{tr})k_i}{k_t} \frac{K_1K_2[AR][M]}{(1 + K_1[AR] + K_2[M])^2} \tag{10-20}$$

Similarly, an expression for $\bar{x}_n$ can be written. These equations are over-simplified for clarity by assuming only one transfer and one termination reaction.

The observed kinetics with Ziegler-Natta catalysts is usually quite complex. A major reason for this is the difficulty in determining the number of active sites and thus K_1 and K_2. In most cases data are presented in terms of reactivity per gram of catalyst, which introduces complications depending on such things as degree of subdivision or surface area.

Other coordination polymerizations

Polymerization with supported metal oxide catalysts (Clark 1969) Catalysts consisting of metals or metal oxides adsorbed on or complexed with the surface of a solid carrier are of commercial interest for the low-pressure polymerization of olefins. Although often used in fixed-bed processes, the catalysts can be slurried and employed in fluid-bed operation like the Ziegler-Natta catalysts.

Common catalyst compositions include oxides of chromium or molybdenum, or cobalt and nickel metals, supported on silica, alumina, titania, zirconia, or activated carbon. The mechanisms and kinetics of the reactions are still largely unknown, although many proposals like those made for Ziegler-Natta polymerization have been presented.

"Alfin" polymerization   The "Alfin" catalyst (Morton 1964; Reich 1966) consists of a suspension in an inert solvent like pentane of a mixture of an alkylenylsodium compound (such as allyl sodium), an alkoxide of a secondary alcohol (such as sodium isopropoxide), and an alkali halide (such as sodium chloride). The catalyst is highly specific for the polymerization of dienes into the 1,4 forms.

Stereoregular polymerization of methyl methacrylate   Initiators and reaction conditions have been found which permit the polymerization of methyl methacrylate to either a highly isotactic or a highly syndiotactic polymer, or a block copolymer of the two, in contrast to the atactic product of free-radical polymerization. Initiation with 9-fluorenyllithium in highly solvating media, such as 1,2-dimethoxyethane, leads to syndiotactic polymer at $-60°C$, whereas use of toluene under the same conditions gives the isotactic form. Addition of small amounts of ethers to the toluene system produces the block copolymer. The mechanism is anionic.

This stereochemical control probably results from the general principle that propagation involving a freely dissociated anionic end group leads to the thermodynamically more stable syndiotactic configuration in which steric and polar repulsions are minimized. The higher-energy isotactic configuration is obtained only under favorable circumstances such as formation of a strong complex between the counterion and the monomer.

GENERAL REFERENCES

Gaylord 1959; Natta 1966; Reich 1966; Boor 1967, 1970; Coover 1967; Ketley 1967; Lenz 1967; Ravve 1967; Cooper 1970; Odian 1970.

E. Ring-Opening Polymerization

Polymers may be prepared by routes involving ring opening, which must be classed stoichiometrically as addition, since no small molecule is split off in the reaction. Some typical examples are given in Table 10-2. The

TABLE 10-2. *Typical ring-scission polymers*

Type	Example	
Lactone	$O(CH_2)_xCO$	$\longrightarrow$ $-[O(CH_2)_xCO-]_y$
Lactam	$HN(CH_2)_xCO$	$\longrightarrow$ $-[HN(CH_2)_xCO-]_y$
Cyclic ether	$(CH_2)_xO$	$\longrightarrow$ $-[(CH_2)_xO-]_y$
Cyclic anhydride	$CO(CH_2)_xCO$ O	$\longrightarrow$ $-[CO(CH_2)_xCOO-]_y$
N-carboxyanhydride	$COCHRNHCO$ O	$\longrightarrow$ $-[COCHRNH-]_y + CO_2$

polymerization of these compounds has some aspects of both chain and step polymerization as far as kinetics and mechanism are concerned. It resembles chain polymerization in that it proceeds by the addition of monomer (but never of larger units) to growing chain molecules. However, the chain-initiating and subsequent addition steps may be similar and proceed at similar rates; if so, these are not chain reactions in the kinetic sense. As in stepwise polymerization, the polymer molecules continue to increase in molecular weight throughout the reaction.

Most of these cyclic compounds polymerize by ionic chain mechanisms in the presence of strong acids or bases when water and alcohols are excluded. These polymerizations are often very rapid. They are of commercial interest in the polymerization of caprolactam (Greenley 1969), ethylene oxide, and 3,3-*bis*(chloromethyl)oxetane. These are true chain reactions.

GENERAL REFERENCES

Furukawa 1963; Lenz 1967; Patsiga 1967; Odian 1970.

BIBLIOGRAPHY

Ashmore 1963. P. G. Ashmore, *Catalysis and Inhibition of Chemical Reactions,* Butterworths, London, 1963.

Beaman 1948. Ralph G. Beaman, "Anionic Chain Polymerization," *J. Am. Chem. Soc.* **70,** 3115–3118 (1948).

Boor 1967. J. Boor, Jr., "The Nature of the Active Site in the Ziegler-Type Catalyst," *Macromol. Revs.* **2,** 115–268 (1967).

Boor 1970. John Boor, Jr., " The Ziegler-Type Catalyst: A Review of Recent Literature *ACS Div. Organic Coatings and Plastics Chem., Papers Presented* **30** (1), 158–1 (1970).

Clark 1969. Alfred Clark, " Olefin Polymerization on Supported Chromium Oxide Cat lysts," *Catalysis Revs.* **3**, 145–174 (1969).

Cooper 1970. W. Cooper, " Aspects of the Mechanism of Coordination Polymerization Conjugated Dienes," *ACS Div. Organic Coatings and Plastics Chem., Papers Present* **30** (1), 191–207 (1970).

Coover 1967. H. W. Coover, Jr., Richard L. McConnell, and F. B. Joyner, " Relationsh of Catalyst Composition to Catalytic Activity for the Polymerization of α-Olefins *Macromol. Revs.* **1**, 91–118 (1967).

Cossee 1964. P. Cossee, " Ziegler-Natta Catalysis. I. Mechanism of Polymerization α-Olefins with Ziegler-Natta Catalysts," *J. Catalysis* **3**, 80–88 (1964).

Eastham 1960. A. M. Eastham, " Some Aspects of the Polymerization of Cyclic Ethers *Fortschr. Hochpolym.-Forsch. (Advances in Polymer Science)* **2**, 18–50 (1960).

Eastham 1965. A. M. Eastham, " Cationic Polymerization," pp. 35–59 in Herman F. Mar Norman G. Gaylord, and Norbert M. Bikales, eds., *Encyclopedia of Polymer Scien and Technology*, Vol. 3, Interscience Div., John Wiley and Sons, New Yor 1965.

Fetters 1969. L. J. Fetters, " Synthesis of Block Polymers by Homogeneous Anion Polymerization," *J. Polymer Sci.* **C26**, 1–35 (1969).

Furukawa 1963. Junji Furukawa and Takeo Saegusa, *Polymerization of Aldehydes a Oxides*, Interscience Div., John Wiley and Sons, New York, 1963.

Gaylord 1959. Norman G. Gaylord and Herman F. Mark, *Linear and Stereoregul Addition Polymers*, Interscience Publishers, New York, 1959.

Greenley 1969. R. Z. Greenley, J. C. Stauffer, and J. E. Kurz, " The Kinetic Equation f the Initiated Polymerization of ε-Caprolactam," *Macromolecules* **2**, 561–567 (1969).

Henderson 1968. J. F. Henderson and M. Szwarc, " The Use of Living Polymers in t Preparation of Polymer Structures of Controlled Architecture," *Macromol. Revs.* 317–401 (1968).

Higginson 1952. W. C. E. Higginson and N. S. Wooding, " Anionic Polymerisation. Pa I. The Polymerisation of Styrene in Liquid Ammonia Solution Catalysed by Potassiu Amide," *J. Chem. Soc.* **1952**, 760–774.

Ketley 1967. A. D. Ketley, ed., *The Stereochemistry of Macromolecules*, Vol. 1, Marc Dekker, New York, 1967.

Lenz 1967. Robert W. Lenz, *Organic Chemistry of Synthetic High Polymers*, Interscien Div., John Wiley and Sons, New York, 1967.

Lenz 1969. Robert W. Lenz, " Applied Reaction Kinetics: Polymerization Reactic Kinetics," *Ind. Eng. Chem.* **61** (3), 67–75 (1969).

Lenz 1970. Robert W. Lenz, " Applied Polymerization Reaction Kinetics," *Ind. En Chem.* **62** (2), 54–61 (1970).

Morton 1964. Avery A. Morton, " Alfin Catalysts," pp. 629–638 in Herman F. Mar Norman G. Gaylord, and Norbert M. Bikales, eds., *Encyclopedia of Polymer Scien and Technology*, Vol. 1, Interscience Div., John Wiley and Sons, New York, 1964.

Morton 1967. Maurice Morton and Lewis J. Fetters, " Homogeneous Anionic Polymeriz tion of Unsaturated Monomers," *Macromol. Revs.* **2**, 71–113 (1967).

Morton 1969. Maurice Morton, " Anionic Polymerization," Chap. 5 in George E. Han ed., *Vinyl Polymerization*, Part I, Vol. 2, Marcel Dekker, New York, 1969.

Mulvaney 1961. J. E. Mulvaney, C. G. Overberger, and Arthur M. Schiller, " Anion Polymerization," *Fortschr. Hochpolym.-Forsch. (Advances in Polymer Science)* 106–138 (1961).

Natta 1955. G. Natta, Piero Pino, Paolo Corradini, Ferdinando Danusso, Enrico Mantica, Giorgio Mazzanti, and Giovanni Moranglio, "Crystalline High Polymers of α-Olefins," *J. Am. Chem. Soc.* **77**, 1708–1710 (1955).

Natta 1960. G. Natta, "Stereospecific Polymerizations," *J. Polymer Sci.* **48**, 219–239 (1960).

Natta 1965. Giulio Natta, "Macromolecular Chemistry: From the Stereospecific Polymerization to the Symmetric Autocatalytic Synthesis of Molecules," *Science* **147**, 261–272 (1965).

Natta 1966. Giulio Natta and Umberto Giannini, "Coordinate Polymerization," pp. 137–150 in Herman F. Mark, Norman G. Gaylord, and Norbert M. Bikales, eds., *Encyclopedia of Polymer Science and Technology*, Vol. 4, Interscience Div., John Wiley and Sons, New York, 1966.

Odian 1970. George Odian, *Principles of Polymerization*, McGraw-Hill Book Co., New York, 1970.

Overberger 1965. C. G. Overberger, J. E. Mulvaney, and Arthur M. Schiller, "Anionic Polymerization," pp. 95–137 in Herman F. Mark, Norman G. Gaylord, and Norbert M. Bikales, eds., *Encyclopedia of Polymer Science and Technology*, Vol. 2, Interscience Div., John Wiley and Sons, New York, 1965.

Patat 1958. F. Patat and Hj. Sinn, "On the Results of Low Pressure Polymerization of α-Olefins (Complex Polymerization I)" (in German), *Angew. Chem.* **70**, 496–500 (1958).

Patsiga 1967. Robert A. Patsiga, "Copolymerization of Vinyl Monomers with Ring Compounds," *J. Macromol. Sci.—Revs. Macromol. Chem.* **C1**, 223–261 (1967).

Plesch 1954. P. H. Plesch, ed., *Cationic Polymerization and Related Complexes*, Academic Press, New York, 1954.

Plesch 1963. P. H. Plesch, ed., *The Chemistry of Cationic Polymerization*, Macmillan, New York, 1963.

Plesch 1968. P. H. Plesch, "Cationic Polymerization," pp. 137–188 in J. C. Robb and F. W. Peaker, eds., *Progress in High Polymers*, Vol. 2, CRC Press, Cleveland, Ohio, 1968.

Ravve 1967. A. Ravve, *Organic Chemistry of Macromolecules*, Marcel Dekker, New York, 1967.

Reich 1966. Leo Reich and A. Schindler, *Polymerization by Organometallic Compounds*, Interscience Div., John Wiley and Sons, New York, 1966.

Robinson 1964. R. E. Robinson and I. L. Meador, "Alkali Metals and Derivatives," pp. 639–658 in Herman F. Mark, Norman G. Gaylord, and Norbert M. Bikales, eds., *Encyclopedia of Polymer Science and Technology*, Vol. 1, Interscience Div., John Wiley and Sons, New York, 1964.

Schildknecht 1947. C. E. Schildknecht, A. O. Zoss, and Clyde McKinley, "Vinyl Alkyl Ethers," *Ind. Eng. Chem.* **39**, 180–186 (1947).

Szwarc 1968a. Michael Szwarc, *Carbanions, Living Polymers and Electron-Transfer Processes*, John Wiley and Sons, New York, 1968.

Szwarc 1968b. M. Szwarc, "'Living' Polymers," pp. 303–325 in Herman F. Mark, Norman G. Gaylord, and Norbert M. Bikales, eds., *Encyclopedia of Polymer Science and Technology*, Vol 8, Interscience Div., John Wiley and Sons, New York, 1968.

Wooding 1952. N. S. Wooding and W. C. E. Higginson, "Anionic Polymerisation. Part III. The Polymerisation of Styrene in Liquid Ammonia Catalysed by Potassium," *J. Chem. Soc.* **1952**, 1178–1180.

Ziegler 1955. Karl Ziegler, E. Holzkamp, H. Breil, and H. Martin, "Polymerization of Ethylene and Other Olefins" (in German), *Angew. Chem.* **67**, 426 (1955).

Ziegler 1964. Karl Ziegler, "Consequences and Development of an Invention" (in German), *Angew. Chem.* **76**, 545–553 (1964).

II

Copolymerization

A. Kinetics of Copolymerization

Although the polymerization of organic compounds has been known fo
over one hundred years, the simultaneous polymerization (*copolymerization*
of two or more monomers was not investigated until about 1911, whe
copolymers of olefins and diolefins were found to have rubbery propertie
For a number of years the study of the kinetics of copolymerization lagge
behind that of the properties of copolymers, which were often found to b
more useful than *homopolymers* made from single monomers.

In the 1930's it was found that monomers differed markedly in thei
tendencies to enter into copolymers. Staudinger (1939) fractionated a viny
chloride-vinyl acetate copolymer made from a mixture of equimolar quanti
ties of the two monomers. He found no polymer containing equal amount
of each monomer, but instead ratios of vinyl chloride : vinyl acetate o
9:3, 7:3, 5:3, and 5:7 among the fractions.

At about the same time, acrylic esters were found to enter copolymer
with vinyl chloride faster than did the second monomer. The first polyme
formed was rich in the acrylate; later, as the amount of acrylate in the mono
mer system was depleted, the polymer became richer in vinyl chloride. I
was found empirically that this change in composition could be reduced b
adding the more active monomer gradually to the reaction mixture through
out the course of the polymerization.

Maleic anhydride and other monomers which homopolymerize wit
great difficulty were found to copolymerize readily with such monomers a

styrene and vinyl chloride. Pairs of monomers, such as stilbene-maleic anhydride and isobutylene-fumaric ester, in which neither monomer polymerizes alone, readily gave high-molecular-weight copolymers in which the monomers appeared in a 1:1 ratio no matter which was in excess in the monomer feed.

The kinetic scheme for chain-reaction copolymerization is developed for radical reactions; extension to ionic systems is straightforward and is treated in the general references. Recent aspects of copolymerization have been reviewed (Lenz 1969, 1970).

The copolymer equation

In 1936 Dostal made the first attack on the mechanism of copolymerization by assuming that the rate of addition of monomer to a growing free radical depends only on the nature of the end group on the radical chain. Thus monomers M_1 and M_2 lead to radicals of types $M_1\cdot$ and $M_2\cdot$. There are four possible ways in which monomer can add:

Reaction			*Rate*	
$M_1\cdot + M_1$	$\longrightarrow$	$M_1\cdot$	$k_{11}[M_1\cdot][M_1]$	
$M_1\cdot + M_2$	$\longrightarrow$	$M_2\cdot$	$k_{12}[M_1\cdot][M_2]$	
$M_2\cdot + M_1$	$\longrightarrow$	$M_1\cdot$	$k_{21}[M_2\cdot][M_1]$	(11-1)
$M_2\cdot + M_2$	$\longrightarrow$	$M_2\cdot$	$k_{22}[M_2\cdot][M_2]$	

Since four independent rate constants were involved, Dostal made no attempt to test his assumption experimentally.

After several unsuccessful approaches, the kinetics of copolymerization was elucidated in 1944 by Alfrey, Mayo, Simha, and Wall. To Dostal's reaction scheme they added the assumption of the steady state applied to each radical type separately, i.e., the concentrations of $M_1\cdot$ and $M_2\cdot$ must each remain constant. It follows that the rate of conversion of $M_1\cdot$ to $M_2\cdot$ must equal that of conversion of $M_2\cdot$ to $M_1\cdot$:

$$k_{21}[M_2\cdot][M_1] = k_{12}[M_1\cdot][M_2] \tag{11-2}$$

The rates of disappearance of the two types of monomer are given by

$$-\frac{d[M_1]}{dt} = k_{11}[M_1\cdot][M_1] + k_{21}[M_2\cdot][M_1]$$

$$-\frac{d[M_2]}{dt} = k_{12}[M_1\cdot][M_2] + k_{22}[M_2\cdot][M_2] \tag{11-3}$$

By defining $r_1 = k_{11}/k_{12}$ and $r_2 = k_{22}/k_{21}$, and combining Eqs. 11-2 and 11-3, it can be shown that the composition of copolymer being formed at any instant is given by

$$\frac{d[M_1]}{d[M_2]} = \frac{[M_1]}{[M_2]} \frac{r_1[M_1] + [M_2]}{[M_1] + r_2[M_2]} \tag{11-4}$$

This is known as the *copolymer equation*; it has been verified by many experimental investigations of copolymer composition.

Monomer reactivity ratios

The *monomer reactivity ratios* r_1 and r_2 are the ratios of the rate constant for a given radical adding its own monomer to that for its adding the other monomer. Thus $r_1 > 1$ means that the radical $M_1\cdot$ prefers to add M_1; $r_1 < 1$ means that it prefers to add M_2. In the system styrene (M_1)-methyl methacrylate (M_2), e.g., $r_1 = 0.52$ and $r_2 = 0.46$; each radical adds the other monomer about twice as fast as its own.

Since the rate constants for initiation and termination do not appear in Eq. 11-4, the composition of the copolymer is independent of over-all reaction rate and initiator concentration. The reactivity ratios are unaffected in most cases by the presence of inhibitors, chain transfer agents, or solvents. Even in heterogeneous systems they remain unchanged unless the availability of the monomers is altered by their distribution between phases. A change from a free radical to an ionic mechanism, however, changes r_1 and r_2 markedly.

A few typical values of monomer reactivity ratios are given in Table 11-1. An extensive compilation is given by Mark (1966).

Types of copolymerization

Ideal A copolymer system is said to be *ideal* when the two radicals show the same preference for adding one of the monomers over the other: $k_{11}/k_{12} = k_{21}/k_{22}$, or $r_1 = 1/r_2$, or $r_1 r_2 = 1$. In this case the end group on the growing chain has no influence on the rate of addition, and the two types of units are arranged at random along the chain in relative amounts determined by the composition of the feed and the relative reactivities of the two monomers. The copolymer equation reduces to $d[M_1]/d[M_2] = r_1[M_1]/[M_2]$.

Alternating Here each radical prefers to react exclusively with the other monomer: $r_1 = r_2 = 0$. The monomers alternate regularly along the chain regardless of the composition of the monomer feed. The copolymer equation simplifies to $d[M_1]/d[M_2] = 1$.

TABLE 11-1.   *Typical monomer reactivity ratios (Mark 1966)*

Monomer 1	Monomer 2	r_1	r_2	$T°C$
Acrylonitrile	1,3-Butadiene	0.02	0.3	40
	Methyl methacrylate	0.15	1.22	80
	Styrene	0.04	0.40	60
	Vinyl acetate	4.2	0.05	50
	Vinyl chloride	2.7	0.04	60
1,3-Butadiene	Methyl methacrylate	0.75	0.25	90
	Styrene	1.35	0.58	50
	Vinyl chloride	8.8	0.035	50
Methyl methacrylate	Styrene	0.46	0.52	60
	Vinyl acetate	20	0.015	60
	Vinyl chloride	10	0.1	68
Styrene	Vinyl acetate	55	0.01	60
	Vinyl chloride	17	0.02	60
Vinyl acetate	Vinyl chloride	0.23	1.68	60

Most actual cases lie between the ideal and the alternating systems; $0 < r_1 r_2 < 1$. A third possibility, with both r_1 and r_2 greater than unity, corresponds to the tendency to form block copolymer (Section D). It has been observed in only a few cases (for example, Gumbs 1969).

Instantaneous composition of feed and polymer

Let F_1 and F_2 be the mole fractions of monomers 1 and 2 in the polymer being formed at any instant:

$$F_1 = 1 - F_2 = \frac{d[M_1]}{d([M_1] + [M_2])} \tag{11-5}$$

If f_1 and f_2 similarly represent mole fractions in the monomer feed,

$$f_1 = 1 - f_2 = \frac{[M_1]}{[M_1] + [M_2]} \tag{11-6}$$

the copolymer equation can be written

$$F_1 = \frac{r_1 f_1^2 + f_1 f_2}{r_1 f_1^2 + 2 f_1 f_2 + r_2 f_2^2} \tag{11-7}$$

It is apparent that in general F_1 does not equal f_1, and both f_1 and F_1 change as the polymerization proceeds. The polymer obtained over a finite range of conversion consists of many increments, each differing in composition (Section *B*).

Equation 11-7 may be used to calculate curves of feed vs. instantaneous polymer composition for various monomer reactivity ratios. Such curves for a series of ideal copolymerizations ($r_1 r_2 = 1$) are shown in Fig. 11-1. Except for pairs of monomers having very similar reactivities, only a small range of feeds gives copolymers containing appreciable amounts of both components.

Several curves for nonideal cases are shown in Fig. 11-2. These curves show the effect of increasing tendency toward alternation. As alternation increases, more and more feeds yield a copolymer containing a good deal of each component. This tendency makes practical the preparation of many important copolymers.

If one of the monomers is very much more reactive than the other, the first polymer formed contains mostly the more reactive monomer. Later in the polymerization this monomer is largely used up, and the last polymer formed consists mainly of the less reactive monomer. Styrene and vinyl acetate form such a system.

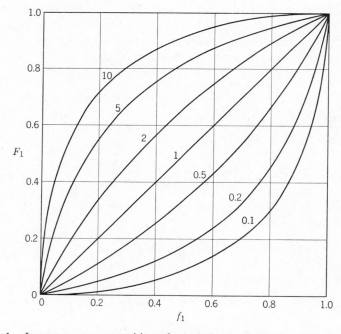

Fig. 11-1. Instantaneous composition of copolymer (mole fraction F_1) as a function of monomer composition (mole fraction f_1) for ideal copolymers with the values of $r_1 = 1/r_2$ indicated.

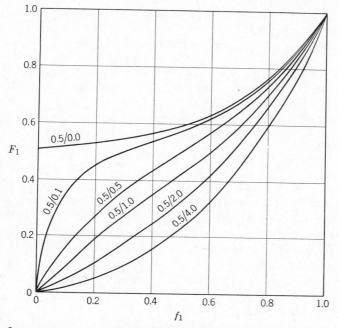

Fig. 11-2.   Instantaneous composition of copolymer F_1 as a function of monomer composition f_1 for the values of the reactivity ratios r_1/r_2 indicated.

For cases in which both r_1 and r_2 are less than unity (or, hypothetically, greater than unity), the curves of Fig. 11-2 cross the line representing $F_1 = f_1$. At the point of intersection polymerization proceeds without a change in the composition of feed or polymer. This is known as *azeotropic copolymerization*. Solution of Eq. 11-4 with $d[M_1]/d[M_2] = [M_1]/[M_2]$ gives the critical composition for the azeotrope:

$$\frac{[M_1]}{[M_2]} = \frac{1 - r_2}{1 - r_1}$$

or, from Eq. 11-7,   (11-8)

$$(f_1)_c = \frac{1 - r_2}{2 - r_1 - r_2}$$

Evaluation of monomer reactivity ratios

The usual experimental determination of r_1 and r_2 involves polymerizing to low conversion for a variety of feed compositions. The polymers are isolated and their compositions measured. Elemental analysis, chemical analysis for reactive groups, or physical analysis (e.g., by refractive index when styrene is one component) may be used.

Four methods of analyzing the data are available:

a. Direct curve fitting on polymer-monomer composition plots. This is a poor method, since the composition curve is rather insensitive to small changes in r_1 and r_2.

b. Plots of r_1 vs. r_2. The copolymer equation may be solved for one of the reactivity ratios:

$$r_2 = \frac{[M_1]}{[M_2]}\left[\frac{d[M_2]}{d[M_1]}\left(1 + \frac{[M_1]}{[M_2]}r_1\right) - 1\right] \tag{11-9}$$

Each experiment with a given feed gives a straight line; the intersection of several of these allows the evaluation of r_1 and r_2 (Fig. 11-3). If the experimental errors are high, the lines may not intersect in a single point; the region within which the intersections occur gives information about the precision of the experimental results.

c. Plots of F_1 vs. f_1. Equation 11-7 can be rearranged to

$$\frac{f_1(1 - 2F_1)}{(1 - f_1)F_1} = r_2 + \frac{f_1{}^2(F_1 - 1)}{(1 - f_1)^2 F_1}r_1 \tag{11-10}$$

This is the equation of a straight line with slope r_1 and intercept r_2. Each experiment gives one point on the line. The least-squares treatment of a

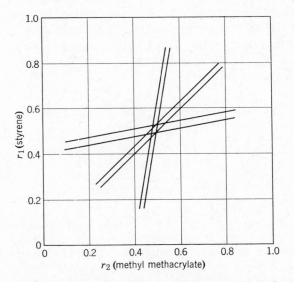

Fig. 11-3. Evaluation of monomer reactivity ratios by graphic solution of the copolymer equation (Burnett 1954, data of Mayo 1944).

series of such points gives the best values of r_1 and r_2 in a straightforward way.

d. Analysis of data giving the copolymer composition as a function of conversion, using the integrated form of the copolymer equation given in Section B (Meyer 1966).

Rate of copolymerization

The discussion so far has considered only the relative rates of the four possible propagation steps in a binary system. The over-all rate of copolymerization depends also on the rates of initiation and termination.

If the initiator releases radicals which have a high efficiency of combination with each monomer, the two types of initiation step need not be considered separately. Steady-state conditions apply both to the total radical concentration and to the separate concentrations of the two radicals. Three types of termination must be considered, involving all possible pairs of radical types. The steady-state condition for all the radicals requires that the initiation and termination rates be equal:

$$v_i = 2k_{t_{11}}[M_1\cdot]^2 + 2k_{t_{12}}[M_1\cdot][M_2\cdot] + 2k_{t_{22}}[M_2\cdot]^2 \qquad (11\text{-}11)$$

Here $k_{t_{11}}$ and $k_{t_{22}}$ are the termination rate constants for pairs of like radicals, and $k_{t_{12}}$ is the constant for the cross reaction. If this condition is applied to the equation for the over-all polymerization rate,

$$-\frac{d([M_1] + [M_2])}{dt} = v_p$$

$$= \frac{(r_1[M_1]^2 + 2[M_1][M_2] + r_2[M_2]^2)(v_i^{1/2}/\delta_1)}{\{r_1^2[M_1]^2 + 2(\phi r_1 r_2 \delta_2/\delta_1)[M_1][M_2] + (r_2\delta_2/\delta_1)^2[M_2]^2\}^{1/2}} \qquad (11\text{-}12)$$

where

$$\delta_1 = \left(\frac{2k_{t_{11}}}{k_{11}^2}\right)^{1/2}, \qquad \delta_2 = \left(\frac{2k_{t_{22}}}{k_{22}^2}\right)^{1/2} \qquad (11\text{-}13)$$

and

$$\phi = \frac{k_{t_{12}}}{2(k_{t_{11}})^{1/2}(k_{t_{22}})^{1/2}} \qquad (11\text{-}14)$$

The δ values are simply related to quantities of the type $k_p^2/2k_t$ as seen in Chapter 9; ϕ compares the cross termination rate constant to the geometric mean of the termination rate constants for like pairs of radicals. A value of ϕ greater than 1 means that cross termination is favored, and conversely.

If r_1 and r_2 are known, measurement of the rate of copolymerization allows ϕ to be determined. A typical result is shown in Fig. 11-4. The value

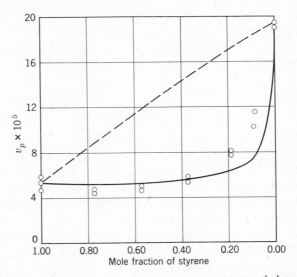

Fig. 11-4. Copolymerization rate data for the system styrene-methyl methacrylate at 60°C (Walling 1949). The broken line assumes $\phi = 1$; the solid line is calculated for $\phi = 13$.

of ϕ greater than unity is typical of most copolymerizations; indeed, cross termination is markedly favored except for nearly ideal copolymerizations. It is likely that differences in polarity are responsible for the preference for cross termination as well as for deviations from ideal copolymerization. As a result, the copolymerization of pairs of monomers of quite different polarity is likely to be slower than the mean of the rates for the separate homopolymerizations.

The above treatment assumes that the termination process is controlled by the chemical nature of the growing-chain radical ends. It is known, however, that termination in radical polymerization is generally diffusion controlled (North 1968); in this case ϕ cannot be interpreted in terms of chemical effects, and a single value of ϕ in Eq. 11-14 is not useful. The situation has been analyzed (Atherton 1962; North 1963) in terms of a single termination constant k_t. The rate equation becomes

$$v_p = \frac{(r_1[M_1]^2 + 2[M_1][M_2] + r_2[M_2]^2)v_i^{1/2}}{k_t^{1/2}(r_1[M_1]/k_{11} + r_2[M_2]/k_{22})} \qquad (11\text{-}15)$$

Ideally, k_t might be expected to be the mole fraction-weighted average of the termination rate constants for the two homopolymerizations, but this appears to be an oversimplification, possibly due to penultimate effects (see below).

Multicomponent copolymerization The mathematics of the copolymer equation has been extended to the cases of three or more monomers (Ham 1964), but the quantitative treatment is quite complex; for example, three monomers require the specification of nine propagation and six termination reactions, and six reactivity ratios. Some simplification results from rather liberal assumptions about the steady state and the probabilities of finding certain sequences of monomers in the resulting polymer. The predicted results are in quite good agreement with experiment.

Penultimate and remote-unit effects Ham (1964) and co-workers have developed the early suggestion of Merz (1946) that units before the last may affect the reactivity of a chain radical. This is often the case, the effect being particularly pronounced with monomers of type $CHX=CHX$, where X is a polar substituent. It has been detected in copolymers of styrene with maleic anhydride and fumaronitrile.

When the penultimate monomer unit does have some influence on the subsequent addition of monomer, four reactivity ratios, corresponding to the four possible types of radicals, must be considered. Effects of units still farther back along the chain have been detected experimentally in some cases.

Depropagation during copolymerization If one or both of the monomers has a tendency to depropagate during a copolymerization, an additional complication is introduced (Lowry 1960). This deviation from the simple treatment can be detected by the dependence of the copolymer composition on the absolute concentration of the monomers even though their relative concentration is invariant.

GENERAL REFERENCES

Alfrey 1952; Ham 1964, 1966; Odian 1970.

B. Composition of Copolymers

The copolymer equation (Eq. 11-4) predicts the average composition of the polymer formed at any instant in the polymerization. Not only may statistical fluctuations of composition about this average occur, but also polymer formed during a finite interval may contain a range of compositions resulting from variation of feed composition. In this case, the over-all polymer composition can be calculated by integrating the copolymer equation.

Statistical fluctuation of instantaneous copolymer composition

Stockmayer (1945) showed that the spread in copolymer composition due to statistical fluctuations is small in practical cases. Alternating copolymers give the minimum spread in composition; even in ideal copolymers, where the spread is greater and is a function of the degree of polymerization, the effect is not large: for a copolymer containing on the average 50% of each monomer, only 12% of the polymer molecules contain less than 43% of either monomer at $\bar{x}_n = 100$, while only 12% of the molecules contain less than 49.3% of either monomer at $\bar{x}_n = 10,000$. The distribution of compositions follows a Gaussian curve. This small spread in composition is neglected in the discussion to follow.

The probability of occurrence of sequences of n similar monomer units can be calculated statistically for an ideal copolymer (Alfrey 1952): the number $N(n)$ of such sequences is

$$N(n) = p^{n-1}(1 - p) \tag{11-16}$$

where p is the probability of addition of that monomer:

$$p = \frac{r_1[M_1]}{r_1[M_1] + [M_2]} \tag{11-17}$$

Except for a monomer present in large excess, the sequences are always short. Any tendency toward alternation clearly leads to shorter sequences than those in an ideal copolymer.

Integration of the copolymer equation

The direct integration of the copolymer equation gives a result which is not convenient for calculations. The most convenient method (Skeist 1946) for calculating copolymer composition and distribution involves the use of Eq. 11-7. For a system in which $F_1 > f_1$, when dM moles of monomer have polymerized, the polymer contains F_1 dM moles of monomer 1, and the feed contains $(f_1 - df_1)(M - dM)$ moles of monomer 1. For a material balance,

$$Mf_1 - (M - dM)(f_1 - df_1) = F_1 \, dM \tag{11-18}$$

Combining this with Eq. 11-7 gives

$$\frac{dM}{M} = \frac{df_1}{F_1 - f_1} \tag{11-19}$$

which can easily be integrated by numerical or graphic means to give the desired composition of polymer as a function of conversion. One convenient closed-form result of this integration is (Meyer 1965, 1967):

$$\log \frac{M}{M_0} = \frac{r_2}{1 - r_2} \log \frac{f_1}{(f_1)_0} + \frac{r_1}{1 - r_1} \log \frac{f_2}{(f_2)_0}$$

$$+ \frac{1 - r_1 r_2}{(1 - r_1)(1 - r_2)} \log \frac{(f_1)_0 - \varepsilon}{f_1 - \varepsilon} \qquad (11\text{-}20)$$

where $\varepsilon = (1 - r_2)/(2 - r_1 + r_2)$.

Variation of copolymer composition with conversion

Ideal system   The system styrene (M_1)-2-vinylthiophene (M_2) is nearly ideal with $r_1 = 0.35$ and $r_2 = 3.10$. The variation of the composition of polymer formed at any instant in this system with conversion and initial feed composition is shown in Fig. 11-5. The greater reactivity of the vinylthiophene causes it to be depleted until at the last only styrene remains, and pure polystyrene is the final product formed. The over-all polymer composition varies less with conversion, and ultimately approaches the initial feed composition as required for a material balance.

The polymer composition distribution for 100% conversion for several initial feeds is shown in the bar graphs in Fig. 11-6. Each block represents a 5% interval in copolymer composition. The U-shaped distribution curves

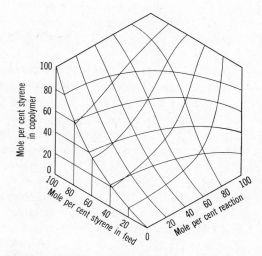

Fig. 11-5.   Variation in instantaneous composition of copolymer with initial feed and conversion for the system styrene-2-vinylthiophene (Mayo 1950).

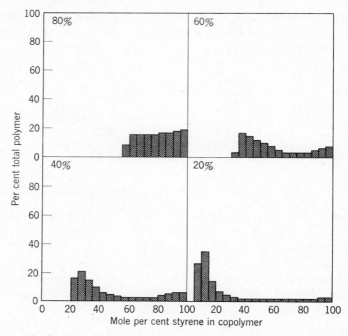

Fig. 11-6. Distribution of copolymer composition at 100% conversion for the indicated values of initial mole percent styrene in the feed, for the system styrene-2-vinylthiophene (Burnett 1954, after Mayo 1950).

arise whenever the monomer reactivity ratios are less than 0.5 and greater than 2, respectively. They indicate that appreciable quantities of two distinct compositions predominate, with little material of intermediate composition.

Alternating system   The system styrene (M_1)-diethyl fumarate (M_2) with $r_1 = 0.30$ and $r_2 = 0.07$ has a strong tendency to alternate and an azeotropic composition at 57 mole percent styrene. The diagrams for this copolymer are more complex than those for the previous system. Feeds near the azeotropic composition remain almost unchanged up to rather high conversion, but eventually all feeds containing more than 57 mole percent styrene drift toward pure styrene and those containing less styrene than the azeotropic amount yield pure diethyl fumarate. Thus (Fig. 11-7), pure polystyrene is formed at the last from all initial feeds higher in styrene than the azeotropic, and pure poly(diethyl fumarate) from all initial feeds containing less than 57 mole percent styrene. The over-all polymer composition shows markedly less change with conversion for all feeds and is of course invariant at the azeotropic composition. The bar graphs of the distribution of copolymer composition (Fig. 11-8) show that feeds containing more styrene than the azeotrope give only polymer containing more styrene than the azeotrope, and vice versa.

Attainment of homogeneity in copolymers   It is often desirable to produce a polymer which is homogeneous in composition rather than one having a broad distribution of compositions. The foregoing discussion suggests two ways of doing this. (*a*) The reaction can be stopped short of complete conversion. For example, the styrene-diethyl fumarate copolymer from feed containing 40% styrene, when stopped at 75% conversion, contains only polymer having between 44 and 52% styrene. (*b*) The polymerization can be started with a feed giving the desired polymer and the feed composition maintained by adding an increment of more reactive monomer as the polymerization proceeds. Both methods find commercial use.

GENERAL REFERENCES

Alfrey 1952; Odian 1970.

C. Chemistry of Copolymerization

The order of reactivity of monomers toward free radicals not only is a function of the reactivity of the monomers, but also depends on the nature of the attacking radical. This is illustrated by the tendency of many monomers

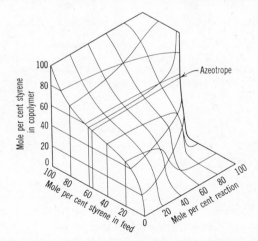

Fig. 11-7.   Variation in instantaneous composition of copolymer with initial feed and conversion for the system styrene-diethyl fumarate (Mayo 1950).

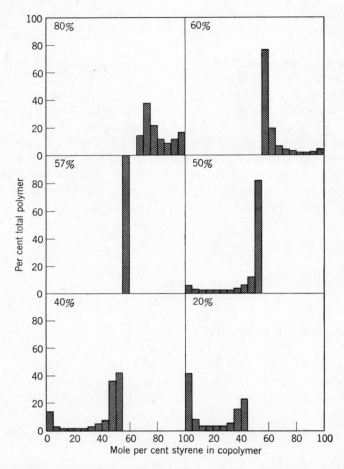

Fig. 11-8. Distribution of copolymer composition at 100% conversion for the indicated values of initial mole percent styrene in the feed, for the system styrene-diethyl fumarate (Burnett 1954, after Mayo 1950).

to alternate in a copolymer chain. The two factors, general reactivity and alternating tendency, are predominant in determining the behavior of monomers in copolymerization.

Reactivity and structure of monomers and radicals

The reactivity of monomers and radicals in copolymerization is determined by the nature of the substituents on the double bond of the monomer. These substituents influence reactivity in three ways: they may activate the double bond, making the monomer more reactive; they may stabilize the resulting radical by resonance; or they may provide steric hindrance at the reaction site.

Reactivity of monomers   The relative reactivities of monomers to a reference radical can be derived from the monomer reactivity ratio. The inverse of this ratio is the rate of reaction of the reference radical with another monomer, relative to that with its own monomer. If the latter rate is taken as unity, relative monomer reactivities can be examined. A few such rates are listed in Table 11-2; more extensive tabulations are given in the general references.

TABLE 11-2.   *Relative reactivities of monomers to reference radicals at 60°C*

| | | Reference Radical | | | |
Monomer	Styrene	Methyl Methacrylate	Acrylo- nitrile	Vinyl Chloride	Vinyl Acetate
Styrene	(1.0)	2.2	25	50	100
Methyl methacrylate	1.9	(1.0)	6.7	10	67
Acrylonitrile	2.5	0.82	(1.0)	25	20
Vinylidene chloride	5.4	0.39	1.1	5	10
Vinyl chloride	0.059	0.10	0.37	(1.0)	4.4
Vinyl acetate	0.019	0.05	0.24	0.59	(1.0)

As a different reference point is taken for each radical, values in different columns are not comparable, although ratios of values can be compared from column to column.

Although there are some exceptions, the effectiveness of substituents in enhancing the reactivity of the monomer lies in about the order

$$-C_6H_5 > -CH{=}CH_2 > -COCH_3 > -CN > -COOR > -Cl$$
$$> CH_2Y > -OCOCH_3 > -OR$$

The effect of a second substituent on the same carbon atom is usually additive.

This order of reactivity corresponds to the resonance stabilization of the radical formed after the addition. In the case of styrene the radical can resonate among forms of the type

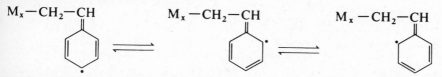

The radical is stabilized with a resonance energy of about 20 kcal/mole. At the other extreme, substituents which have no double bonds conjugated with

the ethylenic double bond give radicals having very low resonance energies (1–4 kcal/mole), since only polar or unbonded forms can contribute to the resonance.

Substituents which stabilize the product radical tend also to stabilize the monomer, but the amount of stabilization is much smaller than for the radical; in styrene the monomer is stabilized to the extent of about 3 kcal/mole. Thus the stabilizing effect of a substituent on the product radical is compensated for only to a limited extent by stabilization of the monomer.

Resonance stabilization   Resonance stabilization of monomers and radicals can be considered in terms of a potential energy diagram of the type shown in Fig. 11-9. The energy changes shown represent situations before and after each of four possible reactions in which a resonance-stabilized (M_s) or unstabilized (M) monomer adds to a stabilized (R_s·) or unstabilized (R·) radical. The set of four curves labeled with the four reaction possibilities represent the combined energy of the growing chain radical and the monomer before reaction, while the two curves labeled " New radical " represent the potential energy of the new bond formed by their combination. Activation energies are represented by solid arrows, heats of reaction by broken arrows. If the entropies of activation are all about the same (a good assumption in the absence of steric hindrance) the order of reaction rates is predicted to be

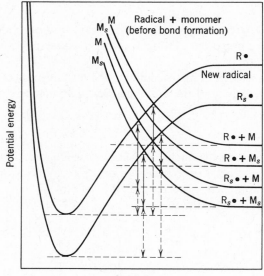

Fig. 11-9.   Potential energy of a radical-monomer pair as a function of their distance apart; see text for further explanation (Walling 1957).

the opposite of that of the activation energies:

$$R \cdot + M_s > R \cdot + M > R_s \cdot + M_s > R_s \cdot + M$$

This series summarizes much of the rate information available for both co-polymerizations and homopolymerizations.

Reactivity of radicals The reactivity of radicals to a reference monomer can be derived by combining the absolute propagation rate constants k_{11} with reactivity ratios to obtain the rate constants k_{12} for radical 1 adding monomer 2. Typical data are listed in Table 11-3.

As expected, resonance stabilization depresses the reactivity of radicals so that their order of reactivity is the reverse of that of their monomers; styrene is one of the least reactive radicals and vinyl acetate one of the most reactive. The effect of a substituent in depressing the activity of a radical is much greater than its effect in enhancing the activity of the monomer. For example, the styrene radical is about 1000 times less reactive than the vinyl acetate radical to a given monomer, if alternation effects are disregarded, but styrene monomer is only about 50 times more reactive than vinyl acetate to a given radical.

Steric effects The effect of steric hindrance in reducing reactivity may be demonstrated by comparing the reactivities of 1,1- and 1,2-disubstituted olefins to reference radicals. As indicated in the preceding discussion, the addition of a second 1-substituent usually increases reactivity three- to tenfold; however, the same substituent in the 2-position usually *decreases* reactivity two- to twentyfold. The extent of reduction in reactivity depends on energy differences between *cis* and *trans* forms and on the possibilities for resonance.

Polarity and alternation

By examining the tendency for alternation as given by the product $r_1 r_2$, it is possible to tabulate monomers in a series arranged so that two monomers farther apart in the series have a greater tendency to alternate. Such an arrangement is shown in Table 11-4. Values of $r_1 r_2$ generally decrease as monomers farther apart in the table are compared. There are some exceptions suggesting specific interactions, probably steric in nature. For example, vinyl acetate alternates more than styrene with vinylidene chloride, and less than styrene with acrylonitrile.

TABLE 11-3. *Radical-monomer propagation rate constants at 60°C (liters/mole sec)*

Monomer	Radical					
	Butadiene	Styrene	Methyl Methacrylate	Methyl Acrylate	Vinyl Chloride	Vinyl Acetate
Butadiene	100	250	2,820	42,000	350,000	—
Styrene	74	145	1,520	14,000	600,000	230,000
Methyl methacrylate	134	278	705	4,100	123,000	150,000
Methyl acrylate	132	206	370	2,090	200,000	23,000
Vinyl chloride	11	8	70	230	12,300	10,000
Vinyl acetate	—	2.6	35	520	7,300	2,300

TABLE 11-4. *Product of reactivity ratios with monomers arranged in order of alternating tendency*

Vinyl acetate	Butadiene	Styrene	Vinyl chloride	Methyl methacrylate	Vinylidene chloride	Acrylonitrile	Diethyl fumarate
	Butadiene						
0.55	0.98	Styrene					
0.39	0.31	0.34	Vinyl chloride				
0.30	0.19	0.24	1.0	Methyl methacrylate			
0.6	<0.1	0.16	0.96	0.61	Vinylidene chloride		
0.21	0.0006	0.016	0.11	0.18	0.35	Acrylonitrile	
0.0049	—	0.021	0.056	—	0.55	—	Diethyl fumarate

There is little doubt that alternation is primarily due to the polarity of the double bond in the monomer: the order of monomers in Table 11-4 closely parallels the order of the tendency of substituents around the double bond to donate electrons to the bond (hydrocarbon, acetoxy) or to withdraw them (carbonyl, cyano).

As might be expected, the tendency toward alternation (if not too great) parallels the tendency toward decreased reactivity induced by substitution of electron-withdrawing groups at the double bond.

The Alfrey-Price treatment (Alfrey 1964)   Alfrey (1947) and Price attempted to express the factors of general reactivity and polarity quantitatively. They wrote the rate constants for copolymerization in the form

$$k_{12} = P_1 Q_2 \exp - e_1 e_2 \tag{11-21}$$

where exp denotes the exponential, P_1 is the general reactivity of the radical $M_1 \cdot$, Q_2 is the reactivity of the monomer M_2, and e_1 and e_2 are proportional to the electrostatic interaction of the permanent charges on the substituents in polarizing the double bond. This form is analogous to Hammett's (1940) equation for the effects of nuclear substituents on the reactivity of aromatic compounds. From the definitions of the monomer reactivity ratios it follows that

$$r_1 = \frac{Q_1}{Q_2} \exp - e_1(e_1 - e_2)$$

$$r_2 = \frac{Q_2}{Q_1} \exp - e_2(e_2 - e_1)$$

(11-22)

and

$$r_1 r_2 = \exp - (e_1 - e_2)^2$$

(11-23)

The Alfrey-Price equation assumes that the same value of e applies to both monomer and radical. This assumption is justified only by the success of the Q-e scheme in interpreting relative reactivities.

It should be possible to assign Q and e values to a series of monomers from the data on a few copolymerization experiments and compute values of r_1 and r_2 for all the possible arrangements of the monomers in pairs (Price 1948). Unfortunately, this practical application of the scheme is hampered by large uncertainties in the values of Q and e which can be assigned with the present experimental error. The scheme is not entirely sound from the theoretical standpoint and is best regarded as semiempirical. In a sense it represents the transcription of the reactivity and polarity series of Tables 11-2 and 11-4 to equation form.

Ionic copolymerization

The order of monomer reactivity in *cationic copolymerization* is quite different from that in free radical polymerization and is distinguished by the high reactivities of vinyl ethers and isobutylene. However, the reactivities correspond to the anticipated effect of substituents upon the reactivity of double bonds toward electrophilic reagents. The differences of reactivity arise from the change in electron availability in the double bond and the resonance stabilization of the resulting carbonium ion.

Compared with radical-induced copolymerization, the carbonium ion reactions show a much wider range of reactivity among the common monomers, and no tendency toward alternation in the copolymer. These characteristics mean that relatively few monomer pairs easily yield copolymers containing large proportions of both monomers.

The data on *anionic copolymerization* suggest that the order of monomer reactivity is much different from both free radical and carbonium ion polymerization. As anticipated, the order of reactivity in the anionic process is determined by the ability of substituents to withdraw electrons from the double bond and to stabilize the carbanion formed. Anionic and co-ordination polymerizations involving monomers of like electronegativity are usually ideal, since the unsubstituted carbon atom at the end of the growing chain has little influence on the course of the reaction. When the electro-negativities of the monomers are widely different, however, the rates of the cross-propagation reactions may be quite different also. In extreme cases, only one monomer can add at all.

The contrast among the three methods of polymerization is brought out very clearly by the results of copolymerization studies using the three modes of initiation. Figure 11-10 shows the composition of the initial polymer formed from a feed of styrene and methyl methacrylate using different initiators. These results verify the different polymerization mechanisms and illustrate the extreme variations of reactivities that monomers exhibit with different type initiators.

Step-reaction copolymerization

Since the reactivities of all functional groups in simple bifunctional stepwise polymerization are essentially identical, irrespective of the length of the molecule to which they are attached, comonomers in such systems are randomly distributed along the chain in amounts proportional to their concentrations in the feed. Thus, the "reactivity ratios" for stepwise

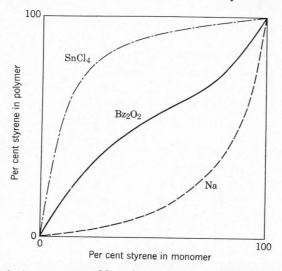

Fig. 11-10. Instantaneous composition of copolymer as a function of monomer composition for the system styrene-methyl methacrylate polymerized by cationic (SnCl$_4$), free radical (benzoyl peroxide, Bz$_2$O$_2$) and anionic (Na) mechanisms (Landler 1950, Pepper 1954).

polymerization are unity. Departures from this generalization are occasionally encountered, however; for example, the esterification rate of succinic acid $HOOC(CH_2)_2COOH$, differs from that of adipic acid, $HOOC(CH_2)_4COOH$, because of the proximity of the carboxyl groups in the former compound. For similar reasons, ethylene glycol and 1,4-butanediol have different esterification rates. Also 1,3-butanediol differs in reactivity from 1,4-butanediol because one of the hydroxyl groups of the former is secondary whereas both of the latter are primary. Where such reactivity differences exist, the tendency is for the more reactive species to enter into the polymer first.

GENERAL REFERENCES

Alfrey 1952; Marvel 1959; Young 1966; Lenz 1967; Odian 1970.

D. Block and Graft Copolymers

It is shown in Section *B* that the probability of finding long sequences of one monomer in an ordinary random copolymer is very small, except in the trivial case where one monomer is present in large excess. Methods of synthesis of polymers containing such long sequences are of interest, however, since they may lead to polymers with properties widely different from those of either homopolymers or random copolymers (as discussed in Chapter 7). Polymers with long sequences of two monomers can have two arrangements of chains: in *block copolymers*, the sequences follow one another along the main polymer chain,

$$——AABB——BBAA——AABB——, \text{ etc.,}$$

whereas in *graft copolymers* sequences of one monomer are "grafted" onto a "backbone" of the second monomer type,

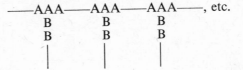

The ultimate aim in preparing either type is to obtain the desired copolymer free from homopolymer or other unwanted species, on a scale which allows evaluation of physical properties. Few of the techniques described in the literature satisfy these requirements.

Block copolymers

Polymers with labile end groups The usual method of preparing block copolymers utilizes a polymer with end groups which can be made to react under different conditions. In stepwise polymerization, the method is trivially simple: two polyesters, e.g., of different type are separately prepared, mixed, and allowed to polymerize further. Ring-scission polymerization can be used effectively.

By means of free radical techniques, block copolymers can be made in several ways. Labile end groups (as for example bromine incorporated through termination by transfer to CBr_4) can be activated by thermal or ultraviolet-light scission of the C—Br bond. Difunctional initiators can be used to produce block copolymers if the two initiator groups can be decomposed independently. Or, polymer radicals can be formed by milling or masticating the polymer. In each case, of course, the polymer radical must be produced in the presence of a second monomer to yield the copolymer.

By far the most important synthesis of block copolymers is that of Szwarc for producing "living" polymers by unterminated anionic polymerization (Chapter 10*C*). Addition of a second monomer to the still active polymer leads to block copolymer uncontaminated with homopolymer and with blocks of accurately known and controlled length. The process can be repeated, with care, to produce multiblock polymers. Several commercial block copolymers are produced in this way and are discussed in Part IV; see also Fetters 1969.

Graft copolymers

Graft copolymerization results from the formation of an active site at a point on a polymer molecule other than its end, and exposure to a second monomer. Most graft copolymers are formed by radical polymerization. The major activation reaction is chain transfer to polymer (Chapter 9). In many instances, the transfer reaction involves abstraction of a hydrogen atom. An important commercial application is the grafting of polystyrene or styrene-acrylonitrile copolymer onto butadiene or acrylonitrile-butadiene copolymer rubber in the production of ABS resins (Chapter 14*A*). Ultraviolet or ionizing radiation, or redox initiation, among other methods, can also be used to produce the polymer radicals leading to graft copolymers.

GENERAL REFERENCES

Burlant 1960; Ceresa 1962, 1965; Hoffman 1964; Battaerd 1967; Aggarwal 1970.

BIBLIOGRAPHY

Aggarwal 1970. S. L. Aggarwal, ed., *Block Polymers*, Plenum Press, New York, 1970.

Alfrey 1944. Turner Alfrey, Jr., and G. Goldfinger, "The Mechanism of Copolymerization," *J. Chem. Phys.* **12**, 205–209 (1944).

Alfrey 1947. Turner Alfrey, Jr., and Charles C. Price, "Relative Reactivities in Vinyl Copolymerization," *J. Polymer Sci.* **2**, 101–106 (1947).

Alfrey 1952. Turner Alfrey, Jr., John J. Bohrer, and H. Mark, *Copolymerization*, Interscience Publishers, New York, 1952.

Alfrey 1964. T. Alfrey, Jr., and L. J. Young, "The *Q-e* Scheme," Chap. II in George E. Ham, ed., *Copolymerization*, Interscience Div., John Wiley and Sons, New York, 1964.

Atherton 1962. J. N. Atherton and A. M. North, "Diffusion-Controlled Termination in Free Radical Copolymerization," *Trans. Faraday Soc.* **58**, 2049–2057 (1962).

Battaerd 1967. H. A. J. Battaerd and G. W. Tregear, *Graft Copolymers*, John Wiley and Sons, New York, 1967.

Burlant 1960. William J. Burlant and Allan S. Hoffman, *Block and Graft Polymers*, Reinhold, New York, 1960.

Burnett 1954. G. M. Burnett, *Mechanism of Polymer Reactions*, Interscience Publishers, New York, 1954.

Ceresa 1962. R. J. Ceresa, *Block and Graft Copolymers*, Butterworths, London, 1962.

Ceresa 1965. R. J. Ceresa, "Block and Graft Copolymers," pp. 485–528 in Herman F. Mark, Norman G. Gaylord, and Norbert M. Bikales, eds., *Encyclopedia of Polymer Science and Technology*, Vol. 2, Interscience Div., John Wiley and Sons, New York, 1965.

Dostal 1936. H. Dostal, "A Basis for the Reaction Kinetics of Mixed Polymerization" (in German), *Monatsh.* **69**, 424–426 (1936).

Fetters 1969. L. J. Fetters, "Synthesis of Block Polymers by Homogeneous Anionic Polymerization," *J. Polymer Sci.* **C26**, 1–35 (1969).

Gumbs 1969. R. Gumbs, S. Penczek, J. Jagur-Grodzinski, and M. Szwarc, "Simultaneous Cationic Homopolymerizations of Vinylcarbazole and Oxetane," *Macromolecules* **2**, 77–82 (1969).

Ham 1964. George E. Ham, "Theory of Copolymerization," Chap. I in George E. Ham, ed., *Copolymerization*, Interscience Div., John Wiley and Sons, New York, 1964.

Ham 1966. George E. Ham, "Copolymerization," *pp.* 165–244 in Herman F. Mark, Norman G. Gaylord, and Norbert M. Bikales, eds., *Encyclopedia of Polymer Science and Technology*, Vol. 4, Interscience Div., John Wiley and Sons, New York, 1966.

Hammett 1940. Louis P. Hammett, *Physical Organic Chemistry*, McGraw-Hill Book Co., New York, 1940.

Hoffman 1964. Allan S. Hoffman and Robert Bacskai, "Block and Graft Copolymerizations," Chap. VI in George E. Ham, ed., *Copolymerization*, Interscience Div., John Wiley and Sons, New York, 1964.

Landler 1950. Yvan Landler, "On Ionic Copolymerization" (in French), *Compt. rend.* **230**, 539–541 (1950).

Lenz 1967. Robert W. Lenz, *Organic Chemistry of Synthetic High Polymers*, Interscience Div., John Wiley and Sons, New York, 1967.

Lenz 1969. Robert W. Lenz, "Applied Reaction Kinetics: Polymerization Reaction Kinetics," *Ind. Eng. Chem.* **61** (3), 67–75 (1969).

Lenz 1970. Robert W. Lenz, "Applied Polymerization Reaction Kinetics," *Ind. Eng. Chem.* **62** (2), 54–61 (1970).

Lowry 1960. George G. Lowry, "The Effect of Depropagation on Copolymer Composition. I. General Theory for One Depropagating Monomer," *J. Polymer Sci.* **42**, 463–477 (1960).

Mark 1966. Herman Mark, B. Immergut, E. H. Immergut, L. J. Young, and K. I. Benyon, "Copolymerization Reactivity Ratios," pp. II–142—II–289, with an Appendix by L. J. Young, pp. II–291—II–340, in J. Brandrup and E. H. Immergut, eds., with the collaboration of H.-G. Elias, *Polymer Handbook*, Interscience Div., John Wiley and Sons, New York, 1966.

Marvel 1959. Carl S. Marvel, *An Introduction to the Organic Chemistry of High Polymers*, John Wiley and Sons, New York, 1959.

Mayo 1944. Frank R. Mayo and Frederick M. Lewis, "Copolymerization. I. A Basis for Comparing the Behavior of Monomers in Copolymerization; The Copolymerization of Styrene and Methyl Methacrylate," *J. Am. Chem. Soc.* **66**, 1594–1601 (1944).

Mayo 1950. Frank R. Mayo and Cheves Walling, "Copolymerization," *Chem. Revs.* **46**, 191–287 (1950).

Merz 1946. E. Merz, T. Alfrey, and G. Goldfinger, "Intramolecular Reactions in Vinyl Polymers as a Means of Investigation of the Propagation Step," *J. Polymer Sci.* **1**, 75–82 (1946).

Meyer 1965. Victor E. Meyer and George G. Lowry, "Integral and Differential Binary Copolymerization Equations," *J. Polymer Sci.* **A3**, 2843–2851 (1965).

Meyer 1966. Victor E. Meyer, "Copolymerization of Styrene and Methyl Methacrylate. Reactivity Ratios from Conversion-Composition Data," *J. Polymer Sci. A-1* **4**, 2819–2830 (1966).

Meyer 1967. Victor E. Meyer and Richard K. S. Chan, "Computer Calculations of Batch-Type Copolymerization Behavior," *Amer. Chem. Soc., Div. Polym. Chem. Preprints* **8** (1), 209–215 (1967).

North 1963. A. M. North, "The ϕ-Factor in Free-Radical Copolymerization," *Polymer* **4**, 134–135 (1963).

North 1968. A. M. North, "Diffusion Control of Homogeneous Free-Radical Reactions," pp. 95–135 in J. C. Robb and F. W. Peaker, eds., *Progress in High Polymers*, Vol. 2, CRC Press, Cleveland, Ohio, 1968.

Odian 1970. George Odian, *Principles of Polymerization*, McGraw-Hill Book Co., New York, 1970.

Pepper 1954. D. C. Pepper, "Ionic Polymerisation," *Chem. Soc. Quart. Revs.* **8**, 88–121 (1954).

Price 1948. Charles C. Price, "Some Relative Monomer Reactivity Factors," *J. Polymer Sci.* **3**, 772–775 (1948).

Simha 1944. Robert Simha and Herman Branson, "Theory of Chain Copolymerization Reactions," *J. Chem. Phys.* **12**, 253–267 (1944).

Skeist 1946. Irving Skeist, "Copolymerization: the Composition Distribution Curve," *J. Am. Chem. Soc.* **68**, 1781–1784 (1946).

Staudinger 1939. H. Staudinger and J. Schneiders, "Macromolecular Compounds. CCXXXI. Polyvinylchlorides" (in German), *Ann.* **541**, 151–195 (1939).

Stockmayer 1945. W. H. Stockmayer, "Distribution of Chain Lengths and Composition in Copolymers," *J. Chem. Phys.* **13**, 199–207 (1945).

Wall 1944. Frederick T. Wall, "The Structure of Copolymers. II," *J. Am. Chem. Soc.* **66**, 2050–2057 (1944).

Walling 1949. Cheves Walling, "Copolymerization. XIII. Over-all Rates in Copolymerization. Polar Effects in Chain Initiation and Termination," *J. Am. Chem. Soc.* **71,** 1930–1935 (1949).

Walling 1957. Cheves Walling, *Free Radicals in Solution,* John Wiley and Sons, Inc., New York, 1957.

Young 1966. Lewis J. Young, "Tabulation of *Q-e* Values," pp. II–341—II–362 in J. Brandrup and E. H. Immergut, eds., with the collaboration of H.-G. Elias, *Polymer Handbook,* Interscience Div., John Wiley and Sons, New York, 1966.

12

Polymerization Conditions and Polymer Reactions

The discussion of polymerization kinetics in Chapters 8–11 has dealt, with the exception of the systems discussed in Chapter 10*D*, with polymerization of pure monomer or of homogeneous solutions of monomer and polymer in a solvent. Certain other types of polymerizing systems are of great interest because they offer practical advantages in industrial applications. Among these are polymerizations in which the monomer is carried in an *emulsion* or in *suspension* in an aqueous phase. Some of the advantages and disadvantages of these systems are itemized in Table 12-1. For purposes of discussion, it is convenient to classify polymerization systems as homogeneous or heterogeneous.

In the chain polymerization of monomers with more than one double bond, crosslinking can occur as in polyfunctional stepwise polymerization. The chain case is important because of the possibility of delaying the crosslinking reaction to a later time and situation, as in the vulcanization of elastomers.

In the same class of postpolymerization reactions are two other phenomena closely related to polymerization: degradation, which frequently involves just the reverse reactions of polymerization, and reactions induced by high-energy radiation, which often involve radicals and lead to crosslinking or degradation.

A. Polymerization in Homogeneous Systems

Bulk polymerization Polymerization in bulk, perhaps the most obvious method of synthesis of polymers, is widely practiced in the manufacture of condensation polymers, where the reactions are only mildly exothermic,

355

TABLE 12-1. *Comparison of polymerization systems*

Type	Advantages	Disadvantages
Homogeneous		
Bulk (batch type)	Minimum contamination. Simple equipment for making castings.	Strongly exothermic. Broadene molecular weight distribution a high conversion. Complex small particles required.
Bulk (continuous)	Lower conversion per pass leads to better heat control and narrower molecular weight distribution.	Requires agitation, material trans fer, separation, and recycling.
Solution	Ready control of heat of polymerization. Solution may be directly usable.	Not useful for dry polyme because of difficulty of complet solvent removal.
Heterogeneous		
Suspension	Ready control of heat of polymerization. Suspension or resulting granular polymer may be directly usable.	Continuous agitation required Contamination by stabilizer pos sible. Washing, drying, possibl compacting required.
Emulsion	Rapid polymerization to high molecular weight and narrow distribution, with ready heat control. Emulsion may be directly usable.	Contamination with emulsifier etc., almost inevitable, leadin to poor color and colo stability. Washing, drying, an compacting may be required.

and most of the reaction occurs when the viscosity of the mixture is stil low enough to allow ready mixing, heat transfer, and bubble elimination Control of such polymerizations is relatively easy.

Bulk polymerization of vinyl monomers is more difficult, since th reactions are highly exothermic and, with the usual thermally decompose initiators, proceed at a rate which is strongly dependent on temperature This, coupled with the problem in heat transfer incurred because viscosit increases early in the reaction, leads to difficulty in control and a tendenc to the development of localized "hot spots" and "runaways." Except i the preparation of castings, e.g., of poly(methyl methacrylate), bulk polymer ization is seldom used commercially for the manufacture of vinyl polymers

Polystyrene and poly(vinyl chloride) are, however, sometimes made in bulk.

Solution polymerization Polymerization of vinyl monomers in solution is advantageous from the standpoint of heat removal (e.g., by allowing the solvent to reflux) and control but has two potential disadvantages. First, the solvent must be selected with care to avoid chain transfer and, second, the polymer should preferably be utilized in solution, as in the case of poly-(vinyl acetate) to be converted to poly(vinyl alcohol) and some acrylic ester finishes, since the complete removal of solvent from a polymer is often difficult to the point of impracticality.

Solid-phase polymerization Although homogeneous in that only a single phase exists, these polymerizations bear much more resemblance to heterogeneous cases, and are treated in Section *B*.

GENERAL REFERENCES

Palit 1965; Ringsdorf 1965; Lenz 1967; Ravve 1967; Odian 1960.

B. Polymerization in Heterogeneous Systems

Gas-phase polymerization

The polymerization of gaseous vinyl monomers with polar substituents takes place rapidly by photoinitiation with the formation of a cloud of polymer. The kinetics of the process is complicated, however, since the polymer separates as a second phase almost immediately after initiation, and most of the propagation and termination must take place in or on the solid particles.

Gas-phase polymerization is little used commercially. The high-pressure polymerization of ethylene (Chapter 13*A*), though technically a vapor-phase reaction, is better considered as taking place in a highly swollen liquid polymer phase with a vapor phase present.

Bulk polymerization with polymer precipitating

In the preparation of a polymer insoluble in its monomer, or in polymerization in the presence of a nonsolvent for the polymer, marked deviations from the kinetics of homogeneous radical polymerization may occur. An example of particular interest is the preparation of polyacrylonitrile, which, like poly(vinyl chloride), poly(vinylidene chloride), and polychloro-trifluoroethylene, is insoluble in its monomer.

The bulk polymerization of acrylonitrile exhibits marked autoaccelera-tion, the rate increasing continuously up to at least 20% conversion. The initial rate is proportional to a power of the initiator concentration only slightly less than 1. These facts, which suggest that the normal bimolecular termination reaction is not effective, are explained by the trapping or occlu-sion of radicals in the unswollen, tightly coiled precipitating polymer. The theory has been confirmed by the demonstration of the presence of radicals in the polymer both by chemical methods and by electron paramagnetic resonance (Chapter 4D). The lifetime of the trapped radicals is many hours at room temperature, and if polymer containing such radicals is heated in the presence of monomer to about 60°C, where the mobility of the radicals is increased, extremely rapid polymerization takes place.

The polymerization of vinyl chloride in bulk or in the presence of a nonsolvent (Mickley 1962) follows a rate equation of the form

$$v_p = k_p \left(\frac{f k_d [\text{I}]}{k_t} \right)^{1/2} \{[\text{M}] + f(\text{P})\} \tag{12-1}$$

where the first term inside the braces arises from the normal rate equation for polymerization in the homogeneous liquid phase, and the second term represents the increment in rate due to polymerization in the precipitated polymer particles. The function $f(\text{P})$ is proportional to the polymer con-centration [P] at low conversion and to $[\text{P}]^{2/3}$ at later times. Radical occlu-sion occurs, but the depth of radical penetration into the polymer particles is small, limiting radical activity to thin surface layers in larger particles present at higher conversions. Termination occurs primarily in the liquid phase, probably as a result of transfer to monomer within the polymer particles, and subsequent diffusion of the monomer radicals to the liquid phase. In contrast, transfer to monomer in the polymerization of acryloni-trile is so slow that permanent radical occlusion occurs.

Popcorn polymerization (Breitenbach 1969)   A troublesome form of hetero-geneous polymerization occurs when tough, insoluble nodules of crosslinked polymer precipitate from systems containing diene monomers. Once formed, popcorn polymer can proliferate rapidly until the supply of monomer is exhausted. It has been a serious cause of the fouling of monomer supply lines in synthetic rubber plants.

Suspension polymerization

The term *suspension polymerization* refers to polymerization in an aqueous system with monomer as a dispersed phase, resulting in polymer as a dispersed solid phase. The process is distinguished from superficially similar emulsion polymerization by the location of the initiator and the kinetics obeyed: in typical suspension polymerization the initiator is dissolved in the monomer phase, and the kinetics is the same as that of bulk polymerization.

The dispersion of monomer into droplets, typically 0.01–0.5 cm in diameter, is maintained by a combination of agitation and the use of water-soluble stabilizers. These may include finely divided insoluble organic or inorganic materials which interfere with agglomeration mechanically, electrolytes to increase the interfacial tension between the phases, and water-soluble polymers to increase the viscosity of the aqueous phase. The tendency to agglomerate may become critical when the polymerization has advanced to the point where the polymer beads become sticky. At the completion of the reaction, the polymer is freed of stabilizer by washing and is dried. For some applications (Chapter 17), the polymer beads can be used directly, whereas for others compaction is required. A typical recipe for suspension polymerization is given in Table 12-2. The method is used com-

TABLE 12-2. *Typical recipes for suspension polymerization (Trommsdorff 1956, Farber 1970)*

	Parts per 100 parts of monomer	
Component	Methyl Methacrylate	Vinyl Chloride
Peroxide initiator	~0.5	0.1–0.5
Water	~350	150–350
Stabilizers*	0.01–1	0.01–1

* Typical stabilizers include methyl cellulose, gelatin, poly(vinyl alcohol), and sodium polyacrylate. Minor amounts of emulsifiers and buffers are also used.

mercially to prepare hard, glassy vinyl polymers such as polystyrene, poly(methyl methacrylate), poly(vinyl chloride), poly(vinylidene chloride), and polyacrylonitrile.

Emulsion polymerization

Emulsion polymerization differs from suspension polymerization in two important respects: the initiator is located in the aqueous phase, and the polymer particles produced are typically of the order of 0.1 μm in diameter, some ten times smaller than the smallest encountered in suspension or "dispersion" polymerization. As demonstrated below, these differences

lead to different kinetics for emulsion polymerization, removing the limi
tation set by Eq. 9-13—for combination,

$$\bar{x}_n = \frac{k_p{}^2}{k_t} \frac{[\text{M}]^2}{v_p}$$

(12-

—on the molecular weight obtainable at a given rate of polymerization. A
a result, emulsion systems allow higher-molecular-weight polymer to b
produced at higher rates than do bulk or suspension systems.

A typical recipe for an emulsion polymerization is given in Table 12-.

TABLE 12-3. *"Mutual" recipe for the emulsion
copolymerization of styrene and
butadiene at 50°C*

Ingredient	Parts by Weight
Butadiene	75
Styrene	25
Water	180
Soap	5.0
Lorol mercaptan*	0.50
Potassium persulfate	0.30

* Crude dodecyl mercaptan.

This is the famous "Mutual" recipe used in the polymerization of styrene
butadiene synthetic rubber during World War II. The ingredients, in addi
tion to monomers and water, are a fatty acid soap, a mercaptan-type chai
transfer agent, and the water-soluble persulfate initiator.

The soap plays an important role in emulsion polymerization. At th
beginning of the reaction, it exists in the form of *micelles*, aggregates c
50–100 soap molecules probably having a layered structure like that show
in Fig. 12–1. Part of the monomer enters the micelles, but most of it exist
as droplets a micrometer or more in diameter.

It can be demonstrated experimentally that no polymer is formed in th
monomer droplets. Polymerization can take place (at a very low rate) i
the homogeneous phase in the absence of soap, but this cannot accoun
for the bulk of the polymer formed. At the beginning of the reaction, poly
mer is formed in the soap micelles; these represent a favorable environmen
for the free radicals generated in the aqueous phase, because of the relativ
abundance of monomer and the high surface/volume ratio of the micelle
compared to the monomer droplets. As polymer is formed, the micelle

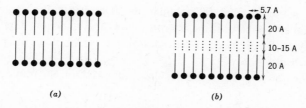

Fig. 12-1.   Idealized structure of a soap micelle (a) without and (b) with solubilized monomer.

grow by the addition of monomer from the aqueous phase (and ultimately from the monomer droplets).

Soon (2–3% polymerization) the polymer particles grow much larger than the original micelles and absorb almost all the soap from the aqueous phase. Any micelles not already activated disappear; further polymerization takes place within the polymer particles already formed. The monomer droplets are unstable at this stage; if agitation is stopped, they coalesce into a continuous oil phase containing no polymer. The droplets act as reservoirs of monomer which is fed to the growing polymer particles by diffusion through the aqueous phase. The polymer particles may contain about 50% monomer up to the point at which the monomer droplets disappear, at 60–80% polymerization. The rate of polymerization is constant over most of the reaction up to this point, but then falls off as monomer is depleted in the polymer particles. Rate increases with increasing soap (and initial micelle) concentration.

Smith-Ewart kinetics (Smith 1948a, b)   In a typical emulsion system, free radicals are generated in the aqueous phase at a rate of about 10^{13} p/cm³/sec. There are about 10^{14} polymer particles per cubic centimeter. Simple calculations show that termination of the free radicals in the aqueous phase is negligible and that diffusion currents are adequate for the rapid diffusion of free radicals into the polymer particles. Thus essentially all the free radicals get into the polymer particles—on the average about one per particle every 10 sec.

It can also be calculated from the known termination rate constants that two free radicals within the same polymer particle would mutually terminate within a few thousandths of a second. Therefore, each polymer particle must contain over most of the time either one or no free radicals. At any time, half of the particles (on the average) contain one free radical, the other half none. The rate of polymerization per cubic centimeter of emulsion is

$$v_p = k_p[\text{M}]\,\frac{N}{2} \tag{12-3}$$

where N is the number of polymer particles per cubic centimeter. Since the monomer concentration is approximately constant, the rate depends principally on the number of particles present and not on the rate of generation of radicals.

The degree of polymerization also depends upon the number of particles:

$$\bar{x}_n = \frac{k_p N [M]}{\rho} \qquad (12\text{-}4)$$

where ρ is the rate of generation of radicals. Unlike v_p, $\bar{x}_n$ is a function of the rate of free radical formation. In bulk polymerization, rate can be increased only by increasing the rate of initiation. This, however, causes a decrease in the degree of polymerization. In emulsion polymerization, the rate may be increased by increasing the number of polymer particles. If the rate of initiation is kept constant, the degree of polymerization increases rather than decreases as the rate rises. Since the number of polymer particles is determined by the number of soap micelles initially present, both rate and molecular weight increase with increasing soap concentration. The Smith-Ewart kinetics require that

$$v_p \propto N,\ [I]^{0.4},\ [E]^{0.6}$$

$$N \propto [I]^{0.4},\ [E]^{0.6} \qquad (12\text{-}5)$$

$$\bar{x}_n \propto N,\ [E]^{0.6},\ [I]^{-0.6}$$

where $[E]$ is the soap or emulsifier concentration.

Deviations from Smith-Ewart kinetics The above kinetic scheme is highly idealized, though valuable for its simplicity. It explains adequately only a small portion of the vast literature on emulsion polymerization, though it works well for monomers such as styrene, butadiene, and isoprene, whose water solubility is very low, less than 0.1 %. Among the circumstances under which the Smith-Ewart kinetics fails to apply quantitatively are:

a. Larger particles (>0.1–0.15 μm diameter), which can accommodate more than one growing chain simultaneously.

b. Monomers with higher water solubility (1–10%), such as vinyl chloride, methyl methacrylate, vinyl acetate, and methyl acrylate. Here initiation in the aqueous phase, followed by precipitation of polymer, becomes important. These particles may absorb emulsifier, decreasing $[E]$ and thus N, or they may serve as sites for polymerization, increasing N.

c. Chain transfer to emulsifier. This often takes place in large enough amount to suggest that growing chains are localized near the surfaces of the particles, where the soap exists.

In fact, a second theory of emulsion kinetics (Sheinker 1954) assumes that the growing chain radicals cannot readily penetrate to the interior of the particles because of their high viscosity, and therefore polymerization occurs only at the particle surfaces. This Medvedev-Sheinker theory predicts that

$$N \propto [E]^3, [I]^0$$

$$v_p \propto N^{0.2}, [E]^{0.5}, [I]^{0.5}$$

(12-6)

This and the Smith-Ewart theory are not mutually exclusive, each being obeyed in certain limiting cases.

Inverse emulsion systems Emulsion polymerization can also be carried out in systems using an aqueous solution of a hydrophilic monomer, such as acrylic acid or acrylamide, emulsified in a continuous oil phase using an appropriate water-in-oil emulsifier. Either oil- or water-soluble initiators can be used. The mechanism seems to be that of normal emulsion polymerization, but the emulsions are often less stable.

"Redox" initiation The decomposition of peroxide-type initiators in aqueous systems is greatly accelerated by the presence of a reducing agent. This acceleration allows the attainment of high rates of radical formation at low temperatures in emulsion systems.

A typical redox system is that of ferrous iron and hydrogen peroxide. In the absence of a polymerizable monomer the peroxide decomposes to free radicals as follows:

$$H_2O_2 + Fe^{+2} \rightarrow HO\cdot + OH^- + Fe^{+3}$$

Another widely used peroxide-type initiator is the persulfate ion. With a reducing agent R the reaction is:

$$S_2O_8^{-2} + R \rightarrow SO_4^{-2} + SO_4^- \cdot + R^+$$

The reducing agent is often the thiosulfate ion:

$$S_2O_8^{-2} + S_2O_3^{-2} \rightarrow SO_4^{-2} + SO_4^- \cdot + S_2O_3^- \cdot$$

or the bisulfite ion:

$$S_2O_8^{-2} + HSO_3^- \rightarrow SO_4^{-2} + SO_4^- \cdot + HSO_3\cdot$$

Many other redox initiator systems have been used.

Solid-phase polymerization

A large number of olefin and cyclic monomers can be polymerized from the crystalline solid state. Among those which react rapidly under these conditions are styrene, acrylonitrile, methacrylonitrile, formaldehyde, trioxane, β-propiolactone, diketene, vinyl stearate, vinyl carbazole, and vinyl pyrrolidone.

Polymerization appears always to be associated with defects in the monomer crystals, most likely line defects (Chapter 5C). Otherwise, it would be required that the monomer and the polymer be isomorphous; that is, that they have the same crystal structure, lattice parameters, etc. This seems extremely unlikely and has not been observed.

Thus, solid-phase polymerization is heterogeneous in the sense that the crystalline monomer and the polymer, whether crystalline or amorphous, are different phases. Many of the features of solid-phase polymerization resemble those of coordination polymerization (Chapter 10C), and the kinetics is correspondingly complex. Even the determination of the nature of the chain carrier is difficult, but by analogy with the behavior of the same monomers in the liquid state it is assumed that both radical and ionic mechanisms can be involved.

General aspects of heterogeneous polymerization have been reviewed (Lenz 1969, 1970).

GENERAL REFERENCES

Palit 1965; Lenz 1967; Ravve 1967; Odian 1970; and specifically:

Polymerization with polymer precipitating: Jenkins 1967.
Suspension polymerization: Trommsdorff 1956, Farber 1970.
Emulsion polymerization: Duck 1966; Vanderhoff 1969; Odian 1970.
Solid-phase polymerization: Lenz 1967; Herz 1968; Bamford 1969.

C. Diene and Divinyl Polymerization

Monomers with two double bonds can arbitrarily be divided into 1,3-*diene* and *divinyl* compounds. With the exception of cases in which the entrance of one double bond in the monomer into a polymer chain leaves the second double bond so unreactive that it will not polymerize, these monomers can produce branched or crosslinked polymers.

By the proper choice of relative reactivity of the two double bonds, it is possible to reduce the reactivity of one just enough so that it will not enter polymerization under the same conditions as the other, but can be made to react under more drastic conditions. This leads to postpolymerization crosslinking reactions of which *vulcanization* reactions are an example.

If the two double bonds are well separated, the reactivity of one is not affected by the polymerization of the other: an example is 2,11-dodeca-diene. If the two double bonds are close enough together that the poly-merization of one can shield the other sterically, or if they are conjugated, as in the dienes, a difference in reactivity can be expected. Some examples are given in Table 12-4.

TABLE 12-4. *Classification of monomers with two double bonds*

Dependency of Reactivity	Symmetrical	Unsymmetrical
Independent	Ethylene glycol dimethacrylate	Allyl acrylate
Intermediate	Divinylbenzene	p-Isopropenylstyrene
Interdependent	Butadiene	Chloroprene

Vulcanization

This section is concerned only with the statistics of vulcanization and the distribution of molecular weights and crosslinks in the resulting poly-mer. Chemical aspects of vulcanization are discussed in Chapter 19.

The random crosslinking of double bonds bears formal resemblance to the formation of three-dimensional networks from polyfunctional monomers by stepwise polymerization. In keeping with previous nomenclature, the fraction of the monomer units on a chain which can be crosslinked is defined as q and the degree of polymerization of the chain as x. The "functionality" of the chain is its total number of vulcanizable groups qx. The extent of reaction p is the fraction of the total available crosslinks that has been formed.

Gelation occurs at a critical value of p, p_c, where there is on the average one crosslink for every two chains:

$$p_c q x = 1 \qquad (12\text{-}7)$$

As in the case of three-dimensional step-reaction polymers, after gelation the finite species constitute only part of the material, the rest being in the form of gel networks. As p increases from p_c to 1, the weight fraction of the finite species drops from unity to zero.

The case above does not correspond to actual vulcanization because it neglects the molecular weight distribution of the chains initially present. If this is considered, the detailed analysis becomes complicated, but some important features remain about the same as for the monodisperse case. The gel point is given by the equation

$$p_c q \bar{x}_w = 1 \qquad (12\text{-}8)$$

where $\bar{x}_w$ is the weight-average degree of polymerization of the starting polymer.

Differences in chemical composition introduced by copolymerization or mixing may have more effect on the type of network structure than does molecular weight distribution. If, e.g., a small amount of high-functionality polymer is mixed with a polymer of low functionality, the resulting vulcanizate will have regions of very tight crosslinking as inclusions in a looser network, leading to pronounced nonuniformity of physical properties. The foregoing treatment also assumes that the points of crosslinking are distributed at random. This is not always true; e.g., if vulcanization is carried out in dilute solution, intramolecular crosslinking may occur before two molecules come near enough to form intermolecular links. Thus gelation is delayed considerably beyond the point predicted by random statistics.

Vinyl-divinyl copolymerization

Consider the copolymerization of monomer A with monomer B—B, where all groups are equally reactive, as in the copolymerization of methyl methacrylate and ethylene glycol dimethacrylate. Let the concentrations of A, B—B, and B groups be [A], [BB], and [B], respectively, where [B] = 2[BB]. Since the A and B groups are equally reactive, their ratio in the copolymer, b/a, is

$$\frac{b}{a} = \frac{[B]}{[A]} \tag{12-9}$$

a simplification of the copolymer equation, Eq. 11-4. At extent of reaction p, here the fraction of the double bonds reacted, there are present $p[A]$ reacted A groups, $(1 - p)[A]$ unreacted A groups, $p^2[BB]$ doubly reacted BB molecules, $2p(1 - p)[BB]$ singly reacted BB's, and $(1 - p)^2[BB]$ unreacted BB's. In terms of B groups, the latter quantities become $(1 - p)^2[B]$ (unreacted) B groups on unreacted BB's, $p(1 - p)[B]$ unreacted or reacted B groups on singly reacted BB's, and $p^2[B]$ (reacted) B groups on doubly reacted BB's.

In this mixture the number of crosslinks is $p^2[BB]$ and the number of chains is derived in terms of the degree of polymerization x: (number of chains) = (total number of reacted units)/x = ([A] + [B])p/x. At the critical extent of reaction p_c for the onset of gelation, the number of crosslinks per chain is $\frac{1}{2}$, whence

$$p_c = \frac{[A] + [B]}{[B]\bar{x}_w} \tag{12-10}$$

since Flory has shown that the appropriate average of x is $\bar{x}_w$. Equation 12-10 holds except where gelation occurs at very low extent of reaction; here the distribution of crosslinks is not random. Similar considerations apply to unsymmetrical divinyl monomers and dienes. In either case, the important practical result of delaying the onset of gelation until high conversions can be achieved in the same ways:

a. By reducing the mole fraction of the divinyl or diene monomer.

b. By reducing the weight-average chain length, possibly by the use of chain transfer agents.

c. By reducing the relative reactivity of the second double bond.

Structure of diene polymers and copolymers

The polymerization behavior of dienes as a class may be illustrated with butadiene as an example. This monomer is quite reactive, since each double bond activates the other by conjugation. After one bond has polymerized, the second is at least 100 times less active because of the lack of the resonance.

The radical formed when a diene has added to a growing chain is of the allylic type, with the odd electron resonating between the 2 and the 4 positions. When the next monomer unit adds onto the radical, these two resonating forms give rise to two possible configurations (Chapter 5*A*) of the resulting chain unit: 1,2 addition, with a pendant vinyl group; and 1,4 addition, with a double bond in the chain. In 1,4 addition, the chain can pass through the residual double bond in either the *cis* or the *trans* configuration. The relative amounts of 1,2 and 1,4 addition can be determined either by chemical means or by infrared spectroscopy (Chapter 4*B*). Since competition among the various propagation steps seems due mainly to differences in energy of activation, an increase in temperature tends to equalize the rates of competing reactions. Thus dienes polymerized at higher temperatures tend to contain more nearly equal numbers of the different types of units, whereas in those polymerized at lower temperatures the preferred form (in this case *trans*-1,4 addition) is favored. This leads to a more regular structure in low-temperature diene polymers.

As discussed in Chapter 10*D*, coordination polymerization can lead to amounts of the various types of addition in diene polymers which differ widely from those obtained in radical polymerization.

Cyclopolymerization (Butler 1966, Gibbs 1967) Nonconjugated dienes which have the double bonds separated by three atoms can be polymerized to give soluble, noncrosslinked polymers by an alternating intra-intermolecular mechanism. Successive propagation steps involve, alternately, the addition of a monomer and the formation of a six-membered ring:

where X is one of a variety of groups. Five- and seven-membered rings form from the appropriate polymers with more difficulty, and some open-chain unsaturated groups occur in the polymer.

GENERAL REFERENCES

Alfrey 1952; Temin 1966.

D. Chemical Reactions of Polymers

The concept of the reactivity of functional groups being independent of molecular weight, used in developing the kinetics and statistics of step-wise polymerization (Chapter 8), applies to all functional groups regardless of their location on the polymer chain. As a result, polymers undergo chemical reactions much as do low-molecular-weight compounds provided that reactants can be supplied to the sites of reaction. To accomplish this, most polymer reactions of importance are carried out in solution. Many of these reactions are discussed elsewhere in this book, as indicated by cross reference.

A well-known sequence of polymer reactions is the conversion of poly(vinyl acetate) through poly(vinyl alcohol) to a poly(vinyl acetal) (Chapter 14C). In the latter reaction, as in the reaction of metals to remove chlorine from poly(vinyl chloride), functional groups react in pairs along the chain, with occasional groups isolated and incapable of reaction.

Polyesters are readily hydrolyzed unless low solubility or steric hindrance interferes. Thus, linear condensation polyesters and polyacrylates hydrolyze readily, whereas poly(ethylene terephthalate) and crosslinked alkyd and "polyester" resins are insoluble. Polymethacrylates are inert because of steric hindrance.

The nitration, sulfonation, and reduction of styrene polymers are widely used to produce ion-exchange resins (Chapter 14A). The acetylation, nitration, and xanthation of cellulose (Chapter 15*C*) are important reactions, as are vulcanization and other reactions of natural and synthetic rubbers (Chapters 13*E–H* and 19).

In addition, unsaturated polymers can undergo such reactions as isomerization, cyclization, addition, epoxidation, and hydrogenation, while saturated polymeric hydrocarbons can be substituted on the main chain, the side chain, or the nucleus if aromatic. Terminally reactive polymers can produce block copolymers whereas branching reactions can lead to graft copolymers (Chapter 11*D*). Coupling reactions can promote increases in molecular weight, as in the polyurethanes (Chapter 15*D*), or crosslinking, as in a variety of thermosetting resins (Chapter 16). The class of oxidation-reduction or redox polymers is discussed by Cassidy (1965). Still other reactions of polymers are described in Sections *E* and *F* of this chapter.

GENERAL REFERENCES

Fettes 1964; Lenz 1967; Ravve 1967; Odian 1970.

E. Degradation of Polymers

In the classical chemical sense, the term degradation means breaking down of structure; in one usage relating to polymers (Grassie 1966) it is taken to mean any process leading to deterioration of properties; but here we take it to mean reduction of molecular weight. There are two general types of polymer degradation processes, corresponding roughly to the two types of polymerization, step-reaction and chain-reaction.

Random degradation is analogous to stepwise polymerization. Here chain rupture or scission occurs at random points along the chain, leaving fragments which are usually large compared to a monomer unit.

Chain depolymerization involves the successive release of monomer units from a chain end in a *depropagation* or *unzippering* reaction which is essentially the reverse of chain polymerization.

These two types may occur separately or in combination, may be initiated thermally or by ultraviolet light, oxygen, ozone, or other foreign agent, and may occur entirely at random or preferentially at chain ends or at other weak links in the chain.

It is possible to differentiate between the two processes in some cases by following the molecular weight of the residue as a function of the extent of reaction. Molecular weight drops rapidly as random degradation proceeds but may remain constant in chain depolymerization, as whole molecules are reduced to monomer which escapes from the residual sample as a gas. Examination of the degradation products also differentiates between the two processes: the ultimate product of random degradation is likely to be a disperse mixture of fragments of molecular weight up to several hundred, whereas chain depolymerization yields large quantities of monomer.

Kinetics of random degradation

The kinetics and statistics of random degradation can be treated in exact analogy with the kinetics of linear stepwise polymerization. If p is defined, as before, as the extent of reaction, then $1 - p =$ (number of broken bonds)/(total number of bonds) is the extent of degradation. If the degradation is random, the number of bonds broken per unit time is constant as long as the total number of bonds present is large compared to the number broken. It follows that the number of chain ends increases linearly with time; hence, $1/\bar{x}_n$ increases linearly with time. The acid-catalyzed homogeneous degradation of cellulose is a random degradation of this type.

An example of random degradation initiated by attack at a " weak link " in a chain is the ozonolysis of the isobutylene-isoprene copolymer butyl rubber. Here the initial attack is at the double bonds of the isoprene residues.

Chain depolymerization

Chain depolymerization is a free radical process which is essentially the reverse of chain polymerization. The point of initial attack may be at the chain end or at a "weak link" which may arise from a chain defect, such as an initiator fragment or a peroxide or ether link arising from polymerization in the presence of oxygen. The slightly higher activity of a tertiary hydrogen atom may be enough to provide a site for the initiation of the degradation process.

Poly(methyl methacrylate) degrades thermally by this process. The yield of monomer is 100 % of the weight of polymer lost over a large fraction of the reaction. Polystyrene shows an intermediate behavior: the depropagation reaction ceases before the chain is completely destroyed. In some cases, such as olefin-SO_2 copolymers, an equilibrium can easily be reached between propagation and depropagation.

Kinetics of chain depolymerization   A general kinetic scheme has been formulated which appears to cover all types of depolymerization. It is based on the concept of inverse chain polymerization and includes the steps of initiation, depropagation, termination, and chain transfer. The important feature of this scheme is the inclusion of the chain transfer step, for it can be shown that the kinetics of random degradation result if the kinetic chain length before transfer is just the breaking of one bond.

The transfer reaction probably occurs rapidly by the abstraction of a hydrogen atom from a polymer. The chain which was attacked is likely to split into a radical and one or more inactive fragments at the elevated temperatures where degradation is rapid. Evidence for the transfer reaction includes the observations that tertiary hydrogen atoms at branch points in polyethylene are preferentially attacked and that the degradation of poly-(methyl methacrylate) which has been copolymerized with a little acrylonitrile is quite different from that of pure poly(methyl methacrylate) because of the activity of the α-hydrogens on the acrylonitrile units.

The two factors which appear to be important in determining the course of degradation are the reactivity of the depropagating radical and the availability of reactive hydrogen atoms for transfer. With the possible exception of styrene, where the radical is stabilized by resonance, all polymers containing α-hydrogens, such as polyacrylates and polyacrylonitrile, give poor yields of monomer.

Conversely, the methacrylates give high yields of monomer because of the active radical and the α-methyl group which blocks the possibility of chain transfer. Polytetrafluoroethylene gives high yields of monomer because the strong C-F bond is not easily broken to allow transfer.

The scheme does not apply to polymers such as poly(vinyl acetate) and poly(vinyl chloride), where degradation results from the removal of side groups rather than from chain scission.

Only two parameters beside the rate constants are important in the kinetic analysis. They are the transfer constant, σ, defined as

$$\sigma = \frac{\text{probability of transfer}}{\text{probability of initiation}} = \frac{k_{tr}[\text{R} \cdot]}{k_i} \tag{12-11}$$

and the kinetic chain length. This is not defined in the same way as in polymerization, but as the number-average number of monomers produced from one chain. The kinetic chain length is $(1/\varepsilon) - 1$, where

$$\frac{1}{\varepsilon} = \frac{\text{probability of (propagation + termination + transfer)}}{\text{probability of (termination + transfer)}} \tag{12-12}$$

The kinetic chain length is approximately measured by the ratio of weight of monomer to that of other volatile fragments when depolymerization is completed. In polyethylene, where almost no monomer is produced, $1/\varepsilon \simeq 1$. In polystyrene or polyisobutylene, $1/\varepsilon \simeq 4$; in poly(methyl methacrylate) $1/\varepsilon$ is at least as high as 200.

Jellinek (1966) has pointed out errors in some earlier treatments of depolymerization.

Degradation products Study of the products of thermal depolymerization in vacuum has shown that the chemical nature and relative amounts of these products are remarkably independent of the temperature and extent of the degradation reaction. Polystyrene, e.g., degraded to 40% styrene, 2.4% toluene, and other products having an average molecular weight of 264, at temperatures between 360° and 420°C, and extents of degradation from 4 to 100%. The amounts of monomer produced by various polymers are shown in Table 12-5.

TABLE 12-5. *Per cent monomer resulting from thermal degradation*

Polymer	Per Cent Monomer	
	Weight	Mole
Poly(methyl methacrylate)	100	100
Poly(α-methylstyrene)	100	100
Polyisobutylene	32	78
Polystyrene	42	65
Polybutadiene	14	57
SBR (butadiene-styrene copolymer)	12	52
Polyisoprene	11	44
Polyethylene	3	21

GENERAL REFERENCES

Madorsky 1964; Grassie 1966; Jellinek 1966, 1968; Pinner 1967; Ravve 1967; Reich 1967, 1968.

F. Radiation Chemistry of Polymers

The interaction of high-energy radiation with molecular substances involves the following sequence of events, regardless of the source of energy (photons, protons, electrons, neutrons, etc.). The molecules are first excited and ionized. Secondary electrons are emitted with relatively low speeds, and produce many more ions along their tracks. Within 10^{-12} sec or so, molecular rearrangements take place in the ions and excited molecules, accompanied by thermal deactivation or the dissociation of valence bonds. As far as subsequent reactions are concerned, bond dissociation is the more important. It leads to the production of ions or radicals whose lifetimes depend on diffusion rates and may be weeks or months in solids at low temperatures.

The major effects in polymers arise from the dissociation of primary valence bonds into radicals, whose existence can be demonstrated by chemical methods or by EPR spectroscopy (Chapter 4*D*). The dissociations of C-C and C-H bonds lead to different results, degradation and crosslinking, which may occur simultaneously.

Degradation The major result of radiation is degradation by chain scission in 1,1-disubstituted polymers, such as poly(methyl methacrylate) and its derivatives, polyisobutylene, and poly(α-methylstyrene), and in polymers containing halogen, such as poly(vinyl chloride), poly(vinylidene chloride), and polytetrafluoroethylene. The tendency toward degradation is related to the absence of tertiary hydrogen atoms, a weaker than average C-C bond (low heat of polymerization), or unusually strong bonds (such as C-F) elsewhere in the molecule.

Degradation is, of course, evidenced by decrease in molecular weight, the weight-average molecular weight being inversely proportional to the amount of radiation received. In polymers with bulky side chains, such as poly(methyl methacrylate), extensive degradation of the side chains to gaseous products also occurs.

Crosslinking Crosslinking is the predominant reaction on the irradiation of polystyrene, polyethylene and other olefin polymers, polyacrylates and their derivatives, and natural and synthetic rubbers. It is accompanied by the formation of gel and ultimately by the insolubilization of the entire specimen. Radiation crosslinking has a beneficial effect on the mechanical properties of some polymers and is carried out commercially to produce polyethylene with enhanced form stability and resistance to flow at high temperatures (Chapter 13*D*).

Other reactions Radiation crosslinking is often accompanied by the formation of *trans*-vinylene unsaturation, both reactions resulting in the formation of hydrogen gas. If the irradiation is carried out in the presence of air, surface oxidation results. The resulting peroxides may be decomposed later in graft copolymerization (Chapter 11*D*).

GENERAL REFERENCES

Bovey 1958; Charlesby 1960; Chapiro 1966, 1969; Lenz 1967; Platzer 1967; Parkinson 1969; Tabata 1969.

BIBLIOGRAPHY

Alfrey 1952. Turner Alfrey, Jr., John J. Bohrer, and H. Mark, *Copolymerization*, Interscience Publishers, New York, 1952.

Bamford 1969. C. H. Bamford and G. C. Eastmond, "Solid-Phase Addition Polymerization," *Quart. Revs. Chem. Soc.* **23**, 271–299 (1969).

Bovey 1958. Frank A. Bovey, *The Effects of Ionizing Radiation on Natural and Synthetic High Polymers*, Interscience Publishers, New York, 1958.

Breitenbach 1969. J. W. Breitenbach, "Proliferous Polymerization," pp. 587–597 in Herman F. Mark, Norman G. Gaylord, and Norbert M. Bikales, eds., *Encyclopedia of Polymer Science and Technology*, Vol. 11, Interscience Div., John Wiley and Sons, New York, 1969.

Butler 1966. George B. Butler, "Cyclopolymerization," pp. 568–599 in Herman F. Mark, Norman G. Gaylord, and Norbert M. Bikales, eds., *Encyclopedia of Polymer Science and Technology*, Vol. 4, Interscience Div., John Wiley and Sons, New York, 1966.

Cassidy 1965. Harold G. Cassidy and Kenneth A. Kun, *Oxidation-Reduction Polymers (Redox Polymers)*, Interscience Div., John Wiley and Sons, New York, 1965.

Chapiro 1962. Adolphe Chapiro, *Radiation Chemistry of Polymeric Systems*, Interscience Div., John Wiley and Sons, New York, 1962.

Chapiro 1969. A. Chapiro, Robert B. Fox, Robert F. Cozzens, W. Brenner, and W. Kupfer, "Radiation-Induced Reactions," pp. 702–783 in Herman F. Mark, Norman G. Gaylord, and Norbert M. Bikales, eds., *Encyclopedia of Polymer Science and Technology*, Vol. 11, Interscience Div., John Wiley and Sons, New York, 1969.

Charlesby 1960. Arthur Charlesby, *Atomic Radiation and Polymers*, Pergamon Press, New York, 1960.

Duck 1966. Edward W. Duck, "Emulsion Polymerization," pp. 801–859 in Herman F. Mark, Norman G. Gaylord, and Norbert M. Bikales, eds., *Encyclopedia of Polymer Science and Technology*, Vol. 5, Interscience Div., John Wiley and Sons, New York, 1966.

Farber 1970. Elliott Farber, "Suspension Polymerization," in press in Herman F. Mark, Norman G. Gaylord, and Norbert M. Bikales, eds., *Encyclopedia of Polymer Science and Technology*, Vol. 13, Interscience Div., John Wiley and Sons, New York, 1970.

Fettes 1964. E. M. Fettes, ed., *Chemical Reactions of Polymers*, Interscience Div., John Wiley and Sons, New York, 1964.

Gibbs 1967. William E. Gibbs and John M. Barton. "The Mechanism of Cyclopolymerization of Nonconjugated Diolefins," Chap. 2 in George E. Ham, ed., *Vinyl Polymerization*, Part I, Marcel Dekker, New York, 1967.

Grassie 1966. N. Grassie, "Degradation," pp. 647–716 in Herman F. Mark, Norman G. Gaylord, and Norbert M. Bikales, eds., *Encyclopedia of Polymer Science and Technology*, Vol. 4, Interscience Div., John Wiley and Sons, New York, 1966.

Herz 1968. J. E. Herz and V. Stannett, "Copolymerization in the Crystalline Solid State," *Macromol. Revs.* **3**, 1–48 (1968).

Jellinek 1966. H. H. G. Jellinek, "Depolymerization," pp. 740–793 in Herman F. Mark, Norman G. Gaylord, and Norbert M. Bikales, eds., *Encyclopedia of Polymer Science and Technology*, Vol. 4, Interscience Div., John Wiley and Sons, New York, 1966.

Jellinek 1968. H. H. G. Jellinek, "Degradation of Stereoregular Polymers," Chap. 10 in A. D. Ketley, ed., *The Stereochemistry of Macromolecules*, Vol. 3, Marcel Dekker, New York, 1968.

Jenkins 1967. A. D. Jenkins, "Occlusion Phenomena in the Polymerization of Acrylonitrile and Other Monomers," Chap. 6 in George E. Ham, ed., *Vinyl Polymerization,* Part I, Marcel Dekker, New York, 1967.

Lenz 1967. Robert W. Lenz, *Organic Chemistry of Synthetic High Polymers,* Interscience Div., John Wiley and Sons, New York, 1967.

Lenz 1969. Robert W. Lenz, "Applied Reaction Kinetics: Polymerization Reaction Kinetics," *Ind. Eng. Chem.* **61** (3), 67–75 (1969).

Lenz 1970. Robert W. Lenz, "Applied Polymerization Reaction Kinetics," *Ind. Eng. Chem.* **62** (2), 54–61 (1970).

Madorsky 1964. Samuel L. Madorsky, *Thermal Degradation of Organic Polymers,* Interscience Div., John Wiley and Sons, New York, 1964.

Mickley 1962. Harold S. Mickley, Alan S. Michaels, and Albert L. Moore, "Kinetics of Precipitation Polymerization of Vinyl Chloride," *J. Polymer Sci.* **60,** 121–140 (1962).

Odian 1970. George Odian, *Principles of Polymerization,* McGraw-Hill Book Co., New York, 1970.

Palit 1965. Santi R. Palit, Tilak Guha, Rajat Das, and Ranjit S. Konar, "Aqueous Polymerization," pp. 229–266 in Herman F. Mark, Norman G. Gaylord, and Norbert M. Bikales, eds., *Encyclopedia of Polymer Science and Technology,* Vol. 2, Interscience Div., John Wiley and Sons, New York, 1965.

Parkinson 1969. W. W. Parkinson, "Radiation-Resistant Polymers," pp. 783–809 in Herman F. Mark, Norman G. Gaylord, and Norbert M. Bikales, eds., *Encyclopedia of Polymer Science and Technology,* Vol. 11, Interscience Div., John Wiley and Sons, New York, 1969.

Pinner 1967. S. H. Pinner, ed., *Weathering and Degradation of Plastics,* Gordon and Breach, New York, 1967.

Platzer 1967. Norbert Platzer, Symposium Chairman, "Irradiation of Polymers," Vol. 66 in Robert F. Gould, ed., *Advances in Chemistry Series,* American Chemical Society, Washington, D.C., 1967.

Ravve 1967. A. Ravve, *Organic Chemistry of Macromolecules,* Marcel Dekker, New York, 1967.

Reich 1967. Leo Reich and David W. Levi, "Dynamic Thermogravimetric Analysis in Polymer Degradation," *Macromol. Revs.* **1,** 173–275 (1967).

Reich 1968. Leo Reich, "Polymer Degradation by Differential Thermal Analysis Techniques," *Macromol. Revs.* **3,** 49–112 (1968).

Ringsdorf 1965. H. Ringsdorf, "Bulk Polymerization," pp. 642–666 in Herman F. Mark, Norman G. Gaylord, and Norbert M. Bikales, eds., *Encyclopedia of Polymer Science and Technology,* Vol. 2, Interscience Div., John Wiley and Sons, New York, 1965.

Scheinker 1954. A. P. Scheinker and S. S. Medvedev, "Investigations of the Kinetics of the Polymerization of Isoprene in Aqueous Solutions of Emulsifiers and in Emulsions" (in Russian), *Doklady Akad. Nauk SSSR* **97,** 111–114 (1954).

Smith 1948(a). Wendell V. Smith, "The Kinetics of Styrene Emulsion Polymerization," *J. Am. Chem. Soc.* **70,** 3695–3702 (1948).

Smith 1948(b). Wendell V. Smith and Roswell H. Ewart, "Kinetics of Emulsion Polymerization," *J. Chem. Phys.* **16,** 592–599 (1948).

Tabata 1969. Yoneho Tabata, "Radiation-Induced Polymerization," Chap. 7 in George E. Ham, ed., *Vinyl Polymerization,* Vol. 1, Part II, Marcel Dekker, New York. 1969.

Temin 1966. Samuel C. Temin and Allan R. Shultz, "Crosslinking," pp. 331–414 in Herman F. Mark, Norman G. Gaylord, and Norbert M. Bikales, eds., *Encyclopedia of Polymer Science and Technology,* Vol. 4, Interscience Div., John Wiley and Sons, New York, 1966.

Trommsdorff 1956. Ernst Trommsdorff and C. E. Schildknecht, "Polymerization in Suspension," Chap. III in Calvin E. Schildknecht, ed., *Polymer Processes*, Interscience Publishers, New York, 1956.

Vanderhoff 1969. John W. Vanderhoff, "Mechanism of Emulsion Polymerization," Chap. 1 in George E. Ham, ed., *Vinyl Polymerization*, Vol 1, Part II, Marcel Dekker, New York, 1969.

IV

Properties of Commercial Polymers

I3

Hydrocarbon Plastics and Elastomers

A. Low-Density (Branched) Polyethylene

The first commercial ethylene polymer was branched polyethylene, commonly designated as low-density or high-pressure material to distinguish it from the essentially linear material described in Section *B*. After a period of relatively slow growth, culminating in the production in the United States of less than 100 million lb in 1952, the annual volume of low-density polyethylene has expanded rapidly. Polyethylene became in 1959 the first plastic with production exceeding 1 billion lb. Over 3.5 billion lb was produced by 14 companies in 1969, at prices near $0.10/lb.

Polyethylene was first produced in the laboratories of Imperial Chemical Industries, Ltd. (ICI), England, in a fortuitous experiment in which ethylene (and other chemicals which remained inert) was subjected to 1400 atm of pressure at 170°C. Traces of oxygen caused polymerization to take place. The phenomenon was first described by E. W. Fawcett in Staudinger 1936.

Polymerization

Ethylene (boiling point − 104°C) can be obtained from the dehydration of ethanol or by the hydrogenation of acetylene. Where supplies of natural or petroleum gas are available, however, the monomer is produced by cracking ethane or propane followed by purification processes.

High polymers of ethylene are made commercially at pressures between 1000 and 3000 atm (15,000–45,000 psi) or possibly higher, and temperatures as high as 250°C.

Traces of oxygen initiate the polymerization of ethylene readily. Rapid exothermic reactions can occur, and violent explosions have taken place.

Many other possible impurities in the monomer, such as hydrogen and acetylene, act as chain transfer agents and must be carefully removed if high-molecular-weight products are to be obtained. Besides oxygen, peroxides (benzoyl, diethyl), hydroperoxides, and azo compounds have been used as initiators.

Ethylene polymerization can be carried out with benzene or chlorobenzene as solvent. At the temperatures and pressures used, both polymer and monomer dissolve in these compounds so that the reactions are true solution polymerizations. Water or other liquids may be added to dissipate the heat of reaction.

Batch polymerizations of ethylene cannot be carried out rapidly with reproducibility and good control. Long reaction times, consistent with good control, are not economical. In addition, chain branching becomes excessive at high conversion and results in poor physical properties of the product. As a result, balanced, continuous polymerization systems are preferred. Emulsion polymerization has had little success.

One continuous process utilizes tubular reactors, which may have diameters of less than 1 in. and lengths up to 100 ft. The stainless steel tube may be filled with water, and ethylene containing initiator and possibly benzene is introduced. Additional initiator and water or benzene can be injected at one or more points along the tube to keep the initiator concentration more nearly constant throughout the reactor. Ten or more per cent of the ethylene is polymerized at the far end of the reactor. Here the gas and liquid phases are taken off continuously, the polymer is separated, and the ethylene is recycled after purification.

Another process utilizes bulk polymerization in a tower-type reactor. Ethylene containing trace amounts of oxygen is charged to the reactor at 1500 atm and 190°C. The reaction is kept essentially isothermal and carried to 10–15% conversion. The effluent from the reactor passes to a separatory vessel in which unconverted ethylene is removed for recycling. The molten polyethylene is chilled below its crystalline melting point and passed through the usual finishing steps.

Low polymers of ethylene (oils, greases, and waxes) can be made at pressures between 100 and 500 atm and at temperatures between room temperature and 250°C.

Structure

Low-density polyethylene is a partially (50–60%) crystalline solid melting at about 115°C, with density in the range 0.91–0.94. It is soluble in many solvents at temperatures above 100°C, but no room-temperature solvents exist.

In 1940 infrared spectroscopy (Chapter 4*B*) revealed that low-density polyethylene contains branched chains. These branches are of two distinct types. Branching due to intermolecular chain transfer, arising from reactions of the type

$$R_1CH_2CH_2\cdot + R_2CH_2CH_2R_3 \xrightarrow[\text{hydrogen transfer}]{\text{intermolecular}} R_1CH_2CH_3 + R_2\overset{\cdot}{C}HCH_2R_3,$$

propagating   dead polymer   dead polymer   propagating chain
chain   molecule   molecule

leads to branches which are, on the average, as long as the main polymer chain. This sort of branching has an observable effect on the solution viscosity of the polymer (Chapter 3*D*) and can be detected by comparing the viscosity of a branched polyethylene with that of a linear polymer of the same molecular weight.

The second branching mechanism in polyethylene is postulated to produce short-chain branching by intramolecular chain transfer:

$$RCH_2CH_2CH_2CH_2CH_2CH_2\cdot \xrightarrow[\text{ring formation}]{\substack{\text{transient} \\ \text{six-membered}}}$$

The transient ring mechanism suggests four carbon atoms as the most probable length of the short branches. Infrared absorption studies and studies of degradation under bombardment with high-energy radiation of polyethylene and substituted polyethylenes of known branch structure suggest that both ethyl and butyl branches are present. A mechanism accounting for the ethyl branches assumes a transfer reaction of the type

$$RCH_2HC\overset{\displaystyle CH_2CH_2\cdot}{\underset{\displaystyle CH_2CH_2CH_2CH_3}{\big\backslash}} \longrightarrow RCH_2HC\overset{\displaystyle CH_2CH_3}{\underset{\displaystyle CH_2\overset{\cdot}{C}HCH_2CH_3}{\big\backslash}}$$

after addition of one monomer unit to the radical resulting from a short-chain branching step by the previous mechanism.

Since the transient ring conformations are relatively probable during propagation, the short-chain branching mechanism accounts for the large majority of the chain ends observed in the infrared. A typical low-density polyethylene molecule may contain fifty short branches and less than one long branch on a number average basis.

The short-chain branched structure of low-density polyethylene can be simulated without long-chain branching by the coordination copolymerization of ethylene with butene-1 and hexene-1. The properties of the copolymer are much like those of low-density polyethylene (Levett 1970), but a tendency toward the formation of block structures leads to some significant differences between the materials.

The molecular-weight distribution of typical polyethylenes has been found experimentally and theoretically to be very much broadened by the long-chain branching mechanism. Weight- to number-average molecular weight ratios $\overline{M}_w/\overline{M}_n$ of 20 to 50 are considered typical. The distribution of long branches among the molecules is also very broad: even for highly branched polymer many molecules contain no long branches, while most of the branches are concentrated on a few very large molecules.

Infrared spectroscopy (Chapter 4*B*) provides much information about the chemical and physical structure of polyethylene. Structural features associated with the crystallinity of polyethylene are described in Chapter 5.

Properties

Effect of structure on properties   The physical properties of polyethylenes are functions of three independent structural variables: molecular weight, molecular weight distribution or long-chain branching, and short-chain branching.

Short-chain branching has a predominant effect on the degree of crystallinity and therefore on the density of polyethylene. (Actually these properties are influenced by total chain branching, but the number of long-chain branch points per molecule in typical polyethylenes is so much less than the number of short-chain branch points that the former can be neglected.) Therefore, as discussed in Chapter 7, properties dependent on crystallinity, such as stiffness, tear strength, hardness, chemical resistance, softening temperature, and yield point, increase with increasing density or decreasing amount of short-chain branching in the polymer, whereas permeability to liquids and gases, toughness, and flex life decrease under the same conditions.

The effect of molecular weight is largely evidenced (Chapter 7*C*) in properties of the melt and properties involving large deformations of the solid. As molecular weight increases, so do tensile strength, tear strength, low-temperature toughness, softening temperature, impact strength, and resistance to environmental stress cracking, while melt fluidity, melt "drawability," and coefficient of friction (film) decrease. These properties are commonly compared on the basis of changes in melt index (Chapter 6*A*), which varies inversely with molecular weight.

The effect of long-chain branching on the properties of polyethylene is often evaluated in terms of the breadth of the molecular weight distribution $\overline{M}_w/\overline{M}_n$. With other structural parameters held constant, a decrease in $\overline{M}_w/\overline{M}_n$ causes a decrease in ease of processing but an increase in tensile strength, toughness and impact strength, softening temperature, and resistance to environmental stress cracking.

The interrelation of density, melt index, and $\overline{M}_w/\overline{M}_n$ in producing physical properties desirable for specific end uses has been described (McGrew 1958). For film uses polymers producing tough and flexible films are needed. Injection-molding applications require polymers characterized by rigidity and good flow; pipe, by strength; and wire insulation, by good processing characteristics and resistance to stress cracking.

General physical properties   The mechanical properties of low-density polyethylene are between those of rigid materials like polystyrene and limp plasticized polymers like the vinyls. Polyethylene has good toughness and pliability over a wide temperature range. Its density falls off fairly rapidly above room temperature, and the resulting large dimensional changes cause difficulty in some fabrication methods. The relatively low crystalline melting point (about 115°C for typical materials) limits the temperature range of good mechanical properties.

The electrical properties of polyethylene are outstandingly good, probably ranking next to those of polytetrafluoroethylene for high-frequency uses. In thick sections polyethylene is translucent because of its crystallinity, but high transparency is obtained in thin films.

Chemical properties (Friedlander 1965)   Polyethylene is very inert chemically. It does not dissolve in any solvent at room temperature, but is slightly swelled by liquids such as benzene and carbon tetrachloride which are solvents at higher temperatures. It has good resistance to acids and alkalis. At 100°C it is unaffected in 24 hr by sulfuric or hydrochloric acid but charred by concentrated nitric acid. It is often used in containers for acids, including hydrofluoric.

Polyethylene ages on exposure to light and oxygen, with loss of strength, elongation, and tear resistance. The probable point of attack is the tertiary hydrogens on the chain at branch points. Stabilizers retard the deterioration, but few are compatible enough with the polymer to do much good. The weathering of carbon black pigmented material is quite good. The polymer also undergoes some crosslinking when heated or worked at elevated temperatures. Few plasticizers or other additives are compatible with polyethylene in amounts larger than 1 % or so.

Film and sheeting   Almost half of the polyethylene produced has gone into film and sheeting uses for the last several years. Few competitive film materials have polyethylene's desirable combination of low density, flexibility without a plasticizer, resilience, high tear strength, moisture and chemical resistance, and little tendency for nicks or cuts to propagate.

Film may be made by extrusion through a slit die at 200–250°C and drawing into a water quench bath. Alternatively, blown film may be produced by extruding a tube of polymer and expanding it by means of internal pressure of inert gas, thus drawing the polymer. The tube may be slit to produce flat film or left as a seamless tube. Film thicknesses are usually 0.001–0.005 in.

Over three-fourths of the polyethylene film produced goes into packaging applications, including bags and pouches and wrappings for produce, textile products, merchandise, frozen and perishable foods, and many other products. Other film uses include drapes and tablecloths, and extensive application in agriculture (greenhouses, ground cover, tank, pond, and canal liners, etc.) and construction (moisture barriers and utility covering material).

Injection molding   Production of housewares and toys by injection molding consumes some 13% of the polyethylene produced.

Wire and cable insulation   Polyethylene has filled a long-standing need for a material which would effectively insulate electrical cables without introducing electrical losses at high frequencies. The nonpolar nature of the polymer makes it ideal for this purpose. Television, radar, and multicircuit long-distance telephony might well have been impossible without such insulating materials. Weathering introduces polar impurities such as carbonyl groups into the polymer and must be carefully guarded against. In addition to the high-frequency uses, polyethylene is being more generally utilized for mechanical protection of wire and cables, where its chemical inertness and light weight are advantageous. About 10% of the polyethylene produced is used for wire and cable insulation.

Coating   About 10% of polyethylene is employed for coatings. These may be applied by dipping from hot solutions, melts, or emulsions, or by flame spraying. The extrusion coating of foils, papers, and other films (Mylar, cellophane) consumes large quantities of polyethylene.

Blow molding   The production of "squeeze" bottles and industrial containers by blow molding accounts for about 2% of polyethylene production.

Flexibility, low cost, and resistance to corrosion and breakage are important advantages in this use.

GENERAL REFERENCES

Raff 1965–66, 1967; Chemplex 1969; Dow 1969; Kresser 1969; Laing 1969; Lasman 1969; Nelson 1969.

B. High-Density (Linear) Polyethylene

Linear polyethylene can be produced in several ways, including radical polymerization of ethylene at extremely high pressures, coordination polymerization of ethylene, and polymerization of ethylene with supported metal oxide catalysts (Chapter 10*D*). Commercial production of linear polyethylene, using the second and third routes named, began in 1957 and reached a volume of close to 1.5 billion lb in 1969, at a price of about $0.15/lb. Eleven companies were producers at the end of 1969.

Polymerization

Coordination polymerization (Chapter 10*D*) A coordination catalyst is prepared as a colloidal dispersion by reacting, typically, an aluminum alkyl and $TiCl_4$ in a solvent such as heptane. Ethylene is added to the reaction vessel under slight pressure, at a temperature of 50–75°C. Heat of polymerization is removed by cooling. Polymer forms as a powder or granules, insoluble in the reaction mixture. At the completion of the reaction, the catalyst is destroyed by the admission of water or alcohol, and the polymer is filtered or centrifuged off, washed, and dried.

Polymerization with supported metal oxide catalysts (Chapter 10*D*) These catalysts can be used in a variety of operating modes, including fixed-bed, moving-bed, fluid-bed, or slurry processes. Ethylene is fed with a paraffin or cycloparaffin diluent, at 60–200°C and around 500 psi pressure. The polymer is recovered by cooling or by solvent evaporation.

Structure

Typical linear polyethylenes are highly (over 90%) crystalline polymers, containing less than one side chain per 200 carbon atoms in the main chain. Melting point is above 127°C (typically about 135°C), and density is in the range of 0·95–0·97. Infrared spectroscopy (Chapter 4*B*) gives detailed in formation on the chemical and physical structure of the polymer. Structure features associated with the crystallinity of linear polyethylene are discusse in Chapter 5.

Most of the differences in properties between branched and linear poly ethylenes can be attributed to the higher crystallinity of the latter polymers Linear polyethylenes are decidedly stiffer than the branched material (modulu of 100,000 psi vs. 20,000 psi), and have a higher crystalline melting point an greater tensile strength and hardness. The good chemical resistance o branched polyethylene is retained or enhanced, and such properties as low temperature brittleness and low permeability to gases and vapors are im proved in the linear material.

Blow molding   The production of bottles and other containers by blo molding accounts for over 40 % of the linear polyethylene made. The adjust ment of structure variables to obtain high resistance to environmental stres cracking, allowing the material to be used in detergent bottles, produced large expansion in this field.

Injection molding   About 20 % of the linear polyethylene produced is use in the injection molding of housewares and toys. The higher stiffness an heat resistance of the linear material have led to its replacement of branche polyethylene in applications where these properties are important.

Other uses   Other major uses of linear polyethylene include film and sheet wire and cable insulation, extrusion coating, and pipe and conduit.

GENERAL REFERENCES

Raff 1965–66, 1967; Chemplex 1969; Dow 1969; Kresser 1969.

C. Polypropylene

With the commercial utilization of coordination polymerization (Chapte 10*D*) in 1957, the production of polypropylene became possible. In the inter vening decade, this has become one of the world's major plastics, wit

United States production at about 1 billion lb in 1969, at a price of about
0.20/lb.

Polymerization

The polymerization of propylene, a by-product gas of alkylate gasoline
refining, is carried out with coordination catalysts essentially as described
in Section *B* for linear polyethylene. Ethylene, propylene, and other α-olefins
can be polymerized in the same equipment with very little modification,
leading to highly flexible operation. Catalysts and operating conditions must
be selected with care to ensure that isotactic polypropylene (see the next
paragraph) is produced.

Structure

Polypropylene can be made in isotactic, syndiotactic, or atactic form
(Chapter 5*A*). The crystallizability of isotactic polypropylene makes it the
sole form with properties of commercial interest. Isotactic polypropylene is
an essentially linear, highly crystalline polymer, with a melting point of
165°C. Its crystal structure is described in Chapter 5*B*.

Properties

Polypropylene is the lightest major plastic, with a density of 0.905. Its
high crystallinity imparts to it high tensile strength, stiffness, and hardness.
The resulting high strength-to-weight ratio is an advantage in many applications. Finished articles usually have good gloss and high resistance to
marring. The high melting point of polypropylene allows well-molded parts
to be sterilized, and the polymer retains high tensile strength at elevated
temperatures.

The low-temperature impact strength of polypropylene is somewhat
sensitive to fabrication and test conditions. This sensitivity results from the
presence of a dominating α-transition (Chapter 6*D*) in polypropylene at about
0°C, resulting in a marked loss in stiffness near this temperature. In high-
density polyethylene, the dominant transition is the lower-temperature β-
transition. Thus, the restriction of molecular motion leading to brittle
behavior takes place not far below room temperature in polypropylene, but
at a much lower temperature in polyethylene.

To overcome brittleness, wide use is made of both random and block
copolymers of propylene with ethylene. The block copolymers are the most
impact resistant, and are used in injection molding applications. To retain
transparency, random copolymers are used for film applications, while the
homopolymer is used almost exclusively for filaments.

Polypropylene has excellent electrical properties and the chemical inert ness and moisture resistance typical of hydrocarbon polymers. It is com pletely free from environmental stress cracking. However, it is inherently les stable than polyethylene to heat, light, and oxidative attack (presumably because of the presence of tertiary hydrogens) and must be stabilized with antioxidants and ultraviolet light absorbers for satisfactory processing and weathering. The resulting formulations are quite satisfactory, even for such applications as indoor-outdoor carpeting, but are more expensive.

Application

Injection molding uses, including wide application in the automotive and appliance fields, account for almost half of the production of polypro pylene. Another third is used as filament (rope, cordage, and webbing) and filament and staple for carpeting. Film uses run well behind.

GENERAL REFERENCES

Raff 1965–66, 1967; Rebenfeld 1967; Frank 1968; Gideon 1969; Howsmon 1969; Jez 1969; Kresser 1969; Walton 1969.

D. Other Olefin Polymers

Poly(α-olefins) and olefin copolymers

Polymers of α-olefins, including propylene, 1-butene, and higher homo logs, and copolymers of these monomers with ethylene, can be prepared by coordination polymerization (Chapter 10D). Homopolymers of olefins higher in chain length than propylene which have been offered commercially include those of 1-butene (Buckley 1965, Rubin 1968) and 4-methyl pentene-1 (Westall 1969). In addition, a number of ethylene copolymers are sold, including those with propylene, 1-butene, and isobutylene. Copolymers of ethylene with vinyl acetate, maleic anhydride, and ethyl acrylate are made commercially by radical polymerization.

The major interest in the copolymers of ethylene with about 5% of 1-butene lies in their improved resistance to stress cracking compared to linear polyethylene. Ethylene-maleic anhydride copolymers have the advantage of potential reactivity through the maleic anhydride groups. Copolymers of 60% ethylene and 40% propylene have elastomeric properties much like those of natural rubber. They are cured by chemical crosslinking (see below) with organic peroxides. Ethylene-vinyl acetate copolymers have softness nd pliability like those of the plasticized vinyl resins. In other com-osition ranges, they are used as wax additives and in wax coatings and dhesives.

Chlorinated olefin products

Chlorinated polyethylene (Schramm 1969)   Polyethylene can be chlorinated a solution in CCl_4 at about 60°C or in suspension. Products containing up o 30% chlorine by weight are softer, more rubbery, and more soluble than olyethylene because of loss of crystallinity. In the range of 40–50% chlorine ne polymer becomes harder and higher softening. As high as 73% chlorine y weight can be incorporated into the chain.

Chlorinated polyethylene can be made to have properties like those of lasticized poly(vinyl chloride), but cannot compete with it in cost. Chlori-ated polyethylene exhibits chemical instability similar to that of poly(vinyl hloride).

Chlorosulfonated polyethylene (Keeley 1959)   When polyethylene is treated vith a mixture of chlorine and sulfur dioxide, some chlorine atoms are ubstituted on the chains and some sulfonyl chloride groups ($—SO_2Cl$) are ormed. The chlorosulfonation can be carried out either on the solid material r in solution. There are two results of these modifications: (*a*) the chlorine toms break up the regularity of the polyethylene chain structure so that rystallization is no longer possible, thus imparting an elastomeric character o the polymer; and (*b*) the sulfonyl chloride groups provide sites for cross-inking. A typical polymer contains 25–30% chlorine (one chlorine for every even carbon atoms) and about 1.5% sulfur (one $—SO_2Cl$ for every ninety arbon atoms).

The elastomer can be crosslinked by a large variety of compounds, in-luding many rubber accelerators. Metallic oxides are recommended for ommercial cures. Fillers are not needed to obtain optimum strength roperties.

Chlorosulfonated polyethylene is resistant to ozone, being better than leoprene and butyl rubber in this respect. Oxidative resistance and heat esistance are good. Chemical resistance is better than that of the common lastomers. The material is poor in "snap" and rebound and has low elonga-ion and some permanent set. Its abrasion resistance, flex life, low-tempera-ure brittleness, and resistance to crack growth are good.

Crosslinked polyethylene

There has been considerable interest in converting polyethylene to a hermosetting material, in order to combine its low cost, easy processing, and good mechanical properties with the enhanced form stability at elevated temperatures, resistance to stress crack, and tensile strength expected in a crosslinked polymer.

Chemical crosslinking   Incorporation of relatively stable peroxides, such as dicumyl peroxide and di-*t*-butyl peroxide, provides a chemical means of crosslinking polyethylene. The peroxides are stable at normal processing temperatures but decompose to provide free radicals for crosslinking at higher temperatures in a postprocessing vulcanization or curing reaction. Chemically crosslinked polyethylene is used in the wire and cable industry and is of interest for pipe, hose, and molded articles. Ethylene-propylene rubbers (Section *G*) are also vulcanized by peroxide curing systems.

Radiation crosslinking   (Chapter 12*F*)   The crosslinking of polyethylene by irradiation with high-energy electrons has been used in the commercial production of films combining the properties typical of polyethylene with form stability up to 200°C and a significant increase in tensile strength. The film can be made heat shrinkable by biaxial stretching. It is used for insulating (by wrapping) electrical power cables, coils, transformers, and motors and generators.

Ionomers

The word ionomer (du Pont 1969*a,b*; Kinsey 1969) has been coined as a generic term for a class of thermoplastics containing ionizable carboxyl groups which can create ionic crosslinks between chains. These substances are produced as copolymers of α-olefins with carboxylic acid monomers, such as methacrylic acid, followed by partial neutralization with a metal cation. Crosslinking thus occurs through metal "bridges." These crosslinks are labile at processing temperatures, allowing the ionomers to be extruded or molded in conventional equipment. The upper use temperature of the ionomers is limited, as might be expected, because the crosslinks begin to "melt out." The primary uses of this new class of materials are centered around their combination of properties such as high transparency, toughness, flexibility, adhesion, and oil resistance. Food packaging, skin and blister packaging, and several shoe uses are examples.

GENERAL REFERENCES

Raff 1965–66; Buckley 1968; Dow 1969; Kresser 1969.

C. Natural Rubber and Other Polyisoprenes

Natural rubber

Natural rubber is a high-molecular-weight polymer of isoprene, in which essentially all the isoprenes have the *cis*-1,4 configuration (Chapter 5*A*). The natural polymer has a number-average degree of polymerization of about 5000 and a broad distribution of molecular weights. About 1.3 billion lb of natural rubber was consumed in the United States in 1969, some 22% of the total consumption of rubber for that period.

Source Natural rubber can be obtained from nearly five hundred different species of plants. The outstanding source is the tree *Hevea brasiliensis*, from which comes the name *Hevea rubber*. Rubber is obtained from a latex which exudes from the bark of the *Hevea* tree when it is cut. Latex is an aqueous dispersion of rubber, containing 25–40% rubber hydrocarbon, stabilized by a small amount of protein material and fatty acids. The latex is gathered, coagulated, washed, and dried. Two different processes are used.

Crepe rubber results if a small amount of sodium bisulfite is added to bleach the rubber. The coagulum is rolled out into sheets about 1 mm thick and dried in air at about 50°C. If *smoked sheets* are to be made, the bleach is omitted and somewhat thicker sheets are rolled. These are dried in smokehouses at about 50°C in the smoke from burning wood or coconut shells.

Mastication It was discovered by Hancock in 1824 that rubber becomes a soft, gummy mass when subjected to severe mechanical working. This process is known as *mastication*. The addition of compounding ingredients is greatly facilitated by this treatment, which is usually carried out on roll mills or in internal mixers or plasticators (similar to extruders). Mastication is accompanied by a marked decrease in the molecular weight of the rubber. Oxidative degradation is an important factor in mastication, since the decrease in viscosity and the other property changes do not take place if the rubber is masticated in the absence of oxygen.

After mastication is complete, compounding ingredients are added, and the rubber mix is prepared for vulcanization. These processes are discussed in Chapter 19.

Properties The properties of vulcanized natural rubber form the model for the ideal elastomeric properties discussed in Chapter 6*B*, including rapid extensibility to great elongations, high stiffness and strength when stretched, and rapid and complete retraction on release of the external stress. The properties of natural rubber are discussed further in comparison with those of the major synthetic elastomers in Chapter 19.

Polymers related to natural rubber

Hard rubber, ebonite   As mentioned in Chapter 19, the final product of the reaction of rubber with an excess of sulfur is a hard, inextensible solid called ebonite, containing about 32% combined sulfur. Ebonite was first made in the 1840's and has been produced on a commercial scale since about 1860.

Ebonite can be machined well and is often produced in bar, tube, or sheet stock for this purpose. Its chief uses depend upon its chemical inertness and corrosion resistance and its electrical and thermal insulating properties. The material softens at about 50°C, and hence is not suitable for high-temperature applications.

Chlorinated rubber   The reaction between chlorine and rubber has been known since 1801. It has traditionally been carried out by adding chlorine gas to a solution of rubber in a chlorinated solvent. To keep the viscosity of the solution low, the rubber is usually masticated beforehand. The chlorination is continued to completion, when the product contains about 65% chlorine, or about 3.5 chlorine atoms per isoprene residue. Newer methods of producing chlorinated rubber involve the direct chlorination of latex or the passage of chlorine over thin sheets of rubber swollen with a solvent such as carbon tetrachloride.

The mechanism of the chlorination reaction and the structure of chlorinated rubber are not known in detail. Several mechanisms have been proposed to account for the final composition involving seven chlorines per two isoprene residues. It is likely that substitution, addition, and cyclization are all involved.

Chlorinated rubber is used chiefly in the production of heat- and chemical-resistant paints, varnishes, and lacquers. Films, impregnating solutions, adhesives, and (with the addition of plasticizers) molding powders can also be made.

Rubber hydrochloride   Hydrogen chloride adds to the double bonds of rubber to give a material of the structure

$$-CH_2-\underset{\underset{Cl}{|}}{\overset{\overset{CH_3}{|}}{C}}-CH_2-CH_2-$$

The chief use of rubber hydrochloride is in film for protective wrapping purposes. It finds extensive application for wrapping foods and precision machines and machine parts.

Oxidized rubber Rubber can be oxidized in a controlled way by mastication in air in the presence of a catalyst. Various grades contain 0.25–1.0 oxygen atom per isoprene residue, distributed among hydroperoxide, acid, ester, ketone, alcohol, and epoxide groups. The material is useful for impregnating paper and cardboard and for protective coatings. It has advantages over chlorinated rubber in that its residual unsaturation can be used for vulcanization or thermal hardening reactions. Varnishes prepared with oxidized rubber have excellent electrical insulating properties.

Cyclized rubber When rubber is heated slowly or treated with acidic reagents, it becomes hard and brittle. Its unsaturation is reduced, but the empirical formula $(C_5H_8)_x$ remains unchanged. The changes are thought to result from the condensation of isoprene residues in pairs to give cyclic structures. Cyclized rubbers are made commercially by treating rubber with either sulfuric acid or various sulfonic acids or sulfonyl chlorides, or chlorostannic acid. The products are nonelastic. They are used primarily as compounding ingredients in shoe soles and heels and for rubber-to-metal bonding adhesives.

Gutta-percha and balata These two natural resins or gums come from trees indigenous to Malaya and Central America, respectively. Both are impure forms of *trans*-polyisoprene. The purified hydrocarbon polymer is useful for cable covering, tissue for adhesive and surgical purposes, and golf-ball covers.

Stereoregular synthetic polyisoprene

In 1955, production began of essentially *cis*-1,4-polyisoprenes with structures closely duplicating that of natural rubber. These polymers can be made by two processes which are almost identical except for the catalyst used: one is based on coordination polymerization (Chapter 10*D*), using a catalyst of titanium tetrachloride and an aluminum alkyl such as triisobutyl aluminum; the other is an anionic polymerization (Chapter 10*C*) with butyl lithium as the catalyst.

In a typical operation, isoprene (derived from petroleum) is mixed with a hydrocarbon solvent such as *n*-pentane. The catalyst is added, and the reaction allowed to take place at about 50°C and moderate pressures until a solids content of about 25% is reached. At this point the reaction mixture is a highly viscous "cement." A catalyst deactivator and an antioxidant are added, and solvent is removed in an extruder or a drum dryer. The polymer made by coordination polymerization has a molecular weight distribution similar to that of masticated natural rubber and requires no further processing before compounding, whereas that made with butyl lithium is higher in molecular weight and requires even longer mastication than natural rubber.

The properties of the *cis*-1,4-polyisoprenes are, commensurate with their structure, very nearly identical with those of natural rubber, and the synthetic polymers are not only complete replacements for the natural product, but are often preferred because of their greater cleanliness and uniformity. Production has, however, remained relatively small, around 175 million lb. in 1969.

GENERAL REFERENCES

Bean 1967; Kennedy 1968–1969; Anderson 1969; Carpenter 1970; Cunneen 1970.

F. Rubbers Derived from Butadiene

Although also made from alcohol during World War II, butadiene is now derived exclusively from petroleum. Fractionation of the products of cracking petroleum, either for producing olefins or for obtaining high-octane gasoline, yields a cut containing largely hydrocarbons of the butane and butene family. 1-Butene is separated and catalytically dehydrogenated in the vapor phase to butadiene.

Styrene-butadiene rubber (SBR)

Production of SBR (then known as GR-S) was begun in the United States during World War II. The product was designed to be similar to the German Buna-S (Chapter 19) but lower in molecular weight for easier processing. The rubber was made by emulsion polymerization using the so-called Mutual recipe (Table 12-3), at 50°C. After the war, product quality was improved by carrying out the polymerization at 5°C (41°F) with some being made at temperatures as low as − 10° or − 18°C. These changes were brought about by the use of more active initiators, such as cumene hydroperoxide and *p*-menthane hydroperoxide, and the addition of antifreeze components to the mixture. The product is known as *cold rubber*.

Recently, anionic solution copolymerization of butadiene and styrene with alkyllithium catalysts has been used to produce so-called *solution SBR*.

This product has a narrower molecular-weight distribution, higher molecular weight, and higher *cis*-1,4-polybutadiene content than emulsion SBR. Tread wear and crack resistance are improved, as is economy because oil extension and carbon-black loading can be increased. It seems likely that as the emulsion plants are replaced, new production of SBR will be by the solution process.

Consumption of SBR in the United States remains at about 50% of all rubber use. Total consumption has risen steadily despite significant drops in percentage use in the past decade. Some 2.9 billion lb. was used in 1969 at prices around $0.17–0.18/lb.

Structure of SBR   By virtue of its free radical polymerization, SBR is a random copolymer. The butadiene units are found to be about 20% in the 1,2 configuration, 20% in the *cis*-1,4, and 60% in the *trans*-1,4 for polymer made at 50°C, with the percentage of *trans*-1,4 becoming higher for polymer made at lower temperatures. In consequence of its irregular structure, SBR does not crystallize.

Branching reactions due to chain transfer to polymer and to polymerization of both double bonds of a diene unit become extensive if conversion is allowed to become too high or a chain transfer agent is not used in SBR polymerization. However, SBR has been shown to have exactly one double bond per butadiene unit. Thus no extensive side reactions occur during its formation, at least up to about 75% conversion.

Processing of SBR   In general, the differences in mastication and vulcanization between SBR and natural rubber are minor. A reinforcing filler is essential to the achievement of good physical properties in SBR. However, some fillers other than carbon black reinforce it moderately well. SBR is compatible with the other major elastomers and can be used in blends. The techniques of oil extension and masterbatching are widely employed. These techniques, as well as the processing of SBR in general, are discussed further in Chapter 19.

Properties of SBR   Tire tread stocks made from regular SBR are inferior in tensile strength to those from natural rubber (3000 vs. 4500 psi), whereas those from "cold rubber" are almost equivalent to *Hevea* (3800 psi). At elevated temperatures, however, regular and "cold" SBR lose almost two-thirds of their tensile strength whereas natural rubber loses only 25%. The ozone resistance of SBR is superior to that of natural rubber, but when cracks or cuts start in SBR they grow much more rapidly. Perhaps the most serious defect of both types of SBR for tire uses, however, is its poorer resilience and greater heat buildup. Tread wear of the synthetic material is at least as good as that of natural rubber. The weatherability of SBR is better than that of natural rubber.

"Cold" SBR is superior to the standard product because it contains less low-molecular-weight (nonreinforceable) rubber, less chain branching and crosslinking, and a higher proportion (70%) of the trans-1,4 configuration around the double bond.

Applications of SBR   Originally, SBR was used in tires only of necessity, but "cold" SBR appears equal in most respects to natural rubber, especially for lighter-duty tire use. It is definitely inferior to the natural product for truck tires. For many mechanical goods SBR is superior to natural rubber and is preferentially used because of its easier processing and good-quality end product. Such items include belting, hose, molded goods, unvulcanized sheet, gum, and flooring. Rubber shoe soles are now almost universally made of SBR. Extruded goods and coated fabrics are other fields in which SBR offers advantages in processability. It is widely used for electrical insulation, although its properties are not as good as those of butyl rubber.

Nitrile rubbers

The nitrile rubbers are polymers of butadiene and acrylonitrile, having ratios of the two monomers similar to the ratio of butadiene to styrene in SBR. They are noted for their oil resistance but are not suitable for tires. The first commercial nitrile rubbers were made in Germany.

The oil resistance of the nitrile rubbers varies greatly with their composition. Several grades are available commercially, ranging in acrylonitrile content between 18% (only fair oil resistance) and 40% (extremely oil resistant).

The nitrile rubbers are prepared in emulsion systems similar to those used for SBR. Because the monomer reactivity ratios are quite different, the compositions of the feed and the polymer differ markedly. This fact is usually taken into account by adjusting the monomer composition during the polymerization to achieve the desired polymer composition.

Nitrile rubber is used primarily for its oil resistance, by which is implied its low solubility, low swelling, and good tensile strength and abrasion resistance after immersion in gasoline or oils. Swelling of nitrile rubbers is greater in polar solvents than in nonpolar solvents, but they can be used in contact with water and antifreeze solutions. Resistance to ethylene glycol is good. The rubbers are inherently less resilient than natural rubber. Their heat resistance is good; and, if properly protected by antioxidants, they show satisfactory resistance to oxidative degradation as well.

Nitrile rubbers are extensively used for gasoline hoses, fuel tanks, creamery equipment, and the like. In addition they find wide application in adhesives and, in the form of latex, for impregnating paper, textiles, and leather. In all, about 150 million lb was used in the United States in 1969, at a price of about $0.55/lb.

Stereoregular polybutadienes

cis-1,4-Polybutadiene This polymer is made by coordination or anionic polymerization in the same processes used for *cis*-1,4-polyisoprene (Section *D*) and has similar properties. It is utilized almost entirely in tires, blended with SBR and natural rubber. A small amount of natural rubber appears to be required to prevent the polybutadiene from crumbling during processing and to improve tack. Polybutadiene has high elasticity, low heat buildup, and good resistance to oxidation. It imparts outstanding abrasion resistance to truck and passenger tires, but cannot be used at levels higher than 40–50%, above which its major deficiency of poor skid resistance becomes too apparent. Use in 1969 was about 570 million lb in the United States.

trans-1,4-Polybutadiene By proper selection of catalyst, the processes leading to stereoregular diene polymers can produce a polybutadiene with about 90% *trans*-1,4 units. This polymer has the toughness, resilience, and abrasion resistance of the natural *trans*-polyisoprenes, balata and gutta-percha.

GENERAL REFERENCES

Saltman 1965; Kennedy 1968–1969; Anderson 1969.

G. Other Synthetic Elastomers

Polyisobutylene and butyl rubber (Buckley 1965; Friedlander 1965)

Butyl rubbers are copolymers of isobutylene with a small amount of isoprene added in order to make them vulcanizable. Since the amount of comonomer is small, the methods of polymerization and the properties of the unvulcanized copolymer are similar to those of polyisobutylene itself.

Polymerization The polymerization of isobutylene and its mixtures with diolefins typifies the industrial application of low-temperature cationic polymerization (Chapter 10B). Isobutylene polymerizes rapidly at $-80°C$ with Friedel-Crafts catalysts. In a bulk system at $-80°C$, rapid polymerization is induced by bubbling BF_3 gas through isobutylene. The heat of reaction can be absorbed by adding solid carbon dioxide to the monomer or by adding a low-boiling diluent such as pentane or ethylene which is refluxed. In a typical process, butyl rubber is manufactured by mixing isobutylene with 1.5–4.5% isoprene and methyl chloride as diluent. This mixture is fed to stirred reactors cooled to $-95°C$ by liquid ethylene. Catalyst solution, made by dissolving anhydrous aluminum chloride in methyl chloride, is added. Polymer forms at once as a finely divided product suspended in the reaction mixture. This slurry is pumped out of the reactor continuously as monomer and catalyst are added. The product mixture is passed into a large volume of hot agitated water in a tank where the volatile components are flashed off and recovered. An antioxidant and some zinc stearate to prevent agglomeration of the polymer particles are added at this point. The polymer is then filtered off, dried, and extruded.

Structure   Polyisobutylene and butyl rubber are amorphous under normal conditions, but crystallize on stretching. Most, if not all, of the isoprene units are present in the 1,4 structure. It is usually assumed that the polymers are linear, although the possibility of a small amount of branching has not been investigated in detail. The molecular weights of the polymers made by low-temperature ionic polymerization can be quite high; chain transfer agents such as diisobutylene are often added to control molecular weight at the 200,000–300,000 level.

Unstabilized polyisobutylenes are degraded by heat or light to sticky low-molecular-weight products. The usual rubber antioxidants or retarders of free radical reactions stabilize the polymers well.

Low-molecular-weight polyisobutylenes are liquids. As molecular weight increases, they change to balsamlike solids, and then to rubberlike polymers. Unless low-molecular-weight material is removed, even polyisobutylene of 100,000 molecular weight is sticky. Unlike natural rubber, polyisobutylene and butyl rubber do not crystallize on cooling and hence remain flexible to as low as $-50°C$.

The response of isobutylene polymers and copolymers to stress is quite different from that of natural rubber. Polyisobutylene is sluggish, showing large viscoelastic and viscous components in its response (Chapter 6C). The polymer has been widely used in the study of viscoelasticity.

The strong tendency toward cold flow in polyisobutylene prevents its direct application as an elastomer. It is used in adhesives, caulking compounds, pressure-sensitive tapes, and coatings for paper.

Properties The properties of butyl vulcanizates are compatible with their structure. The very low residual unsaturation of the rubbers leads to outstanding chemical inertness. The closepacked linear paraffinic chains result in unusually low permeability to gases. Steric hindrance of the methyl groups on the chains causes high internal viscosity and viscoelastic response to stresses.

A property of great importance in elastomers is aging in the presence of oxygen. It has been found that the presence of a double bond in the skeletal structure of a polymer is very important in enhancing the rate of absorption of oxygen, and that the presence of methyl side groups is also significant but less so. Butyl rubber is therefore, as expected, less sensitive to oxidative aging than are most other elastomers except the silicones. Since methyl side groups appear to favor chain scission whereas double bonds favor crosslinking, butyl rubber becomes soft rather than brittle on oxidative degradation.

Butyl has much better ozone resistance than natural rubber. Its solvent resistance is typical of that of hydrocarbon elastomers. Its acid resistance is quite good.

The stress-strain properties of butyl rubber are similar to those of natural rubber. Both show the importance of crystallization in obtaining high tensile strength. Crystallization does not take place in butyl, however, until higher elongations are reached. The tear resistance of butyl is quite good and is retained well at high temperatures and for long times, in contrast to natural rubber. The electrical properties are quite good, as predicted from its nonpolar, saturated nature. The dynamic and elastic properties of butyl are marked by sluggishness over the temperature range $-30°$ to $+40°C$, characteristic of a polymer with high internal friction and high damping power. Rebound is slow and heat buildup is high.

Applications About 75% of the butyl rubber produced is used for inner tubes for tires. This usage was substantially reduced when tubeless passenger tires were introduced in 1953, but to date all nonpassenger tires still use tubes. Other uses for butyl are mainly in the area of mechanical goods.

Consumption of butyl rubber in the United States was about 200 million lb in 1969.

Polychloroprene (neoprene) (Hargreaves 1965)

The generic term neoprene denotes rubberlike polymers and copolymers of chloroprene, 2-chloro-1,3-butadiene. Neoprenes were the first synthetic rubbers developed in the United States. Although they are primarily known for their oil resistance, they are good general-purpose rubbers which can replace natural rubber in most of its uses and are satisfactory in a wide variety of applications. About 260 million lb was used in the United States in 1969, at a price of $0.42–0.45/lb.

Chloroprene is prepared by the catalytic addition of hydrogen chloride to vinylacetylene, which in turn is made by the catalytic dimerization of acetylene. This is an expensive process, however, and interest is turning to the production of chloroprene from butadiene. One process utilizes chlorination to 3,4-dichlorobutene-1, followed by dehydrochlorination.

The neoprenes are produced by emulsion polymerization. Some types are polymerized in the presence of sulfur, which introduces some crosslinking in the polymer. In these cases, the latex is allowed to age in the presence of an emulsion of tetraethylthiuram disulfide, which restores the plasticity of the polymer. The latex is then coagulated by acidification followed by freezing.

Polymerization appears to take place almost entirely in the *trans*-1,4 form. As a result the neoprenes are crystallizable elastomers.

The vulcanization of neoprene is different from that of other elastomers in that it can be vulcanized by heat alone. Zinc oxide and magnesium oxide are the preferred vulcanizing agents. The mechanism by which they cause crosslinking is not known. Sulfur vulcanizes neoprene very slowly, and the usual rubber accelerators are in general not effective—some, in fact, are potent retarders of the cure. A few chemicals are known to accelerate the vulcanization, however, among them antimony sulfide.

Unlike many other elastomers, neoprene vulcanizates have high tensile strength (3500–4000 psi) in the absence of carbon black. No reinforcing effect is found with any filler. Suitably protected neoprene vulcanizates are extremely resistant to oxidative degradation. Weathering resistance and ozone resistance are quite good. Neoprene is slightly inferior to nitrile rubber in oil resistance, but markedly better than natural rubber, butyl, or SBR. The dynamic properties of neoprene are superior to those of most other synthetics and only slightly inferior to those of natural rubber. They are less affected by elevated temperature than those of natural rubber. Neoprene has been shown to make excellent tires but cannot compete with other elastomers in price. Its major uses include wire and cable coatings, industrial hoses and belts, shoe heels, and solid tires. Gloves and coated fabrics are made from neoprene latex.

Ethylene-Propylene-Diene Elastomers (Raff 1967)

Elastomeric terpolymers of ethylene, propylene, and diene monomers (EPDM) were introduced shortly after stereoregular polyisoprene and polybutadiene. They have not yet achieved a major place in the spectrum of synthetic rubbers, United States consumption in 1969 being estimated at about 130 million lb.

Although a small amount of ethylene-propylene copolymer is made for elastomer use with peroxide curing, the addition of a third, diene, monomer is required to give curing rates commensurate with that of SBR. Currently, it is in blends with SBR that EPDM rubbers are thought to stand the best chance of commercial success. A variety of diene monomers is used, with ethylidene norbornene and 1,4-hexadiene most widely used at this time.

EPDM elastomers have several desirable properties, including resistance to weathering and capability of being highly extended with oil and loaded with carbon black. Their ultimate success will probably depend largely on economics; the third monomer is expensive, resulting in a 1969 selling price of about $0.30/lb.

Other elastomers

Several other elastomers have the status of specialty rubbers, at relatively low production volume and high price. Among these are chlorosulfonated polyethylene (Section *D*), the acrylate and various fluorocarbon elastomers (Chapter 14), polyurethane and polysulfide elastomers (Chapter 15), and silicone and epichlorhydrin rubbers (Chapter 16).

GENERAL REFERENCES

Kennedy 1968–1969; Anderson 1969.

BIBLIOGRAPHY

Anderson 1969. Earl V. Anderson, "Rubber," *Chem. & Eng. News* **47**, (29), 39–83 (July 14, 1969).

Bean 1967. Arthur R. Bean, Glenn R. Hines, Geoffrey Holden, Robert R. Houston, John A. Langlon, and Roger H. Mann, "Isoprene Polymers," pp. 782–855 in Herman F. Mark, Norman G. Gaylord, and Norbert M. Bikales, eds., *Encyclopedia of Polymer Science and Technology*, Vol. 7, Interscience Div., John Wiley and Sons, New York, 1967.

Buckley 1965. D. J. Buckley, "Butylene Polymers," pp. 754–795 in Herman F. Mark, Norman G. Gaylord, and Norbert M. Bikales, eds., *Encyclopedia of Polymer Science and Technology*, Vol. 2, Interscience Div., John Wiley and Sons, New York, 1967.

Buckley 1968. D. J. Buckley, B. S. Dyer, M. R. Day, and W. R. Bergenn, "Olefin Polymers," pp. 440–458 in Herman F. Mark, Norman G. Gaylord, and Norbert M. Bikales, eds., *Encyclopedia of Polymer Science and Technology*, Vol. 9, Interscience Div., John Wiley and Sons, New York, 1968.

Carpenter 1970. Ernest L. Carpenter, "Rubber in the 70's," *Chem. & Eng. News* **48** (18), 31–57 (April 27, 1970).

Chemplex 1969. Chemplex Co., "Polyethylene film and sheeting," pp. 348–349 in Sidney Gruss, ed., *Modern Plastics Encyclopedia 1969–1970* (McGraw-Hill Book Co., New York), **46** (10A), October, 1969.

Cunneen 1970. J. I. Cunneen, D. Barnard, F. McL. Swift, A. R. Payne, M. Porter, A. Schallamach, W. A. Southorn, and A. G. Thomas, "Rubber, Natural," in press in Herman F. Mark, Norman G. Gaylord, and Norbert M. Bikales, eds., *Encyclopedia of Polymer Science and Technology*, Vol. 12, Interscience Div., John Wiley and Sons, New York, 1970.

Dow 1969. Dow Chemical Co., "Polyethylene," pp. 181, 184–185 in Sidney Gruss, ed., *Modern Plastics Encyclopedia 1969–1970* (McGraw-Hill Book Co., New York), **46** (10A), October, 1969.

Du Pont 1969a. Du Pont Company, "Ionomer Resin," p. 140 in Sidney Gruss, ed., *Modern Plastics Encyclopedia 1969–1970* (McGraw-Hill Book Co., New York), **46** (10A), October, 1969.

Du Pont 1969b. Du Pont Company, "Ionomer foam," p. 261 in Sidney Gruss, ed., *Modern Plastics Encyclopedia 1969–1970* (McGraw-Hill Book Co., New York), **46** (10A), October, 1969.

Frank 1968. H. P. Frank, *Polypropylene*, Gordon and Breach, New York, 1968.

Friedlander 1965. Henry Z. Friedlander, "Chemically Resistant Polymers," pp. 665–683 in Herman F. Mark, Norman G. Gaylord, and Norbert M. Bikales, eds., *Encyclopedia of Polymer Science and Technology*, Vol. 3, Interscience Div., John Wiley and Sons, New York, 1965.

Gideon 1969. William P. Gideon, "Polyallomer Copolymers," pp. 165–166 in Sidney Gruss, ed., *Modern Plastics Encyclopedia 1969–1970* (McGraw-Hill Book Co., New York), **46** (10A), October, 1969.

Hargreaves 1965. C. A. Hargreaves, II, and D. C. Thompson, "2-Chlorobutadiene Polymers," pp. 705–730 in Herman F. Mark, Norman G. Gaylord, and Norbert M. Bikales, eds., *Encyclopedia of Polymer Science and Technology*, Vol. 3, Interscience Div., John Wiley and Sons, New York, 1965.

Howsmon 1969. John Q. Howsmon, "Polypropylene film," p. 354 in Sidney Gruss, ed., *Modern Plastics Encyclopedia 1969–1970* (McGraw-Hill Book Co., New York), **46** (10A), October, 1969.

Jezl 1969. James L. Jezl and Earl M. Honeycutt, "Propylene Polymers," pp. 597–619 in Herman F. Mark, Norman G. Gaylord, and Norbert M. Bikales, eds., *Encyclopedia of Polymer Science and Technology*, Vol. 11, Interscience Div., John Wiley and Sons, New York, 1969.

Keeley 1959. F. Wayne Keeley, "Hypalon Synthetic Rubber," Chap. 14 in M. Morton, ed., *Introduction to Rubber Technology*, Reinhold Publishing Corp., New York, 1959.

Kennedy 1968–69. Joseph P. Kennedy and Erik Törnquist, eds., *Polymer Chemistry of Synthetic Elastomers*, Interscience Div., John Wiley and Sons, New York; *Part I*, 1968; *Part II*, 1969.

Kinsey 1969. Roy H. Kinsey, "Ionomers, Chemistry and New Developments," *Appl. Poly. Symp.* **11**, 77–94 (1969).

Kresser 1969. Theodore O. G. Kresser, *Polyolefin Plastics*, Van Nostrand-Reinhold, New York, 1969.

Laing 1969. J. S. Laing and D. S. Pavlansky, "Low density polyethylene foam," p. 259 in Sidney Gruss, ed., *Modern Plastics Encyclopedia 1969–1970* (McGraw-Hill Book Co., New York), **46** (10A), October, 1969.

Lasman 1969. Henry R. Lasman and Leo G. Van Beaver, "Crosslinked polyethylene foam," pp. 259–260 in Sidney Gruss, ed., *Modern Plastics Encyclopedia 1969–1970* (McGraw-Hill Book Co., New York), **46** (10A), October, 1969.

Levett 1970. C. T. Levett, J. E. Pritchard, and R. J. Martinovich, "A New Low-Density Polyethylene," *SPE J.* **26** (6), 40–43 (1970).

McGrew 1958. Frank C. McGrew, "The Polyolefin Plastics Field—Present Technical Status," *Modern Plastics* **35** (7), 125, 126, 128, 132–133 (March, 1958).

Nelson 1969. R. E. Nelson, "Foamable ethylene copolymers," pp. 260–261 in Sidney Gruss, ed., *Modern Plastics Encyclopedia 1969–1970* (McGraw-Hill Book Co., New York), **46** (10A), October, 1969.

Raff 1965–66. R. A. V. Raff and K. W. Doak, eds., *Crystalline Olefin Polymers*, Interscience Div., John Wiley and Sons, New York; *Part I*, 1965; *Part II*, 1966.

Raff 1967. R. A. V. Raff, P. E. Campbell, R. V. Jones, E. D. Caldwell, Herbert N. Friedlander, and Peter J. Canterino, "Ethylene Polymers," pp. 275–454 in Herman F. Mark, Norman G. Gaylord, and Norbert M. Bikales, eds., *Encyclopedia of Polymer Science and Technology*, Vol. 6, Interscience Div., John Wiley and Sons, New York, 1967.

Rebenfeld 1967. Ludwig Rebenfeld, "Fibers," pp. 505–573 in Herman F. Mark, Norman G. Gaylord, and Norbert M. Bikales, eds., *Encyclopedia of Polymer Science and Technology*, Vol. 6, Interscience Div., John Wiley and Sons, New York, 1967.

Rubin 1968. I. D. Rubin, *Poly(1-Butene): Its Preparation and Properties*, Gordon and Breach, New York, 1968.

Saltman 1965. W. M. Saltman, "Butadiene Polymers," pp. 678–754 in Herman F. Mark, Norman G. Gaylord, and Norbert M. Bikales, eds., *Encyclopedia of Polymer Science and Technology*, Vol. 2, Interscience Div., John Wiley and Sons, New York, 1965.

Schramm 1969. J. N. Schramm, Sr., "Chlorinated polyethylene," p. 185 in Sidney Gruss, ed., *Modern Plastics Encyclopedia 1969–1970* (McGraw-Hill Book Co., New York), **46** (10A), October, 1969.

Staudinger 1936. H. Staudinger, "The Formation of High Polymers of Unsaturated Substances," *Trans. Faraday Soc.* **32**, 97–121 (1936).

Walton 1969. R. J. Walton, "Polypropylene," pp. 185, 188, 194 in Sidney Gruss, ed., *Modern Plastics Encylopedia 1969–1970* (McGraw-Hill Book Co., New York), **46** (10A), October, 1969.

Westall 1969. J. Westall, "Methylpentene polymer," pp. 150–151, 153 in Sidney Gruss, ed., *Modern Plastics Encyclopedia 1969–1970* (McGraw-Hill Book Co., New York), **46** (10A), October, 1969.

14

Other Carbon-Chain Polymers

This chapter includes discussion of polymers, with a carbon-carbon backbone chain, other than those consisting entirely or almost entirely of hydrocarbons, for which see Chapter 13. Primarily, the polymers discussed in this chapter are made from vinyl monomers, $CH_2 = CHX$, and vinylidene monomers, $CH_2 = CY_2$, plus the fluorocarbons, excepting those thermosetting resins made in part from vinyl monomers, such as "polyester" resins (Chapter 16C). It should be noted that throughout this chapter the term "vinyl" is used in the sense of monosubstituted ethylenic, while in the plastics industry (and in Section D) the phrase "vinyl resins" commonly refers to poly(vinyl chloride) and related plastics.

A. Polystyrene and Related Polymers

The family of styrene polymers includes polystyrene, copolymers of styrene with other vinyl monomers, polymers of derivatives of styrene, and mixtures of polystyrene and styrene-containing copolymers with elastomers. Total production of these polymers exceeded 3.2 billion lb in 1969, including some 510 million lb of ABS resins but not including SBR, described in Chapter 13F. The markets for the styrene plastics (excluding ABS resins) were approximately as follows: packaging (including foams and films), 40%; appliances, 20%; housewares and toys, 10% each; and home furnishings, lighting fixtures, construction materials, and miscellaneous uses, 5% each.

Polystyrene has been a low-cost resin for many years, this fact accounting in no small part for its widespread use. In 1970, it sold for about $0.15/lb. The various copolymers and modified resins described in this section range in price from \$0.17/lb for impact (rubber-modified) polystyrene to \$0.22–0.40/lb for the ABS resins.

Polystyrene is a thermoplastic with many desirable properties. It is clear, transparent, easily colored, and easily fabricated. It has reasonably good mechanical and thermal properties, but is slightly brittle and softens below 100°C.

Monomer   Styrene (vinylbenzene), $\langle\bigcirc\rangle CH{=}CH_2$, is made from benzene and ethylene. In one process, ethylene is passed into liquid benzene under pressure, in the presence of aluminum chloride catalyst:

$$\langle\bigcirc\rangle + CH_2{=}CH_2 \xrightarrow[90°C]{AlCl_3} \langle\bigcirc\rangle CH_2CH_3$$

The resulting ethylbenzene is dehydrogenated to styrene by passing it over an iron oxide or magnesium oxide catalyst at about 600°C. The styrene is then refined by distillation.

Polymerization   Although solution or emulsion polymerization may occasionally be used, most polystyrene is made either by suspension polymerization or by polymerization in bulk. All but the latter process are typically carried out as described in Chapter 12.

The bulk polymerization of styrene is begun in a "prepolymerizer," a stirred vessel in which inhibitor-free styrene is polymerized (usually with a peroxide initiator) until the reaction mixture is as concentrated in polymer as is consistent with efficient mixing and heat transfer. Normally, a solution containing about 30% polymer is as viscous as can be handled.

The syrupy mixture from the prepolymerizer then enters a cylindrical tower (about 40 ft long by 15 ft diameter) maintained essentially full of fluid. By cooling the upper part of the tower and heating the lower part, the polymerization is controlled to prevent runaways but to proceed to essentially pure molten polymer at the bottom. This melt is discharged through spinnerets or into an extruder producing small-diameter rod which is chopped, after cooling, into short lengths to provide the finished molding powder.

Structure and properties   Polystyrene is a linear polymer, the commercial product being atactic and therefore amorphous. Isotactic polystyrene can be produced, but offers little advantage in properties except between the glass transition (about 80°C) and its crystalline melting point (about 240°C), where it is much like other crystalline plastics. Isotactic polystyrene is not available commercially.

Like most polymers, polystyrene is relatively inert chemically. It is quite resistant to alkalis, halide acids, and oxidizing and reducing agents. It can be nitrated by fuming nitric acid, and sulfonated by concentrated sulfuric acid at 100°C to a water-soluble resin (see ion-exchange resins, page 407). Chlorine and bromine are substituted on both the ring and the chain at elevated temperatures. Polystyrene degrades at elevated temperatures to a mixture of low-molecular-weight compounds about half of which is styrene. The characteristic odor of the monomer serves as an identification for the polymer.

As made, polystyrene is outstandingly easy to process. Its stability and flow under injection-molding conditions make it an ideal polymer for this technique. Its optical properties—color, clarity, and the like—are excellent, and its high refractive index (1.60) makes it useful for plastic optical components. Polystyrene is a good electrical insulator and has a low dielectric loss factor at moderate frequencies. Its tensile strength reaches about 8000 psi.

On the other hand, polystyrene is readily attacked by a large variety of solvents, including dry-cleaning agents. Its stability to outdoor weathering is poor; it turns yellow and crazes on exposure. Two of its major defects in mechanical properties are its brittleness and its relatively low heat-deflection temperature of 82–88°C, which means that polystyrene articles cannot be sterilized.

Many of these defects can be overcome by proper formulating, or by copolymerization and blending as described below. For example, the addition of ultraviolet-light absorbers improves the light stability of polystyrene enough to make it useful in lighting fixtures such as fluorescent-light diffusers. Flame-retardant polystyrenes have been developed through the use of additives.

Copolymers of styrene

Butadiene copolymers   The most important of all copolymers of styrene in terms of volume are the styrene-butadiene synthetic rubbers discussed in Chapter 13F. Another group of styrene-butadiene copolymers is widely used in latex paints; these fall in the composition range 60 styrene:40 butadiene by weight.

Block copolymers of styrene and butadiene, prepared by anionic polymerization, form a new and interesting class of materials (Moacanin 1969). These are thermoplastic rubbers, showing the behavior of vulcanized elastomers at room temperature, with hard blocks of styrene acting like crosslinks to prevent creep in the polybutadiene block matrix. At higher temperatures they undergo normal plastic flow.

Heat- and impact-resistant copolymers A number of copolymers of styrene with a minor amount of comonomer have enhanced heat and impact resistance without loss of other desirable properties of polystyrene. Typical comonomers are those which increase intermolecular forces of attraction by introducing polar groups, or those which stiffen the chain and reduce rotational freedom through the steric hindrance of bulky side groups. In the first class are acrylonitrile (CH_2=CHCN), fumaronitrile (*trans*-NCCH=CHCN), and 2,5-dichlorostyrene. Comonomers with bulky side groups include N-vinylcarbazole and N,N-diphenylacrylamide.

The products of major commercial interest in this group contain, typically, 76% styrene and 24% acrylonitrile. Such copolymers have heat-deflection temperatures of 90–92°C and more resilience and impact resistance than polystyrene. Their color, however, is slightly yellow.

Ion-exchange resins Cationic type ion-exchange resins are produced by making a suspension polymerization of styrene with several per cent divinylbenzene. The product, in the form of uniform spheres, is sulfonated to the extent of about one —SO_3H group per benzene ring. Anionic ion-exchange resins are made by copolymerizing styrene with divinylbenzene and vinylethylbenzene. The polymer is treated with chloromethyl ether to put chloromethyl groups on the benzene rings. Quaternary ammonium salts are formed by reaction with tertiary amines such as trimethylamine, methyldiethanolamine, and dimethylpropanolamine.

Copolymers of styrene with methyl methacrylate are discussed in Section *B*.

Polymers of styrene derivatives

Many styrene derivatives have been synthesized, but only a few have found commercial importance. Poly(α-methyl styrene), with better heat resistance than polystyrene, has failed to survive because of brittleness and difficulty in fabrication due to a low ceiling temperature of 61°C. Poly(*p-tert*-butylstyrene) is used as a viscosity improver in motor oils, polychlorostyrene is valuable because of its self-extinguishing characteristics, and poly(sodium styrenesulfonate) is used as a water-soluble flocculating agent.

Rubber-modified polystyrene

Rubber is incorporated into polystyrene primarily to impart toughness. The resulting materials consist of a polystyrene matrix with small inclusions of the rubber (usually 5–10% polybutadiene or copolymer rubber). They are termed *impact polystyrene*, and account for over half of the polystyrene homopolymer produced. Grafting of the rubber to the polysty-rene may occur if the rubber is present during the styrene polymerization; these materials are the most effective in enhanced impact strength, particularly if the rubber is slightly crosslinked; but the mechanical blends are also used.

ABS resins

Like the rubber-modified polystyrenes, ABS resins are two-phase systems consisting of inclusions of rubber in a continuous glassy matrix. In this case the matrix is a styrene-acrylonitrile copolymer, and the rubber a styrene-butadiene copolymer, the name ABS deriving from the initials of the three monomers. Again, development of the best properties requires grafting between the glassy and rubbery phases. The ABS resins have higher temperature resistance and better solvent resistance than the high-impact polystyrenes and are true engineering plastics, particularly suitable for high-abuse applications. They can easily be decorated by painting, vacuum metalizing, and electroplating. In addition to fabrication by all the usual plastics techniques (Chapter 17), the ABS resins can be cold formed, a technique typical of metal fabrication.

Glass reinforcement   Reinforcement of ABS resins, and also high-impact polystyrene and styrene homopolymer and copolymers, leads to further increases in strength, toughness, and stiffness, and significant reduction in creep and relaxation rates. Normally, about 29% glass fibers, 0.25 in. long and 0.00035 in. diameter, is used. Several choices of blending processes for incorporating the glass are available. Fabrication is by injection molding, and many-pound pieces can be produced.

Polystyrene foams

Around 250–300 million lb of styrene homopolymer was used in 1969 to make a wide variety of foamed products. Most of these are based on foamed-in-place beads, made by suspension polymerization in the presence of a foaming agent such as pentane or hexane, liquid at polymerization temperature and pressure. Subsequent heating softens the resin and vola-tilizes the foaming agent.

Coumarone-indene resins

Although totally unrelated to styrene, these resins are considered here because of their aromatic nature and vinyl-like structure. They are produced from coal-tar oils recovered from coking operations. The oils contain coumarone (I), indene (II), and related unsaturated compounds. They are poly-

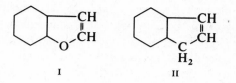

I   II

merized by the addition of sulfuric acid, presumably to a linear polymer with the structure

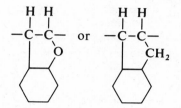

The polymers are low in molecular weight (about 1000 or below) and are completely fluid above their melting points. They are widely used in surface coatings, printing inks, adhesives, waxed papers, and floor tiles, and as softeners and tackifiers in the rubber industry.

Similar resins are made by hydrogenating coumarone-indene resins, or adding cresols, xylenols, or cyclopentadiene to the oil before polymerization. Terpenes and by-products of petroleum cracking and refining operations can be polymerized to similar products.

About 340 million lb of these resins was used in 1969.

GENERAL REFERENCES

Basdekis 1964; Brownell 1964; Abrams 1967; Collins 1969; Dow 1969; Hahn 1969; Hart 1969; Higginbottom 1969; Keskkula 1969; Rathbone 1969; Uniroyal 1969; Boyer 1970.

B. Acrylic Polymers

In this section are discussed the acrylate and methacrylate plastics, of which the most important is poly(methyl methacrylate), and the acrylic fibers, based largely on polyacrylonitrile.

Poly(methyl methacrylate) is a clear, colorless transparent plastic with a higher softening point, better impact strength, and better weatherability than polystyrene. It is available in molding and extrusion compositions, syrups, and cast sheets, rods, and tubes. Production of this polymer for molding and extrusion and in cast form in 1969 was estimated at 350 million lb, at a price of about $0.45/lb for molding compositions.

Monomers The principal commercial processes for the production of acrylate esters are based on ethylene cyanohydrin (much like the acetone cyanohydrin route to methyl methacrylate described below), the carbonylation of acetylene, or the polymerization and subsequent depolymerization of β-propiolactone to give acrylic acid or (in the presence of an alcohol) an acrylate ester.

Methyl methacrylate is made by heating acetone cyanohydrin (from the addition of hydrocyanic acid to acetone) with sulfuric acid to form methacrylamide sulfate. The latter is reacted (without separation) with water and methanol to give methyl methacrylate:

$$CH_3-\underset{\underset{CH_3}{|}}{\overset{\overset{OH}{|}}{C}}-CN \xrightarrow[125°C]{H_2SO_4} CH_2=\underset{\underset{O}{\|}}{\overset{\overset{CH_3}{|}}{C}}CNH_2 \cdot H_2SO_4 \xrightarrow[H_3O]{CH_2OH} CH_2=\underset{\underset{O}{\|}}{\overset{\overset{CH_3}{|}}{C}}COCH_3$$

Polymerization Poly(methyl methacrylate) for molding or extrusion is made by bulk or suspension polymerization, as described in Chapter 12. The production of cast sheets, rods, and tubes is carried out by bulk polymerization, starting in most cases with a syrup of partially polymerized methyl methacrylate with a convenient viscosity for handling. Shrinkage and heat evolution during polymerization are reduced by the use of a syrup.

Sheets of poly(methyl methacrylate) are commonly made by extrusion. Alternatively, they may be cast in cells made up of two sheets of heat-resistant glass separated by a coated rubber gasket. The cell is filled with syrup and sealed, and polymerization is carried out at 60–70°C in an air oven or water bath, with a finishing treatment at 100°C. Peroxide or azo initiators may be used. The sheets are easily removed by separating the glass plates after the cell is chilled in cold water. The plastic sheet may be given a heat treatment to relax residual strains and is protected from scratching by covering with adhesive paper. A few per cent of a plasticizer such as dibutyl phthalate may be used in the polymerization to improve the postforming properties of the sheet.

Poly(methyl methacrylate) rods may be formed by polymerizing syrup under pressure in cylindrical metal tubes. The tubes are heated in successive zones by lowering them vertically into hot water in order to control shrinkage and prevent the formation of bubbles. Tubes of the plastic have been prepared by polymerizing syrup inside heated metal cylinders rotated horizontally on their axis, using just enough syrup to cover the walls to the desired thickness and blanketing the interior with an inert gas.

Properties Poly(methyl methacrylate) is a linear thermoplastic, about 70–75% syndiotactic. Because of its lack of complete stereoregularity and its bulky side groups, it is amorphous. Both isotactic and syndiotactic poly-(methyl methacrylate) have been prepared (Chapter 5A, Chapter 10D) but have not been offered commercially. Poly(methyl methacrylate) is resistant to many aqueous inorganic reagents, including dilute alkalis and acids. It is quite resistant to alkaline saponification, in contrast to the polyacrylates. It undergoes pyrolysis almost completely to monomer by a chain reaction.

Perhaps the outstanding property of poly(methyl methacrylate) is its optical clarity and lack of color. Coupled with its unusually good outdoor weathering behavior, its optical properties make it highly useful in all applications where light transmission is important. These considerations apply to colored compositions as well. An unusually wide range of brilliant, light-fast colors is available in this plastic.

The mechanical and thermal properties of the polymer are good also. Tensile strength ranges as high as 10,000 psi. Impact strength is about equal to that of the impact-resistant styrene copolymers. Heat-deflection temperatures are above 90°C for the heat-resistant grades of poly(methyl methacrylate) molding powder. Electrical properties are good but not outstanding. Fabricability is quite good; only slightly higher temperatures are needed for molding poly(methyl methacrylate) than for polystyrene. Poly-(methyl methacrylate) is less susceptible to crazing than is polystyrene.

A limitation to the optical uses of the material is its poor abrasion resistance compared to glass. Despite considerable effort, attempts to improve the scratch resistance or surface hardness of poly(methyl methacrylate) have so far been accompanied by deterioration in other properties, such as impact strength.

Applications Automotive uses (largely in tail- and signal-light lenses, dials, and medallions) consume more than half the poly(methyl methacrylate) molding powder produced. Other applications of molding compositions include brush backs, jewelry, lenses, and small signs. Major uses for sheets, both cast and extruded, include signs, glazing, skylight, and decorative purposes in the building industry.

Copolymers of methyl methacrylate, ethyl acrylate, and monomers containing reactive functional groups are widely used as thermosetting resins in baked enamel applications. The functionality can be derived from amides (acrylamide, methacrylamide), acids (acrylic acid, methacrylic acid, or others), hydroxyls (hydroxyalkyl acrylates or methacrylates), or oxiranes (glycidyl methacrylate).

Other acrylic plastics

Higher methacrylates   The monomers of higher alkyl methacrylates are most conveniently prepared from methyl methacrylate by alcoholysis. The reaction is carried out at 150°C with an excess of the alcohol whose methacrylate ester is desired and a small amount of sulfuric acid catalyst. As described in Chapter 7B, the higher alkyl methacrylate polymers have glass transition temperatures lower than that of poly(methyl methacrylate). Poly(lauryl methacrylate) is widely used as a pour-point depressant and improver of viscosity-temperature characteristics for lubricating oils.

Polyacrylates   The rubbery, adhesive nature of the lower acrylates makes both suspension polymerization and casting less feasible than with the methacrylates. Consequently, solution and emulsion polymerizations are used commercially. Since the polyacrylates contain an easily removed tertiary hydrogen atom, they undergo some chain transfer to polymer when polymerized to high conversion. This leads to highly branched, less soluble materials.

Since the lower acrylate polymers have glass transition points below room temperature, they are typically soft and rubbery. As the size of the ester group increases, the polymers become harder, tougher, and more rigid. The polyacrylates have been used in finishes, in textile sizes, and in the manufacture of pressure-sensitive adhesives.

Copolymers of ethyl acrylate with a few per cent of a chlorine-containing monomer such as 2-chloroethyl vinyl ether have elastomeric properties. Vulcanization reactions apparently involve the chlorine atom, ester group, and α-hydrogens on the chain. Numerous rubber vulcanization agents and accelerators vulcanize these polymers. They are of interest because of their heat resistance and their excellent resistance to oxidation, allowing their use to over 180°C.

Poly(acrylic acid) and poly(methacrylic acid)   Acrylic acid, $CH_2\!=\!CHCOOH$, can be prepared directly from ethylene cyanohydrin. Methacrylic acid can be prepared from acetone cyanohydrin. Salts of the acids can be polymerized directly, then acidified to give poly(acrylic acid) and poly(methacrylic acid).

Alternatively, poly(methyl acrylate) or poly(methyl methacrylate) may be saponified to form the acids. Or, polymethacrylonitrile may be treated with hydrogen chloride to give poly(methacrylic acid). The polymers are insoluble in their monomers and in most organic liquids, but are soluble in water and very soluble in dilute bases. Poly(acrylic acid) and poly(methacrylic acid) are too water sensitive to serve as plastics. They are brittle when dry, and on heating do not become thermoplastic but crosslink, char, and decompose. In solution they show typical polyelectrolyte behavior, including abnormally high viscosities. Because of this property they are useful as thickening agents for latices and for adhesives.

Cyanoacrylate adhesives Methyl cyanoacrylate monomer is an extremely powerful adhesive. Adhesion occurs when the liquid monomer is spread in a thin layer between the surfaces to be bonded. Traces of bases (even as weak as alcohols or water) on the surfaces catalyze polymerization by an anionic mechanism. Adhesion arises in part from mechanical interlocking between polymer and surface, and in part from strong secondary bond forces.

Polyacrylamide This water-soluble polymer is used as a thickening agent and as a flocculant.

Copolymers and blends Copolymers of styrene and methyl methacrylate are less expensive than acrylic resins but have poorer weathering properties.

A copolymer of α-methylstyrene, 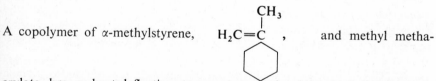 , and methyl metha-

crylate has a heat-deflection temperature and other physical properties similar to those of poly(methyl methacrylate). A modified acrylic polymer [possibly a blend of poly(methyl methacrylate) with a rubber or an olefin polymer] is said to have good impact properties and to be similar to the ABS resins (Section *A*).

Acrylic fibers

The *acrylic* fibers are polymers containing at least 85% acrylonitrile. Other monomers are often used in small amount to make the polymer amenable to dyeing with conventional textile dyes. Common comonomers are vinyl acetate, acrylic esters, and vinyl pyrrolidone. The last may be used as the homopolymer in a blend or grafted onto a copolymer backbone.

A generic class of fibers closely related to the acrylics is the *modacrylic* fibers, containing 35–85% acrylonitrile and, usually, 20% or more vinyl chloride or vinylidene chloride.

Polyacrylonitrile   Acrylonitrile can be made either by the direct catalytic addition of HCN to acetylene, or by the addition of HCN to ethylene oxide to give ethylene cyanohydrin, followed by dehydration. The monomer is soluble in water to the extent of about 7.5% at room temperature, and polymerization is usually carried out in aqueous solution by means of redox initiation. The polymer precipitates from the system as a fine powder.

During the time that other vinyl polymers were being developed commercially as plastics as well as fibers, polyacrylonitrile was considered a useless material because it could not be dissolved or plasticized. It softens only slightly below its decomposition temperature; and, because the polymer is insoluble in the monomer, it could not be polymerized into useful shapes by casting. This intractability was for a time attributed to crosslinking, but better knowledge of the requirements for solubility of polymers led to an extensive search (Houtz 1950) for molecules which might interact with the highly polar —C≡N groups and cause solution of the polymer. Several such solvents were discovered, among them dimethyl formamide and tetramethylene sulfone. Some concentrated aqueous solutions of salts, such as calcium thiocyanate, also dissolve polyacrylonitrile. With these solvents available, both wet- and dry-spinning techniques (Chapter 18) have been used for this polymer.

Properties of acrylic fibers   The acrylic fibers exhibit the properties of high strength, stiffness, toughness, abrasion resistance, resilience, and flex life associated with the synthetic fibers as a class. They are relatively insensitive to moisture and have good resistance to stains, chemicals, insects, and fungi. Their weatherability is outstandingly good. In continuous filament form, they are considered to have superior "feel," while as crimped staple they are noted for bulkiness and a woollike "hand."

In many respects the acrylic fibers are similar to the chlorine-containing fibers described in Section *D*, and to the condensation fibers discussed in Chapter 15. Further information on fibers and fiber technology is found in Chapters 15 and 18.

Production of acrylic fibers in 1969 was about 535 million lb.

GENERAL REFERENCES

Bamford 1964; Davis 1964; Fram 1964; Luskin 1964; Miller 1964; Thomas 1964; Rebenfeld 1967; Beevers 1968; Kennedy 1968; Urquhart 1969; Woodman 1969.

C. Poly(vinyl Esters) and Derived Polymers

The most widely used polymer of a vinyl ester is poly(vinyl acetate). It is utilized not only as a plastic, primarily in the form of emulsions, but also as the precursor for two polymers which cannot be prepared by direct polymerization, poly(vinyl alcohol) and the poly(vinyl acetals). None of these polymers is useful for molding or extrusion, but each is important in certain special applications.

Poly(vinyl acetate)

Monomer Vinyl acetate, $CH_2\!\!=\!\!CH\!-\!O\!-\!\overset{\displaystyle O}{\overset{\displaystyle \|}{C}}\!-\!CH_3$, is prepared by the addition of acetic acid to acetylene:

$$CH\!\equiv\!CH + CH_3COOH \longrightarrow CH_3COOCH\!\!=\!\!CH_2$$

The reaction can be carried out either in the liquid or in the vapor phase.

Polymerization Bulk polymerization of vinyl acetate is difficult to control at high conversions, and in addition the properties of the polymer may deteriorate because of chain branching. Bulk or solution polymerization is usually stopped at low to medium conversion (20–50%), after which the monomer is distilled off. If a solvent (such as methanol, ethanol, ethyl acetate, or benzene) is present, the polymer may be precipitated out or further reacted in solution to one of the derived polymers. The polymerization may be done batchwise or continuously.

Poly(vinyl acetate) may also be made in suspension or emulsion systems. In suspension polymerizations, various additives must be used to coat the beads of polymer to prevent their sticking together during the drying operation. Most of the poly(vinyl acetate) emulsions commercially available are made by emulsion polymerization mechanisms (Chapter 12*B*).

Properties and uses This polymer is atactic and hence amorphous; stereoregular polymers have not been offered commercially. The glass transition temperature of poly(vinyl acetate) is only slightly above room temperature (28°C). As a result the polymer, while tough and form stable at room temperature, becomes sticky and undergoes severe cold flow at only slightly elevated temperatures. Lower-molecular-weight polymers are brittle but become gumlike when masticated and in fact are used in chewing gums.

Poly(vinyl acetate) is water sensitive in certain physical properties, such as strength and adhesion, but does not hydrolyze in neutral systems.

Aside from the production of poly(vinyl alcohol), one major use of poly(vinyl acetate) is in the production of water-based emulsion paints. The low cost, stability, quick drying, and quick recoatability of emulsion paints have led to their wide acceptance. Poly(vinyl acetate) is often co-polymerized with dibutyl fumarate, vinyl stearate, 2-ethylhexyl acrylate, or ethyl acrylate, or is plasticized, to obtain softer compositions for emulsion use.

Poly(vinyl acetate) is also widely used in adhesives, both of the emulsion type and of the hot-melt type.

Poly(vinyl alcohol)

Since vinyl alcohol is unstable with respect to isomerization to acetaldehyde, its polymer must be prepared by indirect methods.

Preparation Poly(vinyl alcohol) is prepared by the alcoholysis of poly-(vinyl acetate) (the less accurate terms "hydrolysis" and "saponification" are also used):

$$-(CH_2CH-)_x + x\ CH_3OH \longrightarrow -(CH_2CH-)_x + x\ CH_3COCH_3$$

$$
\begin{array}{ccc}
| & & | \\
O & & OH \\
| & & \\
C=O & & \\
| & & \\
CH_3 & &
\end{array}
$$

Ethanol or methanol can be used to effect the alcoholysis, with either acid or base as catalyst. Alkaline hydrolysis is much more rapid. The acid hydrolysis is more likely to give some ether linkages in the chain by a mechanism involving the loss of a molecule of water from two adjacent hydroxyl groups. This is an undesirable side reaction. The alcoholysis is usually carried out by dissolving the poly(vinyl acetate) in the alcohol, adding the catalyst, and heating. The poly(vinyl alcohol) precipitates from the solution.

Structure and properties Although poly(vinyl alcohol) is amorphous when unstretched, it can be drawn into a crystalline fiber, the hydroxyl groups being small enough to fit into a crystal lattice despite the atactic chain structure. Poly(vinyl alcohol) does not melt to a thermoplastic, but decomposes by loss of water from two adjacent hydroxyl groups at temperatures above 150°C. Double bonds are left in the chain; and, as more of these are formed in conjugate positions, severe discoloration takes place.

Poly(vinyl alcohol) is water soluble. It dissolves slowly in cold water, but more rapidly at elevated temperatures, and can usually be dissolved completely above 90°C. The aqueous solutions are not particularly stable, especially if traces of acid or base are present. The solutions can undergo a complex series of reversible and irreversible gelation reactions. For example, crosslinking can occur at ether linkages, resulting in increased viscosity through the formation of insoluble products.

Poly(vinyl alcohol) can be reacetylated by heating it with an excess of acetic anhydride in the presence of pyridine. The resulting poly(vinyl acetate) may or may not have the same structure as the parent polymer from which the alcohol was made, owing to the nature of the branched-chain structure in the polymer.

Applications The major uses of poly(vinyl alcohol) fall into two categories. In one type of application, use is made of the water solubility of the polymer. It serves as a thickening agent for various emulsion and suspension systems, and as a packaging film where water solubility is desired. A major outlet is wet-strength adhesives.

In the second type of end use, the final form of the polymer is insoluble in water as a result of chemical treatment. The use of poly(vinyl alcohol) as a textile fiber (*vinal* fiber) is the major example. The polymer is wet spun from warm water into a concentrated aqueous solution of sodium sulfate containing sulfuric acid and formaldehyde. The polymer is insolubilized by the formation of formal groups:

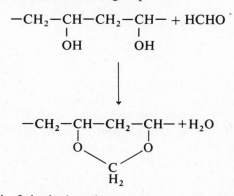

About one-third of the hydroxyl groups is reacted to insolubilize the fiber. Some interchain acetalization is desirable to reduce shrinkage of the fiber, but the amount must be carefully controlled.

Poly(vinyl alcohol) fibers have higher water absorption (about 30%) than other fibers. They can thus replace cotton in uses where the fiber is in contact with the body. The hand of the fabric can be varied from woollike to linenlike. The fiber washes easily, dries quickly, and has good dimensional stability. Tenacity and abrasion resistance are good.

Poly(vinyl acetals)

An important use of poly(vinyl alcohol) is as the starting material for poly(vinyl acetals). By far the most important of these polymers is poly(vinyl butyral), used as the plastic interlayer for automotive and aircraft safety glazings. Lower poly(vinyl acetals), especially poly(vinyl formal), are utilized in enamels for coating electrical wire and in self-sealing gasoline tanks.

Poly(vinyl butyral) This polymer is made by condensing poly(vinyl alcohol) with butyraldehyde in the presence of an acid catalyst, usually sulfuric acid:

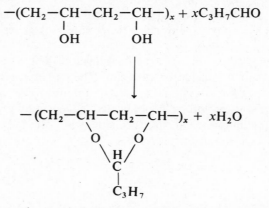

The reaction can be carried out by starting with an aqueous solution of poly(vinyl alcohol) and adding the butyraldehyde and catalyst. The poly(vinyl butyral) precipitates as it is formed. Alternatively, one may start with a suspension of poly(vinyl alcohol) in a water-alcohol mixture, the poly(vinyl butyral) dissolving as it is formed. Poly(vinyl butyral) produced for safety-glass use requires some hydroxyl groups in order to have adequate strength and adhesion to glass. Consequently the condensation reaction is stopped when about one-quarter of the hydroxyl groups is left.

Safety glass The earliest safety-glass interlayer was cellulose nitrate. Its poor weathering qualities led to its general replacement in 1935 by cellulose acetate, and in 1940 by poly(vinyl butyral). Some of the advantages of the latter resin as an interlayer are its superior adhesion to glass, toughness, stability on exposure to sunlight, clarity, and insensitivity to moisture.

Poly(vinyl butyral) must be plasticized for safety-glass use. The usual plasticizers are high-molecular-weight esters such as dibutyl sebacate and triethyleneglycol di-2-ethylbutyrate. About 40–45 parts plasticizer per 100 parts resin by weight is used. The plasticizer may be incorporated in the resin in several ways. One common method is to mix the resin, plasticizer, and a solvent such as ethanol to form a plastic mass. This is extruded in sheet form into a salt-water bath where the solvent is leached out, leaving the plasticized sheet. Since the plasticized resin sticks to itself, it must be powdered with talc or sodium bicarbonate before being rolled up for storage.

Safety-glass laminates are made by washing and drying the poly(vinyl butyral) sheet, and placing it between two pieces of glass. These are subjected to mild heat and pressure to seal the assembly, and are then autoclaved at higher temperature and pressure to complete the lamination process. The usual automotive safety glass contains one plastic layer about 0.030 in. thick.

GENERAL REFERENCES

Lindemann 1970*a,b*; Pritchard 1970.

D. Chlorine-Containing Polymers

Polymers and copolymers of vinyl chloride and vinylidene chloride are widely used as plastics and as fibers. In the trade, these polymers are known as "the vinyls" or "vinyl resins." Once the largest group of thermoplastic materials, the vinyl resins have been surpassed in volume by the olefin polymers in recent years, and about equaled by the styrene-type resins. Consumption of vinyl resins in 1969 was about 2.6 billion lb, at prices of about $0.14–0.24/lb.

Poly(vinyl chloride)

Monomer Vinyl chloride is a gas boiling at $-14°C$. It is currently produced in a two-stage process in which ethylene is first reacted catalytically with HCl and oxygen to yield 1,2-dichloroethane. This is then pyrolized to vinyl chloride and HCl, the latter being recycled. The older process in which HCl was added to acetylene is now largely outmoded.

Polymerization   Suspension polymerization is the most common method of preparing poly(vinyl chloride), and accounts for most of the material used for molding, extrusion, and calendering. Emulsion polymerization, second in importance, is widely used to produce the vinyl resins utilized in plastisols and organisols. Some bulk polymerization has been carried out on a commercial scale in recent years, despite the insolubility of the polymer in its monomer and the sensitivity of the method to radical trapping and local overheating with deterioration of resin properties. A small amount of solution polymerization is also used.

Structure   Poly(vinyl chloride) is a partially syndiotactic material, with sufficient irregularity of structure that crystallinity is quite low. Its structural characterization is complicated by the possibility of chain branching and the tendency of the polymer to associate in solution.

Stability   Poly(vinyl chloride) is relatively unstable to heat and light. Thermal initiation involves loss of a chlorine atom adjacent to some structural abnormality which reduces the stability of the C—Cl bond, such as terminal unsaturation (Grassie 1964). The chlorine radical so formed abstracts a hydrogen to form HCl; the resulting chain radical then reacts to form chain unsaturation with regeneration of a chlorine radical. The reaction can also be initiated by ultraviolet light which is absorbed at unsaturated structures with liberation of an adjacent chlorine atom. In the presence of oxygen, both chain reactions are accelerated, and ketonic structures are formed in the chain.

Stabilizers are almost invariably added to improve the heat and light stability of the polymer. Metallic salts of lead, barium, tin, or cadmium are used. Oxides, hydroxides, or fatty-acid salts are most effective. Epoxy plasticizers (Chapter 16D) aid materially in stabilizing the resin. Free radical acceptance appears to be a prominent mechanism of stabilization. HCl acceptors have been used, but the degradation reactions are not autocatalytic as had previously been supposed.

Vinyl resins

Rigid compounds   The term "rigid vinyls" usually refers to unplasticized poly(vinyl chloride), or compositions with only a few per cent of a plasticizer such as an epoxy resin (Chapter 16D). Often, other polymers are mixed physically with the poly(vinyl chloride) to improve impact resistance (nitrile rubber, chlorinated polyethylene, ABS resins, or methyl methacrylate–butadiene–styrene terpolymer) or processability (styrene-acrylonitrile copolymer or poly(methyl methacrylate)). Fillers (except glass fiber) are usually not used.

Copolymers The advantages in polymer properties resulting from the copolymerization of small amounts of vinyl acetate with vinyl chloride were discovered around 1928. The lower softening point and higher solubility of the copolymers make fabrication very much easier. Stability is improved, if anything, over that of the homopolymer. Color and clarity are also better. Polymerization methods are similar to those for vinyl chloride homopolymer except that emulsion polymerization has had less success. Commercial compositions containing 5–40% vinyl acetate are available.

Polymers containing around 10% vinylidene chloride have better tensile properties than pure poly(vinyl chloride). Copolymers containing 10–20% diethyl fumarate or diethyl maleate have improved workability and toughness and retain the high softening point of poly(vinyl chloride). Acrylic esters have also been used to impart improvements in solubility and workability.

Plasticization Many properties of poly(vinyl chloride) and vinyl chloride-vinyl acetate copolymers are improved by plasticization (Chapter 17). The large majority of the commercial production of vinyl resins is in the form of plasticized compositions. The first important plasticizer for the vinyls was tricresyl phosphate, which has since been replaced by other esters because of its tendency to cause low-temperature brittleness in plasticized compounds. Dibutyl phthalate, dibutyl sebacate, and tributyl phosphate have also been used and in turn replaced. Dioctyl phthalate, trioctyl phosphate, dioctyl sebacate and adipate, and various low-molecular-weight polymers such as poly(propylene glycol) esters are now widely utilized as plasticizers for the vinyls. The plasticizers are usually added to the polymers on hot rolls or in a hot mixer such as a Banbury. The plasticizer content varies widely with the end use of the material, but typically may be around 30% by weight.

Plastisols and organisols These liquid compositions are produced by spray-drying latices obtained from the emulsion polymerization of vinyl chloride. The latex particles are dispersed into plasticizers to make plastisols, or into mixtures of plasticizers and volatile organic liquids to make organisols. Other ingredients, such as stabilizers, fillers, colorants, surfactants, and possibly blowing agents or gelling agents, are also present. The polymer particles do not dissolve in the liquids, but remain dispersed until the mixture is heated. Fusion (plus loss of solvent from the organisols) then yields the final plastic object.

Applications About one-third of the vinyl resins produced is used in calendered products, largely film, sheet, and floor covering. These may be poly(vinyl chloride) or copolymers containing about 5% vinyl acetate, plasticized with 20–30% of a plasticizer like dioctyl phthalate. The film and sheeting are familiar in rainwear, handbags, shower curtains, food covers, etc.

Another third of the vinyls is used for extrusion applications. Much of this material is used for insulation for electrical wire and cable. A typical composition might be poly(vinyl chloride) with about 30% plasticizer, 5% stabilizer, and some filler or pigment. Such resins are quite fire resistant, especially if phosphate plasticizers are used. They are extruded directly around the wire at about 180°C.

Another large field for the vinyls is coated fabrics. The coating is done by hot rolling the fabric with calendered sheets or by coating from solutions. Upholstery and rainwear are common applications.

Another major segment of the vinyl resin production is used for floor covering. In the usual practice, a plastisol coating is applied to a paper base, covered with a protective vinyl film, and then applied to a felt base.

Molding uses, including phonograph records from rigid vinyls, and other plastisol products are also important applications for the vinyls.

Chlorine-containing fibers

Compositions   The use of poly(vinyl chloride) as a fiber was patented as early as 1913, and the material was commercialized in 1931. It was soon withdrawn in favor of modified polymers with better solubility, including chlorinated poly(vinyl chloride) and copolymers of vinyl chloride with vinylidene chloride and with vinyl acetate. Currently, the materials of most interest in the class of chlorine-containing fibers are copolymers of about 60% vinyl chloride and 40% acrylonitrile, and of about 85% vinyl chloride and 15% vinyl acetate. Other comonomers are occasionally used. The generic classes of fibers in this group include *vinyon*, containing more than 85% vinyl chloride; *saran*, containing more than 80% vinylidene chloride; and *modacrylic*, defined in Section *B*.

Properties and uses   The chlorine-containing fibers are usually moderately crystalline and are drawn (typically while hot) five- to fifteenfold to increase strength and reduce extensibility. Like the acrylic (Section *B*) and condensation (Chapter 15) fibers, they are insensitive to moisture and have high wrinkle resistance and good resistance to chemicals, insects, fungi, and the like. They have poorer dimensional stability at high temperatures, and somewhat less strength, elasticity, and abrasion resistance than the condensation fibers.

GENERAL REFERENCES

Anon. 1969; Bulkley 1969; Kaufman 1969; Koleske 1969; Osten 1969; Sarvetnick 1969; Ward 1969; Brighton 1970; Dux 1970; Wessling 1970.

E. Fully Fluorinated Fluorocarbon Polymers

Fluorocarbon polymers represent in many respects the extremes in polymer properties. Within this family are found materials of high thermal stability and concurrent usefulness at high temperatures (in some cases combined with high crystalline melting points and high melt viscosity), and extreme toughness and flexibility at very low temperatures. Many of the polymers are almost totally insoluble and chemically inert, some have extremely low dielectric loss and high dielectric strength, and most have unique nonadhesive and low friction properties.

The prices for fluorocarbon polymers in 1970 ranged from \$3.25–4.00/lb.

Polytetrafluoroethylene

Monomer Tetrafluoroethylene is a nontoxic gas boiling at $-76°C$. It can be made by the pyrolysis of chlorodifluoromethane, the dechlorination of *sym*-dichlorotetrafluoroethane, the decarboxylation of sodium perfluoropropionate, or the decomposition of tetrafluoromethane in an electric arc.

Polymerization The first polymer of tetrafluoroethylene was discovered by R. J. Plunkett (Garrett 1962) when an apparently empty cylinder of the gas was cut open to see why more material had not been obtained from it. The cylinder was partly filled with a waxy white powder shown to be the polymer.

Tetrafluoroethylene is usually polymerized with free radical initiators at elevated pressure in the presence of water. Redox initiation may be used; persulfates and hydrogen peroxide have been employed as initiators.

Structure (Sperati 1962) Polytetrafluoroethylene is a highly crystalline, orientable polymer. These facts indicate a regular structure, which implies the absence of any considerable amount of crosslinking. Branching is presumed to be absent, since branching mechanisms would involve breaking C-F bonds, which are estimated to be too strong to be ruptured easily.

The chances are, therefore, that polytetrafluoroethylene consists of linear —CF_2—CF_2— chains.

The crystal structure and crystalline-phase transitions in polytetrafluoroethylene are discussed in Chapter 5*B*. These transitions, occurring near room temperature, involve a 1.3% volume change having an important effect on the mechanical properties of the polymer for some applications. The degree of crystallinity of the polymer as formed from the monomer is generally quite high, 93–98%. The crystalline melting point is 327°C.

The crystalline density of polytetrafluoroethylene is 2.30 g/cm³. Upon melting the polymer and subsequently cooling it, lower densities are obtained. Subsequent thermal treatment (annealing) generally increases the crystallinity of the solid. A standardized molding and annealing technique under conditions carefully controlled to prevent the formation of voids in the sample gives "standard" specific gravities for various samples ranging between 2.15 and 2.28. This range of densities has been correlated with the molecular weight and thus the melt viscosity of the samples. Higher-molecular-weight polymers have more viscous melts and hence crystallize more slowly and reach lower crystallinities under standardized thermal treatment.

Despite the reputation of polytetrafluoroethylene as a totally insoluble polymer, solvents for it have been found at temperatures not far below its crystalline melting point. Solution viscosities have been measured in perfluorinated kerosene fractions at 300°C, the measurements giving clues as to the molecular weight of the polymer. Other information as to molecular weight is derived from the kinetics of polymerization of tetrafluoroethylene using radioactive initiators, and from the flow properties of the polymers (see the next paragraph). All the evidence points to unusually high number-average molecular weights of many millions for this polymer.

Polytetrafluoroethylene does not flow easily above its crystalline melting point. The viscosity of the polymer is very high because of restricted rotation about the chain bonds and high molecular weight. Upon the continued application of stress to the amorphous polymer above the crystalline melting point, the strength of the polymer is exceeded, causing fracture before the stresses necessary to induce rapid flow are reached. This unusual behavior necessitates unconventional fabrication techniques, as discussed on page 425.

Polytetrafluoroethylene decomposes at elevated temperatures. In vacuum, monomer is the chief product. The vapor pressure of monomer in equilibrium with the polymer at 500°C is 1 mm of mercury. At low temperatures (250–350°C) degradation appears to start at chain ends; at higher temperatures random cleavage becomes more important. In the presence of air the degradation is more complicated.

Properties Polytetrafluoroethylene is extremely resistant to attack by corrosive reagents or solvents. Of many hundreds of reagents tested up to their boiling points, only alkali metals, either molten or dissolved in liquid ammonia, attack the polymer, presumably by removing fluorine atoms from the chain. Fluorine itself degrades the polymer on prolonged contact under pressure. For all practical purposes the polymer is completely una'fected by water. Its thermal stability is such that its electrical and mechanical properties do not change for long intervals (months) at temperatures as high as 250°C.

Molded polytetrafluoroethylene articles have high impact strength but are easily strained beyond the point of elastic recovery. In a tensile test molded articles begin to cold draw at 1500–2000 psi, elongating (with orientation) to 300–450% before breaking at a load of 2500–4500 psi. Highly oriented fibers have tensile strengths as high as 50,000 psi. Polytetrafluoroethylene is subject to cold flow under stress but exhibits some delayed elastic recovery. The polymer is not hard, but is slippery and waxy to the touch, and has a very low coefficient of friction on most substances. Its density is unusually high (2.1–2.3) and its refractive index unusually low (1.375).

Polytetrafluoroethylene has extremely good electrical properties. Its dielectric constant is low (2.0), and its loss factor for all frequencies tested, including those used for television and radar, is one of the lowest known for solids. The properties of polytetrafluoroethylene can be varied widely as a function of molecular structure and fabrication, as described in Chapter 7.

Fabrication Since polytetrafluoroethylene is almost completely insoluble and has impracticably low melt flow rates, most of the fabrication techniques ordinarily used with polymers (Chapter 17) are not suitable for this material. Several unusual techniques have been developed for putting polytetrafluoroethylene into usable shapes. Most of these techniques are variants of two important processing steps: (*a*) pressing the polymer cold into the desired shape, followed by (*b*) sintering at a temperature above the crystalline melting point (say, 380°C) to yield a dense, strong, homogeneous piece.

Techniques for molding polytetrafluoroethylene resemble those of powder metallurgy or ceramics processing more than those of polymer fabrication. Granular polytetrafluoroethylene powder is compressed at room temperature and 2000–10,000 psi to give a preform. Sintering of the preform follows.

Polytetrafluoroethylene may be extruded at slow rates in a ram or screw extruder. In these devices the ram or screw serves merely to compact the molding powder and feed it to the die, within which the material is heated above the crystalline melting point, shaped, and allowed to solidify.

The high degree of cohesion between cold pressed particles of poly-tetrafluoroethylene is utilized in a calendering process for making tape and coating wire. Molding powder is fed to calender rolls, which compress the particles into a solid structure. This passes into a sintering bath or oven.

Polytetrafluoroethylene is available in the form of aqueous dispersions of ultimate particles about 0.2 μm in diameter. They can be used for casting films or for dip coating or impregnating porous structures.

Mechanical properties of polytetrafluoroethylene, including resistance to wear and to deformation under load, stiffness, and compressive strength, can be enhanced by the use of fillers. Most of the desirable properties of the polymer (heat resistance, low friction, weatherability, etc.) are retained.

Applications   The uses of polytetrafluoroethylene are largely those requiring excellent toughness, electrical properties, heat resistance, low frictional coefficient, or a combination of these. Among the electrical applications of the polymer are wire and cable insulation, insulation for motors, generators, transformers, coils, and capacitors, and high-frequency electronic uses. Chemical equipment such as gaskets, pump and valve packings, and pump and valve parts is made from the polymer. Low-friction and antistick applications include nonlubricated bearings, linings for trays in bakeries and other food-processing equipment, mold-release devices, and covers for heat-sealer plates of packaging machines. Low-molecular-weight polymers of tetrafluoroethylene dispersed as aerosols are effective dry lubricants. Uses for the fiber include gasketing, belting, pump and valve packing, filter cloths, and other industrial functions where essentially complete chemical and solvent resistance, combined with heat resistance to above 250°C, is required.

Fluorocarbon copolymers

A copolymer of hexafluoropropylene and tetrafluoroethylene has a crystalline melting point near 290°C. It retains most of the properties of polytetrafluoroethylene, but has melt viscosity in the range allowing fabrication by conventional techniques.

Properties   The copolymer is tough at liquid-air temperatures, yet retains adequate mechanical strength for continuous service at temperatures up to 200°C. Like polytetrafluoroethylene, it is chemically inert and has zero water absorption. It maintains a low dielectric constant and loss factor over the unusually wide range of 60 cycles to 60 megacycles. It has excellent weatherability, nonstick and low friction properties, and very low permeability to gases.

The material can be processed by conventional extrusion, injection molding, compression, transfer, or blow molding, fluidized bed coating, or vacuum forming of sheets or film. It finds application as wire jacketing, as extruded film, rods, tubes, or complex shapes, and as an encapsulating resin. End uses are similar to those for polytetrafluoroethylene.

GENERAL REFERENCES

Osten 1969; McCane 1970.

F. Other Fluorine-Containing Polymers

Polychlorotrifluoroethylene

Monomer   Chlorotrifluoroethylene is made by dechlorination of trichloro-trifluoroethane. The monomer is less subject to spontaneous explosive polymerization than is tetrafluoroethylene. Unlike tetrafluoroethylene, however, chlorotrifluoroethylene is itself toxic.

Polymerization   As in the case of tetrafluoroethylene, the polymerization of chlorotrifluoroethylene is best carried out in an aqueous system using a redox initiator. The details of commercial polymerizations have not been disclosed.

Structure and properties   Polychlorotrifluoroethylene is surpassed only by polytetrafluoroethylene and tetrafluoroethylene-hexafluoropropylene co-polymers in chemical inertness and resistance to elevated temperatures. Differences in the properties of the polymers are a consequence of the lower symmetry of the chlorine-containing polymer. For example, the crystalline melting point of polychlorotrifluoroethylene is 218°C, and the polymer can be quenched to quite clear sheets in which crystallinity is absent. Slower cooling provides opportunity for growth of crystallites and spherulites and results in quite different optical and mechanical properties.

Polychlorotrifluoroethylene is soluble in a number of solvents above 100°C, and is swelled by several solvents at room temperature. The polymer is tough at −100°C and retains useful properties as high as 150°C. Although high compared to that of many other polymers, its melt viscosity is low enough that the usual fabrication techniques, such as molding and extrusion, are practicable. The electrical properties of polychlorotrifluoroethylene are inferior to those of polytetrafluoroethylene, especially for high-frequency applications, because of the more polar nature of the polymer. Typical applications of polychlorotrifluoroethylene include gaskets, valve seats, tubing, and wire and cable insulation. The 1970 price of the polymer was about $4.75/lb.

Copolymers of chlorotrifluoroethylene

Copolymers of chlorotrifluoroethylene and vinylidene fluoride range from tough, flexible thermoplastics to elastomers, depending on composition. One of their outstanding properties is their resistance to the attack and penetration of powerful oxidizing agents such as propellant-grade red-fuming nitric acid and 90% hydrogen peroxide, typical of the fluorocarbons, coupled with the toughness and flexibility required for use as valves, ring seals, caulking compounds, etc. They also find application in automobile transmissions and brake systems as gaskets and seals.

Poly(vinyl fluoride)

Poly(vinyl fluoride) is a highly crystalline plastic which is commercially available in the form of a tough but flexible film. The polymer has the outstanding chemical resistance of the fluorocarbons, and excellent outdoor weatherability. It is extremely resistant to thermal degradation and maintains usable strength above 150°C while remaining tough at −180°C. It has low permeability to most gases and vapors and resists abrasion and staining.

The film has wide use as a protective coating in the building industry. In 0.001–0.002 in. thickness, bonded to wood, metal, or asphalt-based materials, it lasts many times longer than paints, enamels, or other surface coatings.

Poly(vinylidene fluoride)

Poly(vinylidene fluoride) is a crystalline plastic with a melting point of about 170°C. It has good strength properties and resists distortion and creep at both high and low temperatures. It has very good weatherability and chemical and solvent resistance.

This polymer can be processed by extrusion and by compression and injection molding. It is used as a coating, gasketing, and wire- and cable-jacketing material, and in piping for chemical processing.

Vinylidene fluoride-hexafluoropropylene copolymers

These copolymers (in 70%:30% proportions in one composition) are elastomers combining high resistance to heat and to fluids and chemicals with good mechanical properties. They are cured with amine-type systems.

These materials can be compounded to remain serviceable for short periods as high as 300°C, and retain useful properties indefinitely at 200°C. Their resistance to the lubricants, fuels, and hydraulic fluids used in jet aircraft is unequalled by other elastomers. Their mechanical properties are respectable for any elastomer, and excellent among the oil-resistant types.

GENERAL REFERENCES

Brown 1967; Albisetti 1969; Du Pont 1969; Houser 1969; Towler 1969.

BIBLIOGRAPHY

Abrams 1967. Irving M. Abrams and Leo Benzera, "Ion-Exchange Polymers," pp. 692–742 in Herman F. Mark, Norman G. Gaylord, and Norbert M. Bikales, eds., *Encyclopedia of Polymer Science and Technology*, Vol. 7, Interscience Div., John Wiley and Sons, New York, 1967.

Albisetti 1969. C. J. Albisetti, "PVF film," pp. 368–369 in Sidney Gruss, ed., *Modern Plastics Encyclopedia 1969–1970* (McGraw-Hill Book Co., New York), **46** (10A), October, 1969.

Anon. 1969. Anon., "Vinyl film and sheeting," pp. 350, 352–352 in Sidney Gruss, ed., *Modern Plastics Encyclopedia 1969–1970* (McGraw-Hill Book Co., New York), **46** (10A), October, 1969.

Bamford 1964. C. H. Bamford, G. C. Eastmond, Gerald P. Ziemba, and A. Lebovits, "Acrylonitrile Polymers," pp. 374–444 in Herman F. Mark, Norman G. Gaylord, and Norbert M. Bikales, eds., *Encyclopedia of Polymer Science and Technology*, Vol. 1, Interscience Div., John Wiley and Sons, New York, 1964.

Basdekis 1964. C. H. Basdekis, *ABS Plastics*, Reinhold Publishing Corp., New York, 1964.

Beevers 1968. R. B. Beevers, "The Physical Properties of Polyacrylonitrile and Its Copolymers," *Macromol. Revs.* **3**, 113–254 (1968).

Boyer 1970. Raymond F. Boyer, K. E. Coulter, Howard Kedke, Alan E. Platt, Henno Keskkula, John F. Rudd, J. L. Duda, J. S. Vrentas, and K. S. Hyun, "Styrene and Polystyrene," in press in Herman F. Mark, Norman G. Gaylord, and Norbert M. Bikales, eds., *Encyclopedia of Polymer Science and Technology*, Vol. 13, Interscience Div., John Wiley and Sons, New York, 1970.

Brighton 1970. C. A. Brighton, "Vinyl Chloride Polymers," in press in Herman F. Mark, Norman G. Gaylord, and Norbert M. Bikales, eds., *Encyclopedia of Polymer Science and Technology*, Vol. 13, Interscience Div., John Wiley and Sons, New York, 1970.

Brown 1967. Henry C. Brown and Robert P. Bringer, "Fluorine-Containing Polymers," pp. 179–219 in Herman F. Mark, Norman G. Gaylord, and Norbert M. Bikales, eds., *Encyclopedia of Polymer Science and Technology*, Vol. 7, Interscience Div., John Wiley and Sons, New York, 1967.

Brownell 1964. George L. Brownell, "Acids, Maleic and Fumaric," pp. 67–95 in Herman F. Mark, Norman G. Gaylord, and Norbert M. Bikales, eds., *Encyclopedia of Polymer Science and Technology*, Vol. 1, Interscience Div., John Wiley and Sons, New York, 1967.

Bulkley 1969. Charles W. Bulkley, Robert G. Morin, and Alan J. Stockwell, "Vinyl polymers and copolymers," pp. 215–216, 222 in Sidney Gruss, ed., *Modern Plastics Encyclopedia 1969–1970* (McGraw-Hill Book Co., New York), 46 (10A), October, 1969.

Collins 1969. Frederick H. Collins, "Styrene foam sheet," pp. 370–371 in Sidney Gruss, ed., *Modern Plastics Encyclopedia 1969–1970* (McGraw-Hill Book Co., New York), 46 (10A), October, 1969.

Davis 1964. C. W. Davis and Paul Shapiro, "Acrylic Fibers," pp. 342–373 in Herman F. Mark, Norman G. Gaylord, and Norbert M. Bikales, eds., *Encyclopedia of Polymer Science and Technology*, Vol. 1, Interscience Div., John Wiley and Sons, New York, 1964.

Dow 1969. The Dow Chemical Co., "Polystyrene foams," pp. 252–254 in Sidney Gruss, ed., *Modern Plastics Encyclopedia 1969–1970* (McGraw-Hill Book Co., New York) 46 (10A), October. 1969.

Du Pont 1969. Du Pont Co., "Fluoroplastics film and sheet," and K. R. Haberman, "Fluorohalocarbon films," p. 364 in Sidney Gruss, ed., *Modern Plastics Encyclopedia 1969–1970* (McGraw-Hill Book Co., New York) 46 (10A), October, 1969.

Dux 1970. J. P. Dux, "Vinyl Chloride Fibers," in press in Herman F. Mark, Norman G. Gaylord, and Norbert M. Bikales, eds., *Encyclopedia of Polymer Science and Technology*, Vol. 13, Interscience Div., John Wiley and Sons, New York, 1970.

Fram 1964. Paul Fram, "Acrylic Elastomers," pp. 226–246 in Herman F. Mark, Norman G. Gaylord, and Norbert M. Bikales, eds., *Encyclopedia of Polymer Science and Technology*, Vol. 1, Interscience Div., John Wiley and Sons, New York, 1964.

Garrett 1962. Alfred B. Garrett, "The Flash of Genius, 2. Teflon: Roy J. Plunkett," *J. Chem. Educ.* **39**, 288 (1962).

Grassie 1964. N. Grassie, "Thermal Degradation," Chap. VIII-B in E. M. Fettes, ed., *Chemical Reactions of Polymers*, Interscience Div., John Wiley and Sons, New York, 1964.

Hahn 1969. G. Hahn and D. Wiley, "Styrene polymers," pp. 204, 206, 210 in Sidney Gruss, ed., *Modern Plastics Encyclopedia 1969–1970* (McGraw-Hill Book Co., New York) 46 (10A), October, 1969.

Hart 1969. James J. Hart, "Oriented polystyrene film & sheet," p. 354 in Sidney Gruss, ed., *Modern Plastics Encyclopedia 1969–1970* (McGraw-Hill Book Co., New York) 46 (10A), October, 1969.

Higginbottom 1969. Brian Higginbottom and Willis R. Hendricks, "Styrene-butadiene," pp. 241–242 in Sidney Gruss, ed., *Modern Plastics Encyclopedia 1969–1970* (McGraw-Hill Book Co., New York) 46 (10A), October, 1969.

Houser 1969. James H. Houser, "Fluoroplastics: Polyvinylidene fluoride," pp. 137–138 in Sidney Gruss, ed., *Modern Plastics Encyclopedia 1969–1970* (McGraw-Hill Book Co., New York) 46 (10A), October, 1969.

Houtz 1950. R. C. Houtz, "'Orlon' Acrylic Fiber: Chemistry and Properties," *Textile Res. J.* **20**, 786–801 (1950).

Kaufman 1969. Morris Kaufman, *The History of PVC*, Maclaren and Sons, London, 1969.

Kennedy 1968. R. K. Kennedy, "Modacrylic Fibers," pp. 812–839 in Herman F. Mark, Norman G. Gaylord, and Norbert M. Bikales, eds., *Encyclopedia of Polymer Science and Technology*, Vol. 8, Interscience Div., John Wiley and Sons, New York, 1968.

Keskkula 1969. Henno Keskkula, Alan E. Platt, and Raymond F. Boyer, "Styrene Plastics," pp. 85–134 in Raymond E. Kirk and Donald F. Othmer, eds., *Encyclopedia of Chemical Technology*, Vol. 19, 2nd ed., Interscience Div., John Wiley and Sons, New York, 1969.

Koleske 1969. J. V. Koleske and L. H. Wartmen, *Poly(Vinyl Chloride)*, Gordon and Breach, New York, 1969.

Lindemann 1970a. Martin K. Lindemann and Saburo Imoto, "Vinyl Alcohol Polymers," in press in Herman F. Mark, Norman G. Gaylord, and Norbert M. Bikales, eds., *Encyclopedia of Polymer Science and Technology*, Vol. 13, Interscience Div., John Wiley and Sons, New York, 1970.

Lindemann 1970b. Martin K. Lindemann, "Vinyl Ester Polymers," in press in Herman F. Mark, Norman G. Gaylord, and Norbert M. Bikales, eds., *Encyclopedia of Polymer Science and Technology*, Vol. 13, Interscience Div., John Wiley and Sons, New York, 1970.

Luskin 1964. Leo S. Luskin, Robert J. Myers, J. P. Brusie, H. W. Coover, Jr., and T. H. Wicker, Jr., "Acrylic Ester Polymers," pp. 246–342 in Herman F. Mark, Norman G. Gaylord, and Norbert M. Bikales, eds., *Encyclopedia of Polymer Science and Technology*, Vol. 1, Interscience Div., John Wiley and Sons, New York, 1964.

McCane 1970. Donald I. McCane, "Tetrafluoroethylene Polymers," in press in Herman F. Mark, Norman G. Gaylord, and Norbert M. Bikales, eds., *Encyclopedia of Polymer Science and Technology*, Vol. 13, Interscience Div., John Wiley and Sons, New York, 1970.

Miller 1964. M. L. Miller, "Acrylic Acid Polymers," pp. 197–226 in Herman F. Mark, Norman G. Gaylord, and Norbert M. Bikales, eds., *Encyclopedia of Polymer Science and Technology*, Vol. 1, Interscience Div., John Wiley and Sons, New York, 1964.

Moacanin 1969. J. Moacanin, G. Holden, and N. W. Tschoegl, *Block Copolymers (J. Polymer Sci.* **C26**), Interscience Div., John Wiley and Sons, New York, 1969.

Osten 1969. R. A. Osten, "Fluoroplastics: TFE and FEP resins," pp. 136–137 in Sidney Gruss, ed., *Modern Plastics Encyclopedia 1969–1970* (McGraw-Hill Book Co., New York) **46** (10A), October, 1969.

Pritchard 1970. J. G. Pritchard, *Poly(Vinyl Alcohol)—Basic Properties and Uses*, Gordon and Breach, New York, 1970.

Rathbone 1969. W. V. Rathbone, "ABS resins," pp. 76, 83–84 in Sidney Gruss, ed., *Modern Plastics Encyclopedia 1969–1970* (McGraw-Hill Book Co., New York) **46** (10A), October, 1969.

Rebenfeld 1967. Ludwig Rebenfeld, "Fibers," pp. 505–573 in Herman F. Mark, Norman G. Gaylord, and Norbert M. Bikales, eds., *Encyclopedia of Polymer Science and Technology*, Vol. 6, Interscience Div., John Wiley and Sons, New York, 1967.

Sarvetnick 1969. Harold A. Sarvetnick, *Polyvinyl Chloride*, Van Nostrand Reinhold, New York, 1969.

Sperati 1962. C. A. Sperati and H. W. Starkweather, Jr., "Fluorine-Containing Polymers. II. Polytetrafluoroethylene," *Fortschr. Hochpolym.-Forsch. (Advances in Polymer Science)* **2**, 465–495 (1962).

Thomas 1964. W. M. Thomas, "Acrylamide Polymers," pp. 177–197 in Herman F. Mark, Norman G. Gaylord, and Norbert M. Bikales, eds., *Encyclopedia of Polymer Science and Technology*, Vol. 1, Interscience Div., John Wiley and Sons, New York, 1964.

Towler 1969. D. W. Towler, "Fluoroplastics: Chlorotrifluoroethylene (CTFE)," p. 137 in Sidney Gruss, ed., *Modern Plastics Encyclopedia 1969–1970* (McGraw-Hill Book Co., New York) **46** (10A), October, 1969.

Uniroyal 1969. Uniroyal, Inc., "ABS sheet," pp. 359–360 in Sidney Gruss, ed., *Modern Plastics Encyclopedia 1969–1970* (McGraw-Hill Book Co., New York) **46** (10A), October, 1969.

Urquhart 1969. Donald Urquhart, "All-acrylic films," p. 356 in Sidney Gruss, ed., *Modern Plastics Encyclopedia 1969–1970* (McGraw-Hill Book Co., New York) **46** (10A), October, 1969.

Ward 1969. D. W. Ward, "Expanded vinyls," pp. 256–258 in Sidney Gruss, ed., *Modern Plastics Encyclopedia 1969–1970* (McGraw-Hill Book Co., New York) **46** (10A), October, 1969.

Wessling 1970. R. Wessling and F. G. Edwards, "Vinylidene Chloride Polymers," in press in Herman F. Mark, Norman G. Gaylord, and Norbert M. Bikales, eds., *Encyclopedia of Polymer Science and Technology*, Vol. 13, Interscience Div., John Wiley and Sons, New York, 1970.

Woodman 1969. John F. Woodman, "Acrylics," p. 89 in Sidney Gruss, ed., *Modern Plastics Encyclopedia 1969–1970* (McGraw-Hill Book Co., New York) **46** (10A), October, 1969.

15

Heterochain Thermoplastics

Of the various chemical classes of polymers, some, such as polyamides and polyesters, differ from, say, polystyrene and polyethylene in having atoms other than carbon in the chain. Omitting the thermosetting resins discussed in Chapter 16, the majority of the thermoplastic *heterochain* polymers have sufficiently similar properties to warrant their consideration as a group. In this category fall such prominent artificial fibers as the cellulosics, the nylons, and poly(ethylene terephthalate) (but see also Chapters 13, 14, and 18) and strong, tough "engineering" plastics based on similar and related polymers, as well as several important elastomers.

A. Polyamides

The coined word *nylon* has been accepted as a generic term for synthetic polyamides. The nylons are described by a numbering system which indicates the number of carbon atoms in the monomer chains. Amino acid polymers are designated by a single number, as 6 nylon for poly(ω-aminocaproic acid) (polycaprolactam). Nylons from diamines and dibasic acids are designated by two numbers, the first representing the diamine, as 66 nylon for the polymer of hexamethylenediamine and adipic acid, and 610 nylon for that of hexamethylenediamine and sebacic acid. Of greatest commercial importance are 6, 66, 610, and 11 nylon, poly(ω-amino-undecanoic acid), and their copolymers.

The production of nylon fibers in the United States in 1969 was about 1.35 billion lb. Production of nylon for plastics uses was about 90 million lb in 1969. The price of both the plastics and the tire cord and industrial fibers was in the range $0.90–$1.30/lb, with textile fiber prices ranging somewhat higher, depending on form and size.

Development of the nylons

The research leading to the production of synthetic fibers from polyamides began about 1928, when W. H. Carothers (Mark 1940) undertook a series of researches dealing with the fundamentals of polymerization processes. These studies were not aimed at the production of any specific product or process. In his fundamental researches, Carothers studied the condensation reactions of glycols and dibasic acids. He made a large number of polyesters ranging in molecular weight between 2500 and 5000. He also studied polymers of ω-amino acids, obtaining polyamides with a degree of polymerization of about 30. These were hard, insoluble, waxy materials.

Recognizing the equilibria involved in stepwise polymerization, Carothers and J. W. Hill introduced the use of a molecular still to remove the water liberated in the condensation process and to shift the equilibrium toward higher molecular weights. In this way they synthesized materials with molecular weights up to 25,000. These substances had properties so different from those previously studied that Carothers coined the term "superpolymers" to describe those having molecular weights above 10,000.

The superpolyesters obtained from glycols and dibasic acids were tough, opaque solids which melted at moderately elevated temperatures to clear, viscous liquids. Filaments could be pulled from the liquids by touching the surface with a rod and withdrawing it. When cool, these filaments could be drawn to several times their original length. They became tough, transparent, lustrous materials of high strength and elasticity, possessing as high tenacity when wet as dry. Examination by x-ray diffraction showed them to be crystalline and highly oriented.

The properties of the aliphatic superpolyesters made them unsuitable for textile fibers because of their relatively low melting points and high solubility. Mixed aliphatic polyester-polyamides had the same bad features as the polyesters. It was clear that success, if achieved at all, would involve the use of a polyamide fiber.

The next phase of the research was a concentrated effort to find polyamides which could be made into fibers. A large variety of amino acids, diamines, and dibasic acids was studied. Several of the products gave good fibers melting near 200°C and equal to silk in strength and pliability. In 1935, the polyamide of hexamethylenediamine and adipic acid was made. Like the others, it gave fibers by melt spinning or by dry spinning from phenol solutions. The fibers could be cold drawn to obtain high tenacity and elasticity. They melted at about 265°C and were insoluble in all common solvents.

This polymer was selected for commercial development because of its good balance of properties and the possibilities foreseen for the manufacture of the raw materials. At that time, adipic acid was made only in Germany, and hexamethylenediamine was little more than a laboratory curiosity.

Poly(hexamethylene adipamide), 66 nylon

Production Adipic acid can be obtained from a large number of chemicals, including cyclohexane, acetylene, and tetrahydrofuran. One commercial process involves oxidation of cyclohexane by air to a mixture of cyclohexanone and cyclohexanol. The mixture is oxidized to adipic acid by catalytic treatment with nitric acid. Hexamethylenediamine can be made from adipic acid by catalytic dehydration in the presence of ammonia to give adiponitrile, followed by hydrogenation. It can also be made from acetylene, tetrahydrofuran, or butadiene.

Achievement of the stoichiometric balance needed to obtain high-molecular-weight step-reaction polymers (Chapter 8) is simplified by the tendency of hexamethylenediamine and adipic acid to form a 1:1 salt which can be isolated because of its low solubility in methanol. This salt is dissolved in water and added to an autoclave with 0.5–1 mole per cent acetic acid as viscosity stabilizer. As the temperature is raised, the steam generated purges the air from the vessel. Pressure is kept at 250 psi as the temperature is raised to 270–280°C. Pressure is then reduced, and a vacuum may be applied. After a total time of 3–4 hr, nitrogen pressure is used to extrude the nylon as a ribbon through a valve in the bottom of the autoclave. The ribbon is subsequently cut into cubes, the finished product for the plastics-grade material. For fiber use, the polymer (somewhat lower in molecular weight) is melt spun as described in Chapter 18*B*.

Properties Both as a plastic and as a fiber, 66 nylon is characterized by a combination of high strength, elasticity, toughness, and abrasion resistance. Good mechanical properties are maintained up to 150°C, although a more conservative limit for plastics use is 125°. Both toughness and flexibility are retained well at low temperatures.

The solvent resistance of nylon is good; only phenols, cresols, and formic acid dissolve the polymer at room temperature. Strong acids degrade it somewhat. The polymer discolors in air at temperatures of about 130°C and is degraded by hydrolysis at elevated temperatures. Its outdoor weathering properties are only fair unless it is especially stabilized or pigmented with carbon black. Nylon has a moderately low specific gravity, 1.14. Its moisture resistance is fair; moisture acts as a plasticizer to increase flexibility and toughness. The electrical uses of nylon are restricted to low frequencies because of the polar groups in the polymer.

Fiber applications   About half of all the nylon fiber produced goes into tire cord. Most original tires on passenger automobiles have rayon or polyester cord, but nylon cord has most of the replacement market and about half the total volume. Other industrial uses of nylon include rope, thread, and cord, where high tenacity and good elasticity are important, and belting and filter cloths, where resistance to abrasion and to chemical attack is required.

Apparel uses of nylon include ladies' hose, undergarments, and dresses. Low modulus and high elasticity combined with strength and toughness are needed for this application. The ability to give nylon a permanent set by an annealing process at about 100°C has been utilized in the production of permanently pleated garments as well as in preshaped hosiery.

Plastics applications   The most important use of 66 nylon plastic is as an engineering material—a substitute for metal in bearings, gears, cams, etc. Nylon is well suited for such applications because of its high tensile and impact strength, form stability at high temperatures, good abrasion resistance, and self-lubricating bearing properties. The ability to be injection-molded to close dimensional tolerances, augmented by its light weight, frequently gives the plastic a cost advantage in appliances. Rollers, slides, and door latches are often made of nylon. Bearings and thread guides in textile machinery may be made of nylon, which requires no lubricant that could damage yarn or cloth. Electrical wire is jacketed with nylon, which provides a tough, abrasion-resistant outer cover to protect the primary electrical insulation.

Other polyamides

Polycaprolactam, 6 nylon   Caprolactam can be made in a large number of ways. One which has been used commercially involves oxidation of cyclohexane to cyclohexanone, from which the oxime is made. This oxime reacts by the Beckmann rearrangement to give caprolactam. The polymerization of caprolactam is carried out by adding water to open the rings and then removing the water again at elevated temperature, where linear polymer forms. An autoclave or a continuous reactor can be used. Polycaprolactam is in equilibrium with about 10% of the monomer, which must be removed by washing with water before the polymer can be spun. At the spinning temperature, more monomer is formed to restore the equilibrium, and this must again be removed to achieve good properties in the yarn.

Caprolactam can also be polymerized by ionic chain mechanisms, as described in Chapter 10E. The reaction can be carried out below the melting point of the nylon and at atmospheric pressure, making the technique very attractive for the production of large cast articles.

The properties of polycaprolactam are in general similar to those of 66 nylon, but the former has a lower crystalline melting point (225° vs. 265°C for 66 nylon) and is somewhat softer and less stiff. The major use for this polymer is in tire cord.

Poly(hexamethylene sebacamide), 610 nylon   This polymer has not been used as a textile fiber, but has several properties making it suitable for monofilaments (brushes, sports equipment, and bristles). The additional hydrocarbon segments in the sebacic acid give the polymer a lower melting point (215°C) and lower moisture absorption. It thus retains its stiffness and mechanical properties better when wet than does 66 nylon.

Alkoxy substituted nylons   An alkoxy substituted 66 nylon which has especially good resistance to impact and flexural fatigue is used for wristwatch straps, valve seats, and the like. It is alcohol soluble but can be crosslinked by heat or by treatment with acids to an alcohol-insoluble form.

Soluble nylons   In addition to the above polyamides, some nylons soluble in methanol and ethanol are available. They are copolymers of undisclosed composition. More flexible than the other nylons, they are used for wire jacketing, sheeting, and textile coating. Although soluble in alcohol, they are very resistant to hydrocarbons.

Vegetable oil polyamides   Condensation polymers made from dimerized vegetable oil acids and diamines, such as ethylene diamine, are available as low-molecular-weight (6000–9000), film-forming thermoplastics. These materials form high-gloss, flexible films with good adhesion, weatherability, and impact strength. They are also used in thixotropic paints and for paper coatings. Lower-molecular-weight products, liquid at room temperature, serve as curing agents for epoxy resins (Chapter 16D).

11 and 12 nylons   The polymers of the lactam of 12-amino-dodecanoic acid are produced in Europe. They have lower water absorption, and thus better dimensional stability, than the other nylons, and their lower melting points allow them to be processed in ways not achieved with the others, such as fluidized-bed coating.

GENERAL REFERENCES

Rebenfeld 1967; Habermann 1969; Peerman 1969; Schule 1969; Sherwood 1969; Snider 1969; Sweeny 1969.

B. *Other Noncyclic Heterochain Thermoplastics*

This section includes description of heterochain thermoplastics (except the polyamides) which are normally made for commercial use without cyclic or aromatic moieties in the chain. Thus, polycarbonates, polyurethanes, and polyesters based on terephthalic acid are discussed in Section *D*. As mentioned in Section *A*, the linear aliphatic polyesters first made by Carothers, which fall within the scope of this section, are not made commercially.

Ether and acetal polymers

Polyethers, by definition, contain the ether linkage —R—O—R—O— as part of their chain structure. By convention, the polymers in which R is a methylene group, —CH_2— (i.e., polyoxymethylenes), are described by the generic term *acetal resins*, the name polyethers being applied to polymers in which R is more complex. Epoxy resins, also polyethers, are discussed in Chapter 16*D*.

Polyoxymethylene Polyoxymethylene is a polymer of formaldehyde or of trioxane. Although polymeric products of formaldehyde have been known for over 100 years, and were studied in detail by Staudinger (1925), thermally stable polymers of formaldehyde were only recently prepared. The improved stability of these acetal resins allows them to be fabricated into useful articles. Copolymers, thought to contain small amounts of ethylene oxide as the comonomer, are also produced. About 55 million lb of acetal resins was produced in the United States in 1969.

Exceptionally pure formaldehyde (better than 99.9% CH_2O) is polymerized by an anionic mechanism in the presence of an inert solvent (e.g., hexane) at atmospheric pressure and a temperature, preferably between −50 and +70°C, where the solvent is liquid. A wide variety of anionic catalysts is suitable, including amines, cyclic nitrogen containing compounds, arsines, stibines, and phosphines. The polymer is insoluble in the reaction mixture and is removed continuously as a slurry. Thermal stability is improved by acetylation of the hydroxyl end groups of the polymer using acetic anhydride:

$$—CH_2OCH_2OH + CH_3COO^- \longrightarrow$$
$$—CH_2OCH_2O^- + CH_3COOH \xrightarrow{(CH_3CO)_2O}$$
$$—CH_2OCH_2OCOCH_3 + CH_3COO^-$$

Acetal resins have properties characteristic of partially crystalline high-molecular-weight polymers. Typical specimens are about 75% crystalline, with a melting point of 180°C. The impact strength of the polymers is high, and both their stiffness (450,000 psi) and yield stress (10,000 psi) are greater than those of other crystalline polymers. Moisture absorption is negligibly small, and the polymers are insoluble in all common solvents at room temperature. They can be processed by conventional molding and extrusion techniques.

Major uses for the acetal resins are as direct replacements for metals. At their 1969 price of $0.65/lb, they are cheaper than brass per unit volume. Their stiffness, light weight, dimensional stability, and resistance to corrosion, to wear, and to abrasion have led to their replacing brass, cast iron, and zinc in many instances. Typical applications include automobile parts, such as instrument panels, door hardware, and pump housings and mechanisms; pipe, especially for oil field systems; and a wide variety of machine and instrument parts.

Higher aldehydes can be polymerized by both anionic and cationic mechanisms. In some cases stereoregular polymers result. This is the only known case where polymerization involving a double bond other than C=C yields stereoregular polymers.

Polyglycols Polymers made from ethylene oxide and its higher homologs

$$CH_2CHR$$
$$\diagdown O \diagup$$

can be considered to be the condensation products of glycols, although they are in fact produced by ring-scission polymerization of the corresponding oxides (Chapter 10*E*). The product of greatest commercial interest is the polymer of ethylene oxide. This polymer combines the thermoplastic behavior and mechanical properties of a highly crystalline, high-molecular-weight polymer with complete water solubility.

Ethylene oxide polymers have moderate strength and stiffness, similar to the corresponding properties of polyethylene, at room temperature. Their crystalline melting point is 65°C, however, so that their mechanical properties fall off sharply near this temperature. The polymer can be highly oriented by cold drawing. It retains its properties well up to about 75% relative humidity, despite its water solubility. It can be processed by any of the common thermoplastic processing techniques.

The polymers of ethylene oxide are useful where water solubility must be combined with film-forming, fiber-forming, or thickening properties. Typical applications include sizing for cotton and synthetic fibers, thickeners and stabilizers for pastes and adhesives, and binders and film-formers in pharmaceuticals.

Chlorinated polyether   A chlorinated polyether with the structure —CH$_2$C (CH$_2$Cl)$_2$CH$_2$O— is obtained from pentaerythritol by preparing a chlorinated oxetane and polymerizing it by ring scission. The polymer is linear and crystalline, and is much more resistant to thermal degradation than are other halogenated polymers, because of the absence of a hydrogen atom beta to the chlorines.

The polymer has good impact strength, low moisture absorption, high dimensional stability, and good chemical resistance, and retains its properties with increasing temperature well enough to allow use at 100–125°C. It can be processed by injection molding or extrusion, and can be applied as a coating from suspensions or in fluidized bed processes.

The polymer is used in pipes, pumps, and processing equipment where corrosion resistance is important, as a liner for pipe and tubing, and as a material to replace metal die castings.

Polysulfides

Polysulfide elastomers are the reaction products of ethylene dihalides and alkali sulfides. It was discovered in 1920 that a rubberlike material could be made from these reagents. Commercial production began in 1930, and the products were quite successful despite some disadvantages (notably odor) because of their resistance.

The polysulfide rubbers are linear condensation polymers:

$$x\text{RCl}_2 + x\text{Na}_2\text{S}_y \longrightarrow -(\text{R}-\text{S}_y-)_x + 2x\text{NaCl}$$

The physical properties of the materials depend on the length of the aliphatic group and the number of sulfur atoms present. With four sulfurs per monomer all products are rubbery; whereas with only two sulfurs at least four methylene groups are needed in the dihalide to obtain elastomeric properties. The halides ordinarily used are ethylene dichloride, β,β'-dichloroethyl ether, and dichloroethyl formal. Cocondensation can be carried out.

The polysulfide elastomers are outstanding in oil and solvent resistance and in gas impermeability. They have good resistance to aging and ozone. On the other hand, they have disagreeable odors, poor heat resistance, poor abrasion resistance in tire treads, and low tensile strength (1500 psi). Most of their applications as rubbers depend on solvent resistance and low gas permeability. They include gasoline hoses and tanks, diaphragms and gaskets, and balloon fabrics.

Liquid polysulfide elastomers can be cured at room temperature and without shrinkage to tough, solvent-resistant rubbers. These liquid compounds are widely used as gasoline tank sealants and liners for aircraft and in a variety of other sealing and impregnating applications.

These liquid polymers, when combined with oxidizers, burn with great intensity and generate large volumes of gas. They form the basis of solid-fuel rocket propellants. Inorganic oxidizers, in finely ground form, are mixed with the liquid elastomeric binder and cast directly into the rocket motor, binding to the walls of the vessel for support and insulation of the motor case.

Polypeptides

Polypeptides, the basis of natural proteins, are polyamides of α-amino acids. They are characterized by this repeat unit:

$$-NH-CH-C-$$
$$\underset{R}{|} \quad \underset{O}{\|}$$

where R varies widely among the thirty or so amino acids in proteins.

Although synthetic products can be made, the polypeptides of greatest interest are based on naturally occurring materials. They include wool and silk and a variety of fibers and plastics made from proteins extracted from nonfibrous natural products.

The importance of wool and silk is discussed in relation to other fibers in Chapter 18. Other protein-based fibers and plastics are at present of little more than historical interest.

Wool   Wool is a complex polypeptide, a polyamide made up of about twenty α-amino acids. The acids glycine, leucine and isoleucine, proline, cystine, and arginine, glutamic acid, and aspartic acid make up about two-thirds of its weight. About half of wool's mass is in the main polymer chain and half in the side chains. Since some 30% of the side chains have acid or amine groups along their length, wool is a huge polymeric "zwitterion" whose properties and solubility depend markedly upon pH.

Wool can exist in both the α- and β-keratin structures characteristic of the polypeptides: helical and extended sheet structures, respectively. The high reversible extensibility of wool (about 30% compared to 5% for cotton) derives from the fact that it can exist in either of these structures.

The microscopic appearance of wool is easily recognized by the presence of scales in the epidermal or surface layer of the fiber. These scales give wool the ability to felt or pack into dense stable mats.

Wool is noted for its high moisture absorption (up to 10–15%) and good insulating properties, its wrinkle resistance and elastic recovery, and its crimpiness. Its soil resistance, dyeability, and resistance to organic (dry-cleaning) solvents are good. On the other hand, it is attacked by moths and mildew, mild alkalis, and bleaches. It causes allergies. Its strength and stiffness are low compared to those of cotton. It has poor temperature resistance and weatherability in sunlight, and is adversely affected by hot water, with shrinkage and loss of luster and strength.

Despite these drawbacks, the virtues of wool are so great that it probably will continue to find extensive use, as it has to date, in suitings and other wearing apparel, blankets, yarns, carpets, felts, and upholstery. So far, no artificial fiber has equaled it in resilience, hand, insulating properties, dyeability, and flame resistance. The price of wool is notoriously unsteady, however, and its future may well depend upon its price in a competitive market. About 320 million lb was used in the United States in 1969.

Wool is crosslinked through cystine links. This structure gives swelling but not solubility in the natural product. On the addition of thioglycolic acid, however, the crosslinks are broken with the formation of sulfhydryl groups. The tensile strength of the fiber decreases, its stress-strain curve changes to resemble that of a typical elastomer, and it shrinks readily and becomes soluble in polar solvents.

Wool can be recrosslinked by a number of reagents, including alkyl dihalides. Since the thioethers resulting from the treatment are much more stable than the original disulfides, the new fiber has much better resistance to laundering and chemicals. Since moth damage usually begins with rupture of the disulfide links, the treated material has improved moth resistance. The scission and crosslinking can be made so mild that the appearance and hand of the material are not changed.

Silk   Silk is a polypeptide made up of only four amino acids: glycine, H_2NCH_2COOH; alanine, $H_2NCH(CH_3)COOH$; serine; and tyrosine. Glycine and alanine, present in a 2:1 ratio, account for over 75 mole per cent of the material. Silk crystallizes in an anti-parallel-chain pleated sheet β-keratin structure, in which only glycine and alanine residues occur. The other amino acids are present only in the amorphous regions of the fiber. There is no structure in silk equivalent to the α-keratin spiral in wool and other polypeptides. Thus the silk fiber has an elongation (20–25%) lower than that of wool but higher than that of cotton.

The silkworm spins a double fiber from which the two single fibers are easily separated as continuous filaments 400–700 yards long. Silk is thus more nearly analogous to continuous filament that to stable synthetic fibers.

Silk has been noted for thousands of years for its strength, toughness, and smooth, soft feel. It is resilient and shapes well. It is a poor conductor of heat and electricity. Its moisture regain is 11%; it withstands heat and hot water better than wool.

Some disadvantages of silk are those associated with its cultivation and gathering, unevenness of the fibers, loss of strength on dyeing, and poor launderability of the dyed fabrics. It lacks the strength of nylon for hosiery use but would otherwise be preferred for its warmer feel. There is relatively little effort in progress to improve the properties of silk.

GENERAL REFERENCES

Akin 1962; Gaylord 1962, 1963; Furukawa 1963; Bevington 1964; Stone 1967; Berenbaum 1969; Epstein 1969; King 1969; Schar 1969.

C. Cellulosic Polymers

The class of polymers based on cellulose includes the following: native cellulose, including wood and most plant matter, but of interest here primarily because of the widespread popularity of cotton as a fiber; regenerated cellulose, used as a fiber (viscose rayon) and a film (cellophane); chemical derivatives of cellulose, of which the organic esters, particularly the acetate, are most important; and minor polymers with structures similar to that of cellulose.

The importance of the cellulosic fibers has declined steadily with respect to man-made fibers in recent years. In 1970 it was estimated that all natural fibers (mostly, in fact, cotton) accounted for only 39% of the total United States fiber production of some 10.5 billion lb. The estimated 1970 consumption of various cellulosic fibers was: cotton, 4.5 billion lb; viscose rayon, 1.2; acetate rayon, 0.6.

Cellulose

Cellulose is a polysaccharide made up of β-D($+$)-glucose residues joined in linear chains having the structure:

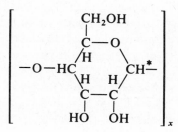

Starch, similarly made up of chains of α-D(+)-glucose, has a quite different chain conformation, resulting from its different steric configuration about the carbon atom starred in the diagram.

With its three hydroxyl groups, cellulose has the opportunity of forming many hydrogen bonds. The resulting high intermolecular forces plus the regular structure of the polymer result in its having an unusually high degree of crystallinity. The crystalline melting point of cellulose is far above its decomposition temperature. The solubility of the polymer is very low; it is doubtful that solution ever takes place unless a chemical derivative is formed. Cellulose does swell, however, in hydrogen-bonding solvents, including water. The swelling is, of course, restricted to the amorphous regions of the structure. When native cellulose is dissolved via chemical reaction and then reprecipitated as pure cellulose, the product is known as *regenerated cellulose*.

The highest-molecular-weight cellulose is found in cotton, which is the best source of very pure cellulose. Molecular weights as high as 570,000 are found, corresponding to a degree of polymerization of 3500. The molecular weight of cellulose from wood and other sources is considerably lower (around 160,000). Almost all chemical reactions of cellulose, including solution, are accompanied by some degradation in chain length. The degree of polymerization of cellulose derivatives is usually below 500.

Cotton

Cotton is about 95% cellulose, with small amounts of protein, pectin, and wax. The fiber can easily be recognized under the microscope because of its flattened, twisted shape.

Cotton fabrics launder well and show excellent resistance to alkalis. Since the strength of cotton is 25% greater wet than dry, it withstands repeated washings well, in contrast to most animal and synthetic fibers which lose strength when wet or moist. Cotton is less stiff than flax, hemp, or jute. Fabrics of cotton can be folded repeatedly without loss of strength. With proper construction and treatment, shrinkage and stretching can be controlled. Cotton is a good conductor of heat, better than silk or wool but not as good as linen. This property and its high moisture absorption give it comfort in wearing apparel. Cotton withstands heat better than the other natural fibers. It can be dyed readily, and the colors are good in lightfastness and washfastness. It is not affected by solvents and usually not attacked by moths.

Cotton is thus a versatile fiber with many desirable properties. These properties and its low price have given it an advantage over the years which has only recently been threatened. Rayon (i.e., viscose) and acetate are now available at prices competitive to cotton with advantages of luster, drape, wrinkle and soil resistance, low shrinkage, rapid drying, and good weathering resistance.

Mercerization The chemical treatment of cotton known as *mercerization* to improve its luster has been practiced for many years. The treatment consists in swelling the cotton by concentrated alkali solution and then washing out the alkali. The chemical changes during the process are similar to those in the formation of alkali cellulose. At the end of the treatment there is no molecular difference in the product, but it is somewhat more amorphous than native cotton and, unless the fibers were held under tension during the process, somewhat less oriented.

As a result of these changes, mercerized cotton is slightly lower in density than the untreated material and has increased water absorption, better dyeability, lower tensile strength, and higher extensibility.

Regenerated cellulose

The term *regenerated cellulose* describes cellulose which has been dissolved by virtue of the production of a soluble chemical derivative, cellulose xanthate, and subsequently reprecipitated. When prepared as a fiber, regenerated cellulose is known as *viscose* or (viscose) *rayon*. Traditionally both viscose and cellulose acetate have been known as rayons, but the term is now restricted to viscose in order to avoid confusion. As a film, regenerated cellulose is known by the generic term *cellophane*.

Manufacture The first step in the production of regenerated cellulose, starting with purified cellulose usually obtained from wood pulp, is the formation of *alkali* or *soda cellulose*. This is produced by the reaction of cellulose with strong alkalis at low temperatures, and is a further step in the reaction of mercerization.

The exact structure of alkali cellulose is not completely understood. It is certain that some of the alkali (say NaOH) is held in the swollen amorphous regions of the cellulose by hydrogen bonding of the hydroxyl radicals to the cellulose molecules. It is also possible that some alcoholates of the type R—ONa are formed, where R is the cellulose residue exclusive of one hydroxyl group.

Alkali cellulose is formed by reacting the cellulose pulp with about 18% aqueous NaOH at room temperature for 15 min to 2 hr. The excess alkali is then pressed out and the mass is shredded into "crumbs" and aged for several days at 25–30°C to promote oxidative degradation of the chains to the desired degree of polymerization.

The alkali cellulose is then converted to cellulose xanthate by the addition of carbon disulfide:

$$R—OH + CS_2 + NaOH \longrightarrow R—O—\overset{\overset{\displaystyle S}{\|}}{C}—S—Na + H_2O$$

where R is a cellulose residue. The average degree of xanthation is about one CS_2 for every two cellulose residues. Since the reaction takes place heterogeneously by the addition of liquid or gaseous CS_2 to the alkali cellulose crumbs, the xanthation in some regions may correspond to the dixanthate or higher.

The mass retains most of its physical form but becomes more sticky or gelatinous and takes on a deep orange color and a characteristic and highly unpleasant odor. It is held at 25–30°C for 1–3 hr with about 10% CS_2 by weight and then evacuated to remove excess CS_2. It is now dissolved in dilute NaOH, becoming completely soluble in this reagent for the first time. The solution is known as viscose. Its color and odor are due to products of side reactions such as Na_2CS_3, NaSH, H_2S, sulfides, and polysulfides.

A fresh viscose solution cannot easily be coagulated and must be allowed to "ripen" for a few days. The changes taking place during ripening are complex, and the reaction must be carefully controlled by time and temperature to avoid premature coagulation.

Cellulose xanthate is essentially unstable and decomposes gradually during the ripening process by hydrolysis and saponification. At the end of the ripening period, enough cellulose residues have been regenerated so that coagulation is imminent. The viscose is then spun into a bath containing sulfuric acid to effect the regeneration of the remaining cellulose residues and coagulate the polymer:

$$R—O—\overset{\overset{\displaystyle S}{\|}}{C}—S—Na + H^+ \longrightarrow R—OH + CS_2 + Na^+$$

If the spinning process places no mechanical stress on the polymer, it can be obtained in a completely amorphous form. This form is unstable, however, and the application of rather small stresses by such processes as stretching, working, spinning, and extruding causes the cellulose to crystallize (usually to a different structure from that of the native material). In this case, the ordering of the chains is not as great as in the native material, and the density and stiffness of the regenerated product are lower. These differences in density, stiffness, strength, and other properties between native and regenerated cellulose can be accounted for by differences in chain length, chain length distribution, crystal structure, degree of crystallinity, orientation, crystallite size, nature and amount of impurities and other flaws, degree of swelling, and chemical reactivity. All these changes on regeneration are in line with the structural differences outlined above.

Viscose rayon

Normal tenacity   Viscose rayon normally has a tenacity of 2–2.5 g/denier, somewhat lower than that of cotton (3–5 g/denier). Its wet tenacity is only about half the dry value but is still adequate for laundering. The moisture absorption of viscose is about twice that of cotton. The elasticity is not high (typically 15% elongation at break), and the fiber shows considerable visco-elastic character.

Viscose of normal tenacity is primarily an apparel fiber, but its popularity is declining. Several one-time major manufacturers have ceased producing the material, and the others are operating well below capacity.

High tenacity   In contrast to apparel rayon, a high-tenacity product (3–6 g/denier) is produced by stretching the fibers just short of their breaking point in the spinning bath. Considerable orientation (and crystallinity) is so introduced. High-tenacity rayon has lower elongation and moisture absorption than the apparel material and is used almost exclusively for tire cord. This market (120 million lb in 1969) is, however, dwindling rapidly with the increasing use of polyester and glass fibers in tires.

Cellophane

In the manufacture of cellophane the viscose solution is extruded as film and then immersed in a bath of ammonium and sodium sulfate and dilute sulfuric acid, which removes the xanthate groups and precipitates the cellulose. The film is passed through various washing, bleaching, and desulfurizing baths, and finally through a bath of glycerol, glucose, or a polyhydric alcohol, which is imbibed and acts as a plasticizer. The film is then dried. For applications in which low moisture permeability is desired, it is coated with a mixture of nitrocellulose and various plasticizers and waxes.

Cellophane is a thin (0.001–0.002 in.) film of fair physical properties. Its tensile strength is good, but its tear strength, impact strength, and resistance to flexing are poor compared to those of newer film materials. The permeability of the uncoated film to water vapor and to water-soluble gases is extremely high. The coated material passes about 0.3 gram of water vapor/sq in./hr at 40°C. Cellophane is widely used as a wrapping and packaging material.

Cellulose sponges are made by stirring lumps of a salt such as sodium sulfate into viscose, coagulating, and washing out the salt to leave a porous product. Sponges may also be made by incorporating a blowing agent in the material.

Cellulose acetate

Manufacture   As a polyhydric alcohol, cellulose can undergo esterification reactions. In the presence of a strong organic acid such as formic acid, equilibrium in the reaction

$$\text{acid} + \text{alcohol} \rightleftharpoons \text{ester} + \text{water}$$

lies partly to the right, and some ester is formed. With other organic acids, including acetic acid, equilibrium is well to the left and ester formation does not take place under normal circumstances. The easiest way to shift the equilibrium to the right is by the removal of water as it is formed in the reaction. To accomplish this, part of the acid is replaced by an acid chloride or anhydride. Sulfuric acid is also utilized in the reaction mixture. It acts as catalyst and also assists in removing water.

Chemical reactions of cellulose must of necessity begin in the (swollen) amorphous regions of the polymer. Unless these heterogeneous reactions are carried to the point where the cellulose is completely dissolved, the chemical homogeneity of the material is usually lower than desired. This is true in the formation of cellulose acetate, and consequently the esterification is carried to the point of complete solution.

Purified cotton linters or wood pulp form the raw material for cellulose acetate. This cellulose is partially dried from its natural moisture content of 5–10% to reduce the water content of the subsequent reaction mixture (complete drying would lead to a product difficult to reswell and low in chemical reactivity). The acetylating mixture of acetic acid, acetic anhydride, and sulfuric acid is added in portions and the reacting mixture is kept at or below 50°C. Since the reaction is exothermic, cooling is required. Higher temperatures are avoided to prevent excessive degradation in molecular weight.

The end of the acetylation is indicated by complete solution of the cellulose. At this point complete acetylation to the triacetate has taken place. The mixture is now held at about 50°C until the desired chain degradation has been achieved, as indicated by the viscosity of the mixture.

To obtain a more easily spinnable composition, it is necessary to reverse the acetylation reaction to the point where the proper solubility relations are obtained. This corresponds to the acetone-soluble diacetate, or an acetyl content of about 30%. This stage of the reaction can be carried out in homogenous solution, thus preserving the maximum amount of chemical homogeneity in the final product.

The acetylation reaction is reversed by the addition of water to the reaction mixture. Since gross addition of water would precipitate the polymer, acetic acid is added to the mixture, which at this point still contains large amounts of acetic anhydride. The final composition of the mixture needed to produce the diacetate corresponds to 90–95% acetic acid and 5–10% water. (Further reversal of the reaction, with a final composition of 60–70% acetic acid, would yield the monoacetate, which is insoluble in acetone and alcohols but is soluble in water.)

At the end of the reaction to produce the diacetate the mixture is cooled and added to a large amount of 25–35% acetic acid. This precipitates the polymer in finely divided form. After the polymer is washed and dried, it is dissolved in acetone for the spinning operation.

Fiber properties and applications The traditional acetate fiber (diacetate) has a tenacity of 1.2–1.5 g/denier and about the same moisture absorption as cotton. It tends to soil less and wash more easily than viscose. Since acetate is thermoplastic, pleats and creases can be permanently set. On the other hand, acetate is less susceptible to creasing and wrinkling in use than viscose because its resilience is greater.

With the advent of spinning processes using methylene chloride, cellulose triacetate has become popular as an apparel fiber. Its properties are improved over those of the diacetate in moisture absorption, wrinkle resistance and creep retention, dyeability, and speed of drying.

Aided by the success of triacetate and by new household, apparel, and industrial applications, including nonwoven fabrics and cigarette filters, the use of cellulose acetate increased to about 600 million lb/year in 1969.

Plastics properties and uses Cellulose acetate was introduced in sheet, rod, and tube form (as well as in lacquers and dopes) around the time of World War I. Its popularity as a substitute for cellulose nitrate (see page 450) was not high, and it was not widely used until injection molding became common in the 1930's. It has been for many years a very popular injection-molding material. The usual molding material has 50–55% acetyl content by weight. It is somewhere between the diacetate and the triacetate in composition. Higher acetyl content gives poorer flow but better moisture resistance; lower acetyl content gives better impact strength. A plasticizer is ordinarily used. Higher-softening grades are preferred for compression molding, and softer grades for extrusion. Cellulose acetate molded and extruded articles are extensively used where extreme moisture resistance is not required.

Cellulose acetate is also widely utilized in sheet form. Thick sheets are made by slicing from blocks and polishing (as with cellulose nitrate). Thinner films can be cast from solution onto polished metal surfaces. The films are also widely used in photography, for wrapping, and for making small enve lopes, bags, and boxes for packaging.

Cellulose acetate has good impact strength and electrical properties, and the advantage of low flammability not possessed by cellulose nitrate. It has however, a rather low softening point and high water absorption.

The plastics uses of cellulose acetate plus the mixed organic esters de scribed below amounted to about 200 million lb in 1969.

Mixed organic esters of cellulose

Cellulose acetate-butyrate   The mixed ester cellulose acetate-butyrate ha several advantages in properties over cellulose acetate. These include lowe moisture absorption, greater solubility and compatibility with plasticizers higher impact strength, and excellent dimensional stability. The materia used in plastics has about 13% acetyl and 37% butyryl content. It is ar excellent injection-molding material. Moldings of cellulose acetate-butyrate are widely used, e.g., in automobile hardware. Motion picture safety film is cellulose acetate-butyrate or cellulose propionate-butyrate. The outdoor weathering properties of cellulose acetate-butyrate are superior to those o the other cellulose esters, but not nearly as good, e.g., as those of the acryli resins.

Cellulose propionate and acetate-propionate   These materials are similar to the acetate-butyrate in properties and uses. The acetate-propionate has some application in surface coatings.

Cellulose nitrate

Although cellulose nitrate accounts today for only a minute fraction of the volume of cellulose plastics, it has considerable historical interest (Chapter 1B). The major use of plastics-grade cellulose nitrate is in the coating field The properties and applications of the material are dependent on its degree of nitration; the plastics and lacquer grades contain 10.5–12% nitrogen corresponding roughly to the dinitrate. Explosives grades contain 12.5–13.5% nitrogen.

Manufacture   The esterification reaction of the hydroxyl group on cellulose with nitric acid is an equilibrium.

$$ROH + HONO_2 \rightleftharpoons RONO_2 + H_2O$$

which can be shifted to the ester side by the removal of water. The usual reagent for this purpose is sulfuric acid; the nitration mixture ordinarily used consists of sulfuric and nitric acids and a limited amount of water.

Fabrication The temperature sensitivity of cellulose nitrate excludes it from all fabrication methods involving heat, such as molding and extrusion. The plastic is handled by so-called "block" methods. Cellulose nitrate, alcohol, and camphor, the plasticizer always used, are mixed in a dough mixer until homogeneous. Here for the first time the fibrous shape of the original cellulose is lost. The colloid is filtered under pressure and rolled on hot rolls (65–80°C) where much of the alcohol is evaporated. It is then sheeted into slabs which are pressed while hot into a homogeneous block about 6″ × 5′ × 2′ in size.

Sheets of any desired thickness are subsequently sliced from the block. They must be aged by storage at about 30°C until the last traces of solvent have diffused out. This may take from 2 hr to 6 months, depending on the thickness of the sheet. The final operation consists of finishing the surface of the sheet by pressing in contact with polished metal sheets or by buffing.

Rods, tubes, and other shapes can be formed by solvent extrusion of the plasticized cellulose nitrate, followed by removal of solvent as above.

Properties and uses Despite its disadvantages of flammability, instability, and poor weathering properties, cellulose nitrate is one of the cheapest and most highly impact-resistant plastics. It still has some applications as a plastic, but is now used mostly for lacquers.

Cellulose ethers

The cellulose ethers are products of the reaction of an organic halide with cellulose swollen by contact with an aqueous base:

$$ROH + R'Cl + NaOH \longrightarrow ROR' + NaCl + H_2O$$

where ROH represents a cellulose residue and one of its hydroxyl groups, and R'Cl is the organic halide. Chlorides are preferred over bromides or iodides, despite their lower reactivity, because of their higher diffusion rate in the heterogeneous reaction system used.

The substitution of the ether groups onto the individual glucose units in the cellulose chain is, of course, a random process. At any over-all degree of substitution a mixture of nonsubstituted and mono-, di-, and trisubstituted glucose units is present. The relative amounts of these can be calculated if the equilibrium and rate constants for all the reactions are known. These considerations apply to the esterification as well as the etherification of cellulose.

Manufacture Cellulose ethers are made by reacting cellulose, alkali, and the organic chloride at about 100°C. The ether content and the viscosity of the product are controlled by temperature, pressure, reaction time, and composition of the reacting mixture.

Ethyl cellulose   The most important of the cellulose ethers is ethyl cellulose. The commercial material has 2.4–2.5 ethoxy groups per glucose residue. It is a molding material which is heat stable and has low flammability and high impact strength. It is flexible and tough at temperatures as low as −40°C. Its electrical, mechanical, and weathering properties are good in comparison with those of other cellulosics, but not generally outstanding. Its softening point is low, and it has a high water absorption (but not as high as cellulose acetate) and some tendency to cold flow.

Other cellulose ethers   Methyl cellulose is an edible, water-soluble material widely used as a thickening agent for foods. It serves also as an emulsifying agent and an adhesive. Carboxymethyl cellulose (available as the sodium salt) has important textile uses.

GENERAL REFERENCES

Ott 1954–55; Bruxelles 1965; Haskell 1965; Hill 1965; Nissan 1965; Savage 1965; Rebenfeld 1967; Eastman 1969; Mitchell 1969; Neff 1969.

D. Aromatic Heterochain Thermoplastics

This section includes the description of those linear polymers having a phenyl group as part of the main chain, excepting the resoles and novolacs which are precursors of the phenolic resins discussed in Chapter 16*A* and polymers which have a phenyl group plus heterocyclic moieties in the main chain, the latter being discussed in Section *E*.

Polyesters

The properties of low melting point and high solubility caused Carothers (see Section *A*) to reject the linear aliphatic polyesters as fiber-forming candidates, and to this day no commercial product is based solely on these polymers. The stiffening action of the *p*-phenylene group in a polymer chain, however, leads to high melting points and good fiber-forming properties, as discussed in Chapter 7*A*. Commercially important polyesters are based on such polymers, of which poly(ethylene terephthalate) is the major product. These polymers are used both for films and for fibers, with a production of just over

1 billion lb of staple fiber and some 325 million lb of monofilament in 1969. The price of the staple fiber was $0.45–0.50/lb.

Polymerization and properties The production of high-molecular-weight polyesters differs somewhat from that of similar polyamides. In the case of the nylons, the chemical equilibrium favors the polyamide under readily achieved polymerization conditions. Stoichiometric equivalence is easily achieved by the use of salts, and amide interchange reactions are slow. In polyester formation, however, the equilibrium is much less favorable, and equivalence is more difficult to achieve, since the salts do not form. Furthermore, the dibasic aromatic acids are very difficult to purify because of low solubility and high melting point.

This situation has been met by taking advantage of the rapidity of ester-interchange reactions. The acid, such as terephthalic acid, is converted to the dimethyl ester, which can easily be purified by distillation or crystallization. This is then allowed to react with the glycol by ester interchange. In practice a low-molecular-weight glycol is used and the reaction takes place in two steps. First, a low-molecular-weight polyester is made with an excess of glycol to ensure hydroxyl end groups. Then the temperature is raised and pressure lowered to effect condensation of these molecules by ester interchange with the loss of the glycol.

In the production of poly(ethylene terephthalate), terephthalic acid is made by the oxidation of *p*-xylene. The polymerization step is similar to that for the polyamides in so far as the equipment and conditions are concerned. The polymer coming from the autoclave is quenched from the molten state to below its glass transition point of about 80°C and is amorphous. Crystallinity develops on heating; the crystalline melting point is 265°C. The polymer is melt spun.

Because of its high crystalline melting point and glass transition temperature, poly(ethylene terephthalate) retains good mechanical properties at temperatures up to 150–175°C. Its chemical and solvent resistance is good, being similar to that of nylon.

Fiber applications The properties of poly(ethylene terephthalate) fiber which influence its uses are its outstanding crease resistance and work recovery and its low moisture absorption. These properties arise from the stiff polymer chain and the resulting high modulus, and the fact that the interchain bonds are not susceptible to moisture. As a result, garments made from the polyester fibers are very resistant to wrinkling and can be washed repeatedly without subsequent ironing. The higher modulus of the polyester fibers and their wrinkle resistance are reminiscent of those of wool, but fully oriented polyester fibers are too stiff to have the crisp but soft hand of wool. It is possible, however, to impart this property to the polyesters by control of crystallinity and orientation so that their stress-strain curve is similar to that of wool. This is done in the production of staple polyester fiber.

The polyester fibers are seldom used alone, but are blended with cotton or wool to make summer- and medium-weight suiting and other goods. They have set new standards for the performance of wearing apparel through the "wash and wear" concept.

Polyester fibers are unique among the synthetics in their ability to form felts. These products are better than wool felts in resistance to temperature, to abrasion, and to further felting.

Polyester fiber is currently popular as a tire cord, with over 20% of the market (about 120 million lb in 1969), largely as a replacement for rayon. This may be a temporary situation, however, if the popularity of glass for the same purpose increases as anticipated.

Film applications   The tensile strength of poly(ethylene terephthalate) film is about 25,000 psi, two to three times that of cellophane or cellulose acetate film (Section *C*). If the area of the specimen at the break point is considered, the tensile strength of this plastic is about twice that of aluminum and almost equal to that of mild steel.

The stiffness of poly(ethylene terephthalate) film is comparable to that of cellophane and other cellulosic films, but its resistance to failure on repeated flexing is unusually high. In one flex test it lasts for over 20,000 cycles compared to a few hundred for cellulosic films. Its tear strength is also better than that of the cellulosics. The impact strength of this material three to four times that of any other plastic film. This toughness is a major advantage in typical applications, such as magnetic recording tape.

Polyurethanes

Polyurethanes are polymers containing the group

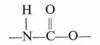

formed typically through the reaction of a diisocyanate and a glycol:

$$xOCNRCNO + xHOR'OH \longrightarrow —[OCONHRNHCOOR'—]_x$$

The polymers formed in this way are useful in four major types of product: foams, fibers, elastomers, and coatings. Polyurethane foams are invariably crosslinked; hence they are discussed with thermosetting resins Chapter 16E.

Elastomers Polyurethane elastomers are made in several steps. A *basic intermediate* is first prepared in the form of a low-molecular-weight (1000–2000) polymer with hydroxyl end groups. This may be a polyester, such as that made from ethylene glycol and adipic acid; a polyether; or a mixed polyester-polyamide.

The basic intermediate, which is here designated **B**, is then reacted with an aromatic diisocyanate to give a *prepolymer*. Typical isocyanates are 2,4-tolylenediisocyanate, 4,4-benzidenediisocyanate, and 1,5-naphthalene-diisocyanate: using the former as an example, a typical prepolymer can be represented as

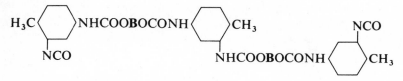

The elastomer is vulcanized through the isocyanate groups by reaction with glycols, diamines, diacids, or amino alcohols. If water is used, carbon dioxide is eliminated during crosslinking, as in the production of urethane foams (Chapter 17*E*).

Polyurethane elastomers are noted for extremely good abrasion resistance and hardness, combined with good elasticity and resistance to greases, oils, and solvents. They make tire treads with unusually long life and are widely used in applications requiring outstanding abrasion resistance, such as heel lifts and small industrial wheels.

About 25 million lb of polyurethane elastomers was produced in 1969.

Fibers Polyurethane fibers with unusually high elasticity are used for lightweight foundation garments and swimsuits. They have replaced rubber latex thread in this use. About 13 million lb was produced in 1969.

Coatings Coatings based on polyurethanes have very good resistance to abrasion and solvent attack plus good flexibility and impact resistance. They can be applied by dip, spray, or brush and adhere well to a wide variety of materials. They are suitable in applications for which unusual impact and abrasion resistance is required, such as gymnasium and dance floors and bowling pins, in magnet wire coatings, and in a variety of outdoor and marine uses because of their good weatherability. About 60 million lb was sold in 1969.

Poromerics Two-layer structures based on polyurethanes have had wide success as leatherlike materials used for high-quality shoes. They consist

of a base substrate in the form of a nonwoven web of poly(ethylene tereph
thalate) fiber impregnated with a porous polyurethane builder. This structure
is punctured with needles to provide vapor transmission. It is covered with
a vapor-permeable coating of a polyurethane with a minor amount of poly
(vinyl chloride). Often, a woven fabric interlayer is used between the web
and the coating (Yuan 1970).

Polycarbonate

A polycarbonate plastic, characterized by the —OCOO— heterochain
unit, can be made either from phosgene and bisphenol A (4,4'-dihydroxy
diphenyl-2,2'-propane) or by ester exchange between bisphenol A and
diphenyl carbonate. It has this structure:

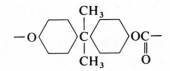

Like the nylon, acetal, and polyether resins, this polymer is a crystalline
thermoplastic of very good mechanical properties. It has unusually high
impact strength, even at low temperatures, ascribed in part to a combination
of relatively high ordering in the amorphous regions and considerable dis
ordering in the crystalline regions. It has low moisture absorption, good
heat resistance (useful to 140°C), and good thermal and oxidative stability
in the melt. It is transparent and self-extinguishing. It can be processed
by conventional injection molding and extrusion. Typical applications
include telephone parts, business machine housings, machinery housings
safety equipment, and graphic arts film bases.

About 30 million lb of the polycarbonate plastics was produced in the
United States in 1969.

High-temperature aromatic polymers

Much interest has been generated in recent years in polymers which
retain good mechanical properties at very high temperatures. Many of those
of current interest are discussed in Section E, but a few fall into the category
of this section.

Poly(phenylene oxide)   The polymer poly(2,6 dimethyl phenylene oxide)
its blends with polystyrene, and both in glass-fiber filled form, have excel
lent dimensional stability at elevated temperatures, good electrical proper
ties, and good resistance to aqueous environments. Water absorption
exceptionally low.

Polysulfones   These materials are characterized by the diphenylene sulfone repeating unit. They have high heat-deflection temperatures and very low creep rates. A polyarylsulfone can be used for extended time periods, retaining good tensile and compressive strength, at 260°C.

Poly(p-oxybenzoate)   This material is stated to resist temperatures of 320°C, to have a self-lubricating surface, and to be virtually inert to solvents and corrosive liquids.

Aromatic polyamide   A poly(phenylene amide) is available commercially as a fiber which retains 50% of its room-temperature strength after 1000 hours at 260°C or 200 hours at 300°C. It does not melt and is ignited only with great difficulty by direct exposure to flame.

Other high-temperature materials in this category are the poly(*p*-phenylenes), polyxylylenes, and the phenoxy polymers, linear analogs of the epoxies discussed in Chapter 16*D*.

GENERAL REFERENCES

Polyesters: Rebenfeld 1967; Anon. 1969; Farrow 1969; Goodman 1969; Hawthorne 1969.
Polyurethanes: Saunders 1962, 1964; Ibrahim 1967; Blokland 1968; Hollowell 1968; Blackfan 1969; Bruins 1969; Carvey 1969; David 1969; Pigott 1969; Wright 1969.
Polycarbonates: Schnell 1964; Bottenbruch 1969; Devin 1969; Rammrath 1969.
High-temperature polymers: Frazer 1968, 1969; Gowan 1969; Hale 1969; Johnson 1969; Kovacic 1969; Walton 1969; Burns 1970.

E. Heterocyclic, Ladder, and Inorganic Polymers

The requirements of modern technology, including those related to the "space age," place increasing demands on the high-temperature behavior of all materials, including polymers. For the last few years there has been an increased effort to define and produce polymeric structures capable of demonstrating good mechanical properties for long periods of time at higher and higher temperatures.

Mark (1967) has pointed out several ways in which the melting or softening points of polymers can be raised, with concomitant improvement in high-temperature properties. These are: stiffening the chains by the addition of ring structures or other stiff elements; crosslinking the chains; and inducing crystallization. With our present capabilities, it has not been possible to devise structures in which all three of these approaches are combined; for example, crystallinity and crosslinking are still largely mutually exclusive. But the application of one or two of these principles has proved fruitful.

The results of adding aromatic rings to polymer chains have been described in Chapter 7*A* and at the end of Section *D* of this chapter. The addition of other ring structures, usually heterocyclic, to form fused-ring groups even larger and stiffer than the phenylene moiety, is a logical extension of this idea. Beyond this, two approaches are suggested: one, attributed (Fraser 1969) to Marvel, is to produce ladder polymers (Chapter 1*A*) whose structure is such that breaking one or a few bonds can not sever the chain; the other is to abandon polymers based on the carbon-carbon bond and go to other structures known collectively as inorganic polymers.

C. S. Marvel has stated (private communication) that the upper temperature limit for the stability of carbon-chain polymers in a nitrogen atmosphere is probably around 550–600°C, unless the structure is very similar to that of graphite itself (as "black orlon" may be—see below). If the polymer is completely free from hydrogen, the stability in air and in nitrogen should be essentially the same. These limits are conventionally determined by dynamic TGA (Chapter 4*E*); isothermal weight loss experiments at high temperatures sometimes yield other results.

Heterocyclic polymer

Polyimide   A polyimide with the structure

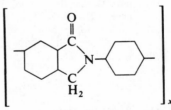

is made by the polycondensation of an aromatic dianhydride and an aromatic amine. The first step is a soluble polyamide-acid, which is converted to the polyimide by further condensation.

This material is available as a film and as fabricated solid parts. The polyamide-acid is also sold as a coating solution. The polyimide retains usable properties at 300°C for months, and at 400°C for a few hours, and withstands exposures of a few minutes to temperatures well over 500°C.

Polybenzimidazole   A polymer with the general structure

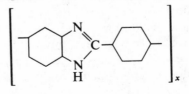

is available in the form of syntactic foams consisting of blends with silica of phenolic microballoons and alumina-silica fibers. The fibers have a maximum continuous-use temperature of over 300°C, but retain some strength even at 500°C.

Other heterocyclic polymers A number of heterocyclic polymer structures has been synthesized and tested for high-temperature use, but none has yet become commercially available. Some of these structures are listed in Table 15-1.

TABLE 15-1. *Structures of some experimental aromatic heterocyclic polymers (Fraser 1969)*

Polymer	Repeat Unit
Polybenzothiazole	
Polybenzoxazole	
Polythiadiazole	
Polyoxadiazole	
Polyphenyltriazole	

Ladder polymers

Ladder polymers are defined as those consisting of an uninterrupted series of rings connected by links around which rotation cannot occur except by bond breaking. If their structure were perfect, the chain of a ladder polymer could be broken only if at least two bonds on the same ring were broken. In a degradation process in which all bonds are equally strong,

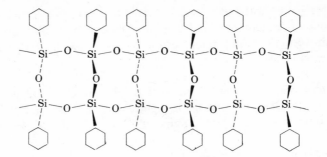

Fig. 15-1.   The ladder polymer polyphenylsilsesquioxane (Brown 1963).

this would be far less probable than a single rupture breaking a single-chain polymer. Thus, ladder polymers should have far greater thermal stability.

Despite the synthesis of many examples of ladder polymers, and the qualitative observation of their good thermal stability, there have been few observations giving direct proof of the above hypothesis. This is no doubt the result in part of difficulties in forcing reactions to completion and maintaining purity required to form perfect ladder structures.

Syntheses and properties   The most nearly perfect and best characterized ladder polymer described in a recent review (Oberberger 1970) is the poly phenylsilsesquioxane described by Brown (1963) (Fig. 15-1). It was produced by an *equilibration* process, in which a prepolymer is subjected to reaction conditions which allow bond reorganization. Here, a phenyl trichlorosilane hydrolysate was equilibrated in the presence of an alkaline rearrangement catalyst and a small amount of solvent for 1 hr at 250°C. The polymers had weight-average molecular weights in the millions, and were markedly more resistant to hydrolysis than linear silicones (Chapter 16F).

A second synthesis method, termed "*zipping up*," involves making linear polymer containing reactive functional groups along the chain, and then linking these groups to form the second part of the ladder. There are several well-known examples: one is the cyclization of poly(methyl vinyl ketone):

$$-CH_2-CH-CH_2-CH-CH_2-CH-$$

with pendant groups:

$$
\begin{array}{ccc}
\text{C=O} & \text{C=O} & \text{C=O} \\
| & | & | \\
\text{CH}_3 & \text{CH}_3 & \text{CH}_3
\end{array}
$$

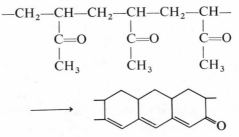

A second is the well-studied production of "black orlon" by the simultaneous cyclization and oxidation at 160–300°C of polyacrylonitrile, originally reported by Houtz (1950). The final material, whose structure may be

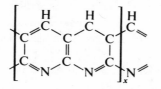

or may be an oxidized and aromatized reaction product of an intermediate of this structure, glows red but maintains its form in a blowtorch flame (as do some of the heterocyclic polymers described above), withstanding temperatures of 700–800°C without loss in properties (which are admittedly not too good).

Multifunctional condensation reactions can lead to ladder polymers if the tendency to form 5- or 6-membered rings (particularly when conjugated and aromatic) rather than crosslinking is used as a driving force. An example of *polyheterocyclization* is the formation of a phenoxazine ladder by the reaction of a substituted quinone with an aminophenol:

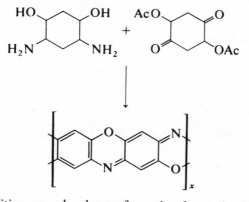

Simpler *cycloaddition* can also be performed, often using the Diels-Alder reaction. These and many other examples are discussed by Overberger (1970).

Inorganic polymers

This subject has been reviewed in several recent books (Stone 1962; Hunter 1964; Andrianov 1965), and by Fraser (1968). The latter author describes three major classes of inorganic polymers on which research has been concentrated.

1. Organic-inorganic polymers in which organic substituents are placed on inorganic chains. The outstanding example, and the only family of inorganic polymers to date, is the silicones, discussed in Chapter 16F. Research on other families of this type has centered on replacing the silicon in silicone-like structures with other elements, such as aluminum, tin, titanium, or boron. The objective is to improve the already good thermal stability of the silicones.

2. Metal chelate polymers, sometimes called coordination polymers (not to be confused with the entirely different conventional organic polymers produced by coordination polymerization—Chapter 10D). These materials can be prepared in several ways: by linking polydentate ligands by metal ions, as in metal acetonylacetonates; by polymer formation in the presence of metals, as in the production of polyphthalocyanines; by incorporation of metal ions in preformed polymers; by reacting chelates containing functional groups, as in the polymerization of basic beryllium carboxylates; and by the preparation of polymers containing ferrocene groups, such as biscyclopentadienyl iron.

3. Completely inorganic linear polymers based on silicon-nitrogen, phosphorous-nitrogen, or boron-nitrogen chains.

GENERAL REFERENCES

Fraser 1968, 1969; and specifically:
Heterocyclic polymers: Levine 1969; Sroog, 1969; Du Pont 1969; Whittaker 1969; Burns 1970.
Ladder polymers: Bailey 1968; Overberger 1970.
Inorganic polymers: Stone 1962; Hunter 1964; Andrianov 1965; Teach 1965; Block 1966; Arledter 1967; Rebenfeld 1967; Venezky 1967; Neuse 1968; Currell 1969; Sander 1969.

BIBLIOGRAPHY

Akin 1962. R. B. Akin, *Acetal Resins*, Reinhold Publishing Corp., New York, 1962.
Andrianov 1965. K. A. Andrianov, *Metalorganic Polymers*, Interscience Div., John Wiley and Sons, New York, 1965.
Anon. 1969. Anon., "Polyester films," pp. 356, 358–359 in Sidney Gruss, ed., *Modern Plastics Encyclopedia 1969–1970* (McGraw-Hill Book Co., New York) **46** (10A) October, 1969.
Arledter 1967. H. F. Arledter, F. L. Pundsack, and W. O. Jackson, "Fibers, Inorganic," pp. 610–690 in Herman F. Mark, Norman G. Gaylord, and Norbert M. Bikales, eds., *Encyclopedia of Polymer Science and Technology*, Vol. 6, Interscience Div., John Wiley and Sons, New York, 1967.
Bailey 1968. William J. Bailey, "Ladder and Spiro Polymers," pp. 97–120 in Herman F. Mark, Norman G. Gaylord, and Norbert M. Bikales, eds., *Encyclopedia of Polymer*

Science and Technology, Vol. 6, Interscience Div., John Wiley and Sons, New York, 1968.

Berenbaum 1969. M. B. Berenbaum, "Polysulfide Polymers," pp. 425–447 in Herman F. Mark, Norman G. Gaylord, and Norbert M. Bikales, eds., *Encyclopedia of Polymer Science and Technology*, Vol. 11, Interscience Div., John Wiley and Sons, New York, 1969.

Bevington 1964. J. C. Bevington and H. May, "Aldehyde Polymers," pp. 609–628 in Herman F. Mark, Norman G. Gaylord, and Norbert M. Bikales, eds., *Encyclopedia of Polymer Science and Technology*, Vol. 1, Interscience Div., John Wiley and Sons, New York, 1964.

Blackfan 1969. C. L. Blackfan, "Thermoplastic polyurethane film and sheet," pp. 364, 368 in Sidney Gruss, ed., *Modern Plastics Encyclopedia 1969–1970* (McGraw-Hill Book Co., New York) **46** (10A), October, 1969.

Block 1966. B. B. Block, "Coordination Polymers," pp. 150–165 in Herman F. Mark, Norman G. Gaylord, and Norbert M. Bikales, eds., *Encyclopedia of Polymer Science and Technology*, Vol. 4, Interscience Div., John Wiley and Sons, New York, 1966.

Blokland 1968. R. Blokland, *Elasticity and Structure of Polyurethane Networks*, Gordon and Breach, New York, 1968.

Bottenbruch 1969. L. Bottenbruch, "Polycarbonates," pp. 710–764 in Herman F. Mark, Norman G. Gaylord, and Norbert M. Bikales, eds., *Encyclopedia of Polymer Science and Technology*, Vol. 10, Interscience Div., John Wiley and Sons, New York, 1969.

Brown 1963. John F. Brown, Jr., "Double Chain Polymers and Nonrandom Crosslinking," *J. Polymer Sci.* **C1**, 83–97 (1963).

Bruins 1969. Paul F. Bruins, ed., *Polyurethane Technology*, Interscience Div., John Wiley and Sons, New York, 1969.

Bruxelles 1965. G. N. Bruxelles and V. R. Grassie, "Cellulose Esters, Inorganic," pp. 307–325 in Herman F. Mark, Norman G. Gaylord, and Norbert M. Bikales, eds., *Encyclopedia of Polymer Science and Technology*, Vol. 3, Interscience Div., John Wiley and Sons, New York, 1965.

Burns 1970. R. L. Burns, "Recent Developments in Thermally Resistant Polymers," *J. Oil Colour Chem. Assoc.* **53**, 52–68 (1970).

Carvey 1969. R. M. Carvey, "Urethane elastomers," p. 215 in Sidney Gruss, ed., *Modern Plastics Encyclopedia 1969–1970* (McGraw-Hill Book Co., New York) **46** (10A), October, 1969.

Currell 1969. B. R. Currell and M. J. Frazer, "Inorganic Polymers," *R.I.C. Reviews* **2**, 13–40 (1969).

David 1969. D. J. David and H. B. Staley, *Analytical Chemistry of the Polyurethanes*, Interscience Div., John Wiley and Sons, New York, 1969.

Devin 1969. Paul E. Devin, "Polycarbonate film and sheet," p. 362 in Sidney Gruss, ed., *Modern Plastics Encyclopedia 1969–1970* (McGraw-Hill Book Co., New York) **46** (10A), October, 1969.

Du Pont 1969. Du Pont Co., "Polyimide film," p. 371 in Sidney Gruss, ed., *Modern Plastics Encyclopedia 1969–1970* (McGraw-Hill Book Co., New York) **46** (10A), October, 1969.

Eastman 1969. Eastman Chemical Products, Inc., "Cellulosics film and sheet," pp. 360–362 in Sidney Gruss, ed., *Modern Plastics Encyclopedia 1969–1970* (McGraw-Hill Book Co., New York) **46** (10A), October, 1969.

Epstein 1969. B. N. Epstein and E. W. Kjellmark, Jr., "Acetal homopolymers," pp. 84–85 in Sidney Gruss, ed., *Modern Plastics Encyclopedia 1969–1970* (McGraw-Hill Book Co., New York) **46** (10A), October, 1969.

Farrow 1969. G. Farrow, E. S. Hill, and P. L. Weinle, "Polyester Fibers," pp. 1–41 in Herman F. Mark, Norman G. Gaylord, and Norbert M. Bikales, eds., *Encyclopedia of Polymer Science and Technology*, Vol. 11, Interscience Div., John Wiley and Sons, New York, 1969.

Frazer 1968. A. H. Frazer, *High Temperature Resistant Polymers*, Interscience Div., John Wiley and Sons, New York, 1968.

Frazer 1969. A. H. Frazer, "High-Temperature Plastics," *Sci. Amer.* **221** (1), 96–100, 103–105 (1969).

Furukawa 1963. Junji Furukawa and Takeo Saegusa, *Polymerization of Aldehydes and Oxides*, Interscience Div., John Wiley and Sons, New York, 1963.

Gaylord 1962. Norman G. Gaylord, ed., *Polyethers. Part 3: Polyalkylene Sulfides and Other Polythioethers*, Interscience Div., John Wiley and Sons, New York, 1962.

Gaylord 1963. Norman G. Gaylord, ed., *Polyethers. Part 1: Polyalkylene Oxides*, Interscience Div., John Wiley and Sons, New York, 1963.

Goodman 1969. I. Goodman, "Polyesters," pp. 62–128 in Herman F. Mark, Norman G. Gaylord, and Norbert M. Bikales, eds., *Encyclopedia of Polymer Science and Technology*, Vol. 11, Interscience Div., John Wiley and Sons, New York, 1969.

Gowan 1969. A. C. Gowan, "Phenylene oxides," p. 164 in Sidney Gruss, ed., *Modern Plastics Encyclopedia 1969–1970* (McGraw-Hill Book Co., New York) **46** (10A), October, 1969.

Habermann 1969. K. R. Habermann, "Nylon films," p. 360 in Sidney Gruss, ed., *Modern Plastics Encyclopedia 1969–1970* (McGraw-Hill Book Co., New York) **46** (10A), October, 1969.

Hale 1969. Warren F. Hale, "Phenoxy Resins," pp. 111–122 in Herman F. Mark, Norman G. Gaylord, and Norbert M. Bikales, eds., *Encyclopedia of Polymer Science and Technology*, Vol. 10, Interscience Div., John Wiley and Sons, New York, 1969.

Haskell 1965. V. C. Haskell, "Cellophane," pp. 60–79 in Herman F. Mark, Norman G Gaylord, and Norbert M. Bikales, eds., *Encyclopedia of Polymer Science and Technology*, Vol. 3, Interscience Div., John Wiley and Sons, New York, 1965.

Hawthorne 1969. J. M. Hawthorne and C. J. Heffelfinger, "Polyester Films," pp. 42–61 in Herman F. Mark, Norman G. Gaylord, and Norbert M. Bikales, eds., *Encyclopedia of Polymer Science and Technology*, Vol. 11, Interscience Div., John Wiley and Sons, New York, 1969.

Hill 1965. Roy O. Hill, Jr., B. P. Rouse, Jr., B. Sheldon Sprague, Lawrence I. Horner and David J. Stanonis, "Cellulose Esters—Organic," pp. 325–454 in Herman F. Mark Norman G. Gaylord, and Norbert M. Bikales, eds., *Encyclopedia of Polymer Science and Technology*, Vol. 3, Interscience Div., John Wiley and Sons, New York, 1965.

Hollowell 1968. J. L. Hollowell, "Leather-like Materials," pp. 210–231 in Herman F Mark, Norman G. Gaylord, and Norbert M. Bikales, eds., *Encyclopedia of Polymer Science and Technology*, Vol. 8, Interscience Div., John Wiley and Sons, New York 1968.

Houtz 1950. R. C. Houtz, "'Orlon' Acrylic Fiber: Chemistry and Properties," *Textil Res. J.* **20**, 786–801 (1950).

Hunter 1964. D. N. Hunter, *Inorganic Polymers*, John Wiley and Sons, New York 1964.

Ibrahim 1967. S. M. Ibrahim and A. J. Ultee, "Fibers, Elastomeric," pp. 573–592 in Herman F. Mark, Norman G. Gaylord, and Norbert M. Bikales, eds., *Encyclopedia of Polymer Science and Technology*, Vol. 6, Interscience Div., John Wiley and Sons New York, 1967.

Johnson 1969. R. N. Johnson, "Polysulfones," pp. 447–463 in Herman F. Mark, Norman G. Gaylord, and Norbert M. Bikales, eds., *Encyclopedia of Polymer Science and Technology,* Vol. 11, Interscience Div., John Wiley and Sons, New York, 1969.

King 1969. N. E. King, "Chlorinated polyether," p. 122 in Sidney Gruss, ed., *Modern Plastics Encyclopedia 1969–1970* (McGraw-Hill Book Co., New York) **46** (10A), October, 1969.

Kovacic 1969. Peter Kovacic and Fred W. Koch, "Poly(phenylenes)," pp. 380–389 in Herman F. Mark, Norman G. Gaylord, and Norbert M. Bikales, eds., *Encyclopedia of Polymer Science and Technology,* Vol. 11, Interscience Div., John Wiley and Sons, New York, 1969.

Levine 1969. H. H. Levine, "Polybenzimidazoles," pp. 188–232 in Herman F. Mark, Norman G. Gaylord, and Norbert M. Bikales, eds., *Encyclopedia of Polymer Science and Technology,* Vol. 11, Interscience Div., John Wiley and Sons, New York, 1969.

Mark 1940. H. Mark and G. Stafford Whitby, eds., *Collected Papers of Wallace Hume Carothers on High Polymeric Substances,* Interscience Publishers, New York, 1940.

Mark 1967. Herman F. Mark, "The Nature of Polymeric Materials," *Sci. Amer.* **217** (3), 148–154, 156 (Sept., 1967).

Mitchell 1969. R. L. Mitchell and G. C. Daul, "Rayon," pp. 810–847 in Herman F. Mark, Norman G. Gaylord, and Norbert M. Bikales, eds., *Encyclopedia of Polymer Science and Technology,* Vol. 11, Interscience Div., John Wiley and Sons, New York, 1969.

Neff 1969. S. B. Neff, "Cellulosics," pp. 117–118 in Sidney Gruss, ed., *Modern Plastics Encyclopedia 1969–1970* (McGraw-Hill Book Co., New York) **46** (10A), October, 1969.

Neuse 1968. Eberhard W. Neuse, "Metallocene Polymers," pp. 667–692 in Herman F. Mark, Norman G. Gaylord, and Norbert M. Bikales, eds., *Encyclopedia of Polymer Science and Technology,* Vol. 8, Interscience Div., John Wiley and Sons, New York, 1968.

Nissan 1965. Alfred H. Nissan, Gunther K. Hunger, and S. S. Sternstein, "Cellulose," pp. 131–226 in Herman F. Mark, Norman G. Gaylord, and Norbert M. Bikales, eds., *Encyclopedia of Polymer Science and Technology,* Vol. 3, Interscience Div., John Wiley and Sons, New York, 1965.

Ott 1954–55. Emil Ott, Harold M. Spurlin, and Mildred W. Grafflin, eds., *Cellulose and Cellulose Derivatives,* Interscience Publishers, New York, Parts I and II, 1954; Part III, 1955.

Overberger 1970. C. G. Overberger and J. A. Moore, "Ladder Polymers," *Fortschr. Hochpolym.-Forsch. (Advances in Polymer Science)* **7,** 113–150 (1970).

Peerman 1969. D. E. Peerman, "Polyamides from Fatty Acids," pp. 597–615 in Herman F. Mark, Norman G. Gaylord, and Norbert M. Bikales, eds., *Encyclopedia of Polymer Science and Technology,* Vol. 10, Interscience Div., John Wiley and Sons, New York, 1969.

Pigott 1969. K. A. Pigott, "Polyurethanes," pp. 506–563 in Herman F. Mark, Norman G. Gaylord, and Norbert M. Bikales, eds., *Encyclopedia of Polymer Science and Technology,* Vol. 11, Interscience Div., John Wiley and Sons, New York, 1969.

Rammrath 1969. H. G. Rammrath, "Polycarbonates," p. 169 in Sidney Gruss, ed., *Modern Plastics Enclyclopedia 1969–1970* (McGraw-Hill Book Co., New York) **46** (10A), October, 1969.

Rebenfeld 1967. Ludwig Rebenfeld, "Fibers," pp. 505–573 in Herman F. Mark, Norman G. Gaylord, and Norbert M. Bikales, eds., *Encyclopedia of Polymer Science and Technology,* Vol. 6, Interscience Div., John Wiley and Sons, New York, 1967.

Sander 1969. Manfed Sander and H. R. Allcock, "Phosphorus-Containing Polymers," pp. 123–144 in Herman F. Mark, Norman G. Gaylord, and Norbert M. Bikales, eds., *Encyclopedia of Polymer Science and Technology*, Vol. 10, Interscience Div., John Wiley and Sons, New York, 1969.

Saunders 1962. J. H. Saunders and K. C. Frisch, *Polyurethanes: Chemistry and Technology. Part 1: Chemistry*, Interscience Div., John Wiley and Sons, New York, 1962.

Saunders 1964. J. H. Saunders and K. C. Frisch, *Polyurethanes: Chemistry and Technology. Part 2: Technology,* Interscience Div., John Wiley and Sons, New York, 1964.

Savage 1965. A. B. Savage, E. D. Klug, Norbert N. Bikales, and David J. Stanonis, "Cellulose Ethers," pp. 459–549 in Herman F. Mark, Norman G. Gaylord, and Norbert M. Bikales, eds., *Encyclopedia of Polymer Science and Technology*, Vol. 3, Interscience Div., John Wiley and Sons, New York, 1965.

Schar 1969. W. C. Schar, "Acetal copolymers," pp. 85–86 in Sidney Gruss, ed., *Modern Plastics Encyclopedia 1969–1970* (McGraw-Hill Book Co., New York) **46** (10A), October, 1969.

Schell 1964. Hermann Schnell, *Chemistry and Physics of Polycarbonates*, Interscience Div., John Wiley and Sons, New York, 1964.

Schule 1969. E. C. Schule, "Polyamide Plastics," pp. 460–482 in Herman F. Mark, Norman G. Gaylord, and Norbert M. Bikales, eds., *Encyclopedia of Polymer Science and Technology*, Vol. 10, Interscience Div., John Wiley and Sons, New York, 1969.

Sherwood 1969. L. T. Sherwood, Jr., "Nylons (polyamides)," pp. 154, 159 in Sidney Gruss, ed., *Modern Plastics Encyclopedia 1969–1970* (McGraw-Hill Book Co., New York) **46** (10A), October, 1969.

Snider 1969. O. E. Snider and R. J. Richardson, "Polyamide Fibers," pp. 347–460 in Herman F. Mark, Norman G. Gaylord, and Norbert M. Bikales, eds., *Encyclopedia of Polymer Science and Technology*, Vol. 10, Interscience Div., John Wiley and Sons, New York, 1969.

Sroog 1969. C. E. Sroog, "Polyimides," pp. 247–272 in Herman F. Mark, Norman G. Gaylord, and Norbert M. Bikales, eds., *Encyclopedia of Polymer Science and Technology*, Vol. 11, Interscience Div., John Wiley and Sons, New York, 1969.

Staudinger 1925. H. Staudinger, "The Constitution of Polyoxymethylenes and Other High-Molecular Compounds" (in German), *Helv. Chim. Acta* **8**, 67–70 (1925).

Stone 1962. F. G. A. Stone and W. A. G. Graham, eds., *Inorganic Polymers*, Academic Press, New York, 1962.

Stone 1967. F. W. Stone, J. J. Stratta, Louis C. Pizzini, John J. Patton, Jr., J. Furukawa, and T. Saegusa, "1,2-Epoxide Polymers," pp. 103–195 in Herman F. Mark, Norman G. Gaylord, and Norbert M. Bikales, eds., *Encyclopedia of Polymer Science and Technology*, Vol. 6, Interscience Div., John Wiley and Sons, New York, 1967.

Sweeny 1969. W. Sweeny and J. Zimmerman, "Polyamides," pp. 483–597 in Herman F. Mark, Norman G. Gaylord, and Norbert M. Bikales, eds., *Encyclopedia of Polymer Science and Technology*, Vol. 10, Interscience Div., John Wiley and Sons, New York, 1969.

Teach 1965. W. C. Teach and Joseph Green, "Boron-Containing Polymers," pp. 581–604 in Herman F. Mark, Norman G. Gaylord, and Norbert M. Bikales, eds., *Encyclopedia of Polymer Science and Technology*, Vol. 2, Interscience Div., John Wiley and Sons, New York, 1965.

Venezky 1967. David L. Venezky, "Inorganic Polymers," pp. 664–691 in Herman F. Mark, Norman G. Gaylord, and Norbert M. Bikales, eds., *Encyclopedia of Polymer Science and Technology*, Vol. 7, Interscience Div., John Wiley and Sons, New York, 1967.

Walton 1969. R. K. Walton, "Polysulfone," pp. 198–204 in Sidney Gruss, ed., *Modern Plastics Encyclopedia 1969–1970* (McGraw-Hill Book Co., New York) **46** (10A), October, 1969.

Whittaker 1969. Whittaker Corp., "Syntactic polybenzimidazole," pp. 261–262 in Sidney Gruss, ed., *Modern Plastics Encyclopedia 1969–1970* (McGraw-Hill Book Co., New York) **46** (10A), October, 1969.

Wright 1969. P. Wright and A. P. C. Cumming, *Solid Polyurethane Elastomers*, Gordon and Breach, New York, 1969.

Yuan 1970. E. L. Yuan, "The Structure and Property Relationships of Poromeric Materials," presented at the 159th National ACS Meeting, Houston, Texas, Feb. 22-27, 1970.

16

Thermosetting Resins

Thermosetting resins are those which change irreversibly under the influence of heat from a fusible and soluble material into one which is infusible and insoluble through the formation of a covalently crosslinked, thermally stable network. In contrast, *thermoplastic* polymers, discussed in Chapters 13–15, soften and flow when heat and pressure are applied, the changes being reversible.

In some of the systems considered in this chapter, network formation occurs with little or no heat required, as in the production of urethane foams and the use of unsaturated polyester resins. Furthermore, vulcanized rubbers consist of covalently crosslinked network polymers, usually formed by the application of heat. However, this network is generated in a separate postpolymerization step. With the exception of the silicone rubbers, thermosetting resins discussed in this chapter are those in which crosslinking occurs simultaneously with the final stages of polymerization, regardless of the amount of heat required in this step.

The most important thermosetting resins, both from a historical standpoint and in current commercial application, are formaldehyde condensation products with phenol (*phenolic resins*) or with urea or melamine (*amino resins*). Other thermosetting types include the epoxy resins, the unsaturated polyester resins, urethane foams, the alkyds widely used for surface coating, and minor types.

A. Phenolic Resins

Phenolic resins have been in commercial use longer than any other synthetic polymer except cellulose nitrate. In contrast to the latter, however, the production of phenolics continues to rise, with some fluctuations, being about 1.2 billion lb in 1969, at a price of $0.22/lb.

Reactions of phenol and formaldehyde

Phenols react with aldehydes to give condensation products if there are free positions on the benzene ring *ortho* and *para* to the hydroxyl group. Formaldehyde is by far the most reactive aldehyde and is used almost exclusively in commercial production. The reaction is always catalyzed, either by acids or by bases. The nature of the product is greatly dependent on the type of catalyst and the mole ratio of the reactants.

Methylolation The first step in the reaction is the formation of addition compounds known as methylol derivatives, the reaction taking place at the *ortho* or *para* position:

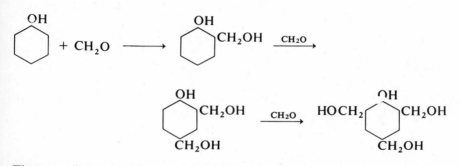

These products, which may be considered the monomers for subsequent polymerization, are formed most satisfactorily under neutral or alkaline conditions.

Novolac formation In the presence of acid catalysts, and with the mole ratio of formaldehyde to phenol less than 1, the methylol derivatives condense with phenol to form, first, dihydroxydiphenyl methane:

and, on further condensation and methylene bridge formation, fusible and soluble linear low polymers called *novolacs* with the structure

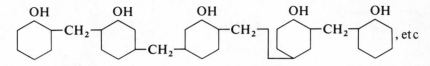

where *ortho* and *para* links occur at random. Molecular weights may range as high as 1000, corresponding to about ten phenyl residues. These materials do not themselves react further to give crosslinked resins, but must be reacted with more formaldehyde to raise its mole ratio to phenol above unity.

Resole formation   In the presence of alkaline catalysts and with more formaldehyde, the methylol phenols can condense either through methylene linkages or through ether linkages. In the latter case, subsequent loss of formaldehyde may occur with methylene bridge formation:

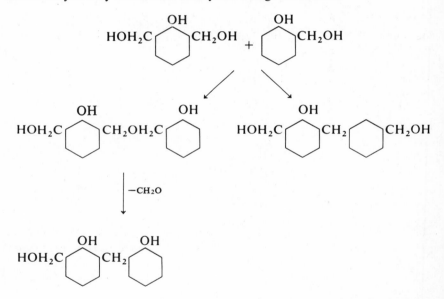

Products of this type, soluble and fusible but containing alcohol groups, are called *resoles*. If the reactions leading to their formation are carried further, large numbers of phenolic nuclei can condense to give network formation.

Summary   The four major reactions in phenolic resin chemistry are as follows: (*a*) addition to give methylol phenols; (*b*) condensation of a methylol phenol and a phenol to give a methylene bridge; (*c*) condensation of two methylol groups to give an ether bridge; and (*d*) decomposition of ether bridges to methylene bridges and formaldehyde, the latter reacting again by the first mode.

Production of phenolic resins

The formation of resoles and novolacs, respectively, leads to the production of phenolic resins by *one-stage* and *two-stage* processes.

One-stage resin In the production of a one-stage phenolic resin, all the necessary reactants for the final polymer (phenol, formaldehyde, and catalyst) are charged into a resin kettle and reacted together. The ratio of formaldehyde to phenol is about 1.25:1, and an alkaline catalyst is used.

Two-stage resin These resins are made with an acid catalyst, and only part of the necessary formaldehyde is added to the kettle, producing a mole ratio of about 0.8:1. The rest is added later as hexamethylenetetramine, which decomposes in the final curing step, with heat and moisture present, to yield formaldehyde and ammonia which acts as the catalyst for curing.

Resin formation The procedures for one- and two-stage resins are similar, and the same equipment is used for both. The reaction is exothermic and cooling is required. The formation of a resole or a novolac is evidenced by an increase in viscosity. Water is then driven off under vacuum, and a thermoplastic *A-stage* resin, soluble in organic solvents, remains. This material is dumped from the kettle, cooled, and ground to a fine powder.

At this point fillers, colorants, lubricants, and (if a two-stage resin) enough hexamethylenetetramine to give a final formaldehyde:phenol mole ratio of 1.5:1 are added. The mixture is rolled on heated mixing rolls, where the reactions are carried further, to the point where the resin is in the *B-stage*, nearly insoluble in organic solvents but still fusible under heat and pressure. The resin is then cooled and cut into final form. The *C-stage*, the final, infusible, crosslinked polymer, is reached on subsequent fabrication— e.g., by molding.

Properties and applications

Molding resins Some 35% of the phenolic resins produced in this country is used in molding applications. In addition to compression molding, the phenolics are now widely injection molded (Chapter 17*A*). The products are outstanding in heat resistance, dimensional stability, and resistance to cold flow. They are widely used for their good dielectric properties in electrical, automotive, radio and television, and appliance parts.

Fillers are almost always used in phenolic molding applications, both to improve properties and to reduce cost. Commonly chosen are wood flour, cotton flock, chopped rags, asbestos, fibrous glass and other fibers, and nitrile rubbers (see also Chapter 17*C*).

The ability of phenolic resins to withstand very high temperatures briefly is important for their use in missile nose cones. Whereas ceramics melt and metals vaporize, phenolics carbonize, maintaining a protective thermal barrier.

Laminating resins   For impregnating paper, wood, and other fillers, about 10–15% of the phenolic resins is produced as alcoholic solutions of one-stage resins (essentially varnishes). These are used to produce decorative laminates for counter tops and wall coverings and industrial laminates for electrical parts, including printed circuits. The impregnated filler is dried in an air oven to remove volatiles and then is hot pressed between polished platens. Shaped products, including cutlery handles and toilet seats, are similarly made.

Bonding resins   A number of industrial applications are based on the excellent adhesive properties and bonding strength of the phenolics. These include the production of brake linings, abrasive wheels and sandpaper, and foundry molds (sand-filled).

Coatings and adhesives   Phenolic resins are widely used in varnishes, electrical insulation, and other protective coatings. Heat-setting adhesives based on phenolics are used in producing most plywood. Some 35% of United States production is utilized in these fields.

Ion-exchange resins   Phenolics are widely used in the production of ion exchange resins, with amine, sulfonic acid, hydroxyl, or phosphoric acid functional groups.

GENERAL REFERENCES

Kentgen 1969; Kuzmak 1969; Leis 1969.

B. Amino Resins

The two important classes of amino resins are the condensation products of urea and of melamine with formaldehyde. They are considered together here because of the similarity in their production and applications. In general, the melamine resins have somewhat better properties but are higher in price. Production of amino resins in the United States in 1969 was about 800 million lb, at $0.32/lb for the urea resins.

Chemistry and production

Both melamine (I), a trimer of cyanamide, and urea react with formal-

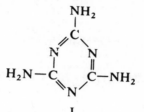

I

dehyde, first by addition to form methylol compounds and then by condensation in reactions much like those of phenol and formaldehyde. The methylol reaction takes the form

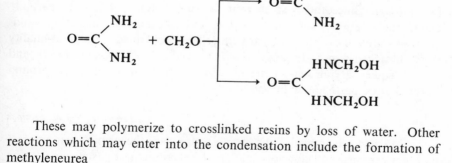

These may polymerize to crosslinked resins by loss of water. Other reactions which may enter into the condensation include the formation of methyleneurea

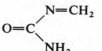

and the following crosslinking reactions:

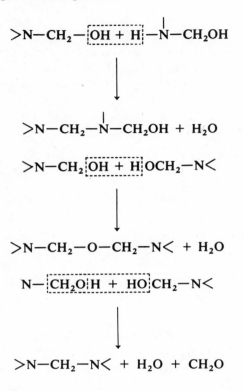

The production of the amino resins is similar to that of phenolic resins (Section *A*). Since the A-stage resin is water-soluble, it is only partially dehydrated, the water solution being used to impregnate the filler. The molding resins are almost always filled with cellulose obtained from good-quality sulfite-bleached paper. Impregnation is carried out in a vacuum mixer, and the subsequent drying step carries the resin to the B-stage. It is then ground to the desired particle size in ball mills.

Properties and applications

A distinct advantage of the amino resins over the phenolics is the fact that they are clear and colorless, so that objects of light or pastel color can be produced. The tensile strength and hardness of the amino resins are better than those of the phenolics, but their impact strength and heat and moisture resistance are lower, although still characteristic of thermosetting resins. The melamine resins have better hardness, heat resistance, and moisture resistance than the ureas.

Molding resins Practically all urea molding compounds are cellulose filled, whereas the melamines, although predominantly cellulose filled, are also used with asbestos, glass or silica, and cotton fabric. Because of their poorer flow characteristics, the urea resins are usually compression molded, but injection molding is now common with both resins. Both resins can be preheated by high-frequency current because of their high polarity.

Because of their colorability, solvent and grease resistance, surface hardness, and mar resistance, the urea resins are widely used for cosmetic container closures, appliance housings, and stove hardware. The production of high-quality dinnerware from cellulose-filled compounds is the largest single use for the melamine resins.

Adhesives The amino resins are widely used for adhesives, largely for plywood and furniture. The melamine resins give excellent, boil-resistant bonds, but for economy are usually blended with the ureas.

Laminating resins Melamine resins are widely used for the production of decorative laminates. These are usually assembled with a core of phenolic-impregnated paper and a melamine-impregnated overlay sheet. They are cured by hot pressing and are widely used for counter, cabinet, and table tops.

Other applications The amino resins modify *textiles* such as cotton and rayon by imparting crease resistance, stiffness, shrinkage control, fire retardance, and water repellency. They also improve the wet strength, rub resistance, and bursting strength of *paper*. Alkylated resins, in which butyl- or amyl-substituted monomethylol ureas or melamines are used, are combined with alkyd resins (Section G) to give *baking enamels*. The urea-based enamels are used for refrigerator and kitchen appliances, and the melamine formulations in automotive finishes.

GENERAL REFERENCES

Widmer 1965; Cordier 1969.

C. Unsaturated Polyester Resins

Polyesters of several diverse types are useful as polymers. This section is concerned only with those in which the dibasic acid or the glycol, or both, contain double bonded carbon atoms. It is further restricted to cases in which radical chain polymerization involving these double bonds and a vinyl monomer, usually styrene, is made to take place in the presence of a fibrous filler, generally glass. Other polyester systems are the fiber-forming saturated polyesters (Chapter 15*D*), the polyester intermediates in the production of urethane elastomers (Chapter 15*D*) or foams (Section *E*), and the alkyd paint and molding resins (Section *G*). The polyester systems discussed here have been known from time to time as *low-pressure laminating resins*, *contact resins*, *polymerizable polyesters*, and *styrenated polyesters*. The term *reinforced plastics*, however, is more general and includes the use of thermosetting resins other than the unsaturated polyesters, among them the phenolic, epoxy, melamine, dialkyl phthalate, and alkyd resins, all discussed elsewhere in this chapter.

The production of unsaturated polyester resins for reinforced plastics use was approximately 650 million lb in 1969, at about $0.21/lb. About 350 million lb of other resins, mostly epoxy and phenolic, was used with fibrous glass to form reinforced plastic articles.

Chemistry of reinforced polyester systems

Much of the versatility of the reinforced polyester systems lies in the wide variation in resin composition and fabrication methods possible, allowing the properties of the product to be tailored to the requirements of the application. This variability is manifested in the large number of resin components used.

Dibasic acids   The unsaturation in the polyester is usually supplied by the inclusion of maleic anhydride or fumaric acid as one component. In addition, a saturated acid or anhydride is often used, such as phthalic anhydride or adipic, azelaic, or isophthalic acid. A higher proportion of unsaturated acid gives a more reactive resin, with improved stiffness at high temperatures, while more of the saturated components give less exothermic cures and less stiff resins, particularly if the aliphatic acids are used.

Dihydric alcohol   Ethylene and propylene glycols are perhaps most popular, but 1,3- and 2,3-butylene, diethylene, and dipropylene glycols are also common.

Monomer   Styrene is by far the most widely used monomer in these systems. Others often encountered are vinyl toluene, methyl methacrylate (leading to improved weatherability), diallyl phthalate (often preferred in molding compounds), and triallyl cyanurate (imparting good heat resistance).

Formulation The ingredients of the polyester resin (a typical formulation is given in Table 16-1) are mixed in a resin kettle and polymerized by step reaction to a molecular weight of 1000–5000, which is in the highly viscous liquid range. After cooling, the mixture is thinned down to a pourable liquid by the addition of the monomer. An inhibitor such as hydroquinone is then added to prevent premature polymerization. When kept cool, the mixture is stable for months to years.

Cure is begun by adding an initiator, usually an organic peroxide, such as benzoyl peroxide, or a hydroperoxide. Typically, promoters or accelerators are used to promote the decomposition of the initiator at room tempera-

TABLE 16-1. *Typical formulation of an unsaturated polyester resin (Sayre 1959)*

Mole	Ingredient	Pounds/100 lb Resin
0.2	Phthalic anhydride	28.86
0.2	Maleic anhydride	19.11
0.2	Propylene glycol	14.83
0.2	Ethylene glycol	12.10
0.3	Styrene	30.00
Trace	Hydroquinone	0.02
		104.92*

* About 5 lb of water is eliminated during esterification.

ture and, thus, rapid low-temperature curing. Common accelerators are cobalt naphthenate or alkyl mercaptans. Cure takes place in two stages: the initial formation of a soft gel is followed by rapid polymerization. The heat evolved can lead to quite high temperatures in large masses of resin.

Fabrication

The fabrication of reinforced plastic articles is usually divided into laminating and molding processes, but the techniques are in some cases quite similar.

Laminating Laminating processes are those in which separate plies of reinforcing material are coated or impregnated with resin and pressed together until cured to a single reinforced structure. The operation can be done batchwise or continuously.

Molding   The simplest molding process for reinforced plastics is *contact molding* or open or hand lay-up molding, a process closely resembling laminating. A single mold is used; the reinforcing material is placed on the mold, impregnated with resin, and allowed to cure in air. Variations include *bag molding*, where a bag or blanket is used to apply low pressure to the open surface of the material, and *spray-up* techniques, in which the resin and sometimes chopped glass are sprayed onto the mold surface.

Matched-die molding includes processes involving two dies which, together, match or conform to all the dimensions of the finished piece. The material can be placed in the die in the form of mats or preforms of reinforcing material with the resin separately added; "pre-pregs," where the preform has been impregnated; or premixes, where the resin, reinforcement, and other extenders have been premixed.

Other fabrication techniques include continuous extrusion, filament winding, casting, and foaming.

Properties and applications

The most important properties of the unsaturated polyester systems include ease of handling, rapid curing with no volatiles evolved, light color, dimensional stability, and generally good physical and electrical properties.

The major applications of glass-reinforced polyester resins fall in the following categories: boat hulls, whose popularity has risen spectacularly; transportation, including passenger car parts and bodies and truck cabs; consumer products, including such diverse items as luggage, chairs, and fishing rods; trays, pipe, and ducts; electrical applications; appliances; construction applications, largely sheet and paneling; and missile and radome uses.

GENERAL REFERENCES

Boenig 1969; Leitheiser 1969; Parker 1970.

D. Epoxy Resins

The epoxy resins are fundamentally polyethers, but retain their name on the basis of their starting material and the presence of epoxide groups in the polymer before crosslinking.

Chemistry of preparation and curing

The epoxy resin used most widely is made by condensing epichlorohydrin with bisphenol A, diphenylol propane. An excess of epichlorohydrin is used, to leave epoxy groups on each end of the low-molecular-weight (900–3000) polymer:

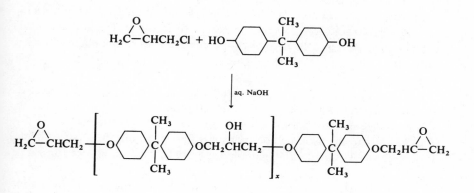

Depending on molecular weight, the polymer is a viscous liquid or a brittle high-melting solid.

Other hydroxyl-containing compounds, including resorcinol, hydroquinone, glycols, and glycerol, can replace bisphenol A. No epoxides other than epichlorohydrin are available at attractive prices, however.

The epoxy resins are cured by many types of materials, including polyamines, polyamides, polysulfides, urea- and phenol-formaldehyde, and acids or acid anhydrides, through coupling or condensation reactions. The reaction with amines involves opening the epoxide ring to give a β-hydroxyamino linkage:

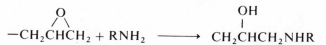

Acids and acid anhydrides react through esterification of the secondary hydroxyl groups on the epoxy resin as well as with the epoxide groups. The phenolic and amino resins may react in several ways, including condensation of methylol groups with the secondary hydroxyls of the epoxy, and reaction of the epoxide groups with phenolic hydroxyls and amino groups.

Epoxy resins can also be cured by cationic polymerization, using Lewis acid catalysts such as BF_3 and its complexes, which form polyethers from the epoxide groups.

Properties and applications

The major use of the epoxy resins is as surface-coating materials, which combine toughness, flexibility, adhesion, and chemical resistance to a nearly unparalleled degree. In addition to curing systems of the types described above, the epoxies can be esterified with drying or nondrying oil fatty acids and then cured by either air drying or baking.

Epoxy resins can be used in both molding and laminating techniques to make glass fiber-reinforced articles with better mechanical strength, chemical resistance, and electrical insulating properties than those obtained with the unsaturated polyesters. Only the higher price (about $0.47/lb) of the epoxies prevents their wider use in this field.

Casting, potting, encapsulation, and embedment are widely practiced with the epoxy resins in the electrical and tooling industries. Liquid resins are often utilized, while hot-melt solids have some application.

Other important uses include industrial flooring, adhesives and solders, foams, highway surfacing and patching materials, and stabilizers for vinyl resins.

About 170 million lb of epoxy resins was used in the United States in 1969.

GENERAL REFERENCES

Lee 1967; Bruins 1968; Madden 1969.

E. Urethane Foams

As indicated in Chapter 15*D*, urethane polymers contain the group —NHCOO—, and are formed through the reaction of a diisocyanate and a glycol. In the production of urethane foams, excess isocyanate groups in the polymer react with water or carboxylic acids to produce carbon dioxide, blowing the foam, at the same time that crosslinking is effected.

Urethane foams can be made in either flexible or rigid form, depending on the nature of the polymer and the type of crosslinking produced. Their popularity has increased very rapidly; production rose from 35 million lb in 1958 to 500 million lb of flexible and 240 million lb of rigid foams in 1969.

Like polyurethane elastomers, urethane foams are made in several steps. A *basic intermediate* of molecular weight around 1000 is a polyether made from poly(1,4-butylene glycol), sorbitol polyethers, or others. The basic intermediate is bifunctional if flexible foams are desired and polyfunctional if rigid foams are to be made.

As in elastomer production, the intermediate is reacted with an aromatic diisocyanate, usually tolylenediisocyanate, to give a *prepolymer* of structure similar to that illustrated in Chapter 16D. Catalysts based on tertiary amines or on stannous soaps are added to achieve rapid production of foam. Crosslinking takes place through the formation of urea linkages:

$$RNCO + H_2O \longrightarrow RNH_2 + CO_2$$

$$R'NCO + RNH_2 \longrightarrow RNHCONHR'$$

The use of low-boiling inert liquids, in particular fluorocarbons, to augment or replace the chemical blowing action described above has led to certain property advantages in the final foams, such as low thermal conductivity characteristic of the entrapped fluorocarbon gas. The ingredients of the foam may be partly expanded (frothed) with an inert gas in a preliminary step, and subsequently expanded to form the final object in a separate second step.

Flexible foams The use of flexible urethane foams for cushions for furniture and automobiles has displaced rubber foam in these applications because of improved strength, lower density, and easier fabrication.

The "one-shot" process, wherein polyether intermediate, tolylenediisocyanate, and catalysts are mixed just before foaming, is widely used for the production of flexible urethane foams. Most of the material is made in the form of slab stock in a continuous process, the foam being up to 8 ft wide and 3–4 ft high. It is cut into 10–60 ft lengths and, after curing 10–24 hr, is further cut up for sale to fabricators.

The foam can also be produced in molding processes, with or without external heating.

Rigid foams Rigid urethane foams are resistant to compression and may be used to reinforce hollow structural units with a minimum of weight. In addition, they consist of closed cells and so have low rates of heat transmission. They develop excellent adhesion when they are formed in voids or between sheets of material. Finally, they are resistant to oils and gasoline and do not absorb appreciable amounts of water. These properties make the rigid foams valuable for sandwich structures used in prefabrication in the building industry, for thermal insulation in refrigerators, portable insulated chests, etc., and for imparting buoyancy to boats.

GENERAL REFERENCES

Saunders 1962, 1964; Bruins 1969; D'Ancicco 1969; David 1969; Pigott 1969; Small 1969

F. Silicone Polymers

Like carbon, silicon has the capability of forming covalent compounds. Silicon hydrides (silanes) up to Si_6H_{14} are known. The Si-Si chain becomes thermally unstable at about this length, however, so that polymeric silanes are unknown. The siloxane link:

is more stable, and is the one found in commercial silicone polymers. Unlike carbon, silicon does not form double or triple bonds. Thus silicone polymers can be formed only by condensation-type reactions.

Silicone polymers became available commercially during World War II. They are particularly noted for their stability at temperatures as high as 150°C. The variety of products available ranges from liquids (lubricants, water repellents, release agents, defoamers) through greases and waxes to resins and rubbers.

Chemistry of the silicones

The study of the chemistry of silicon and its compounds began with the discovery of the element in 1824. Soon afterwards $SiCl_4$ was prepared by reacting silicone and chlorine, and ethyl silicate was first made in 1844 by the reaction of the tetrachloride with ethanol. The compounds of silicon were studied intensively from that time on.

Silicone polymers are made from organosilicon intermediates prepared in various ways from elemental silicon, which is produced by reducing quartz in an electric furnace.

The intermediates ("monomers") are compounds of the type SiR_nX_{4-n}, where R is an alkyl or aryl group and X is a group which can be hydrolyzed to —SiOH, such as chlorine or alkoxy. The intermediates are made by a direct synthesis in which the R and X groups are attached simultaneously to the silicon by a high-temperature reaction of a halide with silicon in the presence of a metal catalyst. The chief reaction is, e.g.

$$2CH_3Cl + Si \longrightarrow Si(CH_3)_2Cl_2$$

but a number of side reactions occur.

Polymerization

Silicone polymers are produced by intermolecular condensation of silanols, which are formed from the halide or alkoxy intermediates by hydrolysis:

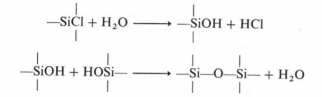

The desired siloxane structure is obtained by using silanols of different functionality, the alkyl (R) groups in the intermediate being unreactive.

Silicone fluids

The silicone fluids are low polymers produced by the hydrolysis reaction mentioned above, in which a predetermined mixture of chlorosilanes is fed into water with agitation. In many cases, the cyclic tetramer predominates in the resulting mixture.

These compounds, not polymers in the sense of this book, are used as cooling and dielectric fluids, in polishes and waxes, as release and antifoam agents, and for paper and textile treatment.

Silicone elastomers

Silicone elastomers are high molecular weight linear polymers, usually polydimethylsiloxanes. They can be cured in several ways:

a. By free-radical crosslinking with, for example, benzoyl peroxide, through the formation of ethylenic bridges between chains.

b. By crosslinking of vinyl or allyl groups attached to silicon through reaction with silylhydride groups:

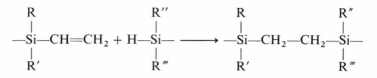

c. By crosslinking linear or slightly branched siloxane chains having reactive end groups such as silanols. In contrast to the above reactions, this yields Si—O—Si crosslinks.

The latter mechanism forms the basis of the curing of room-temperature vulcanizing (RTV) silicone elastomers. These are available as two-part mixtures in which all three essential ingredients for the cure (silanol-terminated polymer, crosslinking agent such as ethyl silicate, and catalyst such as a tin soap) are combined at the time the two components are mixed; and as one-part materials using a hydrolyzable polyfunctional silane or siloxane as crosslinker, activated by atmospheric moisture.

Silicone elastomers must be reinforced by a finely divided material such as silica if useful properties are to be obtained. These materials are outstanding in low-temperature flexibility (to $-80°C$), stability at high temperatures (to $250°C$), and resistance to weathering and to lubricating oils. They are used as gaskets and seals, wire and cable insulation, and hot gas and liquid conduits. They are valuable in surgical and prosthetic devices. The RTV elastomers are very convenient for caulking, sealing, and encapsulating.

Silicone resins

In contrast to the silicone fluids and elastomers, silicone resins contain Si atoms with no or only one organic substituent. They are therefore crosslinkable to harder and stiffer compounds than the elastomers, but many must be handled in solution to prevent premature cure. They are, in fact, usually made by hydrolysis of the desired chlorosilane blend in the presence of a solvent such as mineral spirits, butyl acetate, toluene, or xylene. These materials are usually cured with metal soaps or amines.

The silicone resins are used primarily as insulating varnishes, impregnating and encapsulating agents, and in industrial paints. A part to be coated is typically dipped into the resin solution, and drained or scraped free of excess resin. The solvent is allowed to evaporate, and the resin is cured in an oven.

GENERAL REFERENCES

Stone 1962; Adrianov 1965; Bažant 1965; Kin 1969; Sprung 1970.

G. Miscellaneous Thermosetting Resins

Alkyd resins

Alkyd resins (the name deriving from *al*cohol + a*cid*) are polyesters used primarily in organic paints, with some molding applications. About 600 million lb was produced for paints, varnishes, and lacquers in 1969.

Among many possible compositions for alkyd resins, perhaps the most common is based on phthalic anhydride and glycerol ("glyptal"). Other polyhydric alcohols commonly used are glycols, pentaerythritol, and sorbitol; other acids include maleic anhydride, isophthalic, and terephthalic.

Many alkyd resins are modified by the addition of fatty acids derived from mineral and vegetable oils. If the acids are unsaturated, the resulting resins are the air-drying type which cure by oxidation of the (saponified) acids (*drying oils*). Baking type alkyds cure by heat alone, or by cocondensation with alkylated amino resins.

Surface coatings may be classified as *lacquer* types, in which drying involves only the evaporation of solvent, and *varnish* types, in which chemical reactions also take place on drying. These reactions may involve free radical crosslinking of drying oils, the triglycerides of unsaturated acids such as oleic, linoleic, and ricinoleic.

Alkyd resins are applied in surface coating uses as follows.

Plasticizing resins   These alkyds are of the glycerol-phthalic acid or glycerol-sebacic acid type and are used in lacquers along with natural resins such as shellac to impart flexibility—a plasticizing action.

Drying resins   Drying alkyds contain, in addition to the glycerol phthalate components, some drying oils or acids of drying oils. They are, of course, used in varnish-type coatings. Here phenol-aldehyde or, better, urea- or melamine-aldehyde resins may be added to improve hardness.

Hard resins   The hard resins are usually made from maleic anhydride combined with glycerol and rosin. Their function is to improve film hardness and gloss in both lacquer and varnish coatings.

Allyl resins

Although allylic monomers are not used directly in chain-reaction or addition polymerization because of the stability and consequent low reactivity of the allyl radical, diallyl esters can be crosslinked by polymerization through their double bonds to give thermosetting resins. Two major types are of commercial interest.

Diallyl phthalate   Prepolymers (i.e., partially polymerized but still thermoplastic resins) of diallyl phthalate and diallyl isophthalate are used as molding compounds and in the production of glass fiber-reinforced laminates. They are cured by peroxide catalysts to heat- and chemical-resistant products with good dimensional stability and electrical insulating properties.

Allyl diglycol carbonate   The ester diethylene glycol bisallyl carbonate, $(CH_2{=}CHCH_2OCOOCH_2CH_2)_2O$, is used directly for casting clear glasslike products similar to cast poly(methyl methacrylate) but much harder. The castings are cured with peroxide-type catalysts. They are used for special glazing applications and as spectacle lenses and for other optical purposes.

Furane resins

Furane resins are based on furfuraldehyde,

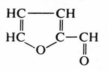

which is derived from waste vegetable matter. It is used in a thermosetting resin in combination with phenol, or converted to furfuryl alcohol, which itself can be thermoset by acids, or reacted with aldehydes or ketones to give polymerizable products. The resins are all dark in color but are strong, have good chemical resistance, and penetrate porous surfaces well.

Lignin

About 30% of woody material consists of lignin, the binder in the growing plant. It is a polymer of various groups having the carbon skeleton of *n*-propylbenzene, linked through oxygen and possibly also carbon bonds. Large quantities of lignin are available as by-products in the manufacture of paper pulp. Little success has been achieved in utilizing the material as a plastic. It lacks the flow properties of a molding material and is used only as a binder or extender.

GENERAL REFERENCES

Patton 1962; Mraz 1964; Schubert 1968; Beacham 1969; Delmonte 1969; Howard 1969.

BIBLIOGRAPHY

Andrianov 1965. K. A. Andrianov, *Metalorganic Polymers,* Interscience Div., John Wiley and Sons, New York, 1965.

Bažant 1965. Vladimir Bažant, Václav Chvalovský and Jiří Rathouský, *Organosilicon Compounds,* Academic Press, New York, 1965.

Beacham 1969. H. H. Beacham, "Allyl resins and monomers," pp. 97–98 in Sidney Gruss, ed., *Modern Plastics Encyclopedia 1969–1790* (McGraw-Hill Book Co., New York) **46** (10A), October, 1969.

Boenig 1969. H. V. Boenig, "Polyesters, Unsaturated," pp. 129–168 in Herman F. Mark, Norman G. Gaylord, and Norbert M. Bikales, eds., *Encyclopedia of Polymer Science and Technology,* Vol. 11, Interscience Div., John Wiley and Sons, New York, 1969.

Bruins 1968. Paul F. Bruins, ed., *Epoxy Resin Technology,* Interscience Div., John Wiley and Sons, New York, 1968.

Bruins 1969. Paul F. Bruins, ed., *Polyurethane Technology,* Interscience Div., John Wiley and Sons, New York, 1969.

Cordier 1969. D. E. Cordier, "Amino resins," p. 104 in Sidney Gruss, ed., *Modern Plastics Encyclopedia 1969–1970* (McGraw-Hill Book Co., New York) **46** (10A), October, 1969.

D'Ancicco 1969. V. V. D'Ancicco, "Integral skin urethane foams," pp. 250, 252 in Sidney Gruss, ed., *Modern Plastics Encyclopedia 1969–1970* (McGraw-Hill Book Co., New York,) **46** (10A), October, 1969.

David 1969. D. J. David and H. B. Staley, *Analytical Chemistry of the Polyurethanes,* Interscience Div., John Wiley and Sons, New York, 1969.

Delmonte 1969. John Delmonte, "Furane resins," p. 138 in Sidney Gruss, ed., *Modern Plastics Encyclopedia 1969–1970* (McGraw-Hill Book Co., New York) **46** (10A), October, 1969.

Howard 1969. J. M. Howard, "Alkyd molding compounds," p. 90 in Sidney Gruss, ed., *Modern Plastics Encyclopedia 1969–1970* (McGraw-Hill Book Co., New York) **46** (10A), October, 1969.

Kentgen 1969. W. A. Kentgen, "Phenolic Resins," pp. 1–73 in Herman F. Mark, Norman G. Gaylord, and Norbert M. Bikales, eds., *Encyclopedia of Polymer Science and Technology,* Vol. 10, Interscience Div., John Wiley and Sons, New York, 1969.

Kin 1969. Myron Kin, "Silicones," pp. 224, 229, 233–234 in Sidney Gruss, ed., *Modern Plastics Encyclopedia 1969–1970* (McGraw-Hill Book Co., New York) **46** (10A), October, 1969.

Kuzmak 1969. George J. Kuzmak, "Phenolics," pp. 160–161 in Sidney Gruss, ed., *Modern Plastics Encyclopedia 1969–1970* (McGraw-Hill Book Co., New York) **46** (10A), October, 1969.

Lee 1967. H. Lee and K. Neville, "Epoxy Resins," pp. 209–271 in Herman F. Mark, Norman G. Gaylord, and Norbert M. Bikales, eds., *Encyclopedia of Polymer Science and Technology*, Vol. 6, Interscience Div., John Wiley and Sons, New York, 1967.

Leis 1969. D. G. Leis and F. T. Yannett, "Phenolic foams," pp. 254, 256 in Sidney Gruss, ed., *Modern Plastics Encyclopedia 1969–1970* (McGraw-Hill Book Co., New York) **46** (10A), October, 1969.

Leitheiser 1969. Robert H. Leitheiser, "Polyesters," pp. 172, 175 in Sidney Gruss, ed., *Modern Plastics Encyclopedia 1969–1970* (McGraw-Hill, Book Co., New York) **46** (10A), October, 1969.

Madden 1969. John J. Madden, "Epoxy resins," pp. 123–124 in Sidney Gruss, ed., *Modern Plastics Encyclopedia 1969–1970* (McGraw-Hill Book Co., New York) **46** (10A), October, 1969.

Mraz 1964. Richard G. Mraz and Raymond P. Silver, "Alkyd Resins," pp. 663–734 in Herman F. Mark, Norman G. Gaylord, and Norbert M. Bikales, eds., *Encyclopedia of Polymer Science and Technology*, Vol. 1, Interscience Div., John Wiley and Sons, New York, 1964.

Parker 1970. Gilbert R. Parker, "Reinforced Plastics 1970–75," *Chem. Eng. News* **48** (4), 57–60, 65–66 (Jan. 26, 1970).

Patton 1962. T. C. Patton, *Alkyd Resin Technology*, Interscience Div., John Wiley and Sons, New York, 1962.

Pigott 1969. K. A. Pigott, "Polyurethanes," pp. 506–563 in Herman F. Mark, Norman G. Gaylord, and Norbert M. Bikales, eds., *Encyclopedia of Polymer Science and Technology*, Vol. 11, Interscience Div., John Wiley and Sons, New York, 1969.

Saunders 1962. J. H. Saunders and K. C. Frisch, *Polyurethanes: Chemistry and Technology. Part 1: Chemistry*, Interscience Div., John Wiley and Sons, New York, 1962.

Saunders 1964. J. H. Saunders and K. C. Frisch, *Polyurethanes: Chemistry and Technology. Part 2: Technology*, Interscience Div., John Wiley and Sons, New York, 1964.

Sayre 1959. James E. Sayre and Paul A. Elias, "Unsaturated Polyesters," *Chem. Eng. News* **37** (51), 56–62 (Dec. 21, 1959).

Schubert 1968. Walter J. Schubert, "Lignin," pp. 233–272 in Herman F. Mark, Norman G. Gaylord, and Norbert M. Bikales, eds., *Encyclopedia of Polymer Science and Technology*, Vol. 8, Interscience Div., John Wiley and Sons, New York, 1968.

Small 1969. F. H. Small, "Urethane foams," pp. 248, 250 in Sidney Gruss, ed., *Modern Plastics Encyclopedia 1969–1970* (McGraw-Hill Book Co., New York) **46** (10A), October, 1969.

Sprung 1970. M. M. Sprung and H. K. Lichtenwalner, "Silicones," in press in Herman F. Mark, Norman G. Gaylord, and Norbert M. Bikales, eds., *Encyclopedia of Polymer Science and Technology*, Vol. 12, Interscience Div., John Wiley and Sons, New York, 1970.

Stone 1962. F. G. A. Stone and W. A. G. Graham, eds., *Inorganic Polymers*, Academic Press, New York, 1962.

Widmer 1965. Gustave Widmer, "Amino Resins," pp. 1–94 in Herman F. Mark, Norman G. Gaylord, and Norbert M. Bikales, eds., *Encyclopedia of Polymer Science and Technology*, Vol. 2, Interscience Div., John Wiley and Sons, New York, 1965.

V

Polymer Processing

17

Plastics Technology

As the earlier sections of this book have suggested, the unique physical and mechanical properties of polymers are responsible in large part for their important place in our modern life. But these properties can be developed and utilized only by fabricating the polymer into useful articles or shapes. The methods by which this fabrication is carried out are described in this chapter.

Fabrication methods are largely determined by the rheological properties of the polymer in question. A primary consideration is whether the material is *thermoplastic*, i.e., retains the ability to flow at elevated temperatures for relatively long times, or *thermosetting*, i.e., subject to (controlled) crosslinking reactions at the temperatures necessary to induce flow, so that the ability to flow is rather quickly lost in favor of form stability. Other considerations of importance in selecting fabrication methods are softening temperature, stability, and, of course, the size and shape of the end product.

In this chapter only fabrication methods pertinent to plastics as distinguished from fibers or rubbers are considered. Techniques specific to the latter classes of material are discussed in Chapters 18 and 19.

A. Molding

Compression molding

Molding processes are those in which a finely divided plastic is forced by the application of heat and pressure to flow into, fill, and conform to the shape of a cavity (mold). One of the oldest methods of polymer processing

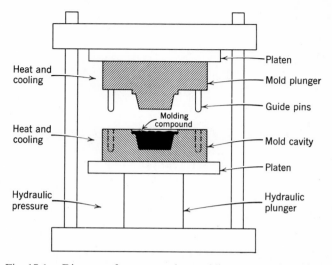

Fig. 17-1.   Diagram of a compression-molding press and mold.

is *compression molding.* Here the polymer is put between stationary and movable members of a mold (Fig. 17-1). The mold is closed, and heat and pressure are applied so that the material becomes plastic, flows to fill the mold, and becomes a homogeneous mass. The necessary pressure and temperature vary considerably depending upon the thermal and rheological properties of the polymer. For a typical compression-molding material they may be near 150°C and 1000–3000 psi. A slight excess of material is usually placed in the mold to insure its being completely filled. The rest of the polymer is squeezed out between the mating surfaces of the mold in a thin, easily removed film known as *flash.*

If a thermoplastic material is being molded, the mold is cooled, the pressure released, and the molded article removed. If a thermosetting material is used, the mold need not be cooled at the end of the molding operation or *cycle,* as the polymer will have set up and can no longer flow or distort.

Injection molding

Most thermoplastic materials are molded by the process of *injection molding.* Here the polymer is preheated in a cylindrical chamber to a temperature at which it will flow and then is forced into a relatively cold, closed mold cavity by means of quite high pressures applied hydraulically, traditionally through a plunger (Fig. 17-2). Injection-molding temperatures are higher than those for compression molding, rising well above 250°C for many materials. Pressures applied to the plunger may range from 10,000 to 30,000 psi. An outstanding feature of injection molding is the speed with which finished articles can be produced. Cycle times of 10–30 sec are common. The method has been used in the past for small articles, with a few ounces of polymer for a single molding. Today, however, large objects such as television and radio cabinets, and door liners for refrigerators, weighing many pounds, are being successfully injection molded.

In the mid-1960's, the pattern of American injection-molding usage changed abruptly. Today well over 80 % of the injection-molding machines used in this country are *reciprocating-screw* machines (Willert 1962), in which the plastic feed is plasticated (i.e., softened by heating and mixing or kneading) and delivered to the mold by a screw, arranged much like that in an extruder (Section *B*). As the screw rotates, it forces the molten polymer forward against a valve, the screw itself moving backwards, until enough material to fill the mold has been processed. The screw then stops; the valve is opened, and the screw is pushed forward by means of a plunger to fill the mold.

The use of the reciprocating-screw molding machine has facilitated several new developments in plastics processing, including the injection molding of thermosets (Ducca 1969), reinforced materials (Conwell 1969, Morrison 1969), and rigid poly(vinyl chloride). Probably no other invention has ever had such a widespread effect on the processing of plastics.

Other molding techniques

Blow molding (Morgan 1969) This operation can be carried out either on an extruder (Section *B*) or a reciprocating-screw injection machine. A section of molten polymer tubing (*parison*) is extruded into an open mold (Fig. 17-3). By means of compressed air or steam the plastic is then blown into the configuration of the mold. This technique is widely used for the manufacture of bottles and similar articles.

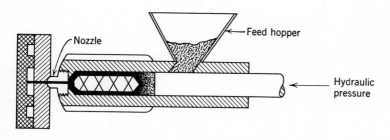

Fig. 17-2. Diagram of a conventional plunger injection-molding machine.

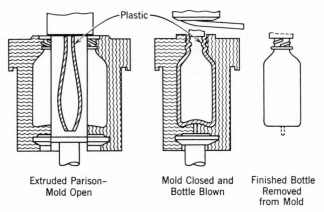

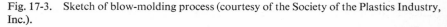

Extruded Parison–   Mold Closed and   Finished Bottle
Mold Open   Bottle Blown   Removed
from Mold

Fig. 17-3.   Sketch of blow-molding process (courtesy of the Society of the Plastics Industry, Inc.).

Rotational molding (Callahan 1969)   In this technique, the powdered polymer is loaded into a relatively inexpensive closed mold, which is intensively heated while being rotated biaxially. The polymer coats the inner walls of the mold to a uniform thickness and is fused there. The method has advantages for producing large hollow parts, and can be used to produce multiwall constructions by successive steps.

GENERAL REFERENCES

McKelvey 1962; Hull 1968; Carley 1969; Elliott 1969.

B. Other Processing Methods

Blowing

Although the blowing of plastics has evolved from standard glass-blowing techniques, procedures have been markedly modified to take advantage of the unique properties of thermoplastic polymers.

Blowing film and tubing   Continuous tubing can be made by extruding the plastic through an annular die, and maintaining gas pressure within the tube. In this fashion, rather thin-walled, uniform-diameter tubing can be manufactured at a rapid rate. By properly controlling the air rate, it is possible to increase the tube diameter to a point where the tube is essentially a cylindrical film. This cylinder can be slit and laid flat, yielding a continuous film of excellent uniformity. In some film-blowing installations, the tube, after blowing and cooling, is nipped between a pair of rolls before being slit; this operation confines the air as a bubble between the extruder and the rolls, making continuous addition of air unnecessary.

Calendering

Calendering is a process used for the continuous manufacture of sheet or film. Granular resin, or thick sheet, is passed between pairs of highly polished heated rolls under high pressure (Fig. 17-4). For the production of thin film, a series of pairs is used, with a gradual reduction in roll separation as the stock progresses through the unit. Proper calendering requires precise control of roll temperature, pressure, and speed of rotation. By maintaining a slight speed differential between a roll pair, it is often possible to impart an exceedingly high gloss to the film or sheet surface. An embossed design can be produced on the surface by means of a calender roll, appropriately engraved. By calendering a mixture of granular resin chips of varying color, it is possible to produce unusual decorative effects (e.g., marblization) in the product; this technique is widely employed in the manufacture of flooring compositions.

Casting

In casting processes, a liquid material is poured into a mold and solidified by physical (e.g., cooling) or chemical (e.g., polymerization) means, and the solid object is removed from the mold. Casting utilizes low-cost equipment (molds can be made out of soft, inexpensive materials such as rubber and plaster) but is a relatively slow process.

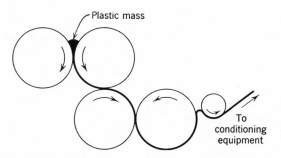

Fig. 17-4.   Diagram of a four-roll calender (Winding 1961).

Casting of thermosetting resins   Thermosetting resins are cast by stopping the polymerization at the A-stage, where the resin is still fusible and fluid. After the mold is filled, the resin is cured in an oven.

Casting of vinyl polymers   Vinyl polymers, primarily the acrylic resins, are cast by preparing a syrup of monomer and polymer, and polymerizing the monomer in the mold. Sheets, rods, and tubes are prepared as described in Chapter 14*B*.

Film casting   Films, including photographic film and cellophane, are made by flowing a solution of the polymer onto an extremely smooth surface in the form of a large polished wheel or, occasionally, a metal belt or band. After the solvent has evaporated (or, in the case of cellophane, the polymer has coagulated) the film is stripped from the casting surface.

Epoxy resins   Cast epoxies are widely used for dies because of their good dimensional stability and high impact strength.

Polyesters   Clear (unreinforced) polyester resins (Chapter 16*C*) are cast in open molds of ceramic, wood, plastic, or metal for both industrial and hobby craft work.

Nylon   Casting from the monomer is used for the anionic polymerization of caprolactam to thick-walled parts of nylon 6.

Urethane elastomers   Liquid urethane prepolymers (Chapter 15*D*) can be used in several casting methods, including hand batch, centrifugal or rotational, solvent, or spray techniques.

Coating

The technology of coating fabrics and paper, a major outlet for some plastics, is a subject whose magnitude and complexity place it outside the scope of this book. The plastic may be utilized as a melt, solution, latex, paste, or enamel or lacquer. It may be applied to the substrate by spreading with a knife, brushing, using a roller, calendering, casting, or extrusion. Coating processes include *dipping*, in which a form (such as for a rubber glove) is dipped into a suspension or latex of polymer and then into a bath of coagulating agent. After several dips to build up the desired thickness, the film is stripped from the form and subjected to heat treatments to cure or crosslink the resin. In "*slush*" *molding*, a hollow mold is filled with a more viscous latex or "slush" of partly plasticized material such as a vinyl plastisol

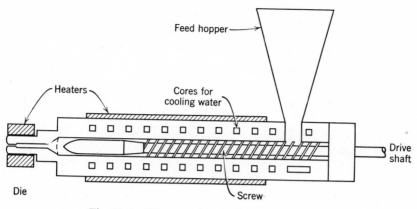

Fig. 17-5. Diagram of a plastics extruder.

(Chapter 14*D*). The excess is poured out, leaving a film which is heat treated and removed. The processes are very flexible and involve inexpensive equipment.

Extrusion

In the extrusion process, polymer is propelled continuously along a screw through regions of high temperature and pressure where it is melted and compacted, and finally forced through a die shaped to give the final object (Fig. 17-5). A wide variety of shapes can be made by extrusion, including rods, channels and other structural shapes, tubing and hose, sheeting up to several feet wide and $\frac{1}{4}$ in. or more thick, and film of similar width down to a few thousandths of an inch in thickness.

The screw of an extruder is divided into several sections, each with a specific purpose. The feed section picks up the finely divided polymer from a hopper and propels it into the main part of the extruder. In the compression section, the loosely packed feed is compacted, melted, and formed into a continuous stream of molten plastic. Some external heat must be applied, but much is generated by friction. The metering section contributes to uniform flow rate, required to produce uniform dimensions in the finished product, and builds up sufficient pressure in the polymer melt to force the plastic through the rest of the extruder and out of the die. Since viscous polymer melts can be mixed only by the application of shearing forces (their viscosity is too high to allow turbulence or diffusion to contribute appreciably to mixing), an additional working section may be needed before the die.

Modern trends in extruder usage include the *twin-screw* or *multiple-screw* extruder, in which two screws turn side by side in opposite directions, providing more working of the melt; and the *vented* extruder, having an opening or vent at some point along the screw which can be opened or led to vacuum to extract volatiles from the polymer melt.

Filament winding

This technique is a form of glass-fiber reinforced polyester technology in which the glass is in the form of long continuous fibers wound in a controlled pattern on a mandrel. Objects with primarily cylindrical symmetry and very high burst strength are obtained.

Film techniques

The production of plastic films by a wide variety of techniques, including blowing and casting, is covered elsewhere. The references cited at the end of this section provide comprehensive reviews.

Foaming

The production of plastic foams is accomplished by generating a gas in a fluid polymer, usually at an elevated temperature. The chemical production of a gas during polymerization to form urethane foams is described in Chapter 16*E*. Thermoplastics are foamed by incorporating either a blowing agent, which decomposes to a gas at elevated temperature, or an inert gas. The references cited at the end of this section provide comprehensive reviews.

Forming

Postforming thermosetting resins   Laminated sheets of thermosetting resins are formed into various shapes by a process resembling that used for sheet metal. The sheet is heated before the final thermosetting reaction, shaped quickly in a mold or around a form, and held in place with light pressure until it sets up.

Forming thermoplastic sheets   Vacuum forming is used widely for the manipulation of cellulose acetate and acrylic resin sheeting. A sheet of the plastic is warmed and laid across a hollow mold cavity, and a vacuum is drawn on the cavity. Atmospheric pressure forces the sheet to conform with the mold; after cooling of the sheet, the vacuum is released, and the formed object removed.

Many modifications of this basic process are in use. In one, an inflatable rubber bag forces the sheet to conform with the mold contour. In another variation the sheeting is first dished by vacuum to a curvature greater than ultimately desired, a *convex* mold is placed into the mold cavity, and the vacuum is released while the sheet is still warm. The sheet recovers sufficiently to conform to the shape of the convex mold, and is frozen in that shape by cooling.

Laminating and low-pressure molding

Both lamination, a high-pressure process, and low-pressure molding involve (*a*) the impregnation of sheets (wood, paper, fabric) with a liquid or dissolved thermosetting resin which acts as an adhesive; (*b*) assembly of the individual sheets; and (*c*) compression and curing.

In low-pressure molding or laminating, the sheets of impregnated material are laid over a mold and held in place by a rubber mattress or bag which is inflated with steam to provide heat and pressure to hold the laminate in place and effect the cure. Many variations of the process are practiced.

Lamination differs from low-pressure molding in that standard shapes are produced in sufficient quantity to allow economical application of hot-press methods. Resins requiring higher temperatures for curing can also be used.

GENERAL REFERENCES

McKelvey 1962; Carley 1969; and specifically:

> *Blowing:* Lasman 1965; Morgan 1969.
> *Calendering:* Meinecke 1965; Perlberg 1969.
> *Casting:* Wallis 1965; Kiernan 1969.
> *Coating:* Higgins 1965, 1967; Pascoe 1969.
> *Extrusion:* Griff 1962; Fisher 1964; Westover 1968; Van Ness 1969.
> *Filament winding:* Rosato 1964, 1967.
> *Film techniques:* Park 1965; Wolinski 1967; Sweeting 1968, 1970.
> *Foaming:* Skochdopole 1965; Benning 1969.
> *Forming:* Monroe 1965; Anon. 1969; McConnell 1969.
> *Laminating:* Power 1968; Skow 1969; White 1969.

C. Fillers, Plasticizers, and other Additives

Many plastics are virtually useless alone but are converted into highly serviceable products by combining them with particulate or fibrous solids. Phenolic and amino resins are almost always compounded, or filled, with substances like wood flour, pure short-fiber cellulose, powdered mica, and asbestos. These materials greatly enhance dimensional stability, impact resistance, tensile and compressive strength, abrasion resistance, and thermal stability. The use of glass fiber as a reinforcing filler for polyester resins is another important illustration. Soft thermoplastics like coumarone-indene and hydrocarbon resins are usually blended with very large amounts (over 80% by weight) of mineral solids such as crushed quartz, limestone, or clay. In compounds of this type, the resin functions as an interparticle adhesive; these products often have poor tensile properties but excellent compressive strength, abrasion resistance, and dimensional stability. A final and extremely important example of the beneficial effect of fillers is the reinforcement of rubber, discussed in Chapter 19*B*.

The principal fillers used in plastics can be divided into two types: particulate and fibrous. Among the particulate fillers are silica products, including sand, quartz, and diatomaceous earth; silicates, including clay, mica, talc, asbestos, and some synthetic silicates; glass, including granules, flakes, and solid and hollow spheres (the latter in syntactic foams); inorganic compounds, including chalk, limestone, alumina, magnesia, and zinc oxide, barytes, silicon carbide, and others; metal powders; and finely divided cellulosics (wood flour) and synthetic polymers (fluorocarbons and others).

Fibrous fillers, some quite old (cellulosics) and some new "space-age" products (metallic fibers, whiskers) include: cellulosic fibers, such as alphacellulose and cotton flock; synthetic fibers, including nylon, polyester, acrylic, and poly(vinyl alcohol); carbon fibers made by pyrolizing materials such as rayon; boron filaments made by depositing boron from a BCl_3-H_2 mixture onto tungsten wire, and having tensile strengths approaching one-half million psi; and single-crystal fibers of aluminum or beryllium oxide, silicon or boron carbide, or others.

Plasticizers

Plasticizers are added to plastics to improve flow and, therefore, processability, and to reduce the brittleness of the product. This is achieved by lowering the glass transition temperature below room temperature, thus achieving a change in properties from those of a hard, brittle, glasslike solid to those of a soft, flexible, tough material (Chapter 7*D*). An example is the plasticization of poly(vinyl chloride) and vinyl chloride-acetate copolymers (Chapter 14*D*). Similar changes in properties can, of course, be brought about by altering the molecular structure of the polymer (e.g., by copolymerization, sometimes called *internal plasticization*).

The basic requirements which must be met by a plasticizer are compatibility and permanence. The plasticizer must be miscible with the polymer.

This implies a similarity in the intermolecular forces active in the two components, and explains why compatibility is difficult to achieve with a non-polar polymer such as polyethylene. Permanence requirements demand low vapor pressure and a low diffusion rate of the plasticizer within the polymer, both of which are obtained by the use of high-molecular-weight plasticizers.

The efficiency of the plasticizer in bringing about the desired changes in properties is important in determining the proportion in which the plasticizer must be added to the resin. Plasticizer efficiency may be evaluated by a number of different semiempirical tests. Some of these measure the amount of nonsolvent needed to cause phase separation when added to polymer-plasticizer solutions (dilution ratio), the viscosity of dilute solutions of the polymer in the plasticizer, polymer-solvent interaction constants measured on these solutions, the depression of the glass transition temperature, the melt viscosity of the plasticized polymer, the electrical or mechanical properties of the plasticized polymer, or the molecular size and shape or viscosity of the plasticizer itself. Not all these tests, needless to say, rate plasticizer candidates in the same order. The selection of a particular plasticizer still depends to a large extent upon empirical results rather than theoretical predictions.

The following types of plasticizers are in common use:

a. Phthalate esters, accounting for over half of the total volume of plasticizers used.

b. Phosphate esters, chiefly tricresyl phosphate, valued primarily for their flameproofing characteristics.

c. Adipates, azelates, oleates, and sebacates, used chiefly in vinyl resins for improving low-temperature flexibility.

d. Epoxy plasticizers, produced by reacting hydrogen peroxide with unsaturated vegetable oils and fatty acids.

e. Fatty acid esters from natural sources, used primarily as extenders to reduce cost (*secondary plasticizers*).

f. Glycol derivatives, employed mainly as lubricants and mold-release agents, and as plasticizers for poly(vinyl alcohol).

g. Sulfonamides, used to plasticize cellulose esters, phenolic and amino resins, and amide and protein plastics.

h. Hydrocarbons and hydrocarbon derivatives, serving as secondary plasticizers.

Other additives

Antioxidants (Maassen 1965; Robin 1969) The role of antioxidant in preventing or inhibiting the oxidation of polymers is usually filled by a substance which itself is readily oxidized, although in some cases the antioxidant may act by combining with the oxidizing polymer to form a stable product. Common antioxidants fall in the classes of phenols, aromatic amines, and salts and condensation products of amines and aminophenols with aldehydes, ketones, and thio compounds.

Colorants (Hopmeir 1969; Zabel 1969)   Colorants for plastics include a wide variety of inorganic and organic materials. A few are molecularly dispersed (oil-soluble dyes) or have small particle size and a refractive index near that of the plastic (organic pigments such as the phthalocyanines) and lead to transparent colored products when incorporated into transparent plastics. Others, including inorganic pigments, impart opacity to the plastic. Common colorants for plastics include, among many others, titanium dioxide and barium sulfate (white), phthalocyanine blues and greens, ultramarine blues, chrome greens, quinacridone reds and magentas, molybdate oranges, cadmium reds and yellows, iron oxide and chrome yellows, carbon black, flake aluminum for a silver metallic effect, and lead carbonate or mica for pearlescence. The coloring of plastics is normally carried out by adding the colorants to the powdered plastic, tumbling, and compounding on hot rolls or in an extruder. Other techniques are occasionally used: e.g., colored castings of poly(methyl methacrylate) are produced by dissolving or dispersing the colorants in the syrup before polymerizing.

Ultraviolet-light absorbers (Miller 1969) are not colorants in the above sense, but have similar light-modifying effects.

Flame retardants (Holderried 1969)   The most useful material imparting flame retardance to plastics is antimony trioxide. It must be used with a source of available chlorine to be effective; it is presumed that antimony oxychloride is the active flame-retarding agent. Aside from this compound and other antimony derivatives, the phosphate ester plasticizers are widely used for reducing flammability, especially in vinyl resins.

The major factors in reducing the flammability of materials appear to be (*a*) elimination of volatile fuel, as by cooling; (*b*) production of a thermal barrier, as by charring, thus eliminating fuel by reducing heat transfer; and (*c*) quenching the chain reactions in the flame, as by adding suitable radical scavengers.

Stabilizers (Thacker 1969)   In addition to antioxidants and flame retardants, certain polymers require stabilizers to achieve and maintain utility. An important example is the vinyl resins, whose stabilization is discussed in Chapter 15D. Other applications of stabilizers include the use of carbon black to prevent photochemical degradation (by excluding light), as in polyethylene, and of ultraviolet light absorbers, such as hydroxybenzophenones, to improve the light stability of both plastics and the dyes which serve as their colorants.

The term *compounding* is applied both to the selection of additives to modify the properties of a polymer, and to their incorporation with the polymer to give a homogeneous mixture, in a form suitable for efficient use in the subsequent processing or fabrication step. It is in the latter sense that compounding is here considered.

The traditional compounding device in the plastics industry is the two-roll mill, which looks and operates much like the top half of a calender (Fig. 17-4). By proper selection of temperatures and speeds of rotation, the plastic is made to adhere to the front roll, except as it is cut off by the operator. The compounding ingredients are added to the plastic mass as it passes between the rolls.

The roll mill has been supplanted in many operations by the compound- or mixer-extruder, an extruder in which the function of the mixing section of the screw is emphasized. The use of the extruder for compounding has many advantages: contamination is reduced, inert atmospheres or vacuum may be utilized, continuous processes are more readily achieved, etc.

Other compounding devices in common use are internal mixers such as kneaders, masticators, and paddle blenders; tumblers; and blenders.

GENERAL REFERENCES

Fillers: Lannon 1965; Frissell 1967; Byrne 1969; Paulus 1969; Seymour 1969 *a, b*.
Plasticizers: Buttery 1960; Lannon 1965; Darby 1969; Wilde 1969.

D. Tables of Plastics Properties

Table 17-1 and 17-2 list comparative properties for typical examples of the major commercial thermoplastic and thermosetting resins, respectively. It should be noted that the properties of a given plastic can vary widely, depending on compounding, fabrication, thermal history, and many other variables. The numbers in the tables may represent only a fraction of the total range of values attainable and should be used only for comparative purposes.

TABLE 17-1. *Typical properties of commercial thermoplastic polymers (Modern Plastics 1969)*

Property	ABS	Acetal Resin	Poly(methyl Methacrylate)	Impact Acrylic	Cellulose Acetate	Cellulose Acetate Butyrate	Chlorinated Polyether
Specific gravity, g/cm³	1.02–1.04	1.42	1.17–1.20	1.08–1.18	1.22–1.34	1.15–1.22	1.4
Refractive index, n_D^{25}	—	1.48	1.49	—	1.46–1.50	1.46–1.49	—
Tensile strength, psi	3500–6200	10,000	7000–11,000	5000–9000	1900–9000	2600–6900	6000
Elongation, %	5–60	25–75	2–10	>15–50	6–70	40–88	60–160
Tensile modulus, 10^5 psi	2–3.5	5.2	4.5	2–4	0.6–4.0	0.5–2	1.6
Impact strength, ft-lb/in. of notch	3–8	1.4–2.3	0.3–0.5	0.5–4.5	0.4–5.2	0.8–6.3	0.4
Heat-deflection temp., °F, 264 psi	200–218	255	155–210	165–215	111–195	113–202	285‡
Dielectric constant, 1000 cycles	2.4–4.5	3.7	3.0–3.6	2.5–3.5	3.4–7.0	3.4–6.4	3.0
Dielectric loss, 1000 cycles	0.004–0.007	0.0048	0.03–0.05	0.02–0.035	0.01–0.07	0.01–0.04	0.01
Water absorption, $\frac{1}{8}$ in. bar, 24 hr, %	0.2–0.45	0.25	0.3–0.4	0.2–0.4	1.7–6.5	0.9–2.2	0.01
Burning rate	Slow	Slow	Slow	Slow	Slow to self-exting.	Slow	Self-exting.
Effect of sunlight	Yellows	Chalks	None	Slight	Slight	Slight	Slight
Effect of strong acids or bases	Attacked, acids	Attacked	Attacked	Attacked, acids	Decomposes	Decomposes	Attacked, acids
Effect of organic solvents	Soluble	Resistant	Soluble	Soluble	Soluble	Soluble	Resistant
Clarity	Opaque	Opaque	Transparent	Opaque	Transparent	Transparent	Opaque

Polymer / Property	Polytetra-fluoroethylene	Ionomer	66 Nylon	Poly(phenylene Oxide)	Polycarbonate	Branched Polyethylene
Specific gravity, g/cm^3	2.14–2.20	0.93–0.96	1.13–1.15	1.06–1.10	1.2	0.910–0.925
Refractive index, n_D^{25}	1.35	1.51	1.53	—	1.586	1.51
Tensile strength, psi	2000–5000	3500–5500	9000–12,000	7800–9600	8000–9500	600–2300
Elongation, %	200–400	350–450	60–300	20–30	100–130	90–800
Tensile modulus, 10^5 psi	0.58	0.2–0.6	1.8–4.2	3.55–3.80	3.5	0.14–0.38
Impact strength, ft-lb/in. of notch	3.0	6–15	1.0–2.0	1.7–1.8†	12–17.5	>16
Heat-deflection temp., °F, 264 psi	250‡	100–120	150–220	212–265	265–285	90–105
Dielectric constant, 1000 cycles	<2.1	2.4–2.5	3.9–4.5	2.64	3.02	2.25–2.35
Dielectric loss, 1000 cycles	<0.0002	0.0015	0.02–0.04	0.0004	0.0021	<0.0005
Water absorption, ⅛ in. bar, 24 hr, %	0.00	0.1–1.4	1.5	0.066	0.15	<0.015
Burning rate	None	Very slow	Self-exting.	Self-exting.	Self-exting.	Very slow
Effect of sunlight	None	Requires protection	Discolors	Slight	Slight	Requires protection
Effect of strong acids or bases	Very resistant	Attacked, acids	Attacked, acids	Resistant	Attacked	Resistant
Effect of organic solvents	Very resistant	Resistant	Resistant	Soluble	Soluble	Resistant below 80°C
Clarity	Opaque	Transparent	Opaque	Opaque	Transparent	Opaque

Table 17-1 (continued)

Property \ Polymer	Linear Polyethylene	Polybutene	Polyimide	Polypropylene	Polymethyl-pentene	Polysulfone	Polystyrene
Specific gravity, g/cm^3	0.941–0.965	0.910–0.915	1.43	0.902–0.906	0.83	1.24	1.04–1.09
Refractive index, n_D^{25}	1.54	1.50	—	1.49	1.465	1.633	1.59–1.60
Tensile strength, psi	3100–5500	3800–4400	10,500	4300–5500	4000	10200§	5000–12000
Elongation, %	20–1000	300–380	5.0–7.0	200–700	15	50–100	1.0–2.5
Tensile modulus, 10^5 psi	0.6–1.8	0.26	4.5	1.6–2.3	2.1	3.6	4–6
Impact strength, ft-lb/in. of notch	0.5–2.0	No break	1.1	0.5–2.0	0.8	1.3*	0.25–0.4*
Heat-deflection temp., °F, 264 psi	110–130	130–140	650	125–140	—	345	220
Dielectric constant, 1000 cycles	2.30–2.35	2.25	3.4	2.2–2.6	2.12	3.13	2.4–2.65
Dielectric loss, 1000 cycles	<0.0005	0.005	0.002	<0.0005–0.0018	0.00003	0.001	0.0001–0.0003
Water absorption, ⅛ in. bar, 24 hr, %	<0.01	<0.01–0.026	0.32	<0.01	0.01	0.22	0.03–0.10
Burning rate	Very slow	Burns	None	Slow	Burns	Self-exting.	Slow
Effect of sunlight	Requires protection	Crazes	None	Requires protection	Requires protection	Slight	Yellows
Effect of strong acids or bases	Resistant	Attacked, acids	Attacked, alkalis	Resistant	Resistant	Resistant	Attacked, acids
Effect of organic solvents	Resistant below 80°C	Resistant	Very resistant	Resistant below 80°C	Soluble	Soluble	Soluble
Clarity	Opaque	Opaque	Opaque	Opaque	Transparent	Transparent	Transparent

Property	Impact Polystyrene	Styrene-butadiene Thermoplastic Elastomer	Rigid Vinyl	Plasticized Vinyl
Specific gravity, g/cm^3	1.04–1.10	0.93–1.10	1.35–1.45	1.16–1.35
Refractive index, n_D^{25}	—	1.52–1.55	1.52–1.55	—
Tensile strength, psi	1500–7000	600–3000	5000–9000	1500–3500
Elongation, %	2–80	300–1000	2.0–40.0	200–450
Tensile modulus, 10^5 psi	1.5–5	0.008–0.5	3.5–6	—
Impact strength, ft-lb/in. of notch	0.5–11	No break	0.4–20	Varies
Heat-deflection temp., °F, 264 psi	210	–150	130–175	—
Dielectric constant, 1000 cycles	2.4–4.5	2.5–3.4	3.0–3.3	4.0–8.0
Dielectric loss, 1000 cycles	0.0004–0.002	0.001–0.003	0.009–0.017	0.07–0.16
Water absorption, $\frac{1}{8}$ in. bar, 24 hr, %	0.05–0.6	0.19–0.39	0.07–0.4	0.15–0.75
Burning rate	Slow	Slow	Self-exting.	Slow to self-exting.
Effect of sunlight	Loses strength	Slight	Slight	Slight
Effect of strong acids or bases	Attacked, acids	Attacked	Resistant	Resistant
Effect of organic solvents	Soluble	Soluble	Soluble	Soluble
Clarity	Opaque	Transparent	Transparent	Transparent

* $\frac{1}{4}$-in.bar; † $\frac{1}{8}$-in. bar; ‡ 66 psi; § at yield.

TABLE 17-2. *Typical properties of commercial thermosetting resins (Modern Plastics 1969)*

Resin and Filler / Property	Epoxy, Cast None	Melamine-Formaldehyde α-cellulose	Phenol-Formaldehyde None	Phenol-Formaldehyde Wood or Cotton	Polyester Cast None	Polyester Glass Cloth	Silicone Cast/RTV None	Urea-Formaldehyde α-cellulose
Specific gravity, g/cm^3	1.11–1.40	1.47–1.52	1.25–1.30	1.34–1.45	1.10–1.46	1.50–2.10	0.99–1.50	1.47–1.52
Refractive index n_D^{25}	1.55–1.61	—	—	—	1.52–1.57	—	1.43	1.54–1.56
Tensile strength, psi	4000–13,000	7000–13,000	7000–8000	5000–9000	6000–13,000	30,000–50,000	350–1000	5500–13,000
Elongation, %	3.0–6.0	0.6–0.9	1.0–1.5	0.4–0.8	<5.0	0.5–2	100–300	0.5–1.0
Tensile modulus, 10^5 psi	3.5	12–14	7.5–10	8–17	3.0–6.4	15–45	0.09	10–15
Impact strength, ft-lb/in. of notch	0.2–1.0	0.24–0.35	0.20–0.36	0.24–0.60	0.2–0.4	5–30	—	0.25–0.40
Heat-deflection temp., °F, 264 psi	115–550	350–370	240–260	260–340	140–400	—	—	260–290
Dielectric constant 1000 cycles	3.5–4.5	7.8–9.2	4.5–6.0	4.4–9.0	2.8–5.2	4.2–6.0	2.7	7.0–7.5
Dielectric loss, 1000 cycles	0.002–0.02	0.015–0.036	0.03–0.08	0.04–0.20	0.005–0.025	0.01–0.06	0.001–0.002	0.025–0.035
Water absorption, ⅛ in. bar, 24 hr, %	0.08–0.15	0.1–0.6	0.1–0.2	0.3–1.2	0.15–0.60	0.05–0.50	0.02	0.4–0.8
Burning rate	Slow	None	Very slow	Very slow	Burns	Burns	Self-exting.	Self-exting.
Effect of sunlight	None	Slight	Darkens	Darkens	Yellows	Slight	None	Slight
Effect of strong acids or bases	Attacked	Attacked	Attacked	Attacked	Attacked	Attacked	Attacked	Attacked
Effect of organic solvents	Resistant	Resistant	Resistant	Resistant	Attacked	Attacked	Attacked	Resistant
Clarity	Transparent	Opaque	Transparent	Opaque	Transparent	Translucent	Transparent	Opaque

BIBLIOGRAPHY

Anon. 1969. Anon., "Cold stamping and warm forging," pp. 618–620 in Sidney Gruss, ed., *Modern Plastics Encyclopedia 1969–1970* (McGraw-Hill Book Co. New York) 46 (10A), October, 1969.

Benning 1969. Calvin James Benning, *Plastic Foams*, John Wiley and Sons, New York, 1969.

Buttery 1960. D. N. Buttery, *Plasticizers*, 2nd ed., Franklin Publishing Co., Palisades, N.J., 1960.

Byrne 1969. Robert E. Byrne, Jr., "Asbestos fibers," pp. 386, 388, 390 in Sidney Gruss, ed., *Modern Plastics Encyclopedia 1969–1970* (McGraw-Hill Book Co., New York) 46 (10A), October, 1969.

Callahan 1969. T. F. Callahan and H. T. MacMillen, "Rotational molding," pp. 568, 570, 575–576, 579 in Sidney Gruss, ed., *Modern Plastics Encyclopedia 1969–1970* (McGraw-Hill Book Co., New York) 46 (10A), October, 1969.

Carley 1969. J. F. Carley, "Processing for High Performance," *Plast. Polym.* 37, 433–447 (1969).

Conwell 1969. Yeates Conwell and D. E. McBournie, "Molding reinforced thermoplastics," pp. 410–412 in Sidney Gruss, ed., *Modern Plastics Encyclopedia 1969–1970* (McGraw-Hill Book Co., New York) 46 (10A), October, 1969.

Darby 1969. J. R. Darby and J. K. Sears, "Plasticizers," pp. 228–306 in Herman F. Mark, Norman G. Gaylord, and Norbert M. Bikales, eds., *Encyclopedia of Polymer Science and Technology*, Vol. 10, Interscience Div., John Wiley and Sons, New York, 1969.

Ducca 1969. F. W. Ducca and E. W. Vaill, "Thermoset molding," pp. 468, 472, 474, 476, 479, 484, 492, 495–496, 500, 505 in Sidney Gruss, ed., *Modern Plastics Encyclopedia 1969–1970* (McGraw-Hill Book Co., New York) 46 (10A), October, 1969.

Elliott 1969. W. O. Elliott, "Injection molding," pp. 419, 421, 426, 428, 430, 433, 436 in Sidney Gruss, ed., *Modern Plastics Encyclopedia 1969–1970* (McGraw-Hill Book Co., New York) 46 (10A), October, 1969.

Fisher 1964. E. G. Fisher, *Extrusion of Plastics*, John Wiley and Sons, New York, 1964.

Frissell 1967. W. J. Frissell, "Fillers," pp. 740–763 in Herman F. Mark, Norman G. Gaylord, and Norbert M. Bikales, eds., *Encyclopedia of Polymer Science and Technology*, Vol. 6, Interscience Div., John Wiley and Sons, New York, 1967.

Griff 1962. A. L. Griff, *Plastics Extrusion Technology*, Reinhold Publishing Corp., New York, 1962.

Higgins 1965. D. G. Higgins and Arthur H. Landrock, "Coating Methods," pp. 765–830 in Herman F. Mark, Norman G. Gaylord, and Norbert M. Bikales, eds., *Encyclopedia of Polymer Science and Technology*, Vol. 3, Interscience Div., John Wiley and Sons, New York, 1965.

Higgins 1967. D. G. Higgins, "Fabrics, Coated," pp. 467–489 in Herman F. Mark, Norman G. Gaylord, and Norbert M. Bikales, eds., *Encyclopedia of Polymer Science and Technology*, Vol. 6, Interscience Div., John Wiley and Sons, New York, 1967.

Holderried 1969. J. A. Holderried, "Flame retardants," pp. 274, 276, 288, 290 in Sidney Gruss, ed., *Modern Plastics Encyclopedia 1969–1970* (McGraw-Hill Book Co., New York) 46 (10A), October, 1969.

Hopmeir 1969. A. P. Hopmeir, L. M. Greenstein, and Anthony J. Petro, "Pigments," pp. 157–219 in Herman F. Mark, Norman G. Gaylord, and Norbert M. Bikales, eds., *Encyclopedia of Polymer Science and Technology*, Vol. 10, Interscience Div., John Wiley and Sons, New York, 1969.

Hull 1968. John L. Hull, Lee J. Zukor, G. E. Pickering, Richard E. Duncan, David R. Ellis, Robert A. McCord, and A. B. Hitchcock, "Molding," pp. 1–157 in Herman F. Mark, Norman G. Gaylord, and Norbert M. Bikales, eds., *Encyclopedia of Polymer Science and Technology*, Vol. 9, Interscience Div., John Wiley and Sons, New York, 1968.

Kiernan 1969. E. F. Kiernan, John F. Woodman, M. Hilrich, George H. Hicks, Philip A. Gianatasio, and F. P. Downing, "Casting of Plastics," pp. 602, 605–606, 611, 614, 616, in Sidney Gruss, ed., *Modern Plastics Encyclopedia 1969–1970* (McGraw-Hill Book Co., New York) **46** (10A), October, 1969.

Lannon 1965. D. A. Lannon and E. J. Hoskins, "Effect of Plasticisers, Fillers, and Other Additives on Physical Properties," Chap. 7 in P. D. Ritchie, ed., *Physics of Plastics*, Van Nostrand, Princeton, N.J., 1965.

Lasman 1965. Henry R. Lasman, "Blowing Agents," pp. 532–565 in Herman F. Mark, Norman G. Gaylord, and Norbert M. Bikales, eds., *Encyclopedia of Polymer Science and Technology*, Vol. 2, Interscience Div., John Wiley and Sons, New York, 1965.

Maassen 1965. G. C. Maassen, R. J. Fawcett, and W. R. Connell, "Antioxidants," pp. 171–197 in Herman F. Mark, Norman G. Gaylord, and Norbert M. Bikales, eds., *Encyclopedia of Polymer Science and Technology*, Vol. 2, Interscience Div., John Wiley and Sons, New York, 1965.

McConnell 1969. William K. McConnell, "Thermoforming," pp. 534, 538, 540, 542, 548–549, 551, 555, 559–560, 563 in Sidney Gruss, ed., *Modern Plastics Encyclopedia 1969–1970* (McGraw-Hill Book Co., New York) **46** (10A), October, 1969.

McKelvey 1962. James M. McKelvey, *Plastics Processing*, John Wiley and Sons, New York, 1962.

Meinecke 1965. Eberhard Meinecke, "Calendering," pp. 802–819 in Herman F. Mark, Norman G. Gaylord, and Norbert M. Bikales, eds., *Encyclopedia of Polymer Science and Technology*, Vol. 2, Interscience Div., John Wiley and Sons, New York, 1965.

Miller 1969. S. B. Miller, G. R. Lappin, and C. E. Tholstrup, "Ultraviolet absorbers," pp. 300, 308, 314 in Sidney Gruss, ed., *Modern Plastics Encyclopedia 1969–1970* (McGraw-Hill Book Co., New York) **46** (10A), October, 1969.

Modern Plastics 1969. Sidney Gruss, ed., *Modern Plastics Encyclopedia 1969–1970* (McGraw-Hill Book Co., New York) **46** (10A), October, 1969.

Monroe 1965. Sam Monroe, "Bag Molding," pp. 300–316 in Herman F. Mark, Norman G. Gaylord, and Norbert M. Bikales, eds., *Encyclopedia of Polymer Science and Technology*, Vol. 2, Interscience Div., John Wiley and Sons, New York, 1965.

Morgan 1969. B. T. Morgan, N. R. Wilson, and D. L. Peters, "Blow molding," pp. 512, 514, 517, 520, 525–526, 529, 532 in Sidney Gruss, ed., *Modern Plastics Encyclopedia 1969–1970* (McGraw-Hill Book Co., New York) **46** (10A), October, 1969.

Morrison 1969. Robert S. Morrison, "Molding reinforced thermosets," pp. 414–415, 417–418 in Sidney Gruss, ed., *Modern Plastics Encyclopedia 1969–1970* (McGraw-Hill Book Co., New York) **46** (10A), October, 1969.

Park 1965. W. R. R. Park and Jo Conrad, "Biaxial Orientation," pp. 339–373 in Herman F. Mark, Norman G. Gaylord, and Norbert M. Bikales, eds., *Encyclopedia of Polymer Science and Technology*, Vol. 2, Interscience Div., John Wiley and Sons, New York, 1965.

Pascoe 1969. Walter R. Pascoe, "Plastic powder coatings," pp. 594, 597, 600 in Sidney Gruss, ed., *Modern Plastics Encyclopedia 1969–1970* (McGraw-Hill Book Co., New York) **46** (10A), October, 1969.

Paulus 1969. H. T. Paulus, "Fibrous glass reinforcements," pp. 392, 394, 396 in Sidney Gruss, ed., *Modern Plastics Encyclopedia 1969–1970* (McGraw-Hill Book, Co., New York) **46** (10A), October, 1969.

Perlberg 1969. S. E. Perlberg and R. C. Seanor, "Calendering," pp. 583, 585–586, 589, 591 in Sidney Gruss, ed., *Modern Plastics Encyclopedia 1969–1970* (McGraw-Hill Book Co., New York) **46** (10A), October, 1969.

Power 1968. G. E. Power, "Laminates," pp. 121–163 in Herman F. Mark, Norman G. Gaylord, and Norbert M. Bikales, eds., *Encyclopedia of Polymer Science and Technology,* Vol. 8, Interscience Div., John Wiley and Sons, New York, 1968.

Robin 1969. Michael Robin, "Antioxidants," pp. 314, 316 in Sidney Gruss, ed., *Modern Plastics Encyclopedia 1969–1970* (McGraw-Hill Book Co., New York) **46** (10A), October, 1969.

Rosato 1964. D. V. Rosato and C. S. Grove, Jr., *Filament Winding,* Interscience Div., John Wiley and Sons, New York, 1964.

Rosato 1967. D. V. Rosato, "Filament Winding," pp. 713–740 in Herman F. Mark, Norman G. Gaylord, and Norbert M. Bikales, eds., *Encyclopedia of Polymer Science and Technology,* Interscience Div., John Wiley and Sons, New York, 1967.

Seymour 1969a. Raymond B. Seymour, "Fillers for polymers," pp. 372–373, 376 in Sidney Gruss, ed., *Modern Plastics Encyclopedia 1969–1970* (McGraw-Hill Book Co., New York) **46** (10A), October, 1969.

Seymour 1969b. Raymond B. Seymour, Kigeu Kawai, A. L. Mlavsky, and S. A. Bortz, "Fibrous reinforcements for polymers," pp. 378–380, 382, 384 in Sidney Gruss, ed., *Modern Plastics Encyclopedia 1969–1970* (McGraw-Hill Book Co., New York) **46** (10A), October, 1969.

Skochdopole 1965. R. E. Skochdopole, "Cellular Materials," pp. 80–130 in Herman F. Mark, Norman G. Gaylord, and Norbert M. Bikales, eds., *Encyclopedia of Polymer Science and Technology,* Vol. 3, Interscience Div., John Wiley and Sons, New York, 1965.

Skow 1969. Norman A. Skow, "High pressure industrial laminates," pp. 398–401, 404–405 in Sidney Gruss, ed., *Modern Plastics Encyclopedia 1969–1970* (McGraw-Hill Book Co., New York) **46** (10A), October, 1969.

Sweeting 1968. Orville J. Sweeting, *The Science and Technology of Polymer Films,* Vol. 1, John Wiley and Sons, New York, 1968.

Sweeting 1970. Orville J. Sweeting, *The Science and Technology of Polymer Films,* Vol. 2, John Wiley and Sons, New York, 1970.

Thacker 1969. George A. Thacker, Jr., "Stabilizers," pp. 290, 294, 297, 300 in Sidney Gruss, ed., *Modern Plastics Encyclopedia 1969–1970* (McGraw-Hill Book Co., New York) **46** (10A), October, 1969.

Van Ness 1969. R. T. Van Ness, G. R. De Hoff and R. M. Bonner, "Extrusion," pp. 438, 442, 446, 448, 452, 456, 460, 462, 466 in Sidney Gruss, ed., *Modern Plastics Encyclopedia 1969–1970* (McGraw-Hill Book Co., New York) **46** (10A), October, 1969.

Wallis 1965. Benedict L. Wallis, "Casting," pp. 1–20 in Herman F. Mark, Norman G. Gaylord, and Norbert M. Bikales, eds., *Encyclopedia of Polymer Science and Technology,* Vol. 3, Interscience Div., John Wiley and Sons, New York, 1965.

Westover 1968. R. F. Westover, "Melt Extrusion," pp. 533–587 in Herman F. Mark, Norman G. Gaylord, and Norbert M. Bikales, eds., *Encyclopedia of Polymer Science and Technology,* Vol. 8, Interscience Div., John Wiley and Sons, New York, 1968.

White 1969. D. Alan White, "Low pressure laminates," pp. 406–407 in Sidney Gruss, ed., *Modern Plastics Encyclopedia 1969–1970* (McGraw-Hill Book Co., New York) **46** (10A), October, 1969.

Wilde 1969. Gene Wilde, David Press, and Charles Cook, "Plasticizers," pp. 286, 288, 290 in Sidney Gruss, Ed., *Modern Plastics Encyclopedia 1969–1970* (McGraw-Hill Book Co., New York) **46** (10A), October, 1969.

Willert 1962. W. H. Willert, "Extruder for Injection Molding," *SPE J.* **18**, 568–572 (1962).
Winding 1961. Charles C. Winding and Gordon D. Hiatt, *Polymeric Materials*, McGraw-Hill Book Co., New York, 1961.
Wolinski 1967. L. E. Wolinski, "Films and Sheeting," pp. 764–794 in Herman F. Mark, Norman G. Gaylord, and Norbert M. Bikales, eds., *Encyclopedia of Polymer Science and Technology*, Vol. 6, Interscience Div., John Wiley and Sons, New York, 1967.
Zabel 1969. Robert H. Zabel "Colorants for plastics," pp. 266–267, 274 in Sidney Gruss, ed., *Modern Plastics Encyclopedia 1969–1970* (McGraw-Hill Book Co., New York) **46** (10A), October, 1969.

18

Fiber Technology

The wide range of materials classified as fibers includes natural and synthetic, organic and inorganic products. Some polymers used as fibers, such as nylon and cellulose acetate, serve equally well as plastics. In the final analysis, the classification of a substance as a fiber depends more on its shape than on any other property. One common definition of a fiber requires that its length be at least 100 times its diameter. Artificial fibers can usually be made in any desired ratio of length to diameter. Among the natural fibers, cotton, wool, and flax are often found with lengths 1000–3000 times their diameter: coarser fibers such as jute, ramie, and hemp have lengths 100–1000 times their diameter.

To be useful as a textile material, a synthetic polymer must have suitable characteristics with respect to several physical properties. These include a high softening point to allow ironing, adequate tensile strength over a fairly wide temperature range, solubility or meltability for spinning, a high modulus or stiffness, and good textile qualities such as those defined on page 515. In addition to these primary requirements, many other properties of the material are important if it is to be suitable for textile applications. Some of these are listed in Table 18-1. The desirable values of these properties and the molecular structure needed to obtain them are discussed in the following sections of this chapter.

The textile industry is quite complex, and many of the technologies which comprise it are intricate and obscure. A new fiber must conform to a set of desirable properties which have been largely decided upon in advance. Manipulation of molecular structure and synthesis variables does not allow freedom to achieve all the desired properties, so compromise is inevitable. Even a successful compromise in physical properties does not ensure success for a new fiber. Other factors which must be favorable include all the

TABLE 18-1. *Fiber properties important in textile uses*

Chemical	Physical	Biological	Fabric Properties
Stability toward	Mechanical	Toxicological	Appearance
Acids	Tenacity	Dermatological	Drape
Bases	Elongation	Resistance to	Hand
Bleaches	Stiffness	Bacteria	Luster
Solvents	Flex life	Molds	Comfort
Heat	Abrasion resistance	Insects	Warmth
Sunlight	Work recovery		Water sorption
Aging	Tensile recovery		Moisture retention
Flammability	Thermal		Wicking
Dyeability	Melting point		Stability
	Softening point		Shape
	Glass transition		Shrinkage
	temperature		Felting
	Decomposition		Pilling
	temperature		Crease resistance
	Electrical		Crease retention
	Surface resistivity		

processes leading to some knitted or woven cloth ready for wearing trials, a sound economic situation and competitive price structure, a good supply of raw materials, and, eventually, a good record of customer acceptance.

Information on price and production volume of the major fibers is found in the sections in Part IV describing the fiber-forming polymers.

A. Textile and Fabric Properties

Definitions of textile terms

Types of fibers   Most artificial fibers may be obtained either as a very long *continuous filament*, or as *staple*, made by cutting continuous filament into relatively short lengths. Natural fibers, with the exception of silk, are obtained only in the form of staple.

Denier   The *denier* of a fiber, a measure of its size, is defined as the weight in grams of 9000 meters of the fiber. It is thus proportional to the density of the fiber and to its cross-sectional area. At least ten other measures of the size of fibers are in use, but the denier is the most widely accepted.

Tenacity The tensile strength of a fiber is usually expressed in terms of *tenacity*. Tenacity is defined as the strength per unit size number, such as the denier, where the size number is expressed as a weight per unit length. Tenacity is thus a function of the density of the fiber as well as its tensile strength. Two fibers having the same denier and the same breaking strength have the same tenacity, usually expressed in grams per denier; if they have the same density, they have the same tensile strength as well. Tensile strength (pounds per square inch) = tenacity (grams per denier) × density × 12,791.

Moisture content and moisture regain Most textile fibers absorb some moisture from their surroundings. If the amount of moisture present at equilibrium under standard conditions (65% relative humidity and 70°F) is expressed as a percentage of the total weight of the (moist) fiber, it is known as the *moisture content*; if it is expressed as a percentage of the oven-dry weight (110°C) of the fiber, it is known as the *moisture regain*.

Crimp Crimp is the waviness of a fiber, a measure of the difference between the length of the unstraightened and that of the straightened fiber. Some naturally occurring fibers, notably certain wools, have a natural crimp. Crimp can be artificially produced in fibers by suitable heat treatment or by rolling them between heated, fluted rolls.

Fabric property terms The esthetic qualities of fabrics are defined in terms such as *appearance*, *hand* (or *handle*), and *drape*. Although the qualitative meanings of these terms are obvious, their quantitative definitions and interpretations in terms of fiber properties are quite difficult. Hand and drape are largely determined by the tensile and elastic behavior of the fiber.

Properties of textile fibers

Electrical properties Since fibers are not normally used in electrical applications, their only electrical property of great interest is their resistivity. Too high resistivity leads to the development of static electrical charges, which cause the fabric to cling unpleasantly and to be difficult to clean. Some of the synthetic fibers, including the nylons, polyesters, and acrylics, are poor in this respect.

Mechanical properties The mechanical properties of fibers are quite complex and have been the subject of much experimental work. A stressed textile fiber is a complicated viscoelastic system, in which a number of irreversible processes can take place. Some typical fiber stress-strain curves are shown in Figs. 18-1 and 18-2. They may be divided roughly into two groups: the silklike curves, featuring almost constant high modulus to the break point, and the woollike curves, featuring a sharp drop in modulus at low stress followed by a long period of elongation at almost constant stress. Intermediate characteristics are shown by some of the synthetics such as nylon: the first region of reversible elasticity is followed by an irreversible region in which drawing takes place. When orientation is complete, another reversible elastic region is found. This may end at the point of failure, or plastic flow may take place.

The tenacity at break of typical fibers ranges from about 1 g/denier for acetate rayon up to 5 or more g/denier for nylon and some of the natural fibers. A fiber weaker than acetate is not a practical textile, and one stronger than nylon would find use for its great strength only in exceptional cases.

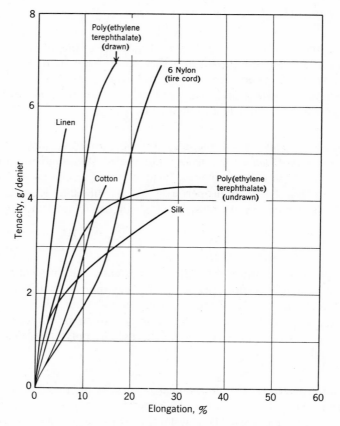

Fig. 18-1.   Stress-strain curves of silklike fibers (Heckert 1953).

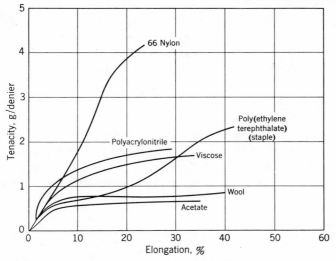

Fig. 18-2. Stress-strain curves of woollike fibers (Heckert 1953).

Moisture regain The moisture regain of the synthetic polyesters, nylons, and acrylics is so much lower than that of any of the natural fibers that the synthetics may be classed as hydrophobic materials. This is an advantage for rapid drying.

Dyeability Hydrophobic fibers are in general difficult to dye. The acrylics are particularly bad in this respect, and modification of the polymer structure by copolymerization is resorted to in order to obtain materials which can be dyed. Nylon is intermediate in dyeability, while cellulose and the cellulosic fibers are eminently dyeable.

Chemical stability All textile fibers must be stable to water, dry-cleaning solvents, and dilute acids, alkalis, and bleaches, and all are reasonably adequate in this respect. Silk and nylon are the least satisfactory fibers from the standpoint of weatherability, whereas the acrylics are outstanding in this respect.

Fabric properties

Esthetic factors (See preceding definitions of textile terms.) Silk is the outstanding example of a fiber with good esthetic properties. Filament acetate, which possibly ranks next, is replacing filament viscose rayon in many applications for which luster, hand, and drape are more important than good mechanical properties. Among the staple fibers, the polyesters appear to be most nearly woollike and most pleasing.

Comfort   The comfort of a fabric is a highly important but little understood property. It is related to the structure of the fabric in determining ventilation and heat-insulating characteristics and wicking ability. Comfort is not, however, dependent upon high moisture absorption in the fiber.

Crease resistance and crease retention   These two properties are related in that a fabric which can be creased only with difficulty tends to retain its crease. The thermoplastic fibers such as nylon, polyesters, and acrylics are very good in both respects. Creasing in wear becomes more important at high moisture contents. Cellulosics are naturally poor in crease resistance, although they may be modified by resin finishes to give good crease resistance.

Fabric stability   The stability of shape and dimensions of the synthetic fibers is outstandingly good. Wool retains its shape well in garments, but felts and shrinks under wet treatment.

The phenomenon of "pilling" has become important in some of the synthetics. A small nodule or pill can be formed on some fabrics by gentle rubbing, as the surface fibers are raised and tangled. In wool these pills are not important because they wear off rapidly, but in nylon and the polyesters the pills do not break away and may become quite unsightly.

Wear resistance   There is no laboratory test for abrasion in a fabric which correlates with actual use. There is evidence that two or more independent parameters are required to define wear in a fabric. In general, acetate rayon is most easily abraded, with viscose, cotton, and wool following as a group (except that viscose when wet is not much better than acetate), while the polyesters, acrylics, and nylon are progressively better.

GENERAL REFERENCES

Mark 1967–1968; McIntyre 1968.

B. Spinning

The conversion of bulk polymer to fiber form is accomplished by *spinning*. In most cases, spinning processes require solution or melting of the polymer (an exception is the spinning of fibers from an aqueous dispersion, as in the case of polytetrafluoroethylene).

If a polymer can be melted under reasonable conditions, the production of a fiber by *melt spinning* is preferred over solution processes. When melt spinning cannot be carried out, a distinction as to type of process is made depending upon whether the solvent is removed by evaporation (*dry spinning*) or by leaching out into another liquid which is miscible with the spinning solvent but is not itself a solvent for the polymer (*wet spinning*). Dry spinning has some advantages over wet spinning. The three processes have many features in common.

The conversion of the spun polymer melt or solution to a solid fiber involves cooling, solvent evaporation, or coagulation, depending on the type of spinning used. The rates of these processes decrease in the order listed. Cooling of a fine filament is normally very rapid and can be controlled within relatively narrow limits. Solvent evaporation involves simultaneous outward mass transfer and inward heat transfer, the rate-controlling step invariably being outward diffusion of solvent. Coagulation involves two-way mass transfer, the coagulating agent (e.g., acid) diffusing inward, and the products of coagulation (e.g., salts, H_2S) diffusing out. As a consequence of these facts, it is very easy to obtain a melt-spun fiber which possesses uniform properties throughout its cross section, but almost impossible to do so with solvent-spun or coagulated fibers. In addition, the rapid cooling of melt-spun fibers tends to produce an almost circular cross section, whereas with solvent-spun fibers the cross sections are usually elliptical, and with coagulated fibers (because of the absence of strong surface-tension effects) the cross sections are ordinarily highly convoluted, as shown in Fig. 18-3.

It is found, however, that convoluted fibers have desirable esthetic properties, particularly hand and luster, so that an effort is made to produce the effect at will. This can be done in melt spinning by the use of noncircular orifices in the spinneret (see below), or in several types of spinning by extruding two different polymers through the same orifice to produce a single *conjugate* fiber (Hicks 1967), as also depicted in Fig. 18-3.

Melt spinning

The process of melt spinning is inherently simple. Molten polymer is pumped at a constant rate under high pressure through a plate called a *spinneret* containing a large number of small holes. The liquid polymer streams emerge downward from the face of the spinneret, usually into air. They solidify and are brought together to form a thread and wound up on bobbins. A subsequent drawing step is necessary to orient the fibers.

The polymer is melted by contacting a hot grid in the form of steel tubing, which is heated by electric current or by hot vapors (Fig. 18-4a). It is usually necessary to protect the polymer melt from oxygen by blanketing

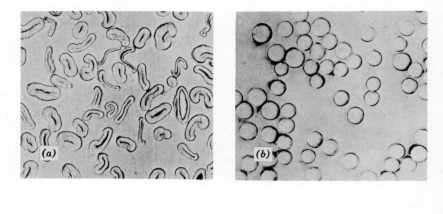

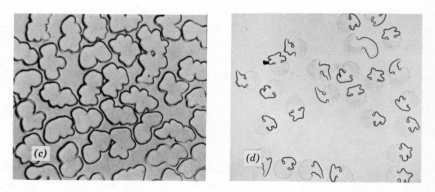

Fig. 18-3.   Photomicrographs, 250 ×, showing cross-sectional shapes of fibers: (*a*) cotton; (*b*) melt-spun polyester; (*c*) dry-spun acetate; (*d*) wet-spun conjugate viscose fiber. [(*a–c*), Riley 1956; (*d*), Hicks 1967.]

it with steam or an inert gas such as carbon dioxide or nitrogen. If the viscosity of the molten polymer is low, it may pass directly to the metering (constant-rate) pump. For melts of higher viscosity a booster pump may be used. Other methods of melting have been proposed, including the use of an extrusion-type screw, one section of which can serve as its own metering pump. Methods in which the metering pump is replaced, e.g., by a source of gas pressure or other device, do not appear to offer sufficiently precise control to hold the denier of fine yarns constant.

The spinneret may consist of a 2- to 3-in. diameter steel disk about $\frac{1}{4}$ in. thick with 50–60 countersunk holes 0.010 in. or less in diameter. The denier of the filament is determined not by the diameter of the holes but by the

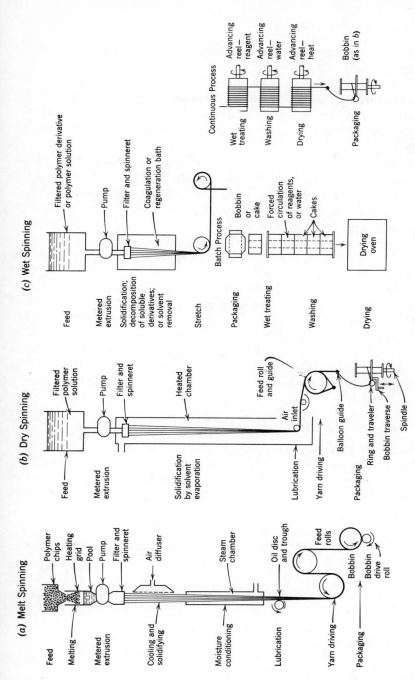

Fig. 18-4. Schematic diagrams of the three principal methods of spinning fibers (Riley 1956).

521

rate at which polymer is pumped through the spinneret and the rate at which the filaments are wound up.

The filaments emerge from the spinneret face into air and begin to cool. An air blast may be used to speed up the cooling process. After the filaments have traveled far enough to become solid (about 2 ft) they are brought together and wound up. Speeds of about 2500 ft/min are usually employed.

The filaments as spun are almost completely unoriented. Most of the stretching that occurs between the spinneret and windup does so while the filament is still molten, and there is sufficient time for molecular orientation to relax before the fiber cools and crystallizes. Consequently a separate drawing step is necessary to produce the orientation of the crystallites necessary for optimum physical properties. In the drawing step somewhat lower speeds are required than can be achieved in spinning, so the two steps are usually done separately. The drawing machine consists of two sets of rolls (Fig. 18-5), one to feed the undrawn yarn from a supply package at velocity v_1 and the other, moving about four times as fast, to collect the drawn yarn at velocity v_2. The filaments may pass over a metal pin between the two sets of rolls; drawing is localized in the neighborhood of the pin. The yarn is then collected on a strong metal bobbin. Freshly drawn yarn has a tendency to contract somewhat in length. Heat is evolved during the drawing process, since work is done on the polymer.

Dry spinning

In the dry-spinning process, the filament is formed by the evaporation of solvent from the polymer solution into air or an inert gas atmosphere. The solvent must, of course, be volatile. Dry spinning has been used for many years for spinning cellulose acetate from acetone solution, and is now

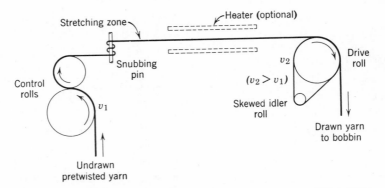

Fig. 18-5. Schematic diagram of a typical drawing process (Riley 1956).

employed for several vinyl fibers including poly(vinyl chloride) and its copolymers, and polyacrylonitrile spun from solution in organic solvents.

So far no process is used in which the polymer is made in the same solvent from which it is dry spun. Therefore, the first step in dry spinning is always the preparation of the polymer solution. The choice of solvent for spinning is based on considerations of solvent power, boiling point, heat of evaporation, stability, toxicity, ease of recovery, etc. Nonpolar solvents are preferred because of convenient boiling points and nonhygroscopic nature, but may cause hazardous buildup of static charge. Low-boiling solvents with high heats of evaporation may cause polymer precipitation or coagulation on the surface of the fiber, in part due to condensation of moisture as the fiber cools when solvent evaporates. This leads to loss of luster and strength. Acetone is preferred for many dry-spinning operations. High solution concentrations are desired, and elevated temperatures are often used to keep viscosities from becoming too high. Cellulose acetate is spun from 20–45% solutions with 400–1000 poise viscosity at 40°C. As in other types of spinning, polymer solutions are usually filtered just before dry spinning.

Spinning is done in a vertical tubular cell (Fig. 18-4*b*), jacketed for temperature control, in which air, steam, or inert gas may be passed through either concurrently or countercurrently as required. Downward spinning is preferred for small-denier fibers and upward spinning for high deniers, to control draw better by eliminating the influence of gravity. Dry spinning is carried out at rates as high as 2500–3000 ft/min, about the same as melt spinning.

Wet spinning

In the wet-spinning process, a solution of a polymer or polymer derivative is spun through a spinneret into a liquid which can coagulate the polymer or derivative (Fig. 18-4*c*). The chemical reaction to convert a derivative to the final polymer can take place simultaneously or later. Wet spinning is commonly used for viscose, cellulose, and some synthetic fibers such as polyacrylonitrile spun from salt solutions.

The essential feature in wet spinning is the transfer of the mass of the solvent from the polymer to the coagulating bath. This transfer is not accompanied by heat of solidification of the polymer, as in melt spinning, or heat evaporation of the solvent, as in dry spinning. A heat of chemical reaction may be present but is not an essential part of the process.

As in other types of spinning, the surface or interfacial tension forces around the filament are quite strong so that it is difficult to produce at the spinneret a filament with a cross section other than circular. These forces also tend to make the filament break up into drops. This tendency is opposed by the viscosity of the polymer; hence more viscous solutions are easier to spin.

It is also advantageous to work with as high concentration of polymer as possible, for obvious reasons. This again leads to high viscosity, and the practical limit may be reached above which the polymer cannot be filtered, pumped, or extruded. For this reason some polymers are wet spun at elevated temperatures where viscosity is lower. Higher temperatures usually promote crystallization and lead to denser, stronger fibers. Cellulose fibers may be spun at 50°C, acrylonitrile fibers as high as 160–185°C.

During coagulation several processes take place simultaneously, including diffusion, osmosis, and salting out. Because of the interplay of these processes, coagulation occurs in rather different ways for different fiber-solvent systems. Usually, coagulation is rapid and the fibers cannot be stretched greatly during this step. The skin of the fiber sets to a gel, and later the volume of material in the core is reduced as the rest of the solvent is removed. The skin then has to fold to accommodate the reduced volume. This gives a characteristic wrinkled cross section to some fibers, such as viscose rayon. It is difficult to produce by wet spinning fibers whose final cross section is circular, despite their initial production at the spinneret in this shape. As a result of the shrinkage, the skin of the fibers is oriented much more than the core.

In other cases, coagulation is much slower and the filaments may be stretched to as much as 30 times their original length before coagulation (as in the spinning of cellulose from cuprammonium solutions into water).

The need for sufficient time for coagulation and other treatments depending on diffusion, and the considerable viscous drag of the coagulating baths, limit the rates of wet spinning to 150–300 ft/min in the usual cases. These speeds are low enough that drawing can be done immediately after spinning in a continuous operation.

Wet-spun yarn may be collected on a bobbin or as a loose cake, which is subjected to chemical treatments, washing, and drying. Alternately, these steps may be carried out continuously as indicated in Fig. 18-4c.

Emulsion and suspension spinning

A certain degree of success has been obtained in spinning insoluble and nonfusible polymers, such as polytetrafluoroethylene, by the following technique: to a solution of ripened cellulose xanthate, an aqueous dispersion of polytetrafluoroethylene was added until the polymeric constituents were about 95% fluoro polymer and 5% cellulosic. This mixture was wet spun. The cellulose was then decomposed completely, and the polytetrafluoroethylene sintered into a continuous fiber, by contact with a metal roll heated to around 390°C. After drawing, the polytetrafluoroethylene fiber (about 2–6 denier) had a tenacity of 1.5–2 g/denier.

GENERAL REFERENCES

Corbiére 1967; Mark 1967; Siclari 1967; Smith 1967; Ziabicki 1967; McIntyre 1968.

C. Fiber After-Treatments

Although it is beyond the scope of this book to treat in detail the production of fabrics from fibers, there are certain intermediate steps between spinning and weaving which are ultimately dependent on the physical, chemical, and molecular structure of the fiber.

All natural and some synthetic fibers must be washed or scoured to free them of natural oils, dirt, chemicals, and other foreign impurities. They must be lubricated or sized or both for proper processing into cloth. They will very likely be dyed for pleasing appearance, and they may have any of a variety of treatments applied to impart or control crease resistance, softening, water repellency, slipping, dimensional stability, shrinkage, or many other properties.

After some or all of these treatments, the fiber is ready to be transformed into a fabric. The most important method of doing this is weaving, in which a set of yarns running lengthwise (warp) is interlaced with a second set at right angles (filling). Other methods of producing fabrics include knitting, in which a series of yarns is looped together, and the manufacture of nonwoven fabrics, papers, and felts, whereby the fibers are bonded together into flat sheets by heat, pressure, and, possibly, bonding agents. Variations of the weaving process lead to pile fabrics, laces, braids, etc. The properties of the fabric depend on the fiber properties, the yarn construction (lengths and diameters of the fibers, size of yarns, amount of twist), the fabric construction (number of yarns per unit area and patterns in which they are combined), and the finishes applied to the fibers and the fabric.

Scouring

The removal of impurities from textile materials, called *scouring*, is carried out by the use of surface-active agents such as soaps and synthetic detergents. These materials have the property of reducing the surface tension of water, and their main function in scouring is to reduce the interfacial tension of their solution toward fats and oils. Most textile fibers contain such fats, oils, or waxes, either naturally or inadvertently or purposely added during various operations. The principal task of scouring is to remove these substances, since most of the other ingredients of the soil are embedded in the fatty material. On repeated laundering, however, some fine soil may be redeposited on the fat-free surface of the fiber, find its way into cavities on the surface, and become extremely resistant to removal. In contrast to initial scouring, one important function of detergents in household and laundry use is to prevent this redeposition of soil.

Lubrication

Lubrication of fibers is necessary to reduce their friction against themselves and against elements of the processing machinery. In the lubrication of yarns, it is necessary to preserve a high static coefficient of friction to keep the yarn in place on its spool or cone, while reducing the dynamic coefficient so as to obtain high speed of movement of the yarn without the generation of heat.

Lubricants may be vegetable or mineral oils or suitably refined petroleum products. Vegetable oils have been preferred because of easy and complete removal by saponification. The best lubricants are water soluble, such as the poly(alkylene glycols).

Another function of the lubricant, which is especially important in the case of the moisture-resistant synthetic fibers, is to reduce the static electric charge on the fibers by lowering their surface resistivity.

Sizing

A *size* is a surface coating used to protect the yarn during weaving. It makes the yarn smooth by binding protruding fibers onto the core of the yarn. Since the fabric is usually dyed after weaving, the size must be easily removed.

Starch is used almost exclusively to size cotton and, to some extent, rayon and wool. It is unsuitable for continuous-filament yarns of rayon, acetate, or the synthetics because of poor adhesion. Gelatin, poly(vinyl alcohol), or other polymeric materials are used to size these yarns.

Dyeing

Dyeing consists in placing a fiber in an (aqueous) solution of a dye and leaving it there until an equilibrium is established in which most of the dye is adsorbed on the fiber and only a small part remains in the dye bath. When equilibrium is reached, the dye concentration is uniform throughout the cross section of the fiber. To attain these conditions, the dye must be preferentially adsorbed in the fiber by means of some type of intermolecular bond such as a van der Waals or hydrogen bond. It is clear also that because of the size of the dye molecules they can penetrate only the amorphous regions of the polymer. Two requirements for successful dyeing are sites to which dye molecules can bond and amorphous regions in the fiber.

Natural cellulosic and protein fibers were known for centuries to dye well with acid dyes which bonded to the hydroxyl or amino groups in the fibers. When cellulose acetate was introduced, a new problem arose which was solved by the use of so-called dispersed dyes containing terminal amine or hydroxyl groups. They are water insoluble and are used as dispersions, hence the name. It seems probable that these dyes undergo hydrogen bonding to the carbonyl oxygen of the acetate groups.

Synthetic fibers have often been very difficult to dye. Polyamide, polyester, and acrylic fibers contain sites for bonding dye molecules, but have such compact structures that dye absorption is extremely slow. Rate of dyeing has been increased by going to higher temperatures, or by utilizing swelling agents for the polymers. The dye sites can be modified, as, e.g., in the treatment of acrylic fibers with copper salts so as to form sites having affinity for anionic dyes. Such fibers as poly(vinyl chloride), poly(vinylidene chloride), and polyethylene contain no sites for dye absorption. It is necessary to copolymerize these monomers with small amounts of materials which do provide dyeing sites in order to obtain satisfactory coloration.

Finishing

Finishes which affect hand   Under the heading of finishes which affect the feel or hand of fabrics come those which increase or decrease the natural friction between the fibers. Thus lubricants make the fabrics feel softer or have increased pliability. On the other hand, antislip finishes such as rosin, carboxymethyl cellulose, and hydroxyethyl cellulose impart a harsh feel to the fabric.

Conditioners   Ethylene and propylene glycols and their low polymers, with the ability to absorb and retain moisture, act as softening, plasticizing, and antistatic agents for the hydrophilic textile fibers, and can counteract to some extent the harshness or stiffness which accompanies the use of antislip agents.

Water-repellent finishes Treatment with waxes such as paraffin emulsions imparts water repellency to cellulosic fabrics, but the effect is not permanent. Permanence can be enhanced by subsequent treatment of the fabric with aluminum or zirconium salts in solution, but the best water repellents at present are those which use stearoylamide pyridinium chloride. This molecule contains the stearoyl radical to give the desired water repellency, the amidomethyl radical which can form a semiacetal with a hydroxyl group of cellulose, and the pyridinium group to solubilize the molecule in water so as to bring it into intimate contact with the fiber. Baking the fiber causes the molecule to split off pyridine hydrochloride, leaving the stearoyl amidomethyl group attached to the cellulose.

Silicones also impart water repellency to fabrics.

Wash-wear finishes One of the major advances in textile technology in recent years has been the introduction of resin finishes which impart the wrinkle- and crease-resistant properties associated with "wash and wear" fabrics. Ordinarily used with cotton and viscose rayon, resin finishes are currently applied to the majority of shirts, blouses, dresses, trousers, and retail piece goods produced.

The resin finishes consist of aqueous solutions of urea-formaldehyde or melamine-formaldehyde precondensates, or of cyclic ureas such as dimethylol ethylene urea. It is thought that wrinkle resistance is imparted through the crosslinking of adjacent cellulose chains in the fiber, rather than from resin formation in or between the fibers. The deposition of the crosslinked polymer between fibers is specifically avoided, since it leads to stiffness and increased body of the fabric.

The fabric is treated with an aqueous solution of the monomers containing an acid or acid-generating catalyst. Excess liquid is removed, and the fabric is dried and cured by heat treatment. Crosslinking is demonstrated by the reduced swelling of the fibers. The treatments function by reducing the irreversible elongation of the fiber: a greater force is needed to stretch the fiber, but a larger percentage of the elongation is recovered.

Textured yarns

The ability to stabilize the physical form of the thermoplastic fibers by heat setting has been utilized in the production of a number of "textured" yarns which have unusually high elasticity. These yarns are made by imparting a tight twist to continuous-filament yarn (usually nylon), heat treating to "set" the twist, and then untwisting the yarn. The resulting coils and twists in the individual fibers can be reversibly pulled out to 200–400%

elongation. Variations of the process exist, and other textured yarns are made by many different processes.

Nonwoven fabrics

A nonwoven fabric is a web or continuous sheet of staple-length fibers, laid down mechanically. The fibers may be deposited in a random manner or preferentially oriented in one direction. The sheet is then bonded together with an adhesive.

Almost all fibers can be used in the production of nonwoven fabrics, those most widely employed being cotton, rayon, and nylon. Binders include poly(vinyl acetate), rubber latices, dispersions of other vinyl polymers, amino resins, and water-soluble cellulose derivatives.

The major use for nonwoven fabrics was once in disposable products, such as tea bags, diapers, and sanitary goods, and industrial wiping cloths. However, their application in household items such as draperies and in interfacings for garments, backings for coated fabrics, etc., has increased rapidly.

GENERAL REFERENCES

Bell 1965; Valko 1965; Ellis 1966; Lewis 1966; Steele 1966; Bikales 1968; Dangel 1969; Hearle 1969; Wray 1970.

D. Table of Fiber Properties

Table 18-2 lists comparative properties for typical examples of the major textile fibers. It should be noted that the properties of fibers of the same chemical type can vary widely depending on precise composition, heat treatment, yarn structure, and many other variables. The values in the table should be used only for comparative purposes.

BIBLIOGRAPHY

Bell 1965. T. E. Bell and H. de V. Partridge, "Bleaching," pp. 438–484 in Herman F. Mark, Norman G. Gaylord, and Norbert M. Bikales, eds., *Encyclopedia of Polymer Science and Technology*, Vol. 2, Interscience Div., John Wiley and Sons, New York, 1965.
Bikales 1968. Norbert M. Bikales, "Nonwoven Fabrics," pp. 345–355 in Herman F. Mark, Norman G. Gaylord, and Norbert M. Bikales, eds., *Encyclopedia of Polymer Science and Technology*, Vol. 9, Interscience Div., John Wiley and Sons, New York, 1968.

TABLE 18-2. Properties of fibers (Modern Plastics 1961)

Property	Acetate	Triacetate	Viscose Regular	Viscose High Tenacity	Nylon 66 Continuous Filament	Nylon 66 Staple	Vinyl Chloride-Acetate Staple	Polyester Continuous Filament	Polyester Staple	Vinylidene Chloride Continuous Filament	Acrylic Staple	Cotton	Silk	Wool
Specific gravity g/cm^3	1.30	1.30	1.50	1.50	1.14	1.14	1.33	1.38	1.38	1.70	1.14	1.50	1.25	1.30
Tensile strength, 10^3 lb/in.^{2}	18–23	20–23	29–47	58–97	66–88	58–105	10–12	77–88	56–78	25–60	33–38	42–125	45–83	17–28
g/denier	1.1–1.4	1.2–1.4	1.5–2.4	3.0–5.0	4.5–6.0	4.0–7.2	0.6–0.7	4.4–5.0	3.2–4.4	1.1–2.9	2.3–2.6	2.1–6.3	2.8–5.2	1.0–1.7
Wet strength, % dry	60–70	67–70	44–54	55–75	85–90	85–90	100	100	100	100	80	110–130	75–95	76–97
Ultimate elongation, %	25–45	25–40	15–30	9–20	26	16–45	10–13	19–23	30–36	20–35	20–28	3–10	13–31	20–50
Recovery, % from elongation, %	23	43	30	37	100	—	—	80	—	95	—	45	33	63
	20	10	20	15	8	—	—	8	—	10	—	5	20	20
Stiffness, 10^{10} dynes/cm^2	3.0–4.8	3.5–5.2	6.4–9.1	9.9–23.6	3.0	1.5	3.3–4.5	12.2	6.1	0.3–1.7	5.0	5.7–11.2	8.4–12.9	2.7–3.9
g/denier	26–41	35–45	48–68	74–176	30	1.5	28–38	100	50	2–11	50	42–82	76–117	24–34
Moisture regain, % at 21°C, 65% R.H.	6.3–6.5	3.2	11.5–16.6	11.5–16.6	4.0–4.5	4.0–4.5	0	0.4	0.4	0	1.5	8.5	11.0	17.0
Melting or softening temperature, °C	260M	300M	—	—	265M	265M	127M	265M	265M	125S	245S	—	—	—
Loss in strength in sun, prolonged exposure	Yes	Slight	Yes	Yes	Yes	Yes	No	Yes	Yes	No	No	Yes	Yes	Yes

Corbiére 1967. J. Corbiére, "Fundamental Aspects of Solution Dry-Spinning," pp. 133–167 in H. F. Mark, S. M. Atlas, and E. Cernia, eds., *Man-Made Fibers, Science and Technology*, Vol. 1, Interscience Div., John Wiley and Sons, New York, 1967.

Dangel 1969. Phoenix N. Dangel, "Nonwoven fabrics," p. 374 in Sidney Gruss, ed., *Modern Plastics Encyclopedia 1969–1970* (McGraw-Hill Book Co., New York) **46** (10A), October, 1969.

Ellis 1966. J. R. Ellis and G. M. Gantz, eds., and Emery I. Valko, "Dyeing," pp. 235–375 in Herman F. Mark, Norman G. Gaylord, and Norbert M. Bikales, eds., *Encyclopedia of Polymer Science and Technology*, Vol. 5, Interscience Div., John Wiley and Sons, New York, 1966.

Hearle 1969. John W. S. Hearle, Percy Grosberg, and Stanley Backer, *Structural Mechanics of Fibers, Yarns and Fabrics*, Interscience Div., John Wiley and Sons, New York, 1969.

Heckert 1953. W. W. Heckert, "Synthetic Fibers from Condensation Polymers," Chap. 5 in Samuel B. McFarlane, ed., *Technology of Synthetic Fibers*, Fairchild Publications, New York, 1953.

Hicks 1967. E. M. Hicks, E. A. Tippetts, J. V. Hewett, and R. H. Brand, "Conjugate Fibers," pp. 375–408 in H. F. Mark, S. M. Atlas, and E. Cernia, eds., *Man-Made Fibers, Science and Technology*, Vol. 1, Interscience Div., John Wiley and Sons, New York, 1967.

Lewis 1966. Charles E. Lewis, "Dyes," pp. 376–405 in Herman F. Mark, Norman G. Gaylord, and Norbert M. Bikales, eds., *Encyclopedia of Polymer Science and Technology*, Vol. 5, Interscience Div., John Wiley and Sons, New York, 1966.

Mark 1967. H. F. Mark and S. M. Atlas, "Principles of Spinning in Emulsion and Suspension," pp. 237–240 in H. F. Mark, S. M. Atlas, and E. Cernia, eds., *Man-Made Fibers, Science and Technology*, Vol. 1, Interscience Div., John Wiley and Sons, New York, 1967.

Mark 1967–1968. H. F. Mark, S. M. Atlas, and E. Cernia, eds., *Man-Made Fibers, Science and Technology*, Interscience Div., John Wiley and Sons, New York; Vol. 1, 1967; Vols. 2 and 3, 1968.

McIntyre 1968. J. E. McIntyre, "Man-Made Fibers, Manufacture," pp. 374–404 in Herman F. Mark, Norman G. Gaylord, and Norbert M. Bikales, eds., *Encyclopedia of Polymer Science and Technology*, Vol. 8, Interscience Div., John Wiley and Sons, New York, 1968.

Modern Plastics 1961. Sidney Gruss, ed., *Modern Plastics Encyclopedia*, Breskin Publications, New York, 1961.

Riley 1956. J. L. Riley, "Spinning and Drawing Fibers," Chap. XVIII in Calvin E. Schildknecht, ed., *Polymer Processes*, Interscience Publishers, New York, 1956.

Siclari 1967. F. Siclari, "Fundamental Aspects of Wet-Spinning Solutions," pp. 95–132 in H. F. Mark, S. M. Atlas, and E. Cernia, eds., *Man-Made Fibers, Science and Technology*, Vol. 1, Interscience Div., John Wiley and Sons, New York, 1967.

Smith 1967. Arthur L. Smith, "Fibers, Identification," pp. 593–609 in Herman F. Mark, Norman G. Gaylord, and Norbert M. Bikales, eds., *Encyclopedia of Polymer Science and Technology*, Vol. 6, Interscience Div., John Wiley and Sons, New York, 1967.

Steele 1966. Richard Steele and D. D. Gagliardi, "Crease Resistance," pp. 307–330 in Herman F. Mark, Norman G. Gaylord, and Norbert M. Bikales, eds., *Encyclopedia of Polymer Science and Technology*, Vol. 4, Interscience Div., John Wiley and Sons, New York, 1966.

Valko 1965. Emery I. Valko and Giuliana C. Teroso, "Antistatic Agents," pp. 204–229 in Herman F. Mark, Norman G. Gaylord, and Norbert M. Bikales, eds., *Encyclopedia of Polymer Science and Technology*, Vol. 2, Interscience Div., John Wiley and Sons, New York, 1965.

Wray 1970. G. R. Wray, "Textile Processing," in press in Herman F. Mark, Norman G. Gaylord, and Norbert M. Bikales, eds., *Encyclopedia of Polymer Science and Technology*, Vol. 13, Interscience Div., John Wiley and Sons, New York, 1970.

Ziabicki 1967. Andrzej Ziabicki, "Principles of Melt-Spinning," pp. 169–236 in H. F. Mark, S. M. Atlas, and E. Cernia, eds., *Man-Made Fibers, Science and Technology*, Vol. 1, Interscience Div., John Wiley and Sons, New York, 1967.

19

Elastomer Technology

The final class of high polymers to be considered in Part V is elastomers. Like fibers, elastomers are considered apart from other polymeric materials because of their special properties. Unlike fibers, elastomers do not in general lend themselves to plastics uses; elastomers must be amorphous when unstretched and must be above their glass transition temperature to be elastic, whereas plastics must be crystalline or must be used below this temperature to preserve dimensional stability.

About 5.7 billion lb of rubber was used in the United States in 1969, with this level expected to increase to 7.5 billion lb by 1975. Synthetics accounted for 77% of the total consumed, with SBR in largest volume by a wide margin, as indicated in Table 19-1. Automobile tires (some 220 million in 1969) remain the largest single end use of rubber, consuming 68% of the total United States production.

History of synthetic rubber   The first attempts to produce synthetic rubbers centered around the homopolymerization of dienes, particularly isoprene because it was known to be the monomer for natural rubber. It was found in the late nineteenth century that rubberlike products could be made from isoprene by treating it with hydrogen chloride or allowing it to polymerize spontaneously on storage. These materials could be vulcanized with sulfur, becoming more elastic, tougher, and more heat resistant.

Around 1900 it was discovered that other dienes such as butadiene and 2,3-dimethylbutadiene could be polymerized to rubberlike materials spontaneously by alkali metals or by free radicals. Application was made of these facts during World War I in Germany, where 2,3-dimethylbutadiene was polymerized spontaneously.

After World War I this research on rubberlike products continued, with the emphasis shifting to butadiene because of its more ready availability, at that time from acetaldehyde via the aldol synthesis. Alkali metals were used as initiators, and the products were called *buna rubbers* from the first letters of butadiene and the symbol Na for sodium.

Improvements in processing and properties were made in two ways: copolymers of butadiene with vinyl monomers, notably styrene, were introduced, and emulsion polymerization was adopted. The significance of initiators, reduction activators, and oxygen became known. The importance of maintaining low conversion or utilizing modifiers such as CCl_4

TABLE 19-1. *Approximate United States consumption of rubber in 1969 (Carpenter 1970)*

Type	Consumption, million lb
SBR	2850
Polybutadiene	600
Neoprene	260
Butyl	200
Polyisoprene	190
Nitrile	140
EPDM	100
Other synthetics	60
Total synthetics	4400
Natural rubber	1300
Total	5700

and long-chain mercaptans was discovered. By the beginning of World War II, acceptable polymers (buna-S) were being produced in Germany containing 68–70% butadiene and 30–32% styrene.

In review of earlier chapters, it may be recalled that the unique properties of elastomers include their ability to stretch and retract rapidly, exhibit high strength and modulus while stretched, and recover fully on release of the stress. To obtain these properties, certain requirements are placed upon the molecular structure of the compounds: they must be high polymers, be above their glass transition temperatures, be amorphous in the unstretched state (but preferably develop crystallinity on stretching), and contain a network of crosslinks to restrain gross mobility of the chains.

A. Compounding and Elastomer Properties

The process by which a network of crosslinks is introduced into an elastomer is called *vulcanization*. The chemistry of vulcanization (Section *B*) is complex and has not been well understood throughout the century of practice of the process since its discovery by Goodyear in 1839. The profound effects of vulcanization, however, are clear: it transforms an elastomer from a weak thermoplastic mass without useful mechanical properties into a strong, elastic, tough rubber (Table 19-2). The tensile strength, stiffness,

TABLE 19-2. Properties typical of raw, vulcanized, and reinforced natural rubber

Property	Raw Rubber	Vulcanized Rubber*		Reinforced Rubber
Tensile strength, psi	300	3000		4500
Elongation at break, %	1200	800		600
Modulus, psi†	—	400		2500
Permanent set	Large		Small	
Rapidity of retraction (snap)	Good		Very good	
Water absorption	Large		Small	
Solvent resistance (hydrocarbons)	Soluble		Swells only	

* Not reinforced.
† Tensile stress at 400% elongation; a measure of stiffness.

and hysteresis (representing loss of energy as heat) of natural rubber before and after vulcanization are shown in Fig. 19-1.

For many uses, however, even vulcanized rubbers do not exhibit satisfactory tensile strength, stiffness, abrasion resistance, and tear resistance. Fortunately, these properties can be enhanced by the addition of certain fillers to the rubber before vulcanization. Fillers for rubber can be divided into two classes: *inert fillers*, such as clay, whiting, and barytes, which make the rubber mixture easier to handle before vulcanization but have little effect on its physical properties; and *reinforcing fillers* (Section *C*), which do improve the above-named unsatisfactory properties of the vulcanized rubber. Carbon black is the outstanding reinforcing filler for both natural and synthetic rubbers. The added effects of reinforcement with carbon black on the

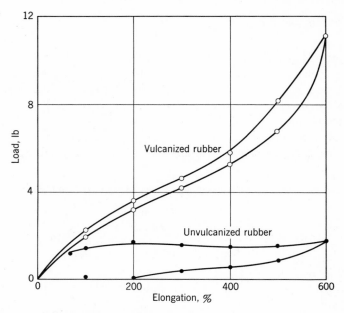

Fig. 19-1. Stress-strain curves to 600% elongation and back, typical of unvulcanized and vulcanized natural rubber.

properties of rubbers over the effects of vulcanization alone are also illustrated in Table 19-2. Although the nature of reinforcement is not completely understood, it appears to add a network of many relatively weak "fix points" to the more diffuse network of strong primary bond crosslinks introduced by vulcanization. Vulcanization restrains the long-range movements of the polymer molecules but leaves their local segmental mobility high; reinforcement stiffens the mass and improves its toughness by restricting this local freedom of movement.

Mechanical properties of elastomers

The stress-strain behavior of pure gum (i.e., nonreinforced) vulcanizates of the various elastomers depends markedly on the molecular structure, polarity, and crystallizability of the polymers. Nitrile rubber and SBR, being copolymers of complex and irregular structure, do not crystallize at all, but natural rubber and butyl rubber crystallize on stretching at room temperature. In these elastomers the crystallites have a stiffening effect like that of a reinforcing filler with the result that their stress-strain curves turn up markedly at higher elongations. The similarity of these stress-strain curves to those predicted from the kinetic theory of rubber elasticity (Chapter 6*B*) should be noted.

Unvulcanized rubber undergoes a very large amount of mechanical conditioning when first exposed to tensile stress. Figure 19-1 shows the first cycle of loading and unloading of a tensile specimen of unvulcanized natural rubber. As successive cycles take place, the changes in resistance to stretching, tensile strength, energy absorption, and permanent set become smaller. This mechanical conditioning is not accompanied by molecular weight degradation. Such degradation does occur, however, in the process of milling raw rubber to reduce its melt viscosity to the desired range before the incorporation of fillers, etc. Here the process is complicated by the presence of oxygen. Vulcanized rubber undergoes a similar but much less extensive change in stress-strain properties on successive cycles of test.

The mechanical properties of elastomers depend greatly upon the rate of testing. Since service conditions for automobile tires (by far the most important end use of elastomers) involve rapid cyclic stresses, dynamic test methods are essential if the results are to correlate with the performance of the finished article.

Oxidative aging of elastomers

Rubbers which retain double bonds in their vulcanized structure, such as natural rubber, SBR, and nitrile rubber, are sensitive to heat, light, and particularly oxygen. Unless protected with antioxidants (Maassen 1965) (of which the phenyl α- and β-naphthylamines are the most common) the rubbers age by an autocatalytic process accompanied by an increase in oxygen content. The results of the aging may be either softening or embrittlement, suggesting competing processes of degradation and crosslinking of the polymer chains. The first step in the process is presumed to be the same as that in the oxidation of any olefin, i.e., the formation by free radical attack of a peroxide at a carbon atom next to a double bond:

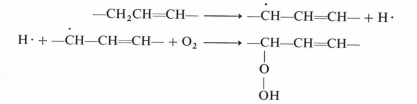

Subsequent steps may include crosslinking or chain cleavage or involvement of the double bond, probably initially through an epoxide group. Many of the steps in the reaction resemble closely those of certain types of vulcanization.

Many rubbers are sensitive also to attack by ozone, requiring protection by antiozonants (Cox 1965). These are often derivatives of *p*-phenylene diamine, and are thought either to react with the ozone before it can undergo reaction with the rubber surface, or to aid in reuniting chains severed by ozone.

Natural rubber is far more sensitive than most of the other elastomers to both oxygen and ozone attack.

Compounding

The term *compounding* describes the selection of additives and their incorporation into a polymer so as to give a homogeneous mixture ready for subsequent processing steps. Except for differences in the nature of the compounding ingredients, especially as required for vulcanization and reinforcement and described under those headings, the compounding of rubbers is similar to that of most plastics. Typical equipment described in Chapter 17 is used for rubber as well as plastics compounding.

In the case of natural rubber, SBR, and the other synthetic rubbers produced by emulsion polymerization, the polymer is first available in the form of a latex. The normal procedure in compounding is to coagulate and dry the latex, and then to masticate and compound the rubber on mills or in other equipment. With SBR, however, two important compounding steps, *oil extending* and *masterbatching*, are commonly carried out before coagulation of the latex.

Oil extending   The use of hydrocarbon oils to dilute, or extend, rubbers has been known for many years. Since 1950, this practice has been applied to SBR polymerized to high molecular weight. The oil serves as a plasticizer and softener, reducing the melt viscosity of the rubber to that normally required in compounding. The process, which leads to a less expensive final product, is widely used: about 90% of the tire tread stock produced in this country is oil extended to some extent. The oil is added as an emulsion to the latex before coagulation.

Masterbatching   A convenient method of mixing rubber and carbon black is the coprecipitation of a mixture of the rubber latex and an aqueous slurry of the black. This method gives adequate mixing when the particle size of the black is similar to that of the latex particle and the number of each type of particle per cubic centimeter before the coagulation can be made nearly the same. This situation holds for SBR and the most useful blacks. Latex masterbatching has now become an established technique in the rubber industry.

The technology of elastomers has been reviewed (Weissert 1969).

GENERAL REFERENCES

Morton 1959; Bateman 1963; Cooper 1966; Stagg 1968; Anderson 1969; Carpenter 1970; Cunneen 1970; Garvey 1970.

B. *Vulcanization*

Ever since Goodyear's first experiments in heating rubber with small amounts of sulfur, this process has been the best and most practical method for bringing about the drastic property changes described by the term *vulcanization*, not only in natural rubber but also in the diene synthetic elastomers such as SBR, butyl, and nitrile rubbers. It has been found since, however, that neither heat nor sulfur is essential to the vulcanization process. Rubber can be vulcanized or *cured* without heat by the action of sulfur chloride, for example. A large number of compounds which do not contain sulfur can vulcanize rubber; these fall generally into two groups, oxidizing agents (selenium, tellurium, organic peroxides, nitro compounds) and generators of free radicals (organic peroxides, azo compounds, many accelerators, etc.). Thus it is clear that there is no single method or chemical reaction of vulcanization.

Since the chemical reactions associated with vulcanization are varied and involve only a few atoms in each polymer molecule, a definition of vulcanization in terms of the physical properties of the rubber is necessary. In this sense, vulcanization may be defined as any treatment that decreases the flow of an elastomer, increases its tensile strength and modulus, but preserves its extensibility. There is little doubt that these changes are due primarily to chemical crosslinking reactions between polymer molecules. As might be expected, such properties as tensile strength are relatively insensitive to the onset of these crosslinking reactions; tensile strength does indeed change tenfold during curing, but this is evidence of the profound alteration of polymer properties by the process. Tests based on melt flow are more sensitive to initial crosslinking reactions and are widely used in the rubber industry.

Although vulcanization takes place by heat in the presence of sulfur alone, the process is relatively slow. It can be speeded many-fold by the addition of small amounts of organic or inorganic compounds known as *accelerators*. Many accelerators require the presence of still other chemicals known as *activators* or *promoters* before their full effects are realized. These activators are usually metallic oxides, such as zinc oxide. They function best in the presence of a rubber-soluble metallic soap, which may be formed during the curing reaction from the activator and a fatty acid. The most efficient combination of chemicals for sulfur vulcanization includes sulfur, an organic accelerator, a metallic oxide, and a soap.

Chemistry of vulcanization

It is convenient to group vulcanization reactions into two categories: nonsulfur vulcanizations, in which peroxides, nitro compounds, quinones, or azo compounds are the curing agents; and vulcanizations brought about by sulfur, selenium, or tellurium.

The understanding of vulcanization has been considerably advanced by the recognition of the following two concepts which arise from the analogy of vulcanization with the crosslinking and (oxidative) degradation of simple olefins:

a. The point of initial attack in most vulcanization reactions is a hydrogen atom on a methylene group alpha to the double bond itself.

b. Free radical mechanisms are involved in almost all cases.

Nonsulfur vulcanization From the two concepts just given, the following scheme has been developed for nonsulfur vulcanizations:

a. A free radical R· is formed by the decomposition or oxidation of the curing agent, or as a step in the oxidative degradation of the rubber.

b. This free radical initiates vulcanization by abstracting a hydrogen atom from one of the α-methylene groups (in natural rubber the methyl side group directs the attack to the methylene group nearest it):

$$\text{R·} + \underset{\overset{|}{\text{H}}}{-\text{CH}}-\overset{\overset{\text{CH}_3}{|}}{\text{C}}=\text{CH}-\text{CH}_2- \longrightarrow \text{RH} + -\underset{\bullet}{\text{CH}}-\overset{\overset{\text{CH}_3}{|}}{\text{C}}=\text{CH}-\text{CH}_2-$$

c. The rubber free radical then attacks a double bond in an adjacent polymer chain. This results in the formation of a crosslink and the regeneration of a free radical in a reaction analogous to propagation in an addition polymerization (Chapter 9):

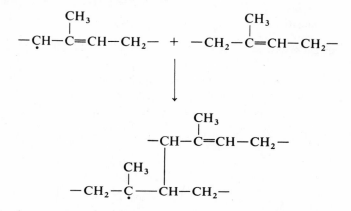

Vulcanization may continue by several such propagation steps.

Chain transfer may also occur. Termination probably occurs by reaction of the rubber free radical with a free radical fragment of the curing agent. In contrast to addition polymerization in a fluid system, the termination reaction between two rubber free radicals is considered unlikely because of the low probability of two such radicals coming into position to react, owing to the high viscosity of the medium.

Sulfur vulcanization Not all the reactions which take place in sulfur curing are well understood. It is clear that the sulfur can undergo a variety of reactions. Some, but not all, of it goes to form sulfide or disulfide crosslinks between chains, e.g.,

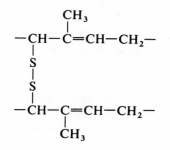

Some double bonds are saturated by reactions analogous to step *c* of the mechanism above, some by the addition of hydrogen sulfide (a known product of sulfur vulcanization—see next paragraph), and some by the formation of cyclic sulfides such as

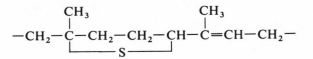

Some sulfur reacts with an activator such as zinc oxide to form the corresponding sulfide.

Most accelerators are either free radical formers or strong proton acceptors (as is sulfur itself). Thus C—C crosslinks may also be formed by the free radical mechanism described above, or by the removal of hydrogens by the proton acceptors, e.g.,

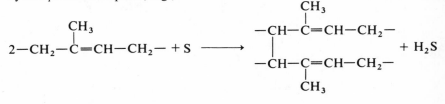

Many other possibilities exist.

Other types of vulcanization   Some reagents, such as mercaptans, thiol acids, and mercapto acids, can add directly to diene double bonds. Bisthiol acids can produce direct interchain crosslinks by this mechanism.

Physical aspects of vulcanization

Vulcanization of the type just described takes place when 0.5–5 parts (by weight) of sulfur is combined with 100 parts of rubber. If the reactions are allowed to continue until considerably more sulfur has combined (say 30–50 parts per 100 parts rubber), a rigid, nonelastomeric plastic known as *hard rubber* or *ebonite* is formed. Tensile strength reaches a maximum, with elongation remaining high, in the early *soft cure* region; then both fall to much lower values with increasing amount of combined sulfur.

The region of maximum tensile properties in the neighborhood of 5–10 parts combined sulfur is known as the region of *optimum cure* for this *mix*. Beyond this point the *stock* (i.e., mixture) is said to be *overcured* and is likely to be leathery, i.e., stiffer and harder than at the optimum cure but weaker and less extensible. At high proportions of combined sulfur the strength of the material increases again, and the elongation becomes very low as the hard rubber region is reached.

Many properties of soft vulcanizates, including tensile and tear strength, stiffness, and hardness, go through a shallow maximum as the cure time is increased. Others, such as elongation and permanent set, drop continuously.

The effect of increasing temperature is, as expected, to move the curves to shorter times. Experiments in which the composition of the mix is varied show that the initial rise in tensile strength is accompanied by the rapid incorporation of sulfur in the rubber, and that, as the rate of sulfur addition decreases, the rate of increase of tensile strength begins to decline. It is also known that the total amount of combined sulfur does not measure the extent of change in physical properties during cure, and that the ratio of sulfur combined to double bonds used up varies widely from one mix to another.

These facts suggest that processes of crosslinking and degradation occur simultaneously during vulcanization. Additional evidence for this view arises from the fact that badly overcured stocks may undergo *reversion*, i.e., become soft and tacky with notable loss in strength. Such stocks resemble oxidatively degraded rubber. Reversion differs from overcuring in that it usually takes place on prolonged heating when not enough sulfur is present to get into the region of leathery cures.

Accelerators

Until about 1900, rubber vulcanization was a slow process. Large excesses of sulfur over the amount required for optimum cure were used. The remaining free sulfur diffused to the surface of the stock and caused *bloom*. The curing time was several hours, despite the fact that lead, calcium, magnesium, or zinc oxides (inorganic accelerators) speeded the reaction slightly.

The rubber industry was revolutionized by the introduction of organic accelerators for vulcanization. These include sulfur-containing compounds such as thioureas, thiophenols, mercaptans, dithiocarbamates, xanthates, trithiocarbamates, dithio acids, mercaptothiazoles and mercaptobenzothiazoles, and thiuram sulfides, plus a few nonsulfur types such as ureas, guanidines, and aldehydeamines.

Except for the generalization that accelerators are likely to contribute to vulcanization by initiating free radical chains or by abstracting protons, little is known about their specific action in speeding up vulcanization. Some accelerators, such as thiuram disulfides, are known to dissociate thermally at vulcanization temperatures:

$$\underset{\substack{\| \\ R_2N-C}}{\overset{S}{}}-S-S-\underset{\substack{\| \\ C}}{\overset{S}{}}-NR_2 \longrightarrow CS_2 + S + R_2N\underset{\substack{\| \\ }}{\overset{S}{C}}NR_2$$

The sulfur is consumed in promoting vulcanization. Since the mechanism may involve free radical steps, further action is possible. Zinc salts of mercaptobenzothiazole and the dithiocarbamic acids may act similarly. On the other hand, an accelerator such as mercaptobenzothiazole itself, which is known not to decompose or to react with olefins at curing temperatures, must always be used with an activator.

Activators

The presence of basic metallic oxides or salts of lead, calcium, zinc, or magnesium appears to be necessary to obtain the full effect of almost all organic accelerators. Solubility of the compound is important; for this reason the oxides are used with organic acids (stearic acid, rosin), or the soaps (stearates, laurates) of the metals are utilized instead. Aside from the specific need for an activator with an acidic accelerator such as mercaptobenzothiazole, as just noted, the presence of the oxide appears to be important in determining the type of crosslinking reactions which take place. These appear to be strong ionic bridges of the type

$$\text{rubber—(accelerator + S)}^- \text{ Zn}^{+2} \text{ }^-\text{(accelerator + S)—rubber}$$

formed during vulcanization. In cases in which the activator is important, these bridges may contribute substantially to the crosslinking in the vulcanizate.

Typical vulcanization mix

The majority of sulfur-vulcanized products of natural rubber are based on a mix with the following composition range: rubber, 100 parts by weight; sulfur, 0.25–1.5 parts; an accelerator, 0.25–1.5; an activator such as zinc oxide, 1–10; a soap such as stearic acid or zinc laurate, 1–5; and an antioxidant, 0–1.5. If no other ingredients are added, the product is known as a *pure-gum vulcanizate. Loaded stocks*, made by adding carbon black or other fillers to this recipe, are discussed in Section *C*.

GENERAL REFERENCES

Morton 1959; Garvey 1970; Wolfe 1970.

C. Reinforcement

Pure-gum vulcanizates are relatively soft, pliable, and extensible, and are most useful for such items as rubber bands, tubing, and gloves. Rubber would be of little value in modern industry were it not for the fact that certain fillers enhance such mechanical properties as tensile strength, stiffness, tear resistance, and abrasion resistance of the stocks into which they are incorporated. Such loading agents are called *reinforcing fillers*, and the improvement in service life of a heavy-duty rubber item, such as a belt or tire tread, resulting from the enhancement of these properties, is called *reinforcement*.

Types of fillers

Only a few fillers are reinforcing for even one or two of the important mechanical properties of natural rubber, while some weaken the vulcanizates in one or more important respects. The latter are known as *inert fillers*. In general it is found that the reinforcing action of a filler depends upon its nature, the type of elastomer with which it is used, and the amount of filler present.

Carbon black is the only important reinforcing filler for most elastomers, although silica and silicate fillers can be used in some cases, as with the silicone elastomers (Chapter 16*F*). Important nonreinforcing fillers include talc, zinc oxide, and magnesium carbonate.

Nature of the filler   The chemical composition of the filler is a primary determinant of its reinforcing action. Carbon blacks are always more efficient reinforcing agents than, e.g., talc, even though particle size, surface condition, and other properties can be varied widely for both. Within a single chemical class, particle size and surface condition are important variables.

Nature of the elastomer   Although a good reinforcing filler such as carbon black increases the tear and abrasion resistance of all the current major elastomers, it has no effect in enhancing the tensile strength of neoprene or butyl rubber, in contrast to the marked tensile reinforcement it causes in natural rubber, SBR, nitrile rubber, and even the polysulfide elastomers. These differences are ascribed to a conflict between two effects of the filler. (*a*) A reinforcing filler stiffens and strengthens the structure by introducing a network of many relatively weak fix points. (*b*) Simultaneously, it may interfere with the ability of the polymer to crystallize at high elongations, simply by its bulkiness in the system. Since stiffness and tensile strength are greatly enhanced by crystallization, the specific effect of the filler depends upon which of these actions predominates.

Amount of filler   Increasing amounts of reinforcing filler cause continuing improvement in properties until a maximum is reached, representing the optimum loading for this composition. Beyond this point, additional filler merely acts as diluent and the properties of the vulcanizate again deteriorate.

Various colloidal forms of carbon constitute the most important reinforcing filler for rubber. Carbon blacks are the only inexpensive materials that reinforce all three of the important properties—tensile strength, tear resistance, and abrasion resistance. Many different carbon blacks are used, differing mainly in particle size, surface condition, and degree of agglomeration. Roughly, the degree of reinforcement increases with decreasing particle size of the black down to the practical lower limit at which blacks can be made, about 100 A diameter.

Types of carbon black   *Channel blacks* are made by burning natural gas in a limited supply of air and allowing the carbon to deposit on cold iron channels. They are sometimes designated by processability, as easy-processing channel, etc. Particle diameters range from 200 to 300 A. *Furnace blacks* are replacing channel blacks in many uses. They are made by burning gas or oil in a furnace in limited air and removing the carbon from the off-gases by a centrifugal and electrostatic precipitation. They range from 300 to 800 A in particle diameter, and are often designated by their effects on the vulcanizates, as high-abrasion, high-modulus, semi-reinforcing, or conductive furnace blacks. Coarser blacks, such as *thermal blacks* (from hydrocarbons decomposed in the absence of air), *acetylene blacks*, and *lamp blacks*, are less commonly used as reinforcing fillers.

Structure of carbon blacks   The electron microscope shows that the carbon blacks are essentially spherical, with a tendency in some cases to align in continuous chainlike structures. These chains form electrically conducting paths through the rubber, hence the names conductive channel and conductive furnace blacks. Surface areas calculated from electron microscope measurements of particle diameter and by nitrogen absorption agree well, indicating that the particles are relatively nonporous. They have a crystal structure similar to that of graphite except that the sheets of hexagonally close-packed carbon atoms are arranged randomly above one another. This suggests that unsatisfied bonds are present which may be important in reinforcement. The surfaces of the particles contain much adsorbed material, including hydrogenated and oxygenated structures. As a result the pH of a water slurry of the black may range from 2.5 to 11.

Effects of carbon black structure on reinforcement   The three parameters—particle size, pH, and *structure index*, i.e., tendency to agglomerate or form chains—appear to be predominant in determining the reinforcing behavior of blacks. Tensile strength, abrasion and tear resistance, hardness, and toughness increase with decreasing particle size, whereas rebound and ease of processing become poorer. Blacks with a high structure index are difficult to disperse and give high stiffness and hardness but low tensile strength, toughness, and electrical resistivity. The pH influences the rate of vulcanization, depending upon the acidic or basic nature of the accelerator.

The stiffness of a reinforced vulcanizate rises steeply with the loading of carbon black and is nearly independent of the particle size of the black. These two features are unique for this property. Other properties, such as tensile strength, go through a maximum as loading is increased. They also depend upon particle size, as mentioned above.

The tensile strength of natural rubber can be increased about 40% by reinforcement. In SBR, however, the properties of the pure-gum vulcanizates are very poor, since the elastomer does not crystallize. The tensile strength of SBR can be raised about tenfold by reinforcement, making it equivalent to natural rubber for fully reinforced stocks. The tear resistance of SBR is also poor for unloaded vulcanizates, but is equivalent to that of natural rubber after reinforcement.

Abrasion resistance is highly important for heavy-duty rubber such as tire tread stock. The abrasion resistance of both natural rubber and SBR can be improved at least fivefold by proper reinforcement. Unfortunately, the resilience of the rubber decreases with increasing loading of filler. As a consequence hysteresis loss and heat buildup increase. Thus the use of a reinforcing filler represents a compromise between adequate abrasion and tear resistance and abnormal heat buildup.

Origin of reinforcement   Only two conditions must be met for significant reinforcement of a rubber by carbon black. First, the particle size of the black must be small, usually between 200 and 500 A. This insures a large surface area and thus a large filler-rubber interface. Second, the rubber must " wet " the carbon black. This not only assures dispersion of the black in the rubber, but implies that the rubber-black adhesion approaches in strength the cohesion of the rubber to itself. This is necessary if the rubber-black bonds are to survive the large strains associated with high elongations. This adhesion is produced in part by secondary bond forces, which alone are enough to account for reinforcement, and in part by chemical grafting of rubber on to the surface of the black particle, almost certainly as the result of radical processes.

GENERAL REFERENCES

Morton 1959; Burgess 1965; Kraus 1965, 1970; Garvey 1970.

TABLE 19-3. *Typical properties of commercial elastomers (Whitby 1954, Morton 1959)*

Property	Natural Rubber	SBR Normal	SBR Cold	Neoprene	Butyl	Urethane	Nitrile	Polysulfide	Acrylate	Silicone	Fluorocarbon	cis-1,4-Isoprene
Tensile strength, psi	4500	3000	3800	4000	3000	2000	2500	1200	2500	1500	1000	4500
Elongation, %	600	500	550	800	400	300	550	400	400	600	600	600
Modulus, psi, 300–400% elongation	2500	2000	2500	1000	1000	1200	1500	1400	—	—	250	2500
Dynamic properties	Excel.	Fair	Good	Fair	Poor	Good	Poor	Poor	Good	Poor	Poor	Excel.
Permanent set	Low	Low	Low	Moder.	Moder.	Moder.	Moder.	High	Low	High	High	Low
Tear resistance	Good	Poor	Fair	Good	Excel.	Good	—	—	Good	Poor	—	Good
Abrasion resistance	Fair	Good	Good	Good	Good	Excel.	—	Poor	Good	Poor	—	Fair
Gas Permeability	High	High	High	Moder.	Low	—	Moder.	Low	Low	—	—	High
Upper use temp., °C	100	100	100	80–100	120	100	140	80–150	200	250	250	100
Lower use temp., °C	−60	−55	−55	−45	−50	−45	−15 to −55	−40	−25	−90	−15 to −40	−60
Oxidative and ozone resistance	Poor	Fair	Fair	Good	Good	Good	Good	Good	Excel.	Excel.	Excel.	Fair
Solvent resistance	Poor	Poor	Poor	Good	Poor	Good	Good	Good	Good	Excel.	Excel.	Poor

D. Table of Elastomer Properties

Table 19-3 lists values of some physical and chemical properties of major commercial elastomers. In comparison with the corresponding tables in Chapters 17 and 18 it should be noted that the properties of elastomers depend in very large extent on the details of vulcanization, reinforcement, and compounding. It is, consequently, almost futile to assign "typical" values of properties according to polymer type. Therefore, most of the data in Table 19-3 are of no more than qualitative significance, and comparisons should be made with extreme caution.

BIBLIOGRAPHY

Anderson 1969. Earl V. Anderson, "Rubber," *Chem. & Eng. News* **47** (29), 39–83 (July 14, 1969) (21 pp).

Bateman 1963. L. Bateman, ed., *The Chemistry and Physics of Rubber-like Substances,* John Wiley and Sons, New York, 1963.

Burgess 1965. K. A. Burgess, F. Lyon, and W. S. Stoy, "Carbon," pp. 820–836 in Herman F. Mark, Norman G. Gaylord, and Norbert M. Bikales, eds., *Encyclopedia of Polymer Science and Technology,* Vol. 2, Interscience Div., John Wiley and Sons, New York, 1965.

Carpenter 1970. Ernest L. Carpenter, "Rubber in the 70's," *Chem. & Eng. News* **48** (18), 31–57 (April 27, 1970).

Cooper 1966. W. Cooper, "Elastomers, Synthetic," pp. 406–482 in Herman F. Mark, Norman G. Gaylord, and Norbert M. Bikales, eds., *Encyclopedia of Polymer Science and Technology,* Vol. 5, Interscience Div., John Wiley and Sons, New York, 1966.

Cox 1965. William L. Cox., "Antiozonants," pp. 197–203 in Herman F. Mark, Norman G. Gaylord, and Norbert M. Bikales, eds., *Encyclopedia of Polymer Science and Technology,* Vol. 2, Interscience Div., John Wiley and Sons, New York, 1965.

Cunneen 1970. J. I. Cunneen, D. Barnard, F. McL. Swift, A. R. Payne, M. Porter, A. Schallamach, W. A. Southorn, and A. G. Thomas, "Rubber, Natural," in press in Herman F. Mark, Norman G. Gaylord, and Norbert M. Bikales, eds., *Encyclopedia of Polymer Science and Technology,* Vol. 12, Interscience Div., John Wiley and Sons, New York, 1970.

Garvey 1970. B. S. Garvey, "Rubber Compounding and Processing," in press in Herman F. Mark, Norman G. Gaylord, and Norbert M. Bikales, eds., *Encyclopedia of Polymer Science and Technology,* Vol. 12, Interscience Div., John Wiley and Sons, New York, 1970.

Kraus 1965. Gerard Kraus, ed., *Reinforcement of Elastomers,* Interscience Div., John Wiley and Sons, New York, 1965.

Kraus 1970. Gerard Kraus, "Reinforcement," in press in Herman F. Mark, Norman G. Gaylord, and Norbert M. Bikales, eds., *Encyclopedia of Polymer Science and Technology,* Vol. 12, Interscience Div., John Wiley and Sons, New York, 1970.

Maassen 1965. G. C. Maassen, R. J. Fawcett, and W. R. Connell, "Antioxidants," pp. 171–197 in Herman F. Mark, Norman G. Gaylord, and Norbert M. Bikales, eds., *Encyclopedia of Polymer Science and Technology*, Vol. 2, Interscience Div., John Wiley and Sons, New York, 1965.

Morton 1959. Maurice Morton, ed., *Introduction to Rubber Technology*, Reinhold Publishing Corp., New York, 1959.

Stagg 1968. Richard Stagg, "Latexes," pp. 164–195 in Herman F. Mark, Norman G. Gaylord, and Norbert M. Bikales, eds., *Encyclopedia of Polymer Science and Technology*, Vol. 8, Interscience Div., John Wiley and Sons, New York, 1968.

Weissert 1969. Fred C. Weissert, "Elastomer Technology," *Ind. Eng. Chem.* **61** (8), 53–59 (1969).

Whitby 1954. G. S. Whitby, C. C. Davis, and R. F. Dunbrook, eds., *Synthetic Rubber*, John Wiley and Sons, New York, 1954.

Wolfe 1970. James R. Wolfe, Jr., "Vulcanization," in press in Herman F. Mark, Norman G. Gaylord, and Norbert M. Bikales, eds., *Encyclopedia of Polymer Science and Technology*, Vol. 13, Interscience Div., John Wiley and Sons, New York, 1970.

Appendix I

List of Symbols

Symbols in dimensional formulas have the following significance:

M = mass	T = time	Q = electric charge
L = length	θ = temperature	

Symbol	Dimension	Common Unit	Definition	Chapter Where First Defined
A	L	...	Angstrom unit	
A	ML^2T^{-2}	kcal	Helmholz free energy, work content	$1C$
A	$ML^{-1}T^{-1}$	poise	Frequency factor for viscous flow	$6A$
A	T^{-1}	sec^{-1}	Collision frequency factor (first-order reaction)	$9F$
A	$M^{-1}L^3T^{-1}$	liter/mole sec	Collision frequency factor (second-order reaction)	$9F$
A_1	...	mole/g	First virial coefficient	$3B$
A_2	$M^{-1}L^3$	ml mole/g^2	Second virial coefficient	$2C$
A_3	$M^{-2}L^6$	ml^2 mole/g^3	Third virial coefficient	$3B$
a	...	...	Arbitrary constant	
a	...	...	Exponent in modified Staudinger equation	$3D$
a	L	A	Length of crystal unit cell axis	$5B$
a_T	...	...	"Shift factor" for time-temperature superposition	$6C$

551

Symbol	Dimension	Common Unit	Definition	Chapter Where First Defined
b	...	...	Arbitrary constant	
b	L	A	Length of crystal unit cell axis	$5B$
b	L	cm or A	Size parameter of Gaussian distribution	$6B$
C	...	...	Arbitrary constant	
C	...	...	Transfer constant in chain polymerization	$9C$
C_m	...	$\text{mole}^{1/2}/\text{g}^{1/2}$	Thermodynamic constant of the Flory-Huggins dilute solution theory	$2C$
C_p	$L^2T^{-2}\theta^{-1}$	cal/g °C	Specific heat	$6B$
c	ML^{-3}	g/cm^3	Concentration	
c	ML^{-3}	g/deciliter	Concentration	$3D$
c	L	A	Length of crystal unit cell axis	$5B$
D	L^2T^{-1}	cm^2/sec	Diffusion constant	$3E$
d	...	...	Signifies the total derivative	
d	L	cm	Diameter of spherical particle	$3C$
E	ML^2T^{-2}	kcal	Energy content	$1C$
E	$M^{-1/2}L^{5/2}T^{-1}$	$(\text{cal cm}^3)^{1/2}$ cm/mole	Molar attraction constant	$2A$
E	ML^2T^{-2}	kcal	Activation energy	$6A$
e	...	...	Base of natural logarithms	
e	...	...	Polarity factor in the Alfrey-Price equation	$11C$
F_1,F_2	...	...	Mole fractions in copolymer	$11A$
f	...	...	Denotes a functional relationship	
f	...	...	Fraction of polymer in given phase	$2D$

Symbol	Dimension	Common Unit	Definition	Chapter Where First Defined
f	MT^{-1}	dyne sec/cm	Frictional coefficient	3E
f	MLT^{-2}	dyne	Force	6B
f	...	...	Volume fraction	6D
f	...	...	Functionality	8C
f	...	...	Initiator efficiency	9C
f_1, f_2	...	...	Mole fractions in monomer feed	11A
G	ML^2T^{-2}	kcal	Gibbs free energy	1C
G	$ML^{-1}T^{-2}$	psi or dyne/cm^2	Modulus of elasticity	6B
g	...	...	Ratio of sizes of branched and unbranched molecules	2B
g	...	...	Coefficient of $\Gamma^2 c^2$ in expansion of osmotic pressure	3E
g	...	...	Symmetry tensor in EPR analysis	4D
H	ML^2T^{-2}	kcal	Heat content or enthalpy	1C
H	$M^{-1}L^2$	mole cm^2/g^2	Light scattering calibration constant	3C
H	$L^{-1}T^{-1}Q^{-1}$	gauss	Magnetic field strength	4D
h	ML^2T^{-1}	erg sec	Planck's constant	
I	ML^2T^{-2}	lumen	Transmitted light intensity	3C
I_0	ML^2T^{-2}	lumen	Incident light intensity	3C
i	...	...	Summation index	
i	$ML^{-1}T^{-2}$	lumen/cm^3	Scattered light intensity per unit volume of scatterer	3C
J	$M^{-1}LT^2$	cm^2/dyne	Elastic compliance	6C
j	...	...	Summation index	
K	$M^{-1}L^2$	mole cm^2/g^2	Light scattering calibration constant	3C
K	$M^{-1}L^3$	deciliter mole$^{1/2}$/g$^{3/2}$	Constant in Flory viscosity equation	3D
K	$M^{-1}L^3T^{-1}$	liter/mole sec	Rate constant for initiation (second order)	10B

Symbol	Dimension	Common Unit	Definition	Chapter Where First Defined
K', K''	...	...	Empirical constants in modified Staudinger viscosity equations	3D
k, k'	...	...	Arbitrary constants	
k	$ML^2T^{-2}\theta^{-1}$	erg/deg molecule	Boltzmann's constant	2C
k	$M^{-1}L^3T^{-1}$	liter/mole sec	Rate constant (second order)	8C
k_d	T^{-1}	sec^{-1}	Rate constant for decomposition of initiator (first order)	9C
k', k''	...	...	Coefficients in Huggins viscosity equation	3D
l	L	cm or A	Length	
ln	...	...	Abbreviation for natural logarithm	
log	...	...	Abbreviation for logarithm to the base 10	
M	...	g/mole	Molecular weight	
$\overline{M}_n$	...	g/mole	Number-average molecular weight	1A
$\overline{M}_v$	...	g/mole	Viscosity-average molecular weight	3D
$\overline{M}_w$	...	g/mole	Weight-average molecular weight	1A
$\overline{M}_z$	...	g/mole	z-Average molecular weight	3E
m	L^{-2}	cm^{-2}	Number of elastic chains per unit area	6B
N	...	...	Number of items	
N_o	M^{-1}	$mole^{-1}$	Avogadro's number	2C
n	...	...	Mole fraction	2C
n	...	...	Refractive index	3C
n	...	...	Exponent in Avrami equation	5E
$\tilde{n}$	...	...	Refractive increment	3E

Symbol	Dimension	Common Unit	Definition	Chapter Where First Defined
P	...	...	Reactivity of radical in Alfrey-Price equation	11C
$P(\theta)$	...	...	Angular light-scattering function	3C
p	$ML^{-1}T^{-2}$	psi or dyne/cm^2	Pressure	
p	...	...	Extent of reaction	8C
$\tilde{p}$	...	...	Reduced pressure	2C
Q	...	...	General reactivity factor in Alfrey-Price equation	11C
q	...	...	Fraction of cross-linkable monomers on a chain	12C
R	$ML^2T^{-2}\theta^{-1}$	erg/mole deg or cal/mole deg	Gas constant	
R	...	...	Ratio of volumes of dilute and precipitated phases	2D
R_θ	L^{-1}	cm^{-1}	Rayleigh ratio	3C
r	L	cm or A	Distance from origin	2B
r	...	...	Ratio of number of A and B groups present in stepwise polymerization	8C
$r,(\overline{r^2})^{1/2}$	L	cm or A	Root-mean-square end-to-end distance	2B
r_1, r_2	...	...	Monomer reactivity ratios	11A
S	$ML^2T^{-2}\theta^{-1}$	kcal/deg	Entropy	1C
$s,(\overline{s^2})^{1/2}$	L	cm or A	Radius of gyration or root-mean-square distance of chain segments from center of gravity	2B
s	T	sec	Sedimentation constant	3E

Symbol	Dimension	Common Unit	Definition	Chapter Where First Defined
s	$ML^{-1}T^{-2}$	psi or dyne/cm^2	Stress or applied force per unit area	6A
T	θ	K or °C	Temperature	
T_g	θ	K or °C	Glass transition temperature	6D
T_m	θ	K or °C	Crystalline melting point	2A
$\tilde{T}$	...	...	Reduced temperature	2C
t	T	sec	Time, efflux time	3D
u	...	...	Size parameter for light scattering of spheres	3C
V	L^3	cm^3	Volume, specific volume, molar volume	
$\tilde{V}$	...	...	Reduced volume	2C
v	...	...	Volume fraction	2A
v	...	...	Size parameter for light scattering of random coils	3C
v	LT^{-1}	cm/sec	Velocity	6A
v	$ML^{-3}T^{-1}$	mole/liter sec	Rate of reaction	9C
$\bar{v}$	$M^{-1}L^3$	cm^3/mole	Partial molar or specific volume	2C
W	...	...	Probability function	2C
w	...	...	Weight, weight fraction	
x	L	cm or A	Distance, space coordinate	
x	...	...	Chain length, number of segments in chain, degree of polymerization	2B
x	...	...	Size parameter for light scattering of rods (Fig. 3-7 only)	3C
$\bar{x}_n$	...	...	Number-average degree of polymerization	8C

Symbol	Dimension	Common Unit	Definition	Chapter Where First Defined
$\bar{x}_w$	...	...	Weight-average degree of polymerization	8C
Y	$ML^{-1}T^{-2}$	psi or dyne/cm^2	Yield stress	6E
y	...	...	Space coordinate	
y	...	...	Size parameter for light scattering of discs (Fig. 3-7 only)	3C
y	...	...	Parameter of distribution functions	9E
Z	...	...	Number of atoms in polymer chain	6A
z	...	...	Space coordinate	
z	...	...	z-Average (see $\bar{M}_z$)	
z	...	...	Dissymmetry of scattered light	3C
z	...	...	Parameter of distribution functions	9E
α	...	...	Denotes first in a series	
α	...	...	Expansion factor of dissolved polymer chain	2B
α	L	cm or A	Dimension	6B
α	...	...	Branching coefficient	8D
α	$L^3\theta^{-1}$	cm^3/deg	Volume expansion coefficient	6D
β	...	...	Denotes second in a series	
β	$ML^3T^{-1}Q^{-1}$	Bohr magneton	Magnetic moment of electron spin	4D
β	(angle)	degree	Angle of crystal unit cell	5B
β	L	cm or A	Dimension	6B
Γ	$M^{-1}L^3$	cm^3/g	Coefficient of c in expansion of osmotic pressure	3B
Γ	...	...	Gamma function	9E
γ	...	...	Activity coefficient	3E
γ	L	cm or A	Dimension	6B

Symbol	Dimension	Common Unit	Definition	Chapter Where First Defined
γ	...	...	Strain or deformation in response to applied stress	6C
γ	T^{-1}	sec^{-1}	Shear rate	6A
Δ	...	...	Signifies a change in or difference between the values of a thermodynamic function for two states	
δ	$M^{1/2}L^{-1/2}T^{-1}$	$(calorie/cm^3)^{1/2}$	Solubility parameter	2A
δ	...	...	Chemical shift (NMR)	4D
δ	(angle)	degree	Phase difference due to energy absorption	4G
δ	...	...	Ratio of termination to propagation rate constants in copolymerization	11A
δ^2	$ML^{-1}T^{-2}$	$calorie/cm^3$	Cohesive energy density	2A
ε	...	...	Function of reactivity ratios	11B
ε	...	...	Kinetic chain length parameter in polymer degradation	12E
ζ	...	...	Size parameter of crystallization theory	5F
η	$ML^{-1}T^{-1}$	dyne sec/cm^2, poise	Viscosity	
η_{inh}	$M^{-1}L^3$	deciliter/g	Inherent viscosity	3D
η_r	...	...	Relative visocity	3D
η_{red}	$M^{-1}L^3$	deciliter/g	Reduced viscosity	3D
η_{sp}	...	...	Specific viscosity	3D
$[\eta]$	$M^{-1}L^3$	deciliter/g	Intrinsic viscosity	2E, 3D
Θ	θ	K	Flory temperature, where polymer-solvent interactions are zero	2B

Symbol	Dimension	Common Unit	Definition	Chapter Where First Defined
θ	(angle)	degree	Angle	2C
θ	...	...	Fraction of surface sites occupied	10D
κ	...	...	Thermodynamic constant of Flory-Huggins dilute solution theory	2C
κ	$M^{-1}LT^2$	$cm^2/dyne$	Compressibility of liquid	3C
λ	L	cm or A	Wavelength of light in air	3C
λ_s	L	cm or A	Wavelength of light in solution $= \lambda/n$	3C
μ	$ML^3T^{-1}Q^{-1}$	Bohr magneton	Magnetic moment of nucleus	4D
μ	...	...	Parameter of distribution functions	9E
μ_h	...	...	Huggins polymer-solvent interaction parameter	2C
μm	L	micrometer	Abbreviation for micrometer	
μ_0	MLQ^{-2}	henry/meter	Magnetic permeability of vacuum	4D
ν	L^{-3}	cm^{-3}	Number of particles per cm^3	3C
ν	...	...	Number of items	
ν	T^{-1}	sec^{-1}	Frequency	4A
ν	...	...	Kinetic chain length	9C
π	$ML^{-1}T^{-2}$	atmosphere	Osmotic pressure	2C
ρ	ML^{-3}	g/cm^3	Density	2A
ρ	...	...	Ratio of number of functional groups on branch units to total number of such groups	8D
ρ	...	...	Parameter of distribution functions	9E

Symbol	Dimension	Common Unit	Definition	Chapter Where First Defined
ρ	T^{-1}	sec^{-1}	Rate of generation of radicals	12B
Σ	...	...	Summation sign	
σ	...	...	Parameter of fractionation theory	2D
σ	...	...	Parameter of distribution functions	9E
σ	...	...	Chain transfer constant in polymer degradation	12E
τ	L^{-1}	cm^{-1}	Turbidity	3C
τ	...	...	Chemical shift (NMR)	4D
τ	T	sec	Relaxation time, retardation time	6C
τ_s	T	sec	Mean lifetime of a free radical	9E
Φ	M^{-1}	deciliter/mole cm^3	Universal constant of Flory viscosity theory	3D
ϕ	...	...	Cross termination probability constant in copolymerization	11A
χ	...	...	Polymer-solvent interaction constant	2C
ψ	...	...	Thermodynamic constant of Flory-Huggins dilute solution theory	2C
ω	...	...	Denotes last in a series	
ω	T^{-1}	sec^{-1}	Angular velocity	3E
∂	...	...	Signifies partial derivative	
[ ]	...	...	Denote concentration of substance within brackets (except in symbol [η])	
°	...	...	Degree sign	
×	...	...	Multiplication sign	

Appendix II

Table of Physical Constants

Symbol	Name	Value
A	Angstrom unit	$1 \text{ A} = 10^{-10}$ m (not a preferred unit)
e	Base of natural logarithms	2.7183
h	Planck's constant	6.6255×10^{-27} erg sec
k	Boltzmann's constant	1.3805×10^{-16} erg/deg
N_0	Avogadro's number	6.0226×10^{23} mole^{-1}
R	Gas constant	8.3143×10^{7} erg/mole K
		1.99 cal/mole K
		82.1 cm^3 atm/mole K
K	Absolute temperature	$0°C = 273.16$ K (Kelvin)
μm	Micrometer	$1 \ \mu\text{m} = 10^{-3}$ mm $= 10^{-6}$ m
nm	Nanometer	$1 \text{ nm} = 10^{-3} \ \mu\text{m} = 10^{-6}$ mm $= 10^{-9}$m

Author Index[*]

Abe, A., 145, *181*
Abrams, I. M., 409, *429*
Adams, E. T., Jr., 95, *96*
Aggarwal, S. L., 351, *352*
Akin, R. B., 443, *462*
Albisetti, C. J., 429, *429*
Alexander, L. E., 114, *133*
Alexander, P., 95, *103,* 126, *134*
Alfrey, T., Jr., 204, 206, 207, 213, 214, 217, *217,* 227, 231, 234, *250,* 329, 337, 338, 341, 347, 350, *352, 353,* 368, *374*
Allcock, H. R., 462, *466*
Allen, P. W., 26, 47, 48, 52, *57, 59,* 75, 84, 90, *97, 102*
Altgelt, K. H., 56, *57*
Aminco, 81, *97*
Anderson, E. V., 394, 397, 401, *401,* 539, *549*
Anderson, F. R., 161, *181*
Andrianov, K. A., 461, 462, *462,* 485, *487*
Anonymous, 423, *429,* 457, *462,* 499, *509*
Arakawa, T., 161, *181*
Archibald, W. J., 93, *97*
Arledter, H. F., 462, *462*
Armstrong, J. L., 72, *97*
Armstrong, R. W., 96, *97*
Ashmore, P. G., 323, *325*
Asmussen, F., 26, *57, 61*
ASTM, 126, 133, *133*

Atherton, J. N., 336, *352*
Atkins, J. T., 82, *98*
Atlas, S. M., 165, 177, 178, 180, *182,* 518, 519, 520, 525, *531, 532*
Aubrey, D. W., 146, *180*
Ayrey, G., 106, *134*

Backer, S., 529, *531*
Bacskai, R., 351, *352*
Badgley, W. J., 74, *97*
Baer, E., 133, *134,* 207, 217, *218, 219,* 250, *251*
Baer, M., 123, *134*
Bagchi, S. N., 113, *135,* 161, *181*
Bailey, W. J., 462, *462*
Baker, C. A., 49, *57*
Baker, W. O., 16, *20,* 217, *218*
Baldwin, R. L., 95, *97, 104*
Bamford, C. H., 44, *57,* 297, 299, 301, 304, *307,* 364, *374,* 414, *429*
Bannerman, D. G., 223, *250*
Barb, W. G., 297, 301, 304, *307*
Barclay, R. K., 145, *181*
Bareiss, R., 68, *99*
Barnard, D., 394, *402,* 539, *549*
Barton, J. M., 367, *374*
Basdekis, C. H., 409, *429*
Bateman, L., 539, *549*
Battaerd, H. A. J., 351, *352*

[*]Page numbers in *italics* refer to references at end of chapters.

563

Subject Index